HOW do you learn?

Read • Reflect • Watch • Listen • Connect • Discover • Interact

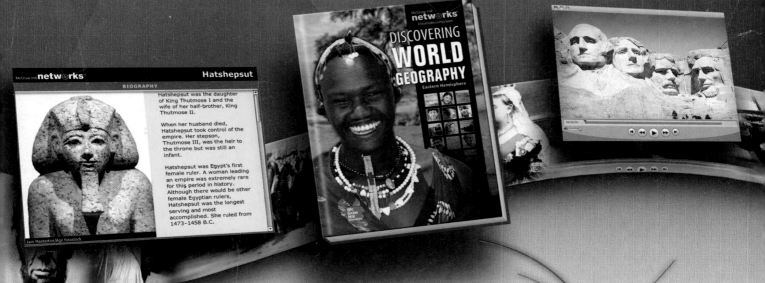

start netw**o**rking

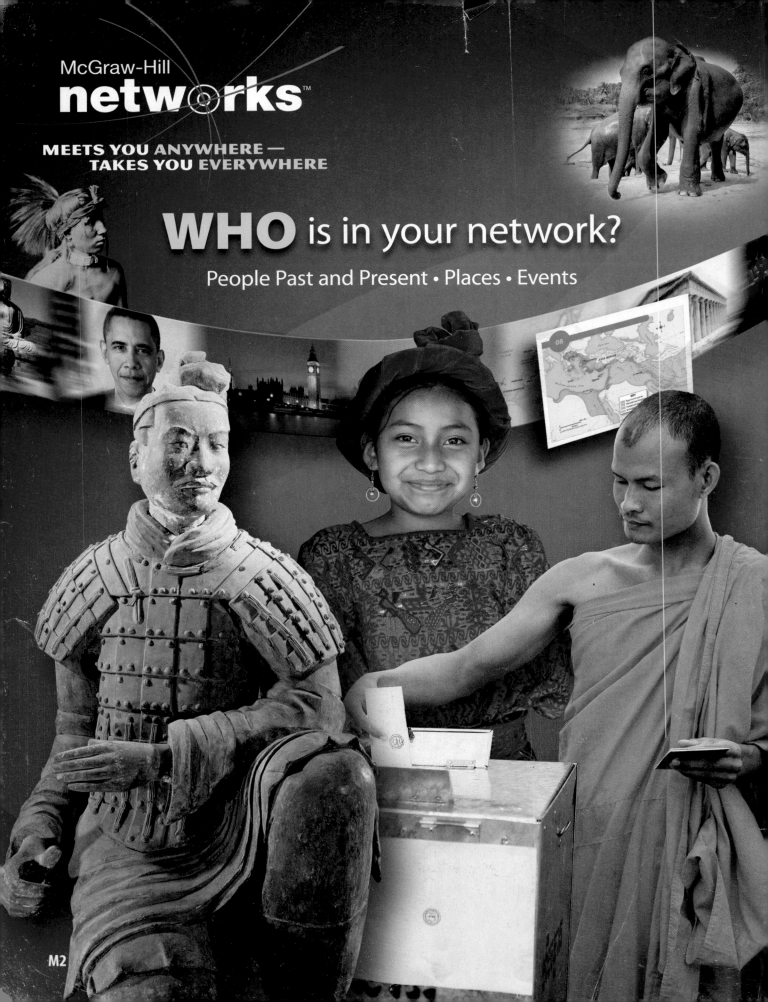

WHAT do you learn?
History • Geography • Economics • Government • Culture

start **networking**

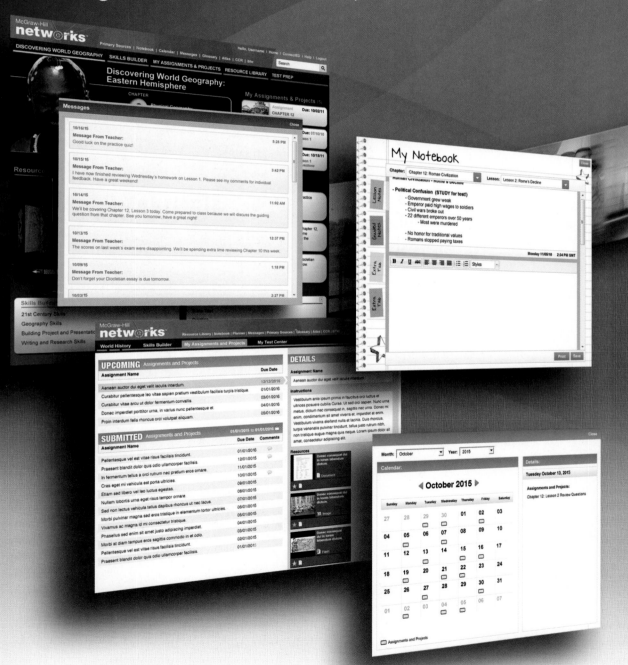

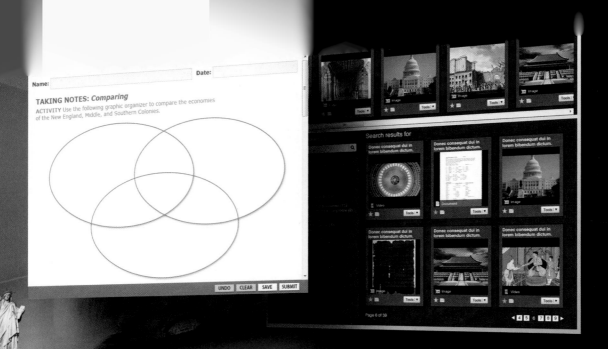

WHAT do you use?
Graphic Organizers • Primary Sources • Videos • Games • Photos

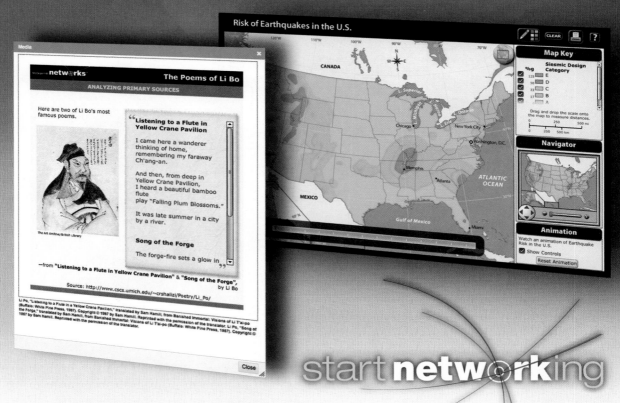

start **networking**

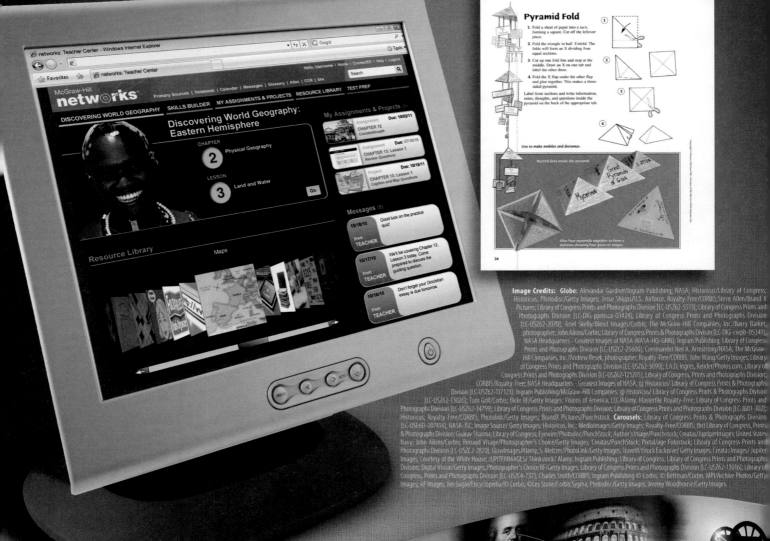

DISCOVERING
WORLD
GEOGRAPHY

Eastern Hemisphere

Richard G. Boehm, Ph. D.

Bothell, WA • Chicago, IL • Columbus, OH • New York, NY

About the Cover: Young warrior of Samburu people wearing traditional bracelets and headdress; Tiras Mountain landscape in country of Namibia in Africa
Cover Photo Credits: ©Roy Toft/National Geographic Stock; (bkgd)Christian Heinrich/Getty Images; (l to r, t to b)Anja Fleig/age fotostock; ©Brand X Pictures/PunchStock; Lissa Harrison; ©Image Source, all rights reserved.; NASA/NOAA/SPL/Getty Images; Fancy Collection/SuperStock; George Clerk/Getty Images; Ingram Publishing; ©Pete Atkinson/Getty Images; Getty Images; Author's Image/PunchStock; D. Normark/PhotoLink/Getty Images; ©Image Source/Getty Images

www.mheonline.com/networks

Copyright © 2014 McGraw-Hill Education

All rights reserved. No part of this publication may be reproduced or distributed in any form or by any means, or stored in a database or retrieval system, without the prior written consent of McGraw-Hill Education, including, but not limited to, network storage or transmission, or broadcast for distance learning.

Send all inquiries to:
McGraw-Hill Education
8787 Orion Place
Columbus, OH 43240

ISBN: 978-0-07-663609-9
MHID: 0-07-663609-7

Printed in the United States of America.

2 3 4 5 6 7 8 9 RJC 17 16 15 14 13

AUTHORS

SENIOR AUTHOR

Richard G. Boehm, Ph.D., was one of the original authors of *Geography for Life: National Geography Standards,* which outlined what students should know and be able to do in geography. He was also one of the authors of the *Guidelines for Geographic Education*, in which the Five Themes of Geography were first articulated. Dr. Boehm has received many honors, including "Distinguished Geography Educator" by the National Geographic Society (1990), the "George J. Miller Award" from the National Council for Geographic Education (NCGE) for distinguished service to geographic education (1991), "Gilbert Grosvenor Honors" in geographic education from the Association of American Geographers (2002), and the NCGE's "Distinguished Mentor Award" (2010). He served as president of the NCGE, has twice won the *Journal of Geography* award for best article, and also received the NCGE's "Distinguished Teaching Achievement." Presently, Dr. Boehm holds the Jesse H. Jones Distinguished Chair in Geographic Education at Texas State University in San Marcos, Texas, where he serves as director of The Gilbert M. Grosvenor Center for Geographic Education. His most current project includes the production of the video-based professional development series, Geography: Teaching With the Stars. Available programs may be viewed at www.geoteach.org.

CONTRIBUTING AUTHORS

Jay McTighe has published articles in a number of leading educational journals and has coauthored 10 books, including the best-selling *Understanding by Design* series with Grant Wiggins. McTighe also has an extensive background in professional development and is a featured speaker at national, state, and district conferences and workshops. He received his undergraduate degree from the College of William and Mary, earned a master's degree from the University of Maryland, and completed postgraduate studies at the Johns Hopkins University.

Dinah Zike, M.Ed., is an award-winning author, educator, and inventor recognized for designing three-dimensional, hands-on manipulatives and graphic organizers known as Foldables®. Foldables are used nationally and internationally by parents, teachers, and other professionals in the education field. Zike has developed more than 150 supplemental educational books and materials. Her two latest books, *Notebook Foldables®* and *Foldables®, Notebook Foldables®, & VKV®s for Spelling and Vocabulary 4th–12th,* were each awarded *Learning Magazine's* Teachers' Choice Award for 2011. In 2004 Zike was honored with the CESI Science Advocacy Award. She received her M.Ed. from Texas A&M, College Station, Texas.

CONSULTANTS AND REVIEWERS

ACADEMIC CONSULTANTS

William H. Berentsen, Ph.D.
Professor of Geography and European Studies
University of Connecticut
Storrs, Connecticut

David Berger, Ph.D.
Ruth and I. Lewis Gordon Professor of Jewish History
Dean, Bernard Revel Graduate School
Yeshiva University
New York, New York

R. Denise Blanchard, Ph.D.
Professor of Geography
Texas State University–San Marcos
San Marcos, Texas

Brian W. Blouet, Ph.D.
Huby Professor of Geography and International Education
The College of William and Mary
Williamsburg, Virginia

Olwyn M. Blouet, Ph.D.
Professor of History
Virginia State University
Petersburg, Virginia

Maria A. Caffrey, Ph.D.
Lecturer, Department of Geography
University of Tennessee, Knoxville
Knoxville, Tennessee

So-Min Cheong, Ph.D.
Associate Professor of Geography
University of Kansas
Lawrence, Kansas

Alasdair Drysdale, Ph.D.
Professor of Geography
University of New Hampshire
Durham, New Hampshire

Rosana Nieto Ferreira, Ph.D.
Assistant Professor of Geography and Atmospheric Science
East Carolina University
Greenville, North Carolina

Eric J. Fournier, Ph.D.
Associate Professor of Geography
Samford University,
Birmingham, Alabama

Matthew Fry, Ph.D.
Assistant Professor of Geography
University of North Texas
Denton, Texas

Douglas W. Gamble, Ph.D.
Professor of Geography
University of North Carolina, Wilmington
Wilmington, North Carolina

Gregory Gaston, Ph.D.
Professor of Geography
University of North Alabama
Florence, Alabama

Jeffrey J. Gordon, Ph.D.
Associate Professor of Geography
Bowling Green State University
Bowling Green, Ohio

Alyson L. Greiner, Ph.D.
Associate Professor of Geography
Oklahoma State University
Stillwater, Oklahoma

William J. Gribb, Ph.D.
Associate Professor of Geography
University of Wyoming
Laramie, Wyoming

Joseph J. Hobbs, Ph.D.
Professor of Geography
University of Missouri
Columbia, Missouri

Ezekiel Kalipeni, Ph.D.
Professor of Geography and Geography Information Science
University of Illinois at Urbana–Champaign
Urbana, Illinois

Pradyumna P. Karan, Ph.D.
University Research Professor of Geography
University of Kentucky
Lexington, Kentucky

Christopher Laingen, Ph.D.
Assistant Professor of Geography
Eastern Illinois University
Charleston, Illinois

Jeffrey Lash, Ph.D.
Associate Professor of Geography
University of Houston–Clear Lake
Houston, Texas

Jerry T. Mitchell, Ph.D.
Research Professor of Geography
University of South Carolina
Columbia, South Carolina

Thomas R. Paradise, Ph.D.
Professor, Department of Geosciences and the King Fahd Center for Middle East Studies
University of Arkansas
Fayetteville, Arkansas

David Rutherford, Ph.D.
Assistant Professor of Public Policy and Geography
Executive Director, Mississippi Geographic Alliance
University of Mississippi
University, Mississippi

Dmitrii Sidorov, Ph.D.
Professor of Geography
California State University, Long Beach
Long Beach, California

Amanda G. Smith, Ph.D.
Professor of Education
University of North Alabama
Florence, Alabama

Jeffrey S. Ueland, Ph.D.
Associate Professor of Geography
Bemidji State University
Bemidji, Minnesota

Fahui Wang, Ph.D.
Professor of Geography
Louisiana State University
Baton Rouge, Louisiana

TEACHER REVIEWERS

Precious Steele Boyle, Ph.D.
Cypress Middle School
Memphis, TN

Jason E. Albrecht
Moscow Middle School
Moscow, ID

Jim Hauf
Berkeley Middle School
Berkeley, MO

Elaine M. Schuttinger
Trinity Catholic School
Columbus, OH

Mark Stahl
Longfellow Middle School
Norman, OK

Mollie Shanahan MacAdams
Southern Middle School
Lothian, MD

Sara Burkemper
Parkway West Middle Schools
Chesterfield, MO

Alicia Lewis
Mountain Brook Junior High School
Birmingham, AL

Steven E. Douglas
Northwest Jackson Middle School
Ridgeland, MS

LaShonda Grier
Richmond County Public Schools
Martinez, GA

Samuel Doughty
Spirit of Knowledge Charter School
Worcester, MA

CONTENTS

Reference Atlas Maps RA1

World: Political.....................RA2
World: Physical.....................RA4
North America: Political.............RA6
North America: Physical.............RA7
United States: Political.............RA8
United States: Physical.............RA10
Canada: Physical/Political..........RA12
Middle America: Physical/Political...RA14
South America: Political............RA16
South America: Physical.............RA17
Europe: Political...................RA18

Europe: Physical....................RA20
Africa: Political...................RA22
Africa: Physical....................RA23
Middle East: Physical/Political.....RA24
Asia: Political.....................RA26
Asia: Physical......................RA28
Oceania: Physical/Political.........RA30
World Time Zones....................RA32
Polar Regions.......................RA34
A World of Extremes.................RA35
Geographic Dictionary...............RA36

Scavenger Hunt RA38

UNIT ONE Our World: The Eastern Hemisphere 1

Earth's Land, People, and Environments 15

ESSENTIAL QUESTION
How does geography influence the way people live?

LESSON 1 **How Geographers Think**.....18

LESSON 2 **Geographers and Their Tools**..................26

Thinking Like a Geographer Relief..........29
Think Again The Height of Mount Everest.....31
What Do You Think? Is Globalization Destroying Indigenous Cultures?..................34

The Physical World 39

ESSENTIAL QUESTION
How does geography influence the way people live?

LESSON 1 **The Earth-Sun Relationship**...............42

LESSON 2 **Forces Shaping Earth**........52

LESSON 3 **Earth's Land and Water**.....58

v

CONTENTS

The Human World 69

ESSENTIAL QUESTIONS
How do people adapt to their environment? • What makes a culture unique? • Why do people make economic choices?

LESSON 1 Earth's Population 72
Think Again Immigration.................78

LESSON 2 The World's Cultures 82
Global Connections Rain Forest Resources.....90

LESSON 3 Economies of the World 94
Think Again Command Economy 96

UNIT TWO Asia 105

East Asia 113

ESSENTIAL QUESTIONS
How does geography influence the way people live? • What makes a culture unique? • How do cultures spread? • Why do people trade? • How does technology change the way people live?

LESSON 1 Physical Geography of East Asia 116
Thinking Like a Geographer Rice............120

LESSON 2 History of East Asia 124
Think Again The Great Wall of China........125

LESSON 3 Life in East Asia 132
Global Connections The Fury of a Tsunami....140

Southeast Asia 147

ESSENTIAL QUESTIONS
How does geography influence the way people live? • What makes a culture unique? • Why does conflict develop?

LESSON 1 Physical Geography of Southeast Asia 150
Think Again Hurricanes, Typhoons, and Cyclones.........................155

LESSON 2 History of Southeast Asia 156

LESSON 3 Life in Southeast Asia 164

(t) Aurora Photos/Alamy; (c) Jon Arnold/Alamy; (b) ©ADREES LATIF/X90022/Reuters/Corbis

There's More Online! Videos • Animations • Games • Interactive Maps, Images, Graphs, Charts...and more

CHAPTER 6 South Asia 173

ESSENTIAL QUESTIONS
How does geography influence the way people live? • How do governments change? • What makes a culture unique?

LESSON 1 Physical Geography of South Asia 176

LESSON 2 History of South Asia 182

LESSON 3 Life in South Asia 188
Thinking Like a Geographer Place Names 191

CHAPTER 7 Central Asia, the Caucasus, and Siberian Russia 197

ESSENTIAL QUESTIONS
How does geography influence the way people live? • How do governments change?

LESSON 1 Physical Geography of the Regions 200

LESSON 2 History of the Regions 206

LESSON 3 Life in the Regions 212
Thinking Like a Geographer World Time Zones 213

CHAPTER 8 Southwest Asia 223

ESSENTIAL QUESTIONS
How does geography influence the way people live? • Why do civilizations rise and fall? • How does religion shape society?

LESSON 1 Physical Geography of Southwest Asia 226
Thinking Like a Geographer Seas 228

LESSON 2 History of Southwest Asia 232
Think Again The Middle East and Southwest Asia 239

LESSON 3 Life in Southwest Asia 240
What Do You Think? Are Trade Restrictions Effective at Changing a Government's Policies? 248

CONTENTS

UNIT THREE Africa 253

Chapter 9 North Africa 261

ESSENTIAL QUESTIONS
How do people adapt to their environment? • How does religion shape society? • Why do conflicts develop?

- **LESSON 1** The Physical Geography of North Africa 264
  - Think Again The Sahara Desert 269
- **LESSON 2** The History of North Africa 272
- **LESSON 3** Life in North Africa 280

Chapter 10 East Africa 291

ESSENTIAL QUESTIONS
How does geography influence the way people live? • Why do people trade? • Why does conflict develop?

- **LESSON 1** Physical Geography of East Africa 294
  - Think Again Great Migration 301
- **LESSON 2** History of East Africa 302
- **LESSON 3** Life in East Africa 310
  - Global Connections Sudan Refugees and Displacement 318

Chapter 11 Central Africa 325

ESSENTIAL QUESTIONS
How do people adapt to their environment? • How does technology change the way people live? • Why do people make economic choices?

- **LESSON 1** Physical Geography of Central Africa 328
- **LESSON 2** History of Central Africa 334
- **LESSON 3** Life in Central Africa 340
  - What Do You Think? Has the United Nations Been Effective at Reducing Conflict in Africa? 346

There's More Online! Videos • Animations • Games • Interactive Maps, Images, Graphs, Charts...and more

CHAPTER 12 West Africa 351

ESSENTIAL QUESTIONS
How does geography influence the way people live? • How do new ideas change the way people live? • What makes a culture unique?

LESSON 1 **Physical Geography of West Africa** 354

LESSON 2 **The History of West Africa** 360
Thinking Like a Geographer Religion 367

LESSON 3 **Life in West Africa** 368

CHAPTER 13 Southern Africa 377

ESSENTIAL QUESTIONS
How does geography influence the way people live? • How do new ideas change the way people live?

LESSON 1 **Physical Geography of Southern Africa** 380

LESSON 2 **History of Southern Africa** 388

LESSON 3 **Life in Southern Africa** 394
Think Again Madagascar 396

UNIT FOUR Oceania, Australia, New Zealand, and Antarctica 405

CHAPTER 14 Australia and New Zealand 413

ESSENTIAL QUESTIONS
How does geography influence the way people live? • Why does conflict develop? • What makes a culture unique?

LESSON 1 **Physical Geography** 416

LESSON 2 **History of the Region** 424

LESSON 3 **Life in Australia and New Zealand** 432
Global Connections Unfriendly Invaders 440

(t) Ariadne Van Zandbergen/Alamy; (c) Gallo Images - LKIS/Getty Images; (b) Penny Tweedie/The Image Bank/Getty Images

CONTENTS

CHAPTER 15 Oceania 447

ESSENTIAL QUESTIONS
How does geography influence the way people live? • What makes a culture unique? • Why do people make economic choices?

LESSON 1 Physical Geography of Oceania 450

LESSON 2 History and People of Oceania 456

LESSON 3 Life in Oceania 462

CHAPTER 16 Antarctica 471

ESSENTIAL QUESTIONS
How does geography influence the way people live? • How do people adapt to their environment?

LESSON 1 The Physical Geography of Antarctica 474

LESSON 2 Life in Antarctica 480
Thinking Like a Geographer Lake Vostok 485
What Do You Think? Is Global Warming a Result of Human Activity? 486

FOLDABLES® LIBRARY 491

GAZETTEER 499

ENGLISH-SPANISH GLOSSARY 507

INDEX 524

This icon indicates where reading skills and writing skills from the *Common Core State Standards for English Language Arts & Literacy in History/Social Studies, Science, and Technical Subjects* are practiced and reinforced.

FEATURES

Think Again

The Height of Mount Everest	31
Immigration	78
Command Economy	96
The Great Wall of China	125
Hurricanes, Typhoons, and Cyclones	155
The Middle East and Southwest Asia	239
The Sahara	269
Great Migration	301
Madagascar	396

Explore the Continent

Explore the World	2
Asia	106
Africa	254
Oceania, Australia, New Zealand, and Antarctica	406

Thinking Like a Geographer

Relief	29
Rice	120
Place Names	191
World Time Zones	213
Seas	229
Religious Rivalry	367
Pans	383
Lake Vostok	485

GLOBAL CONNECTIONS

Rain Forest Resources	90
The Fury of a Tsunami	140
Sudan: Refugees and Displacement	318
Unfriendly Invaders	440

What Do You Think?

Is Globalization Destroying Indigenous Cultures?	34
Are Trade Restrictions Effective at Changing a Government's Policies?	248
Has the UN Been Effective at Reducing Conflict in Africa?	346
Is Global Warming a Result of Human Activity?	486

xi

MAPS

REFERENCE ATLAS MAPS

World: Political	**RA2**
World: Physical	**RA4**
North America: Political	**RA6**
North America: Physical	**RA7**
United States: Political	**RA8**
United States: Physical	**RA10**
Canada: Physical/Political	**RA12**
Middle America: Physical/Political	**RA14**
South America: Political	**RA16**
South America: Physical	**RA17**
Europe: Political	**RA18**
Europe: Physical	**RA20**
Africa: Political	**RA22**
Africa: Physical	**RA23**
Middle East: Physical/Political	**RA24**
Asia: Political	**RA26**
Asia: Physical	**RA28**
Oceania: Physical/Political	**RA30**
World Time Zones	**RA32**
Polar Regions: Physical	**RA34**
A World of Extremes	**RA35**
Geographic Dictionary	**RA36**

UNIT 1: OUR WORLD: THE EASTERN HEMISPHERE

The World: Physical	**4**
The World: Political	**6**
The World: Population Density	**8**
The World: Economic	**10**
The World: Climate	**12**
The World	**16**
Tectonic Plate Boundaries	**40**
Earth's Wind Patterns	**47**
Human Geography	**70**
Global Impact: The World's Rain Forests	**93**

UNIT 2: ASIA

Asia: Physical	**108**
Asia: Political	**109**
Asia: Population Density	**110**
Asia: Economic Resources	**111**
Asia: Climate	**112**
East Asia	**115**
The Silk Road	**127**
Korean War, 1950–1953	**130**
Southeast Asia	**149**
Independence of Southeast Asian Countries	**160**
South Asia	**175**
South Asia: Seasonal Rains	**179**
Central Asia, the Caucasus, and Siberian Russia	**198**
Trans-Siberian Railroad	**207**
World Time Zones	**213**
Population Density	**214**
Southwest Asia	**225**
The Spread of Islam	**235**
Territorial Changes	**237**
Language Groups of Southwest Asia	**242**

UNIT 3: AFRICA

Africa: Physical	**256**
Africa: Political	**257**
Africa: Population Density	**258**
Africa: Economic Resources	**259**
Africa: Climate	**260**
North Africa	**262**
Ancient Egypt	**273**
North African Independence	**277**
East Africa	**293**
Desertification in the Sahel	**299**
Trade in East Africa	**304**
Sudan and South Sudan	**321**
Central Africa	**327**
Independence for Central African Countries	**338**
West Africa	**352**
Trading Kingdoms of West Africa	**362**
Triangular Trade Routes	**365**
Southern Africa	**379**
Population of Southern Africa	**395**

There's More Online! Videos • Animations • Games • Interactive Maps, Images, Graphs, Charts...and more

UNIT 4: OCEANIA, AUSTRALIA, NEW ZEALAND, AND ANTARCTICA

The Region: Physical.....................................**408**
The Region: Political....................................**409**
The Region: Population Density............................**410**
The Region: Economic Resources...........................**411**
The Region: Climate.....................................**412**
Australia and New Zealand................................**415**
Global Impact: Australia's Invasive Species..............**443**
Oceania...**448**
Antarctica..**473**

CHARTS, GRAPHS, DIAGRAMS, AND INFOGRAPHICS

UNIT 1: OUR WORLD: THE EASTERN HEMISPHERE

Diagram: Latitude and Longitude **21**
Chart: The Six Essential Elements **24**
Diagram: Earth's Hemispheres **27**
Diagram: Earth's Layers **43**
Diagram: Seasons **45**
Diagram: Rain Shadow **48**
Infographic: Salt Water vs. Freshwater **60**
Graph: Population Pyramid **73**
Chart: Major World Religions **84**
Infographic: Making a New Drug **95**
Diagram: Factors of Production **98**
Graph: GDP Comparison **99**

UNIT 2: ASIA

Chart: Dynasties of China **126**
Graph: U.S. Trade Deficit With China, 2001–2011 .... **139**
Diagram: Below Ground in Siberia **203**
Infographic: Oil: Reserves and Consumption **230**

UNIT 3: AFRICA

Infographic: Pyramids of Egypt **274**
Infographic: West African Energy **358**

UNIT 4: OCEANIA, AUSTRALIA, NEW ZEALAND, AND ANTARCTICA

Diagram: Atoll Formation **453**
Diagram: Ozone Hole **483**

networks ONLINE RESOURCES

Videos

Every lesson has a video to help you learn more about your world!

Infographics

Chapter 2
Lesson 3 Fresh and Salt Water in the World

Chapter 5
Lesson 2 The Spice Islands

Chapter 6
Lesson 3 Population of India

Interactive Charts/Graphs

Chapter 2
Lesson 1 Rain Shadow; Climate Zones

Chapter 3
Lesson 1 Understanding Population
Lesson 3 Economic Questions

Chapter 6
Lesson 1 Water Wells

Chapter 7
Lesson 3 Living in a Yurt

Chapter 8
Lesson 1 Oases
Lesson 2 Ziggurats

Chapter 11
Lesson 3 Social Issues in Central Africa

Chapter 14
Lesson 2 Australian Gold Rush

Animations

Chapter 1
Lesson 1 The Earth; Regions of Earth
Lesson 2 Elements of a Globe

Chapter 2
Lesson 1 Earth's Daily Rotation; Earth's Layers; Seasons on Earth
Lesson 3 How the Water Cycle Works

Chapter 3
Global Connections: Social Media

Chapter 4
Global Connections: How Tsunamis Form

Chapter 6
Lesson 1 Rivers in South Asia

Chapter 8
Lesson 1 Why Much of the World's Oil Supply Is in Southwest Asia

Chapter 9
Lesson 2 How the Pyramids Were Built

Chapter 10
Lesson 2 East African Independence
Global Connections: Refugees in Sudan

Chapter 14
Global Connections: Aussie Invasive Species

Chapter 15
Lesson 1 How Volcanoes Form Islands

Chapter 16
Lesson 1 How Icebergs Form

Slide Shows

Chapter 1
Lesson 1 Spatial Effect; Places Change Over Time
Lesson 2 Special Purpose Maps; History of Mapmaking

Chapter 2
Lesson 1 Effects of Climate Change
Lesson 2 Human Impact on Earth

Chapter 3
Lesson 1 Different Places in the World

networks ONLINE RESOURCES

Chapter 4
Lesson 1 Chinese Mountain Ranges
Lesson 3 Influence of Japanese Anime

Chapter 5
Lesson 1 Wildlife of Southeast Asia

Chapter 6
Lesson 3 Culture Groups in Southeast Asia

Chapter 9
Lesson 2 Egyptian Artifacts
Lesson 3 Artisans of North Africa

Chapter 10
Lesson 1 The Nile's River Source
Lesson 2 Ancient Africa
Global Connections: Effects of Climate on Refugee Camps

Chapter 11
Lesson 1 Rain Forest and Savanna
Lesson 3 Types of Housing in Central Africa

Chapter 12
Lesson 1 West Africa's Variety of Land
Lesson 2 The Salt Trade

Chapter 13
Lesson 1 Diamonds
Lesson 3 Cities in Southern Africa

Chapter 14
Lesson 1 Views of Australia and New Zealand

Chapter 15
Lesson 2 Life in Antarctica
Global Connections: Invasive Plant Species

Interactive Maps

Chapter 4
Lesson 2 The Silk Road and Trade

Chapter 5
Lesson 1 Volcanoes in Southeast Asia
Lesson 2 History of Southeast Asian Civilizations

Chapter 6
Lesson 1 Monsoons
Lesson 2 The British in India

Chapter 7
Lesson 2 Trans-Siberian Railroad
Lesson 2 The Soviet Union

Chapter 8
Lesson 1 Bodies of Water in Southwest Asia
Lesson 2 Islamic Expansion

Chapter 9
Lesson 2 360° View: Ancient Egypt
Lesson 2 The Punic Wars

Chapter 10
Lesson 1 Desertification of the Sahel; Resources in East Africa
Lesson 2 African Trade Routes and Goods
Global Connections: Crisis in Darfur

Chapter 12
Lesson 2 The First Trading Kingdoms

Chapter 13
Lesson 3 Population: Southern Africa

Chapter 14
Lesson 1 The Great Barrier Reef
Lesson 2 The History of Australia

Chapter 15
Global Connections: How Invasive Species Get to Australia

Interactive Images

Chapter 1
Lesson 1 360° View: Times Square, New York City

Chapter 2
Lesson 3 Ocean Floor; Garbage

Chapter 3
Lesson 1 Refugee Camps
Lesson 2 Cultural Change

Chapter 4
Lesson 1 Mount Fuji in Art
Lesson 2 360° View: China's Great Wall; Samurai
Lesson 3 Baseball in Japan

Chapter 5
Lesson 1 Indonesian Tsunami
Lesson 3 Minority Groups in Southeast Asia

Chapter 7
Lesson 1 Taiga Trees
Lesson 2 360° View: Samarkand, Uzbekistan
Lesson 3 The Aral Sea

Chapter 8
Lesson 3 360° View: Khalifa, Dubai, United Arab Emirates; Foods of Ramadan

xvi

Chapter 9
- Lesson 1 360° View: Berber Homes; 360° View: Cairo on the Nile; The Suez Canal; Mediterranean Climate and Agriculture
- Lesson 2 Islam

Chapter 10
- Lesson 1 360° View: Lake Bogoria in the Great Rift Valley; Glaciers in East Africa
- Lesson 2 British at Omdurman; Ethiopian Freedom and the Battle of Adwa
- Lesson 3 360° View: Masai and Kenya; Animal Poaching

Chapter 11
- Lesson 1 Congo River; Slash-and-Burn Agriculture; Mining for Gold
- Lesson 2 Triangular Trade
- Lesson 3 Brazzaville

Chapter 12
- Lesson 1 Lake Chad
- Lesson 2 Kwame Nkrumah
- Lesson 3 360° View: Lagos, Nigeria; West African Art

Chapter 13
- Lesson 1 Drakensberg Mountains; 360° View: Etosha Pan; Kalahari Dunes
- Lesson 2 Boer War

Chapter 14
- Lesson 1 Fjords; 360° View: Australia's Outback
- Lesson 2 Australia in World War II
- Lesson 3 Wool Processing

Chapter 15
- Lesson 2 Colonizing Oceania; 360° View: Bora Bora in the South Pacific
- Lesson 3 360° View: A Beach in Polynesia

Chapter 16
- Lesson 1 360° View: Antarctica
- Lesson 2 Earth's Ozone Layer

Games

Chapter 1
- Lesson 1 True or False
- Lesson 2 Concentration

Chapter 2
- Lesson 1 Fill in the Blank
- Lesson 2 Tic-Tac-Toe
- Lesson 3 Columns

Chapter 3
- Lesson 1 Crossword
- Lesson 2 Identification
- Lesson 3 Flashcard

Chapter 4
- Lesson 1 Identification
- Lesson 2 Flashcard
- Lesson 3 Columns

Chapter 5
- Lesson 1 Crossword
- Lesson 2 Tic-Tac-Toe
- Lesson 3 Concentration

Chapter 6
- Lesson 1 Fill in the Blank
- Lesson 2 True or False
- Lesson 3 Columns

Chapter 7
- Lesson 1 Columns
- Lesson 2 Concentration
- Lesson 3 Identification

Chapter 8
- Lesson 1 True or False
- Lesson 2 Flashcard
- Lesson 3 Fill in the Blank

Chapter 9
- Lesson 1 Tic-Tac-Toe
- Lesson 2 Crossword
- Lesson 3 Identification

Chapter 10
- Lesson 1 True or False
- Lesson 2 Columns
- Lesson 3 Fill in the Blank

Chapter 11
- Lesson 1 Concentration
- Lesson 2 Crossword
- Lesson 3 Identification

Games

Chapter 12
Lesson 1 Fill in the Blank
Lesson 2 Columns
Lesson 3 Flashcard

Chapter 13
Lesson 1 Tic-Tac-Toe
Lesson 2 Flashcard
Lesson 3 Concentration

Chapter 14
Lesson 1 Crossword
Lesson 2 True or False
Lesson 3 Fill in the Blank

Chapter 15
Lesson 1 Crossword
Lesson 2 Identification
Lesson 3 Columns

Chapter 16
Lesson 1 Flashcard
Lesson 2 Tic-Tac-Toe

Reference Atlas

Reference Atlas Maps	RA1	Europe: Physical	RA20
World: Political	RA2	Africa: Political	RA22
World: Physical	RA4	Africa: Physical	RA23
North America: Political	RA6	Middle East: Physical/Political	RA24
North America: Physical	RA7	Asia: Political	RA26
United States: Political	RA8	Asia: Physical	RA28
United States: Physical	RA10	Oceania: Physical/Political	RA30
Canada: Physical/Political	RA12	World Time Zones	RA32
Middle America: Physical/Political	RA14	Polar Regions	RA34
South America: Political	RA16	A World of Extremes	RA35
South America: Physical	RA17	Geographic Dictionary	RA36
Europe: Political	RA18		

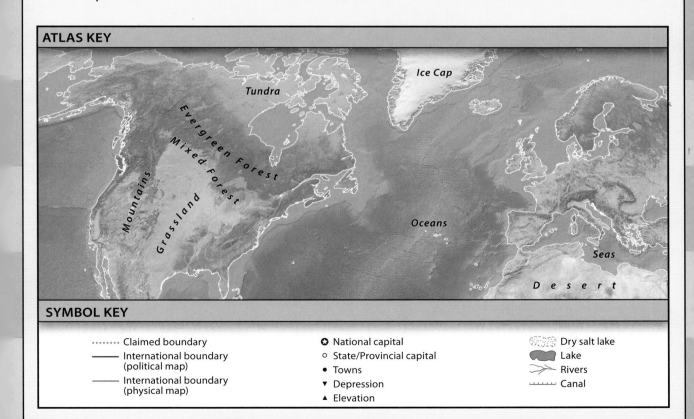

ATLAS KEY

SYMBOL KEY

- .... Claimed boundary
- ——— International boundary (political map)
- ——— International boundary (physical map)
- ⊙ National capital
- ○ State/Provincial capital
- • Towns
- ▼ Depression
- ▲ Elevation
- Dry salt lake
- Lake
- Rivers
- Canal

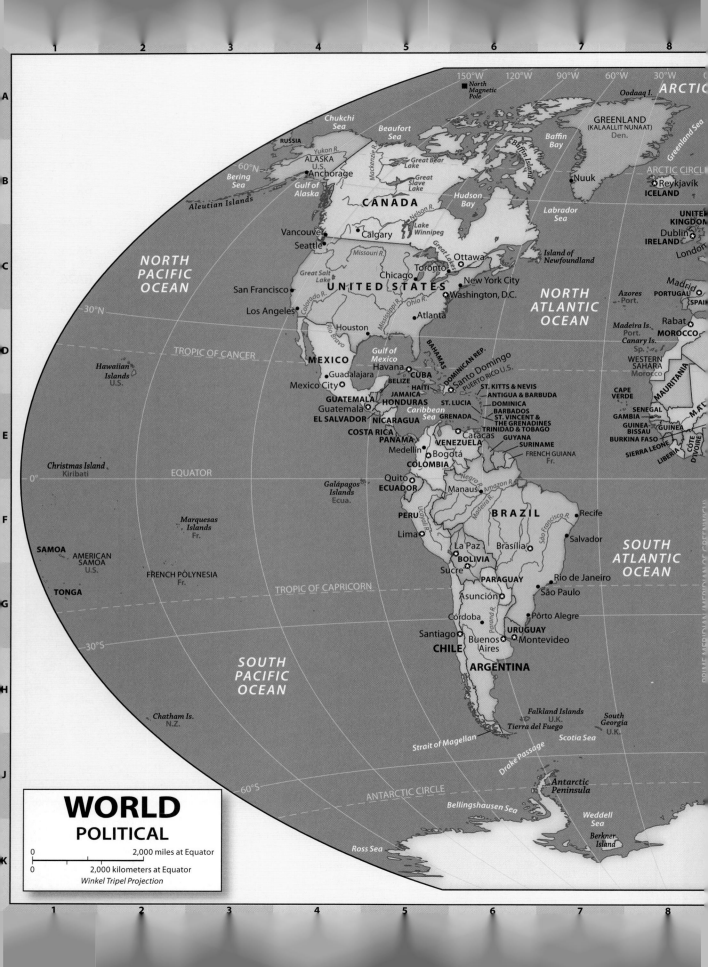

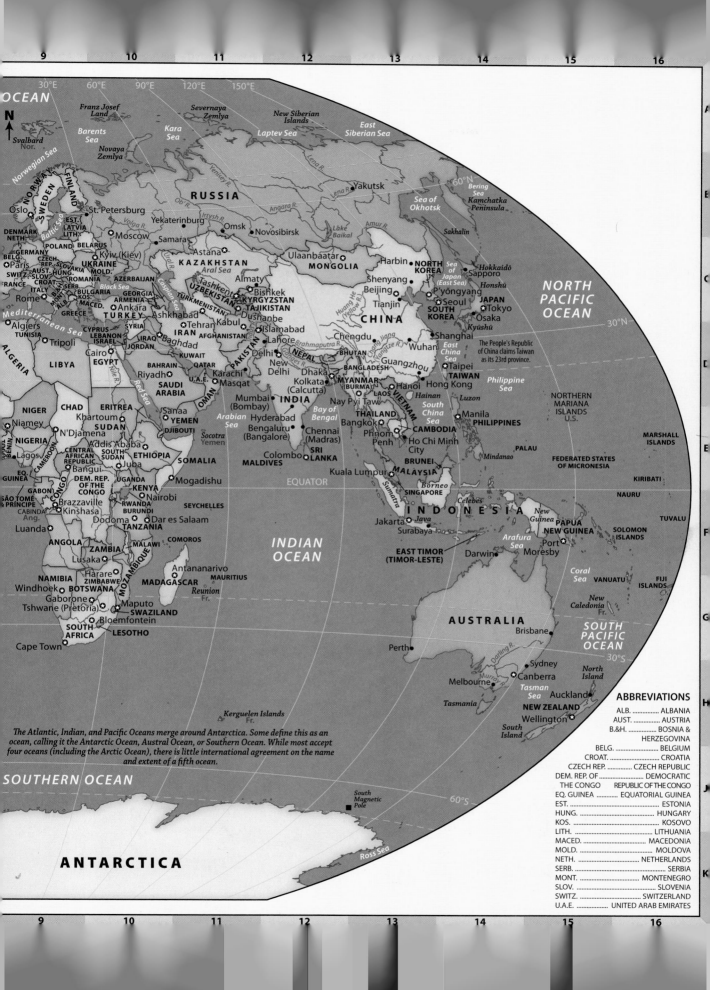

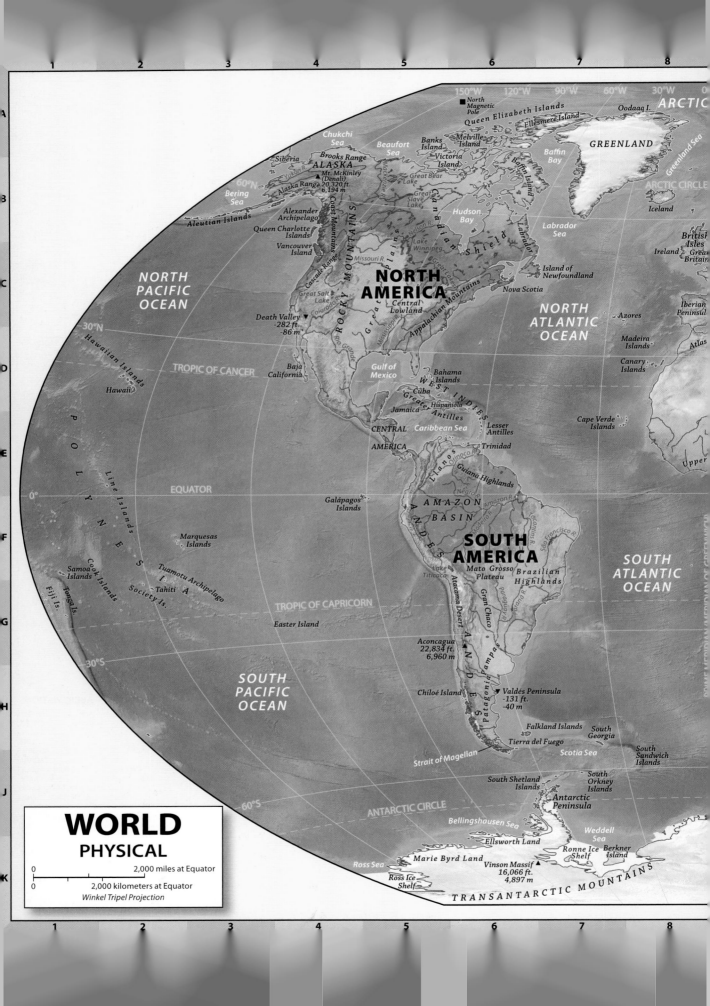

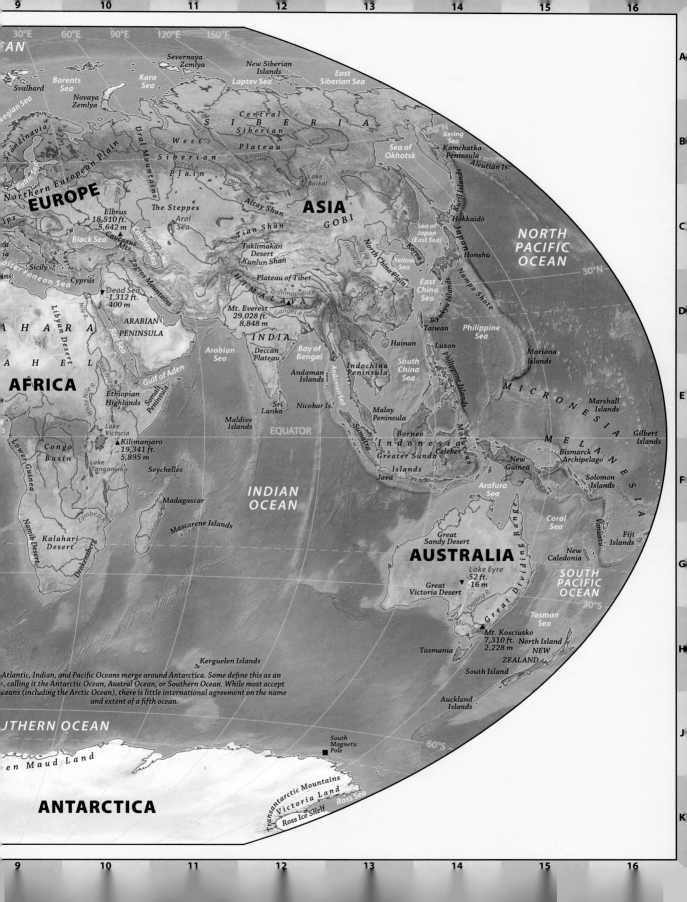

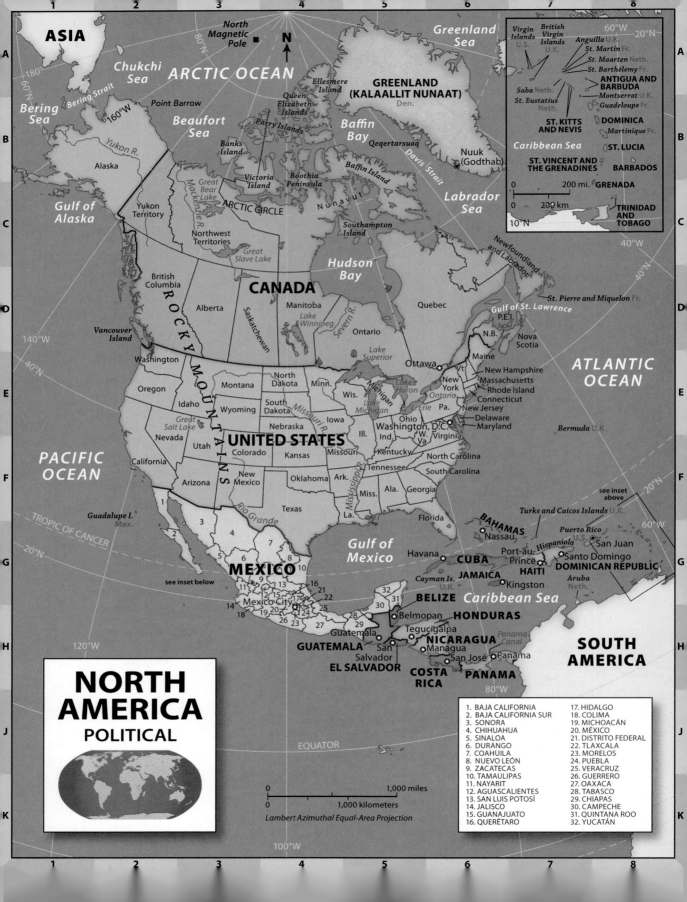

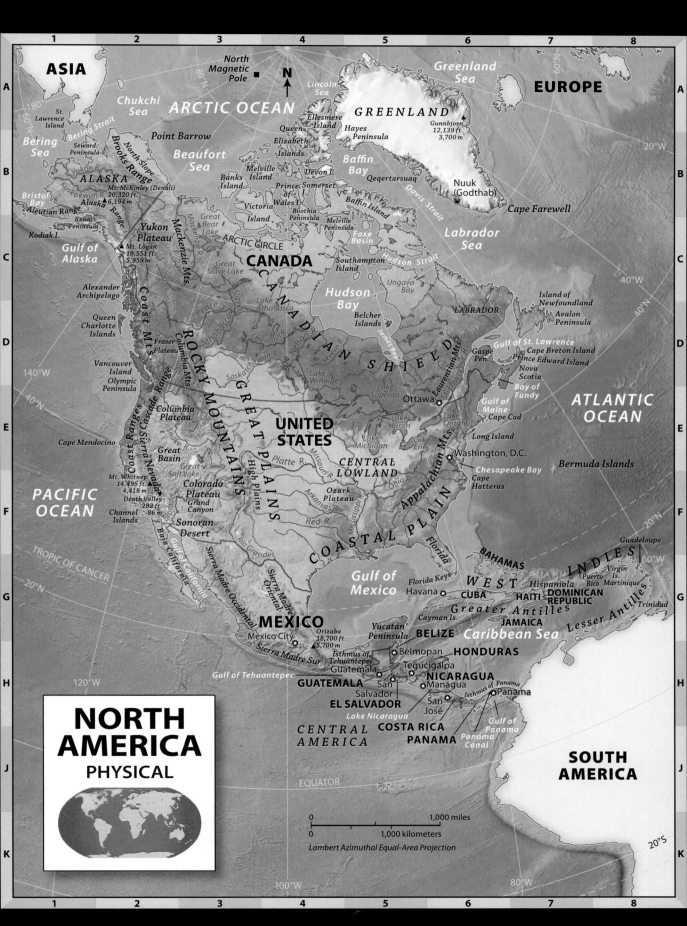

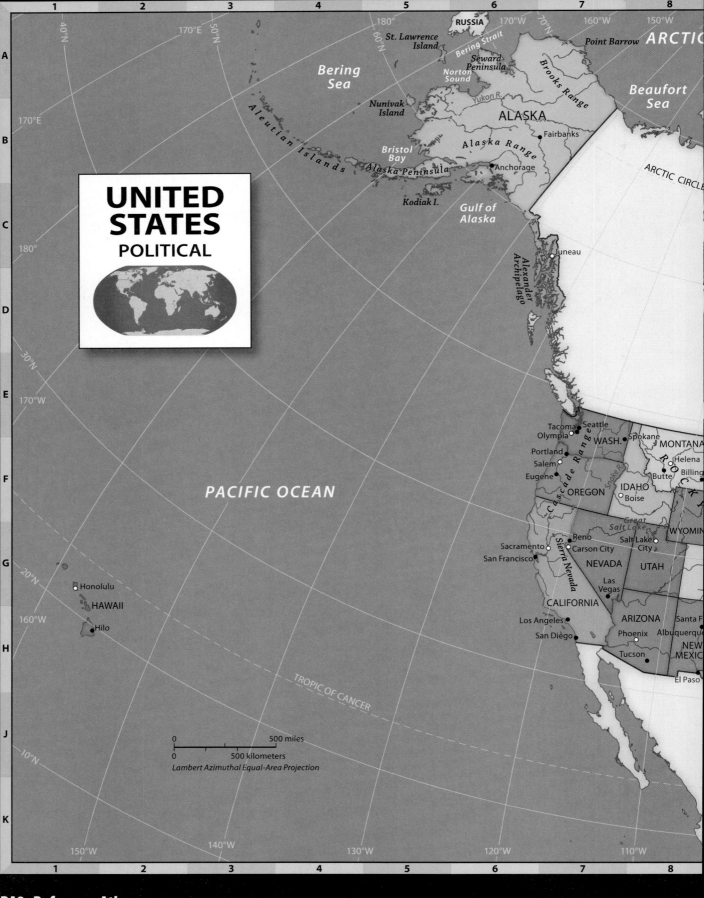

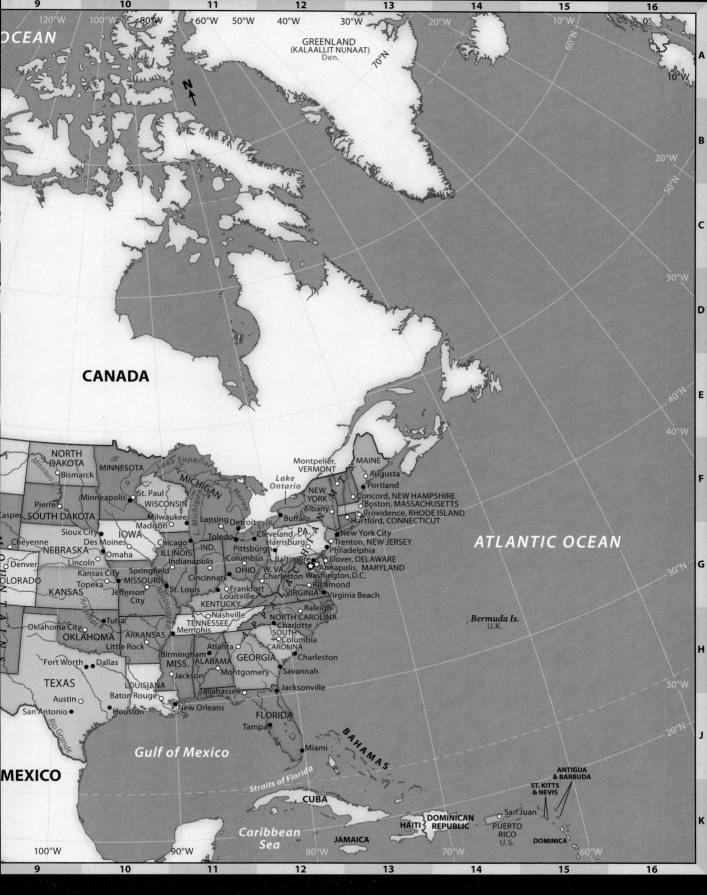

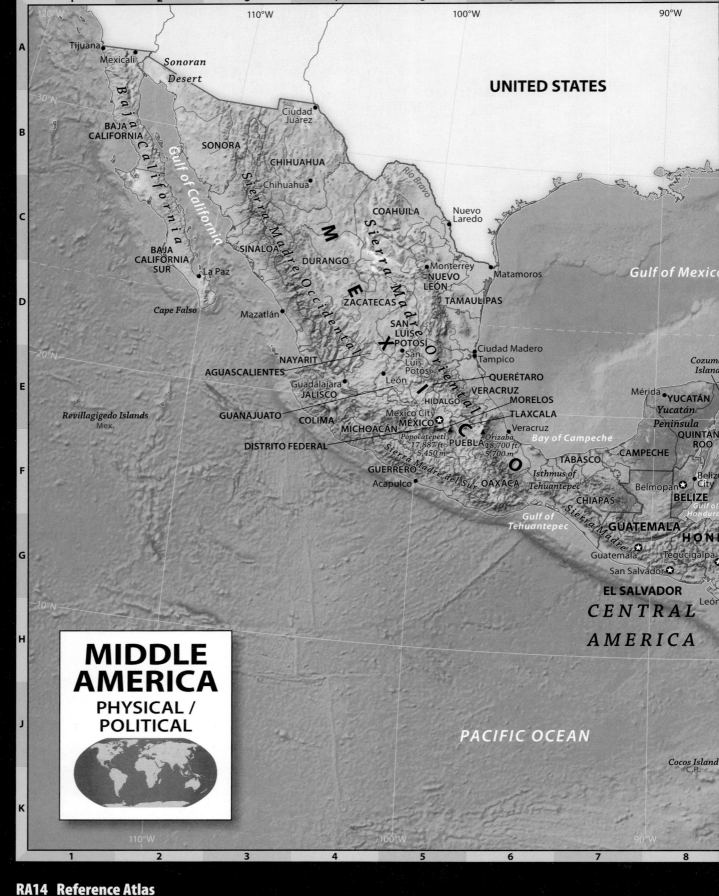

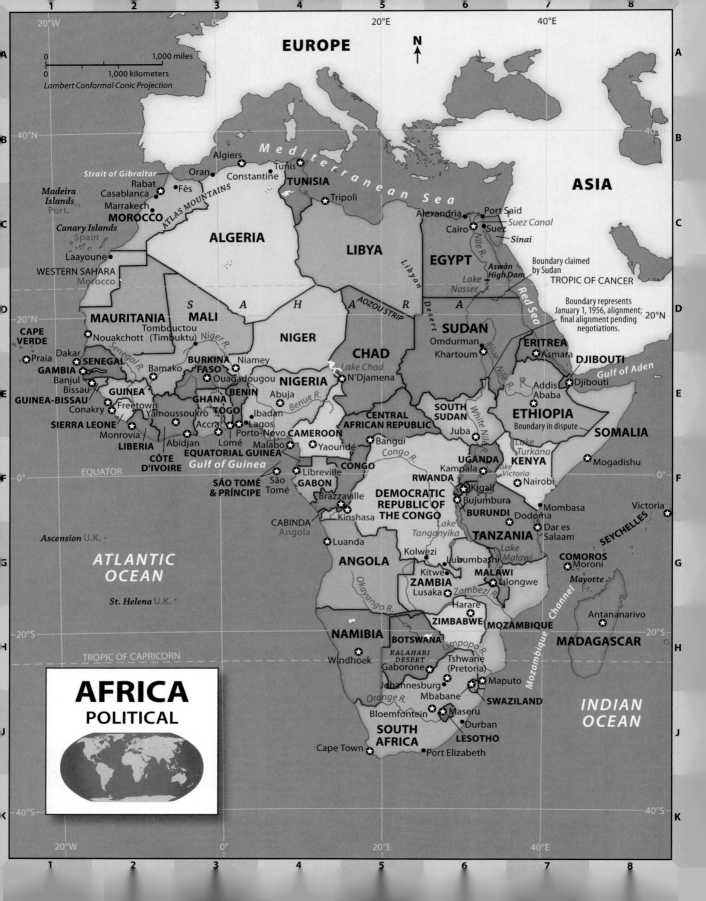

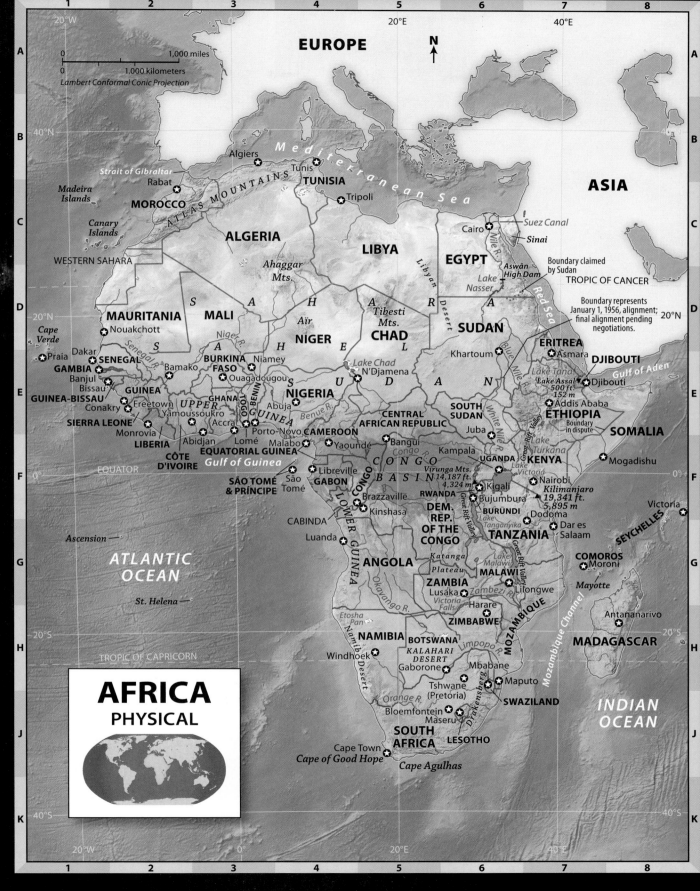

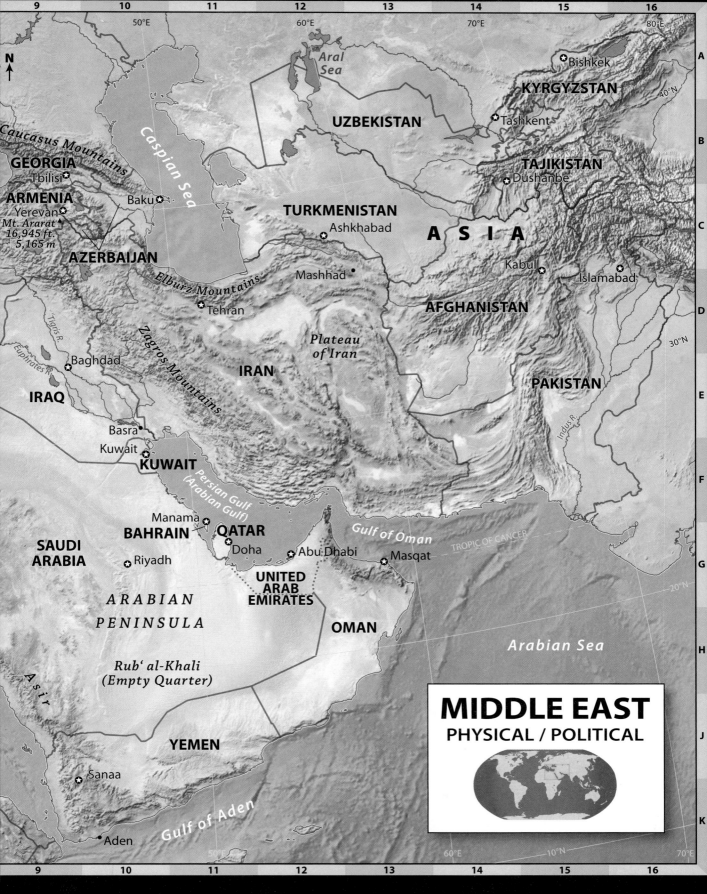

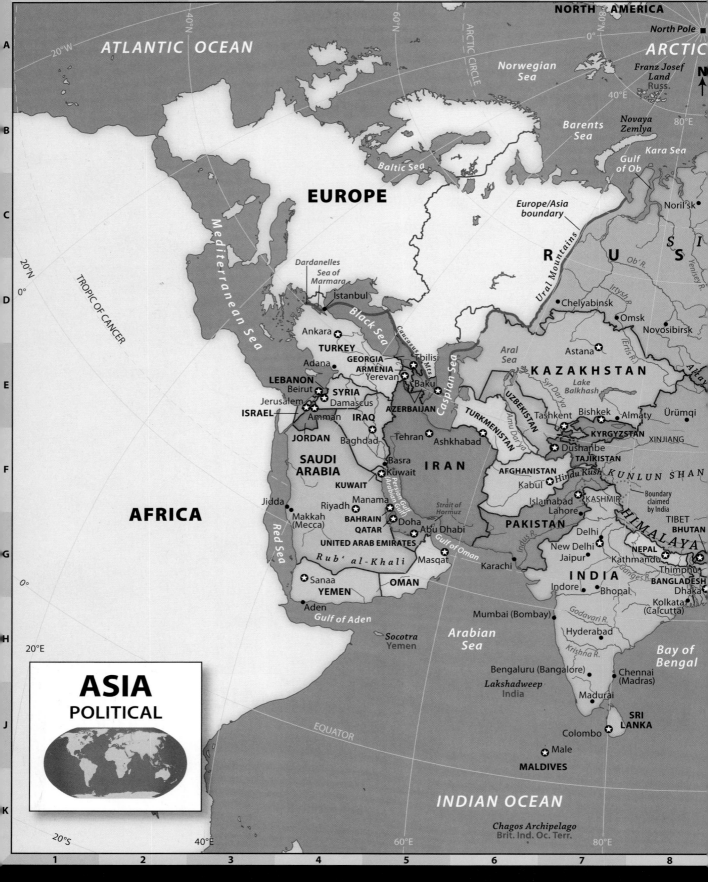

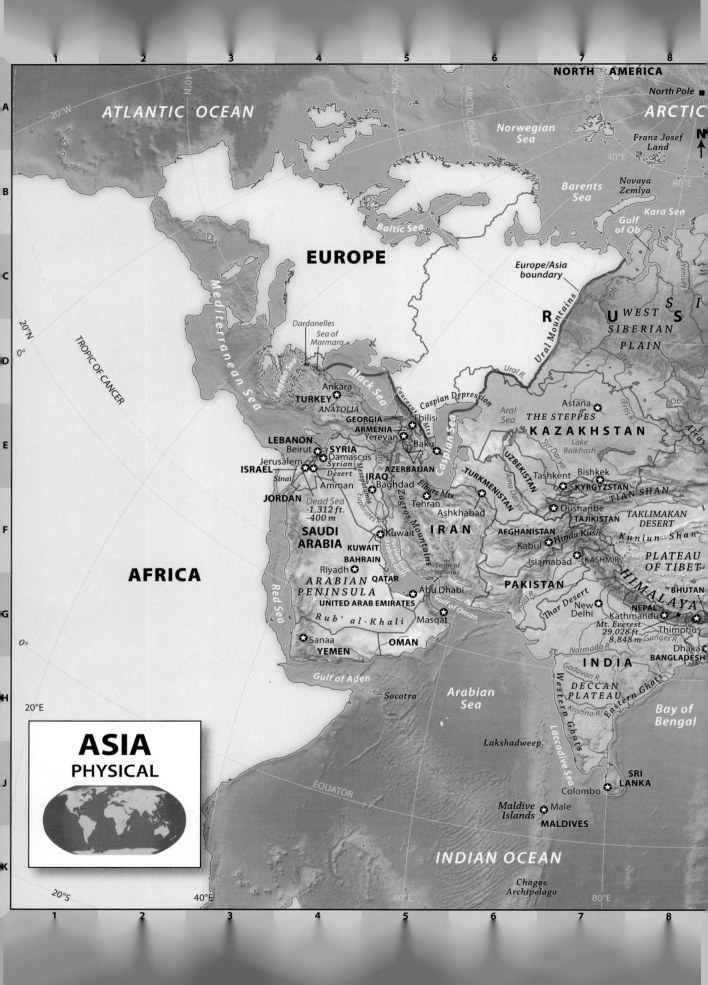

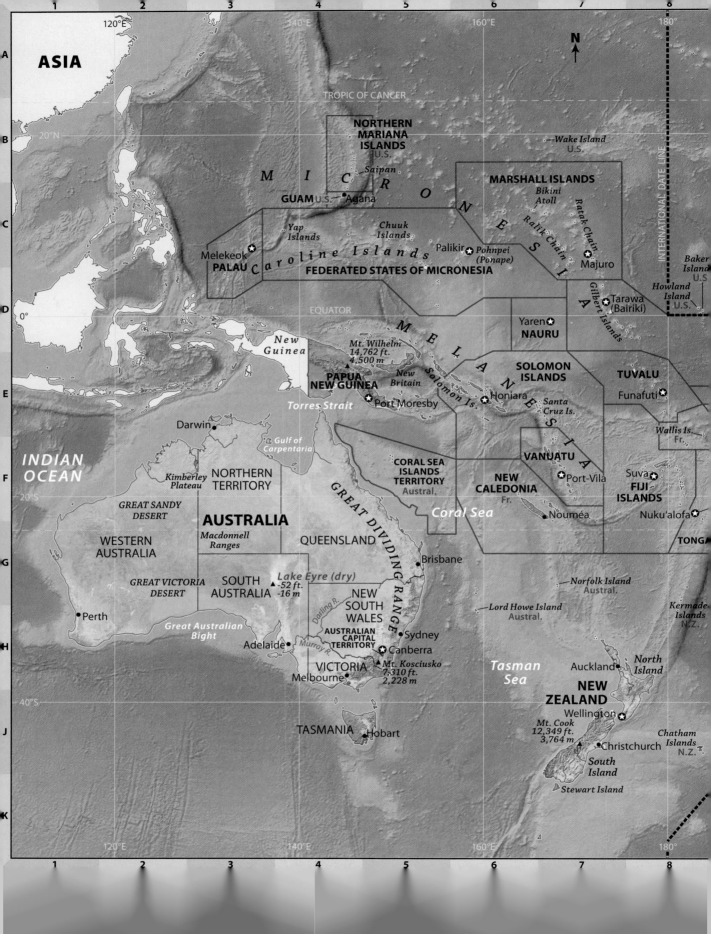

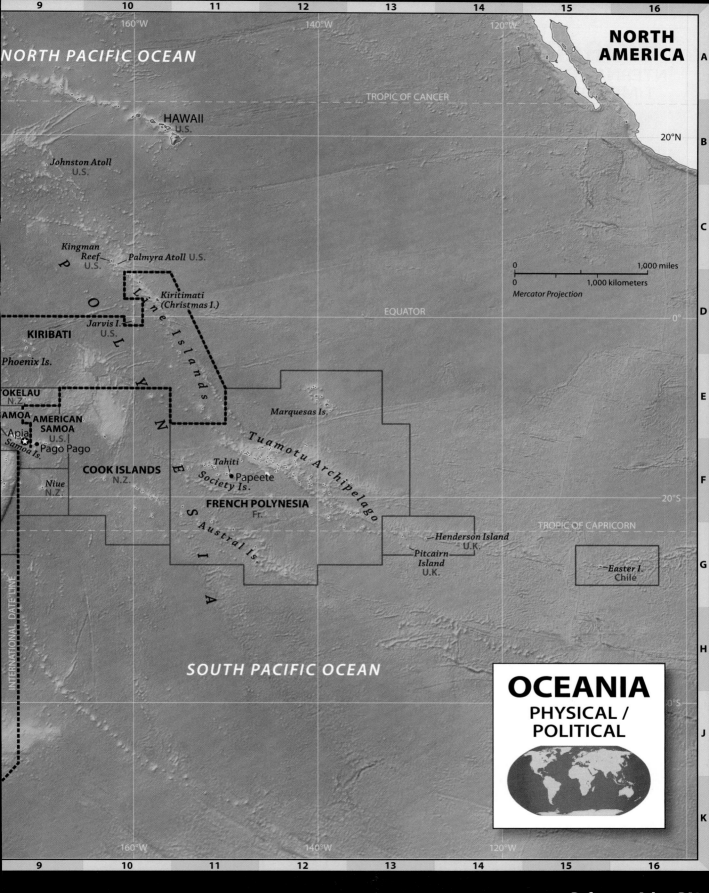

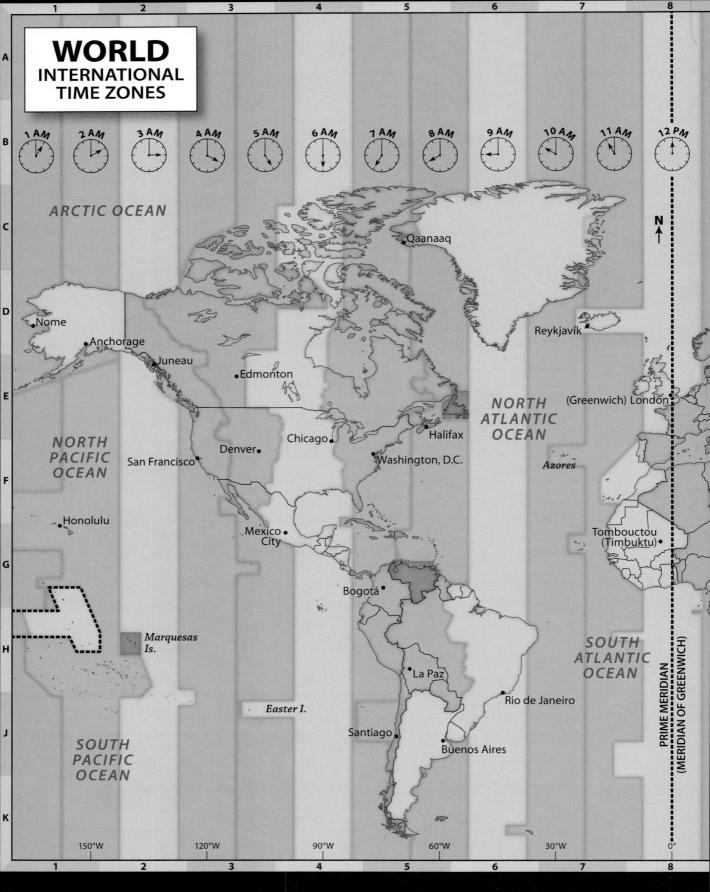

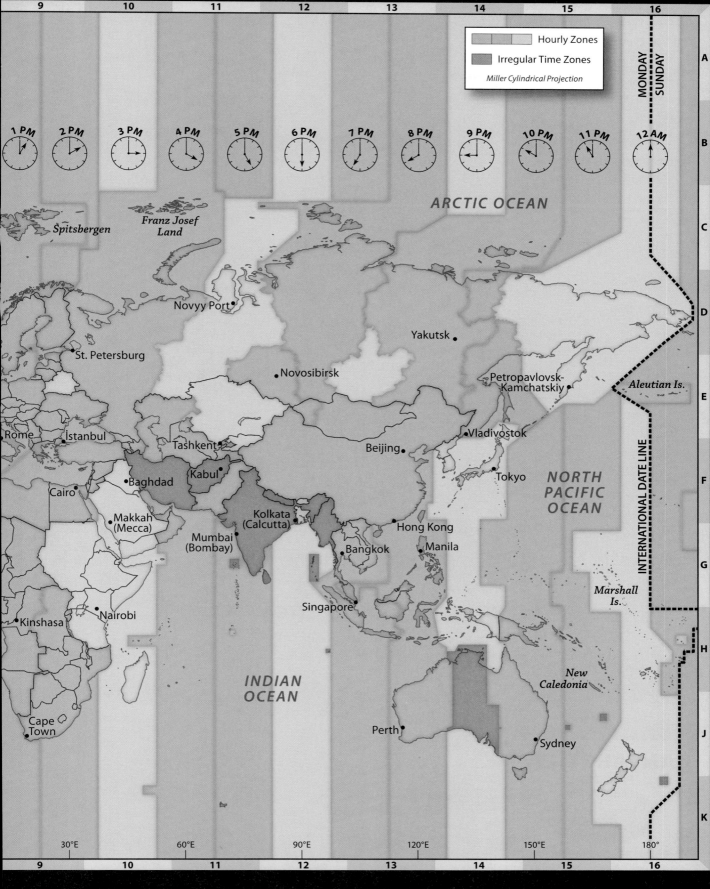

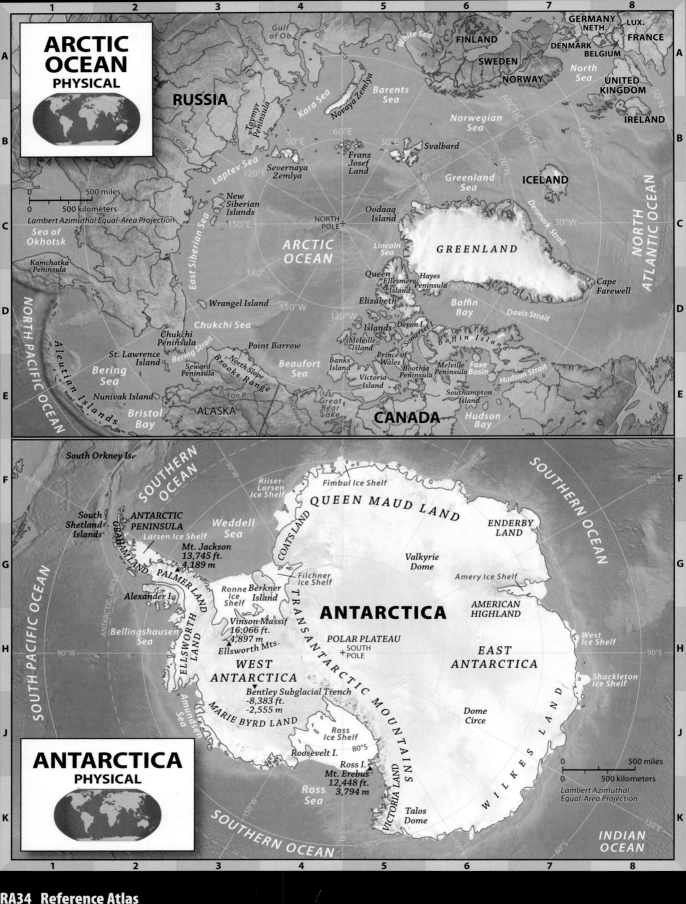

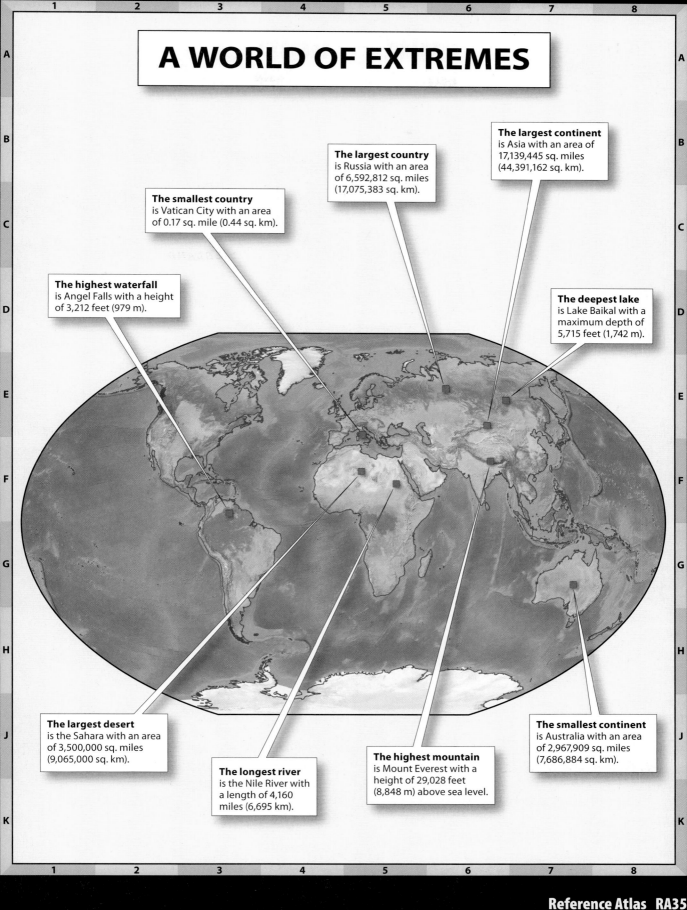

GEOGRAPHIC DICTIONARY

archipelago a group of islands

basin area of land drained by a given river and its branches; area of land surrounded by lands of higher elevations

bay part of a large body of water that extends into a shoreline, generally smaller than a gulf

canyon deep and narrow valley with steep walls

cape point of land that extends into a river, lake, or ocean

channel wide strait or waterway between two landmasses that lie close to each other; deep part of a river or other waterway

cliff steep, high wall of rock, earth, or ice

continent one of the seven large landmasses on the Earth

delta flat, low-lying land built up from soil carried downstream by a river and deposited at its mouth

divide stretch of high land that separates river systems

downstream direction in which a river or stream flows from its source to its mouth

escarpment steep cliff or slope between a higher and lower land surface

glacier large, thick body of slowly moving ice

gulf part of a large body of water that extends into a shoreline, generally larger and more deeply indented than a bay

harbor a sheltered place along a shoreline where ships can anchor safely

highland elevated land area such as a hill, mountain, or plateau

hill elevated land with sloping sides and rounded summit; generally smaller than a mountain

island land area, smaller than a continent, completely surrounded by water

isthmus narrow stretch of land connecting two larger land areas

lake a sizable inland body of water

lowland land, usually level, at a low elevation

mesa broad, flat-topped landform with steep sides; smaller than a plateau

mountain land with steep sides that rises sharply (1,000 feet or more) from surrounding land; generally larger and more rugged than a hill

mountain peak pointed top of a mountain

mountain range a series of connected mountains

mouth (of a river) place where a stream or river flows into a larger body of water

oasis small area in a desert where water and vegetation are found

ocean one of the four major bodies of salt water that surround the continents

ocean current stream of either cold or warm water that moves in a definite direction through an ocean

peninsula body of land jutting into a lake or ocean, surrounded on three sides by water

physical feature characteristic of a place occurring naturally, such as a landform, body of water, climate pattern, or resource

plain area of level land, usually at low elevation and often covered with grasses

plateau area of flat or rolling land at a high elevation, about 300 to 3,000 feet (90 to 900 m) high

reef a chain of rocks, coral or sand at or near the surface

river large natural stream of water that runs through the land

sea large body of water completely or partly surrounded by land

seacoast land lying next to a sea or an ocean

sound broad inland body of water, often between a coastline and one or more islands off the coast

source (of a river) place where a river or stream begins, often in highlands

strait narrow stretch of water joining two larger bodies of water

tributary small river or stream that flows into a large river or stream; a branch of the river

upstream direction opposite the flow of a river; toward the source of a river or stream

valley area of low land usually between hills or mountains

volcano mountain or hill created as liquid rock and ash erupt from inside the Earth

SCAVENGER HUNT

NETWORKS contains a wealth of information. The trick is to know where to look to access all the information in the book. If you complete this scavenger hunt exercise with your teachers or parents, you will see how the textbook is organized and how to get the most out of your reading and studying time. Let's get started!

1 How many lessons are in Chapter 2?

2 What does Unit 1 cover?

3 Where can you find the Essential Questions for each lesson?

4 In what three places can you find information on a Foldable?

5 How can you identify content vocabulary and academic vocabulary in the narrative?

6 Where do you find graphic organizers in your textbook?

7 You want to quickly find a map in the book about the world. Where do you look?

8 Where would you find the latitude and longitude for Dublin, Ireland?

9 If you needed to know the Spanish term for *earthquake*, where would you look?

10 Where can you find a list of all the charts in a unit?

McGraw-Hill **networks**™

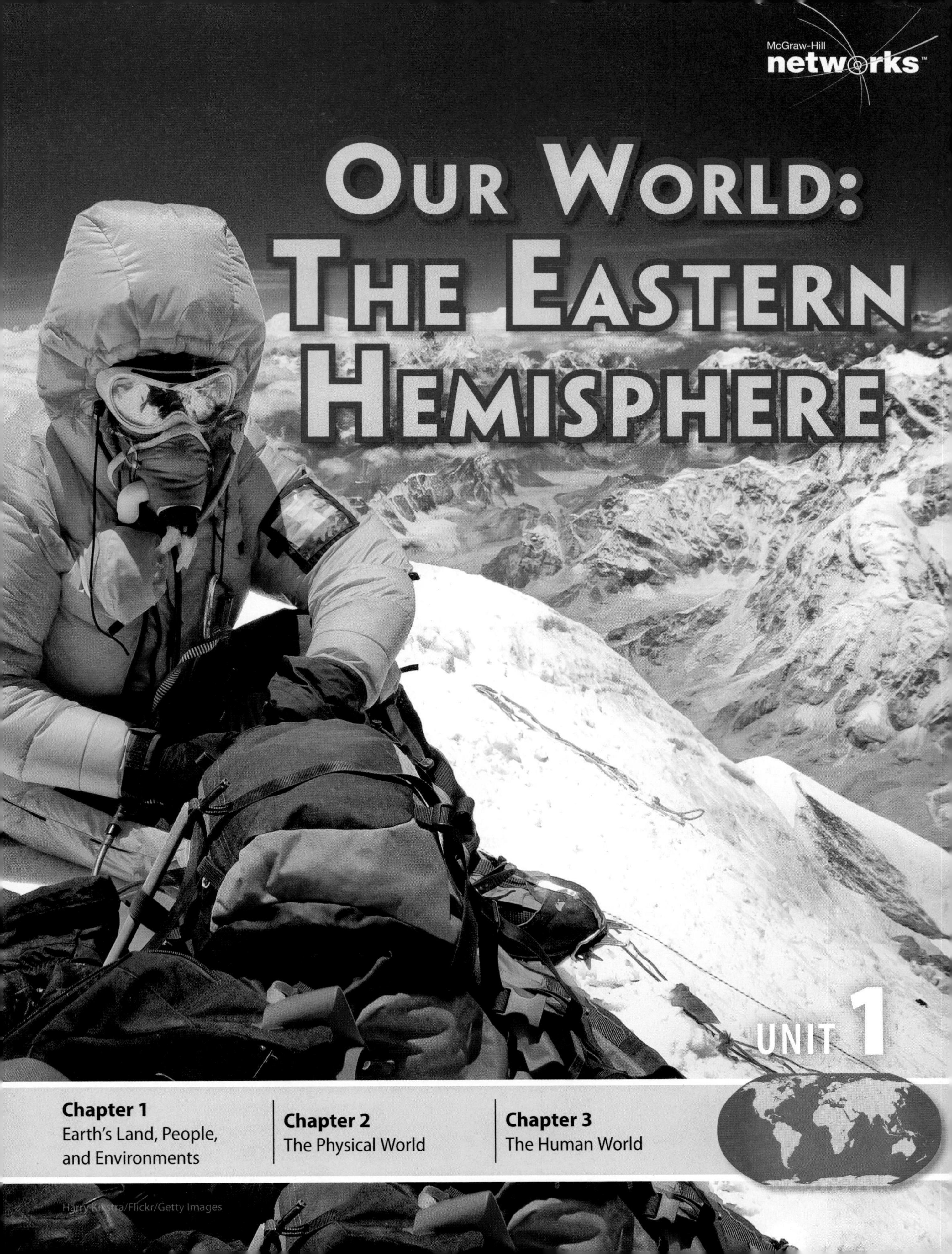

Our World: The Eastern Hemisphere

UNIT 1

Chapter 1
Earth's Land, People, and Environments

Chapter 2
The Physical World

Chapter 3
The Human World

Harry Kikstra/Flickr/Getty Images

EXPLORE the WORLD

Geography is the study of Earth and all of its variety. When you study geography, you learn about the planet's land, water, plants, and animals. Some people call Earth "the water planet." Do you know why? Water—in the form of streams, rivers, lakes, seas, and oceans—covers nearly 70 percent of Earth's surface.

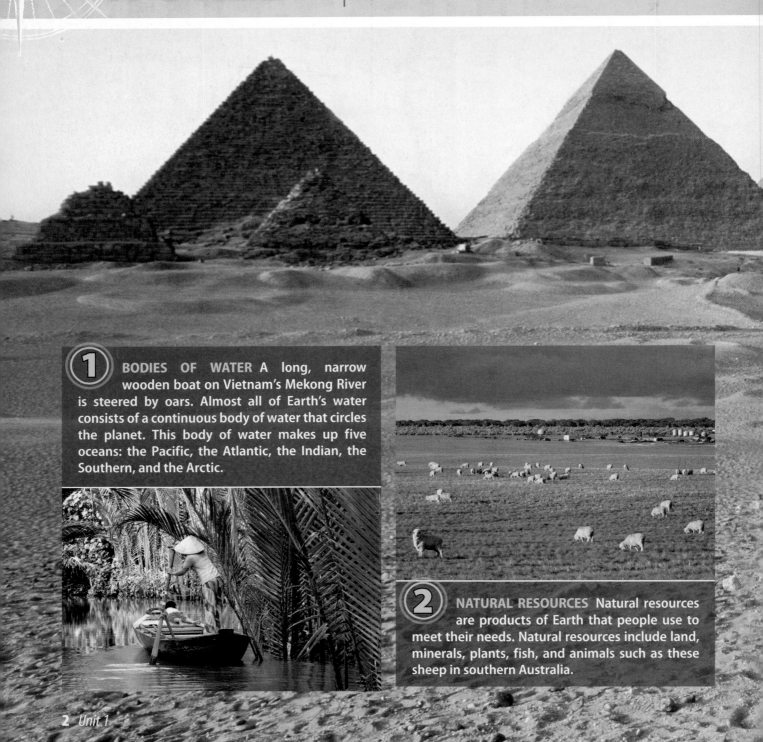

① BODIES OF WATER A long, narrow wooden boat on Vietnam's Mekong River is steered by oars. Almost all of Earth's water consists of a continuous body of water that circles the planet. This body of water makes up five oceans: the Pacific, the Atlantic, the Indian, the Southern, and the Arctic.

② NATURAL RESOURCES Natural resources are products of Earth that people use to meet their needs. Natural resources include land, minerals, plants, fish, and animals such as these sheep in southern Australia.

UNIT 1

3 LANDFORMS Landforms are features of the land, such as the great deserts of Egypt. Landforms influence where people live and how they relate to their environment.

FAST FACT

Earth's longest mountain range is underwater.

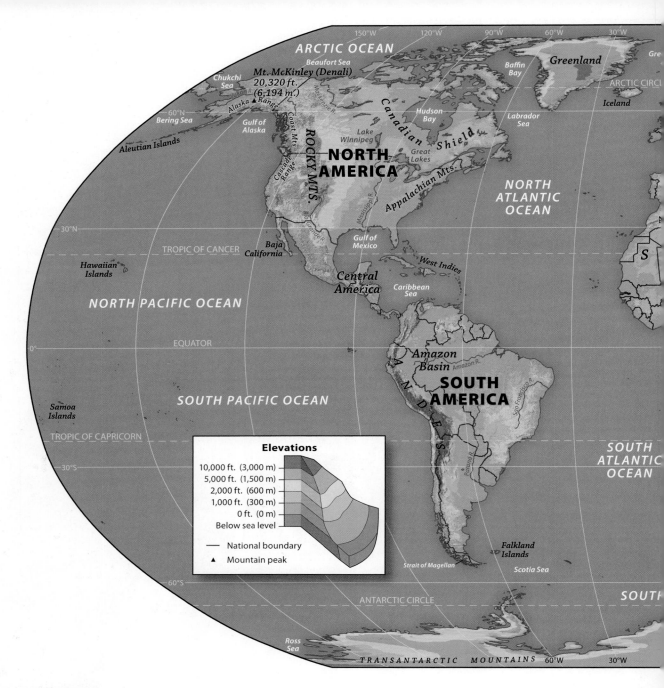

OUR WORLD

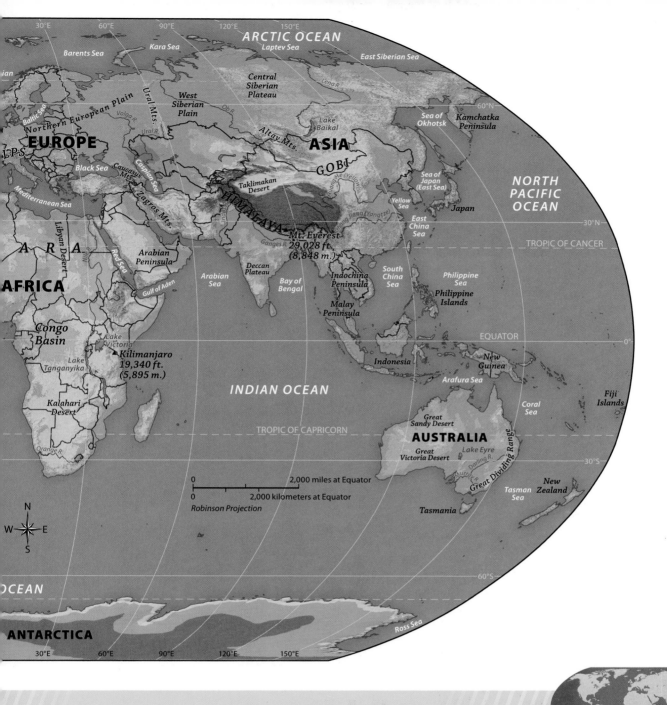

PHYSICAL

MAP SKILLS

1 THE GEOGRAPHER'S WORLD What part of Asia has the highest elevation?

2 THE GEOGRAPHER'S WORLD What body of water is located west of Australia?

3 PLACES AND REGIONS How would you describe Southern Africa?

Unit 1 **5**

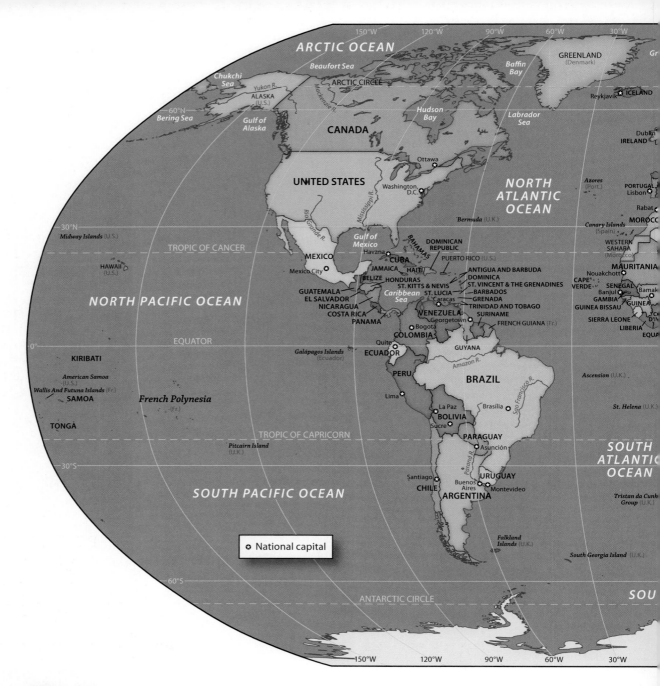

OUR WORLD

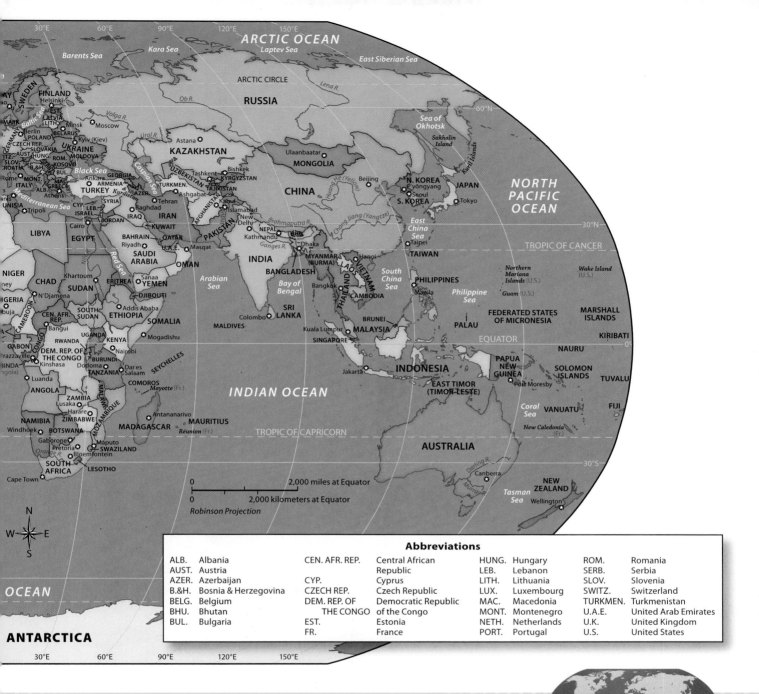

POLITICAL

MAP SKILLS

1. **PLACES AND REGIONS** How would you describe the region north of Australia?

2. **THE GEOGRAPHER'S WORLD** Which country is located west of Egypt?

3. **PLACES AND REGIONS** What is the capital of Mongolia?

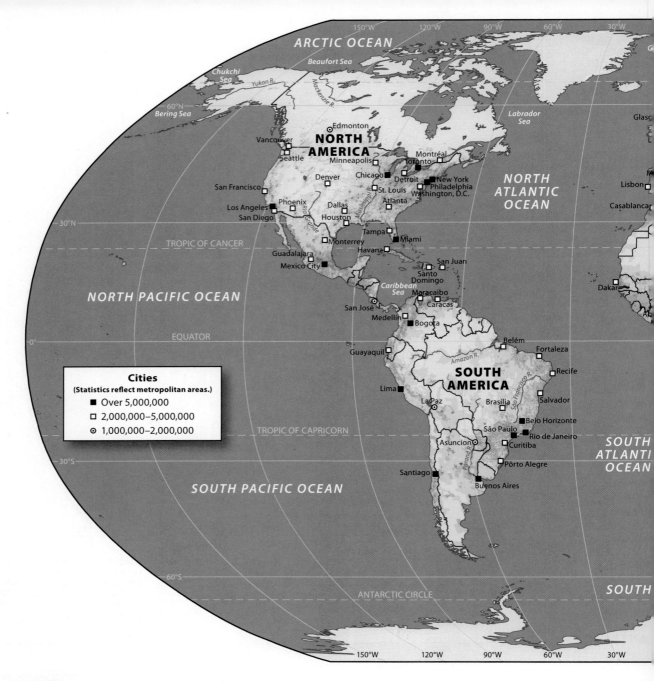

Our World

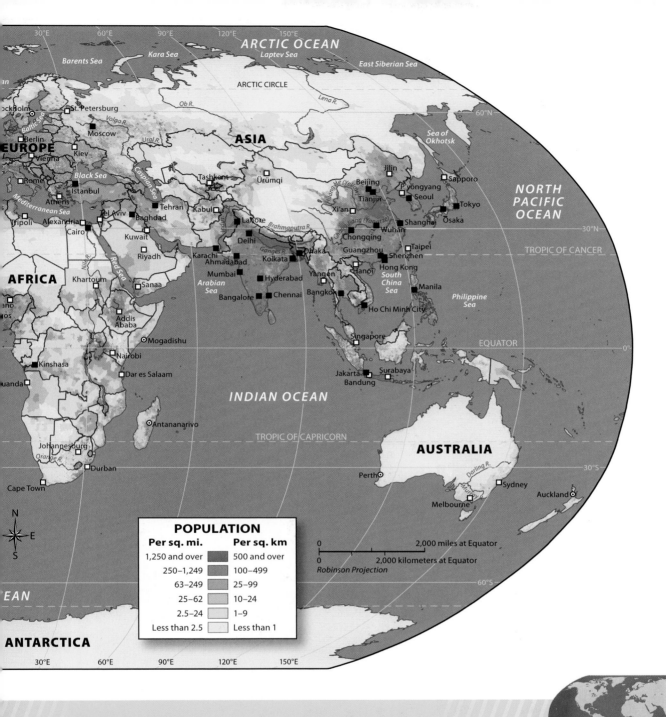

POPULATION DENSITY

MAP SKILLS

1. **PLACES AND REGIONS** What parts of Asia are the most densely populated?

2. **PLACES AND REGIONS** Which part of Africa has the lowest population density?

3. **ENVIRONMENT AND SOCIETY** In general, what population pattern do you see in northern Africa?

Unit 1 9

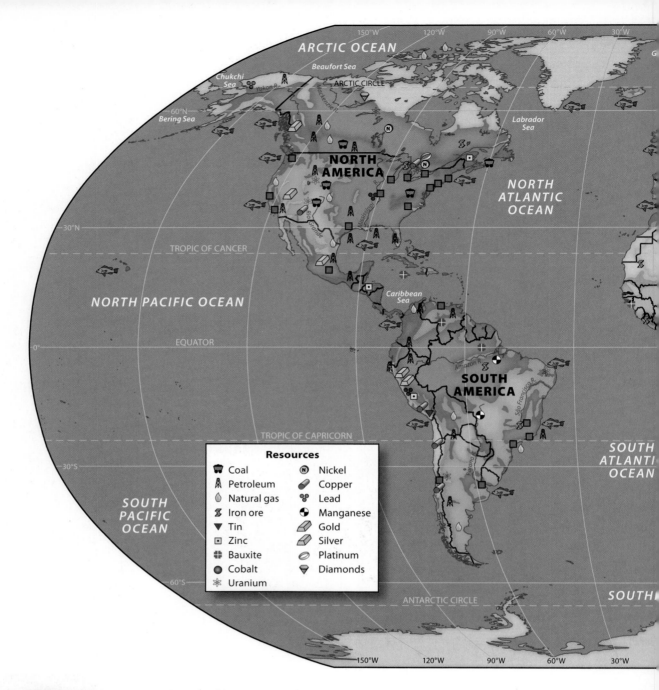

OUR WORLD

Unit 1

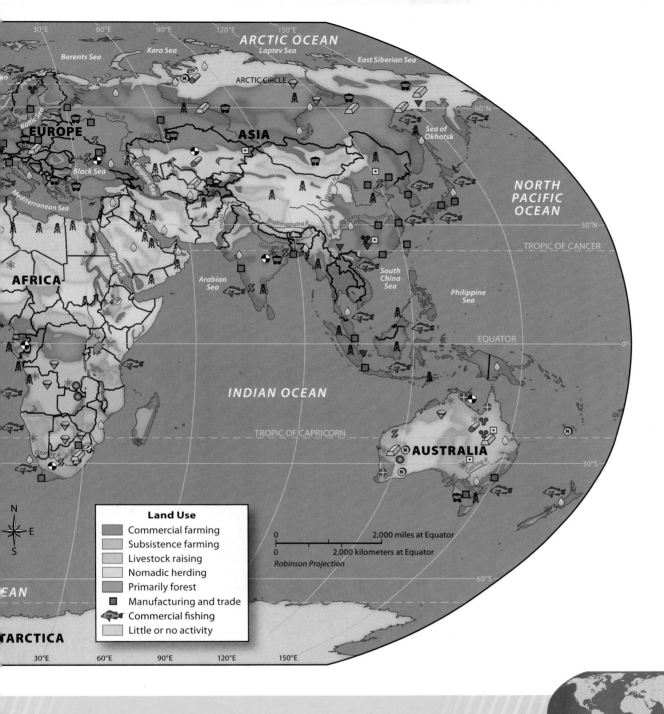

ECONOMIC RESOURCES

MAP SKILLS

1. **ENVIRONMENT AND SOCIETY** What economic activity is found along most coastal regions?

2. **HUMAN GEOGRAPHY** Describe the general use of land in northern Africa.

3. **PLACES AND REGIONS** Which area produces the most oil—Australia or Southwest Asia?

Unit 1 **11**

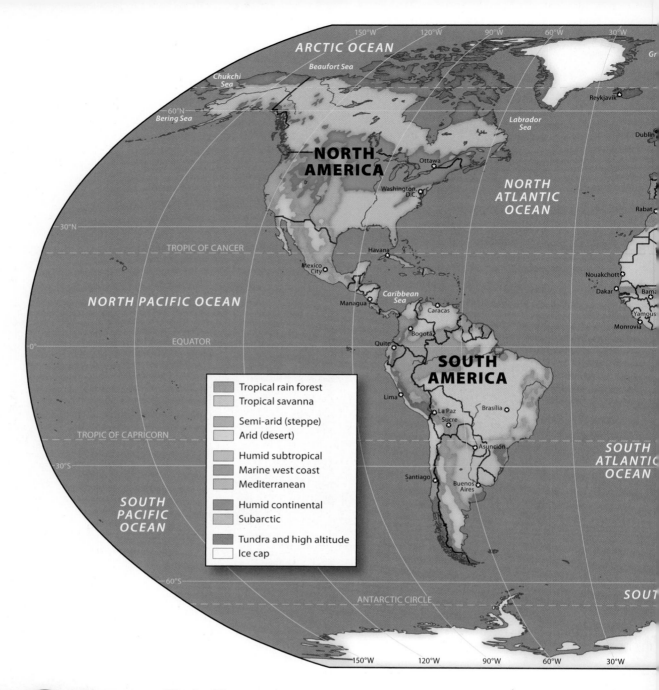

OUR WORLD

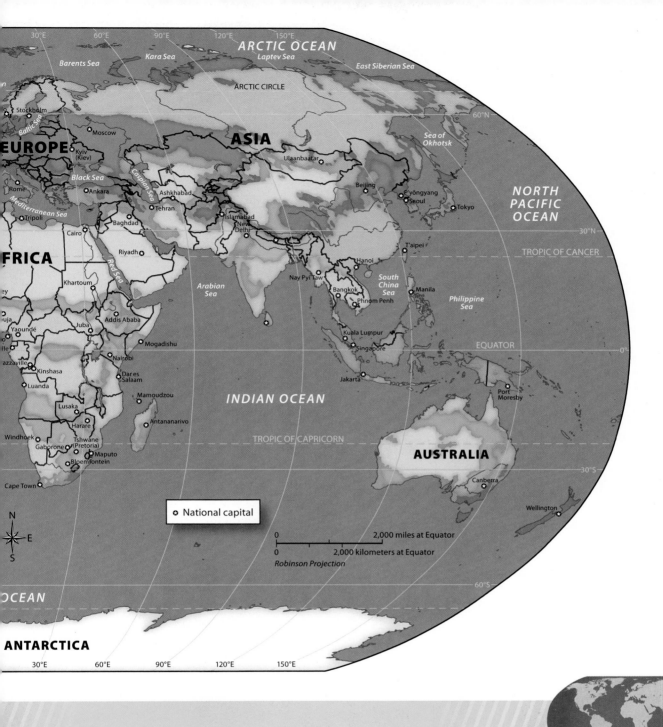

CLIMATE

MAP SKILLS

1. **PHYSICAL GEOGRAPHY** Which climate zones appear in Australia?

2. **THE GEOGRAPHER'S WORLD** Which continent receives more rain—Australia or Asia? Why?

3. **PHYSICAL GEOGRAPHY** In general, how does the climate of Southern Africa compare with the climate of North Africa?

Earth's Land, People, and Environments

ESSENTIAL QUESTION • *How does geography influence the way people live?*

A geographer drills for an ice sample.

networks

There's More Online about Earth's Land, People, and Environments.

CHAPTER 1

Lesson 1
How Geographers Think

Lesson 2
Geographers and Their Tools

The Story Matters...

Since ancient times, people have drawn maps to show their known world. As people explored, they came into contact with different places and people, which expanded their understanding of the world. Today, what we know about the world continues to grow as geographers study the world's environments with the latest technology. More importantly, by understanding the connections between humans and the environment, geographers can find solutions to significant problems.

Go to the Foldables library in the back of your book to make a Foldable that will help you take notes while reading this chapter.

Chapter 1 Earth's Land, People, and Environments

Geography is the study of Earth in all of its variety. When you study geography, you learn about the physical features and the living things—humans, plants, and animals—that inhabit Earth.

Step Into the Place

MAP FOCUS Use the map to answer the following questions.

1. **THE GEOGRAPHER'S WORLD** What are the names of the large landmasses on the map?

2. **THE GEOGRAPHER'S WORLD** What are the names of the large bodies of water on the map?

3. **THE GEOGRAPHER'S WORLD** What do you think the blue lines are that appear within the landmasses?

4. **CRITICAL THINKING**
   Analyzing How is this world map similar to other maps you have seen?

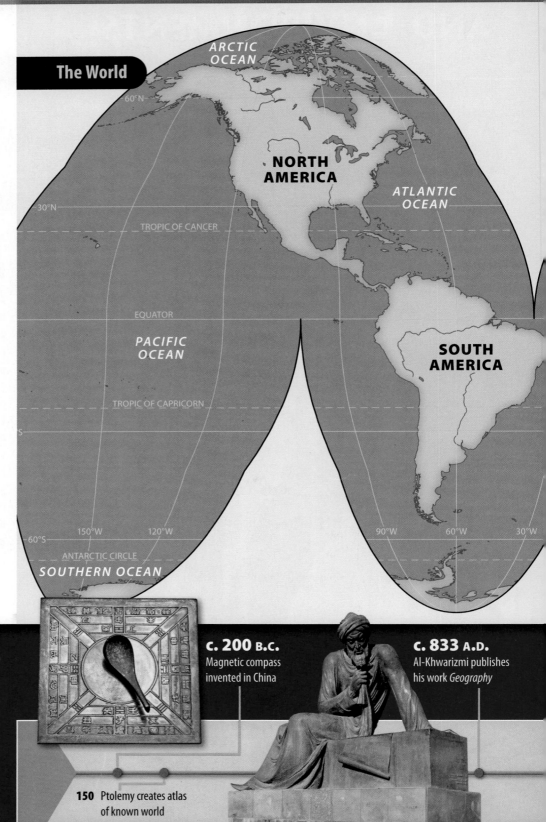

The World

Step Into the Time

DESCRIBING Choose an event from the time line and write a paragraph describing how it might have changed how people understood or viewed the world in which they lived.

c. 200 B.C. Magnetic compass invented in China

150 Ptolemy creates atlas of known world

c. 833 A.D. Al-Khwarizmi publishes his work *Geography*

16 Chapter 1

(l) Hans-Joachim Schneider/Alamy; (r) Egmont Strigl/age fotostock

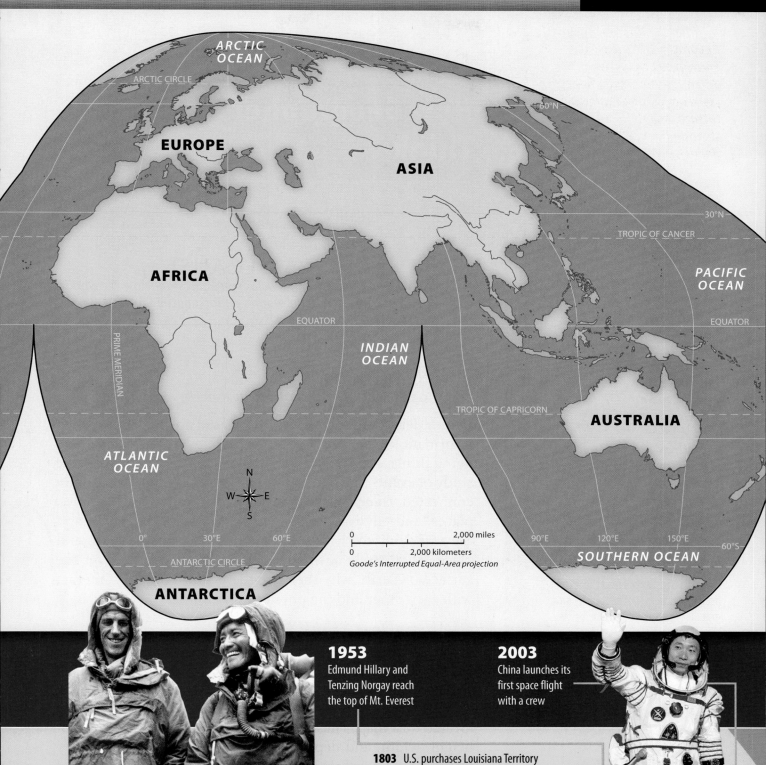

networks

There's More Online!

- ☑ **CHART/GRAPH** Six Essential Elements of Geography
- ☑ **IMAGES** Places Change Over Time
- ☑ **ANIMATION** Regions of Earth
- ☑ **VIDEO**

Reading **HELP**DESK

Academic Vocabulary

- dynamic
- component

Content Vocabulary

- geography
- spatial
- landscape
- relative location
- absolute location
- latitude
- Equator
- longitude
- Prime Meridian
- region
- environment
- landform
- climate
- resource

TAKING NOTES: *Key Ideas and Details*

Identifying As you read the lesson, list the five themes of geography on a graphic organizer like the one below.

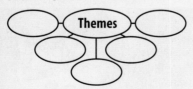

Lesson 1
How Geographers Think

ESSENTIAL QUESTION • *How does geography influence the way people live?*

IT MATTERS BECAUSE
Thinking like a geographer helps you understand how the world works and appreciate the world's remarkable beauty and complexity.

Thinking Spatially

GUIDING QUESTION *What does it mean to think like a geographer?*

Our understanding of the world is based on a combination of information from many sources. Biology is the study of how living things survive and relate to one another. History is the study of events that occur over time and how those events are connected. **Geography** is the study of Earth and its peoples, places, and environments. Geographers look at people and the world in which they live mainly in terms of space and place. They study such topics as where people live on the surface of Earth, why they live there, and how they interact with each other and the physical environment.

Geographers Think Spatially

Geography, then, emphasizes the spatial aspects of the world. **Spatial** refers to Earth's features in terms of their locations, their shapes, and their relationships to one another.

Physical features such as mountains and lakes can be located on a map. These features can be measured in terms of height, width, and depth. Distances and directions to other features can be determined. The human world also has spatial dimensions. Geographers study the size and shape of cities, states, and countries. They consider how close or far apart these human features are to one another. Geographers also think about the relationships between human features and physical features.

But thinking spatially is more than just the study of the location or size of things. It means looking at the characteristics of Earth's features. Geographers ask what mountains in different locations are made of. They examine what kinds of fish live in different lakes. They study the layout of cities and think about how easy or difficult it is for people to move around in them.

The Perspective of Place

Locations on Earth are made up of different combinations of physical and human characteristics. Physical features such as climate, landforms, and vegetation combine with human features such as population, economic activity, and land use. These combinations create what geographers call places.

Places are locations on Earth that have distinctive characteristics that make them meaningful to people. The places where we live, work, and go to school are important to us. Our home is an important place. Even small places such as our bedroom or a classroom often have a unique and special meaning. In the same way, larger locations, such as our hometown, our country, or even Earth, are places that have meaning for people.

One way that geographers learn about places is by studying landscapes. **Landscapes** are portions of Earth's surface that can be viewed at one time and from one location. They can be as small as the view from the front porch of your home, or they can be as large as the view from a tall building that includes the city and surrounding countryside.

The geography theme of *place* describes all of the characteristics that give an area its own special quality. The Taj Mahal was built by an Indian ruler in the 1600s. The white marble dome and the slender towers reflect the Islamic influence at the time.

Whether we visit a landscape or we look at photographs of the landscape, we can tell much about the people who live there. Geographers look at landscapes and try to explain their unique combinations of physical and human features. As you study geography, notice the great variety in the world's landscapes.

The Perspective of Experience

Geography is not something you learn about only in school or just from books. Geography is something you experience every day.

We all live in the world. The seasons change. Birds chirp and car horns honk. We walk on sidewalks and in forests. People ride in cars along streets and highways. We shop in malls and grocery stores. We fly in airplanes to distant places. By surfing the Internet or watching TV, we learn about people and events in our neighborhood, our country, and the world.

This is all geography. By learning about geography in school, we can better appreciate and understand this world in which we live.

A Changing World

Earth is **dynamic**, or always changing. Rivers shift course. Volcanoes suddenly erupt, forming mountains or collapsing the peaks of mountains. The pounding surf removes sand from beaches.

The things that people make change, too. Farmers shift from growing one crop to another. Cities grow larger. Nations expand into new areas.

Geographers, then, study how places change over time. They try to understand what impact those changes have. What factors made a city grow? What effect did a growing city have on the people who live there? What effect did the city's growth have on nearby communities and on the land and water near it? Answering questions like these is part of the field of geography.

☑ READING PROGRESS CHECK

Describing Why do geographers study how places change?

Geography Themes

GUIDING QUESTION *How can you make sense of a subject as large as Earth and its people?*

To help study geography, geographers organize information about the world into five themes. These themes help them view and understand Earth.

Location

Location is where something is found on Earth. There are two types of location. **Relative location** describes where a place is compared to another place. This approach often uses the cardinal directions— north, south, east, and west. A school might be on the east side of

Academic Vocabulary

dynamic always changing

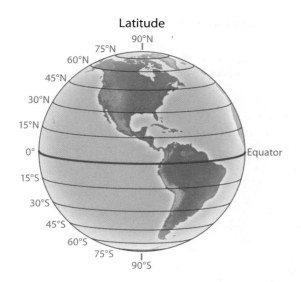
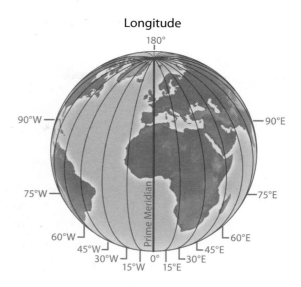

town. Relative location can also tell us about the characteristics of a place. For example, knowing that New Orleans is near the mouth of the Mississippi River helps us understand why the city became an important trading port.

Absolute location is the exact location of something. An address like 123 Main Street is an absolute location. Geographers identify the absolute location of places using a system of imaginary lines called latitude and longitude. Those lines form a grid for locating a place precisely.

Lines of **latitude** run east to west, but they measure distance on Earth in a north-to-south direction. One of these lines, the **Equator**, circles the middle of Earth. This line is equally distant from the North Pole and the South Pole. Other lines of latitude between the Equator and the North and South Poles are assigned a number from 1° to 90°. The higher the number, the farther the line is from the Equator. The Equator is 0° latitude. The North Pole is at 90° north latitude (90° N), and the South Pole is at 90° south latitude (90° S).

Lines of **longitude** run from north to south, but they measure distance on Earth in an east-to-west direction. They go from the North Pole to the South Pole. These lines are also called *meridians*. The **Prime Meridian** is the starting point for measuring longitude. It runs through Greenwich, England, and has the value of 0° longitude. There are 180 lines of longitude to the east of the Prime Meridian and 180 lines to the west. They meet at the meridian 180°, which is the International Date Line.

Geographers use latitude and longitude to locate anything on Earth. In stating absolute location, geographers always list latitude first. For example, the absolute location of Washington, D.C., is 38° N, 77° W.

Lines of latitude circle Earth parallel to the Equator and measure the distance north or south of the Equator in degrees. Lines of longitude circle Earth from the North Pole to the South Pole. These lines measure distances east or west of the Prime Meridian.

▶ **CRITICAL THINKING**

The Geographer's World At what degree of latitude is the Equator located?

Disney World, located in Orlando, Florida, attracts millions of visitors every year.

▶ **CRITICAL THINKING**
Human Geography What effect do you think Disney World has on the surrounding communities?

Place

Another theme of geography is place. The features that help define a place can be physical or human.

Why is Denver called the "Mile High City"? Its location one mile above sea level gives it a special character. Why does New Orleans have the nickname "the Crescent City"? It is built on a crescent-shaped bend along the Mississippi River. That location has had a major impact on the city's growth and how its people live.

Region

Although places are unique, two or more places can share characteristics. Places that are close to one another and share some characteristics belong to the same **region**. For example, Hong Kong and Macau are located in southeastern China. They have some features in common, such as nearness to the ocean. Both cities also have mostly warm temperatures throughout the year.

In the case of those two cities, the region is defined using physical characteristics. Human characteristics can also define regions. For instance, the countries of North Africa are part of the same region. One reason is that most of the people living in these countries follow the same religion—Islam.

Geographers study regions so they can identify the broad patterns of larger areas. They can compare and contrast the features in one region with those in another. They also examine the special features that make each place in a region distinct from the others.

Human-Environment Interaction

People and the environment interact. That is, they affect each other. The physical characteristics of a place affect how people live. Flat, rich, well-watered soil is good for farming. Mountains full of coal can be mined. The environment can present all kinds of hazards, such as floods, droughts, earthquakes, and volcanic eruptions.

People affect the environment, too. They blast tunnels through mountains to build roadways and drain swamps to make farmland. Although these actions can improve life, they can also harm the environment. Exhaust from cars on the roadways can pollute the air, and turning swamps into farms destroys natural ecosystems.

The **environment** is the natural surroundings of a place. It includes several key features. One is **landforms**, or the shape and nature of the land. Hills, mountains, and valleys are types of landforms. The environment also includes the presence or absence of a body of water. Cities located on coastlines, like Cape Town, South Africa, have different characteristics than inland cities, like Johannesburg, South Africa.

Weather and climate also play a role in how people interact with their environment. The average weather in a place over a long period of time is called its **climate**. The climate in Norlisk, Russia, is marked by long, cold, wet winters and short, mild summers. Jakarta's climate in Indonesia is warm year-round. People in Norlisk interact with their environment differently than the people in Jakarta do.

Another **component**, or part, of the environment is **resources**. These are materials that can be used to produce crops or other products. Forests are a resource because the trees can be used to build homes and furniture. Oil is a resource because it can be used as a source of energy.

Movement

Geographers also look at how people, products, ideas, and information move from one place to another. People have many reasons for moving. Some move because they find a better job.

Academic Vocabulary

component part

Climate affects people in many ways. (Below left) Hot, humid weather during a summer day in Seoul, South Korea, makes it possible for children to swim in nearby streams. (Right) The people must also be prepared to deal with bitterly cold winter days.

CHART SKILLS

THE SIX ESSENTIAL ELEMENTS

Element	Definition
The World in Spatial Terms	Geography studies the location and spatial relationships among people, places, and environments. Maps reveal the complex spatial interactions.
Places and Regions	The identities of individuals and people are rooted in places and regions. Distinctive combinations of human and physical characteristics define places and regions.
Physical Systems	Physical processes, like wind and ocean currents, plate tectonics, and the water cycle, shape Earth's surface and change ecosystems.
Human Systems	Human systems are things like language, religion, and ways of life. They also include how groups of people govern themselves and how they make and trade products and ideas.
Environment and Society	Geography studies how the environment of a place helps shape people's lives. Geography also looks at how people affect the environment in positive and negative ways.
The Uses of Geography	Understanding geography and knowing how to use its tools and technologies helps people make good decisions about the world and prepares people for rewarding careers.

Being aware of the six essential elements will help you sort out what you are learning about geography.

▶ **CRITICAL THINKING**
Identifying The study of volcanoes, ocean currents, and climate is part of which essential element?

Sometimes, people are forced to move because of war, famine, or religious or racial prejudice. Movement by large numbers of people can have important effects. People may face shortages of housing and other services. If new arrivals to an area cannot find jobs, poverty levels can rise.

In our interconnected world, a vast number of products move from place to place. Apples from Washington State move to supermarkets in Texas. Clothes produced in Thailand end up in American stores. Oil from Saudi Arabia powers cars and trucks across the United States. All this movement relies on transportation systems that use ships, railroads, airplanes, and trucks.

Ideas can move at an even faster pace than people and products. Communications systems, such as telephone, television, radio, and the Internet, carry ideas and information all around Earth. Remote villagers on the island of Borneo watch American television shows and learn about life in the United States. Political protestors in Egypt use text messaging and social networking sites to coordinate their activities. The geography of movement affects us all.

The Six Essential Elements

The five themes are one way of thinking about geography. Geographers also divide the study of geography into six essential elements. Elements are the topics that make up a subject. Calling them *essential* means they are necessary to understanding geography.

✓ **READING PROGRESS CHECK**
Determining Central Ideas How do people and the environment interact?

Building Geography Skills

GUIDING QUESTION *How will studying geography help you develop skills for everyday life?*

Many Web sites offer maps and directions to find a location. Have you ever used a Web browser to map a location? If you followed that map to your destination, you were using a geography skill.

Interpreting Visuals

Maps are one tool geographers use to picture the world. They use other visual images, as well. These other visuals include graphs, charts, diagrams, and photographs.

Graphs are visual displays of numerical information. Three of the most commonly used graphs are bar graphs, circle graphs, and line graphs. They can help you compare and analyze information. Charts display information in columns and rows. Diagrams are drawings that use pictures to represent something in the world or an abstract idea. A diagram might show the steps in a process or the parts that make up something.

Critical Thinking

Geographers ask analytical questions. For example, geographers might want to know why earthquakes are more likely in some places than in others. That question looks at causes. They might ask, How does climate affect the ways people live? Such questions examine effects.

Geographers might ask how the characteristics of a place have changed over time. That is a question of analysis. Or they could ask why people in different nations use their resources differently. That question calls on them to compare and contrast.

Learning how to ask—and answer—questions like these will help sharpen your mind. In addition to understanding geography better, you will also be able to use these skills in other subjects.

✓ **READING PROGRESS CHECK**

Determining Word Meanings What is an analytical question?

Include this lesson's information in your Foldable®.

LESSON 1 REVIEW CCSS

Reviewing Vocabulary
1. Compare and contrast a *landscape* and a photograph.

Answering the Guiding Questions
2. *Analyzing* To understand the geography of an island, what would a geographer have to consider?
3. *Determining Word Meanings* Explain the difference between place and location.
4. *Identifying* Give one example of a way Earth is dynamic and one example of how people make a change.
5. *Analyzing* Would a chart, a diagram, or a graph be the best way to show a population comparison for three different years?
6. *Informative/Explanatory Writing* Describe the relative location of your school.

There's More Online!

- ☑ **SLIDE SHOW** History of Mapmaking
- ☑ **ANIMATION** Elements of a Globe
- ☑ **VIDEO**

Reading **HELPDESK**

Academic Vocabulary
- sphere
- convert
- distort

Content Vocabulary
- hemisphere
- key
- scale bar
- compass rose
- map projection
- scale
- elevation
- relief
- thematic map
- technology
- remote sensing

TAKING NOTES: *Key Ideas and Details*

Describing As you read the lesson, identify three parts of a map on a graphic organizer. Then, explain what each part shows.

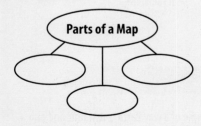

Lesson 2
Geographers and Their Tools

ESSENTIAL QUESTION • *How does geography influence the way people live?*

It Matters Because
The tools of geography help you understand the world.

Visual Representations

GUIDING QUESTION What is the difference between globes and maps?

If you close your eyes, you can probably see your neighborhood in your mind. When you do, you are using a mental map. You are forming a picture of the buildings and other places and where each is located in relation to the others.

Making and using maps is a big part of geography. Of course, geographers make maps that have many parts. Their maps are more detailed than your mental map. Still, paper maps are essentially the same as your mental map. Both are a way to picture the world and show where things are located.

Globes
The most accurate way to show places on Earth is with a globe. Globes are the most accurate because globes, like Earth, are **spheres**; that is, they are shaped like a ball. As a result, globes represent the correct shapes of land and bodies of water. They show distances and directions between places more correctly than flat images of Earth can.

The Equator and the Prime Meridian each divide Earth in half. Each half of Earth is called a **hemisphere**. The Equator divides Earth into two sections called the Northern and Southern Hemispheres. The Prime Meridian, together with the International Date Line, splits Earth into the Eastern and Western Hemispheres. Everything west of the Prime Meridian for 180 degrees is in the Western Hemisphere.

26

Maps

Maps are not round like globes. Instead, maps are flat representations of the round Earth. They might be sketched on a piece of paper, printed in a book, or displayed on a computer screen. Wherever they appear, maps are always flat.

Maps **convert**, or change, a round space into a flat space. As a result, maps **distort** physical reality, or show it incorrectly. This is why maps are not as accurate as globes are, especially maps that show large areas or the whole world.

Despite this distortion problem, maps have several advantages over globes. Globes have to show the whole planet. Maps, though, can show only a part of it, such as one country, one city, or one mountain range. As a result, they can provide more detail than globes can. Think how large a globe would have to be to show the streets of a city. You could certainly never carry such a globe around with you. Maps make more sense if you want to study a small area. They can focus on just that area, and they are easy to store and carry.

Maps tend to show more kinds of information than globes. Globes generally show major physical and political features, such as landmasses, bodies of water, the countries of the world, and the largest cities. They cannot show much else without becoming too difficult to read or too large. However, some maps show these same features. But maps can also be specialized. One map might illustrate a large mountain range. Another might display the results of an election. Yet another could show the locations of all the schools in a city.

☑ **READING PROGRESS CHECK**

Analyzing Why are globes sometimes more accurate than maps?

Academic Vocabulary

sphere a round shape like a ball

convert to change from one thing to another

distort to present in a manner that is misleading

A set of imaginary lines divides Earth into hemispheres.
▶ **CRITICAL THINKING**
The Geographer's World What line divides Earth into Eastern and Western Hemispheres?

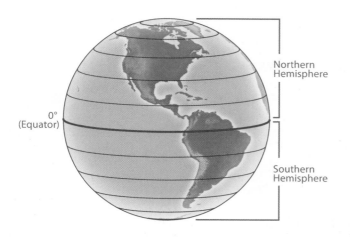

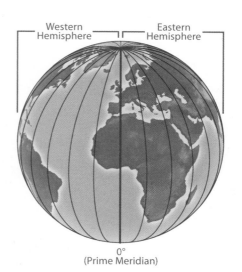

Chapter 1 27

Maps

GUIDING QUESTION *How do maps work?*

Maps are available in many different locations. You can see them in a subway station. Subway maps indicate the routes each train takes. In a textbook, a map might show new areas that were added to the United States at different times. At a company's Web site, a map can locate all its stores in a city. The map of a state park would tell visitors what activities they can enjoy in each area of the park. Each of these maps is different from the others, but they have some traits in common.

Parts of a Map

Maps have several important elements, or features. These features are the tools that convey information.

The map title tells what area the map will cover. It also identifies what kind of information the map presents about that area. The **key** unlocks the meaning of the map by explaining the symbols, colors, and lines. The **scale bar** is an important part of the map. It tells how a measured space on the map corresponds to actual distances on Earth. For example, by using the scale bar, you can determine how many miles in the real world each inch on the map represents. The **compass rose** shows direction. This map feature points out north, south, east, and west. Some maps include insets that show more detail for smaller areas, such as cities on a state map. Many maps show latitude and longitude lines to help you locate places.

Map Projections

To convert the round Earth to a flat map, geographers use **map projections**. A map projection distorts some aspects of Earth in order to represent other aspects as accurately as possible on a flat

Many different kinds of maps are available because maps are useful for showing a wide range of information.
▶ **CRITICAL THINKING**
Describing What is the difference between a large-scale and a small-scale map?

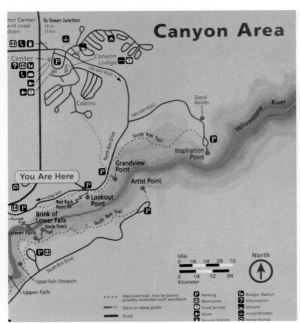

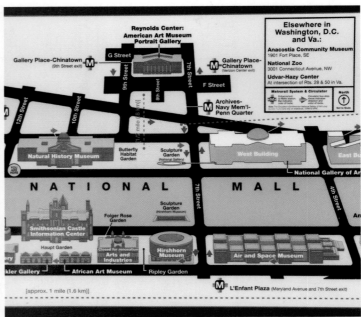

map. Some projections show the correct size of areas in relation to one another. Other map projections emphasize making the shapes of areas as accurate as possible.

Some projections break apart the world's oceans. By doing so, these maps show land areas more accurately. They clearly do not show the oceans accurately, though.

Mapmakers, known as cartographers, choose which projection to use based on the purpose of the map. Each projection distorts some parts of the globe more or less than other parts. Finally, mapmakers think about what part of Earth they are looking at and how large an area they want to cover.

Map Scale

Scale is another important feature of maps. As you learned, the scale bar relates distances on the map to actual distances on Earth. The scale bar is based on the scale at which the map is drawn. **Scale** is the relationship between distances on the map and on Earth.

Maps are either *large scale* or *small scale*. A large-scale map focuses on a smaller area. An inch on the map might correspond to 10 miles (16 km) on the ground. A small-scale map shows a relatively larger area. An inch on a small-scale map might be the same as 1,000 miles (1,609 km).

Each type of scale has benefits and drawbacks. Which scale to use depends on the map's purpose. Do you want to map your school and the streets and buildings near it? Then you need a large-scale map to show this small area in great detail. Do you want to show the entire United States? In that case, you need a small-scale map that shows the larger area but with less detail.

Types of Maps

The two types of maps are general purpose and thematic. The type depends on what kind of information is drawn on the map. General-purpose maps show a wide range of information about an area. They generally show both the human-made features of an area and its natural features.

Political maps are one common type of general-purpose map that shows human-made features. They show the boundaries of countries or of divisions within them, like the states of the United States. They also show the locations and names of cities.

Physical maps display natural features such as mountains and valleys, rivers, and lakes. They picture the location, size, and shape of these features. Many physical maps show **elevation**, or how much above or below sea level a feature is. Maps often use colors to present this information. A key on the map explains what height above or below sea level each color represents.

Physical maps usually show **relief**, or the difference between the elevation of one feature and the elevation of another feature near it.

Thinking Like a Geographer

Relief

Relief is the height of a landform compared to other nearby landforms. If a mountain 10,000 feet (3,048 m) high rises above a flat area at sea level, the relief of the mountain equals its elevation: 10,000 feet. If the 10,000-foot-high mountain is in a highland region that is 4,000 feet (1,219 m) above sea level, its relief is *less than* its elevation—only 6,000 feet (1,829 m). The difference in height between it and the land around it is much less than its absolute height. What would be the relief of a mountain 7,500 feet (2,286 m) high compared to its highest foothill, at 3,000 feet (914 m) high?

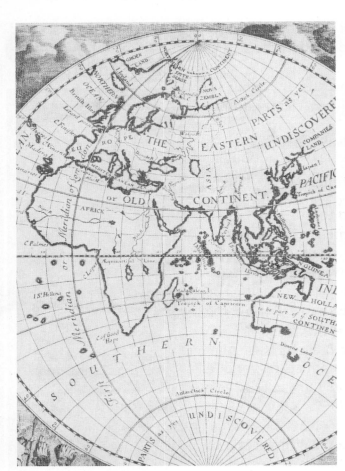
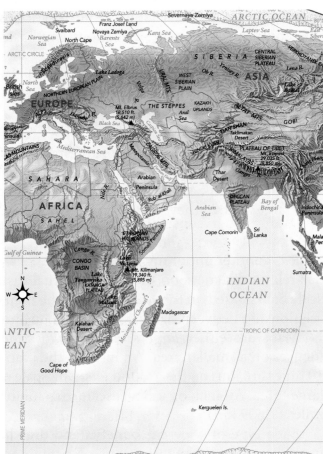

Cartography is the science of making maps. As knowledge of Earth grew, maps became increasingly accurate.

▶ CRITICAL THINKING

The Geographer's World Describe two ways in which the historical map differs from the present-day map.

Elevation is an absolute number, but relief is relative. It depends on other landforms that are nearby. The width of the colors on a physical map usually shows the relief. Colors that are narrow show steep places, and colors that are wide show gently sloping land.

Thematic maps show more specialized information. A thematic map might indicate the kinds of plants that grow in different areas. That kind of map is a vegetation map. Another could show where farming, ranching, or mining takes place. That kind of map is called a land-use map. Road maps show people how to travel from one place to another by car. Just about any physical or human feature can be displayed on a thematic map.

☑ **READING PROGRESS CHECK**

Describing What is the difference between thematic and general purpose maps?

Geospatial Technologies

GUIDING QUESTION *How do geographers use geospatial technologies?*

GPS devices and maps are available in many cars and cell phones. These electronic maps are an example of geospatial technologies. **Technology** is any way that scientific discoveries are applied to practical use. Geospatial technologies can help us think spatially. They provide practical information about the locations of physical and human features.

Global Positioning System

GPS devices work with a network called the Global Positioning System (GPS). This network was built by the U.S. government. Parts of it can be used only by the U.S. armed forces. Parts of it, though, can be used by ordinary people all over the world. The GPS has three elements.

The first element of this network is a set of more than 30 satellites that orbit Earth constantly. The U.S. government launched the satellites into space and maintains them. The satellites send out radio signals. Almost any spot on Earth can be reached by signals from at least four satellites at all times.

The second part of the network is the control system. Workers around the world track the satellites to make sure they are working properly and are on course. The workers reset the clocks on the satellites when needed.

The third part of the GPS system consists of GPS devices on Earth. These devices receive the signals sent by the satellites. By combining the signals from different satellites, a device calculates its location on Earth in terms of latitude and longitude. The more satellite signals the device receives at any time, the more accurately it can determine its location. Because satellites have accurate clocks, the GPS device also displays the correct time.

GPS is used in many ways. It is used to track the exact location and course of airplanes. That information helps ensure the safety of flights. Farmers use it to help them work their fields. Businesses use it to guide truck drivers. Cell phone companies use GPS to provide services. And of course, GPS in cars helps guide us to our destinations.

Geographic Information Systems

Another important geospatial technology is known as a geographic information system (GIS). These systems consist of computer hardware and software that gather, store, and analyze geographic information. The information is then shown on a computer screen. Sometimes it is displayed as maps. Sometimes the information is shown in other ways. Companies and governments around the world use this new tool.

A GIS is a powerful tool because it links data about all kinds of physical and human features with the locations of those features. Because computers can store and process so much data, the GIS can be accurate and detailed.

People select what features they want to study using the GIS. Then they can combine different features on the same map and analyze the patterns.

> **Think Again**
>
> **Geographers know the exact height of Mount Everest.**
>
> **Not true.** Mount Everest, on the border of Nepal and China, is the world's tallest mountain. It is said to be 29,028 feet (8,848 m) tall. Scientists disagree on this measurement, however, and the government of Nepal does not accept it. In 2011 Nepal launched a two-year effort to find the exact height using three GPS devices.

GPS satellites are used to measure, as well as determine, location on Earth. Some cell phones receive this satellite information, allowing people to locate places in a city.

▶ **CRITICAL THINKING**

Analyzing Why is it important for geographers to know exactly where places are located on Earth?

For instance, a farmer might want to compare the amount of moisture in the soil to the health of the plants. At the same time, he or she could add soil types around the farm to the comparison. The farmer could then use the results of the analysis to answer all kinds of questions. What plants should I plant in different locations? How much irrigation water should I use? How can I drive the tractor most efficiently?

Satellites and Sensors

Since the 1970s, satellites have gathered data about Earth's surface. They do so using remote sensing. **Remote sensing** simply means getting information from far away. Most early satellite sensors were used to gather information about the weather. Weather satellites help save lives during disasters by providing warnings about approaching storms. Before satellites, tropical storms were often missed because they could not be tracked over open water.

Satellites gather information in different ways. They may use powerful cameras to take pictures of the land. They can also pick up other kinds of information, such as the amount of moisture in the soil, the amount of heat the soil holds, or the types of vegetation that

are present. In the early 2000s, scientists used satellites and GIS technology to help conserve the plants and animals that lived in the Amazon rain forest. Using the technology, scientists can compare data gathered from the ground to data taken from satellite pictures. Land use planners use this information to help local people make good decisions about how to use the land. These activities help prevent the rain forest from being destroyed.

Some satellites gather information regularly on every spot in the world. That way, scientists can compare the information from one year to another. They look for changes in the shape of the land or in its makeup, spot problems, and take steps to fix them.

Limits of Geospatial Technology

Geospatial technologies allow access to a wealth of information about the features and objects in the world and where those features and objects are located. This information can be helpful for identifying and navigating. By itself, however, the information does not answer questions about why features are located where they are. These questions lie at the heart of understanding our world. The answers are crucial for making decisions about this world in which we live.

It is important to go beyond the information provided by geospatial technologies. We must build understanding of people, places, and environments and the connections among them.

✓ **READING PROGRESS CHECK**

Identifying What are the three elements of the Global Positioning System?

Include this lesson's information in your Foldable®.

LESSON 2 REVIEW

Reviewing Vocabulary
1. How has *remote sensing* changed the way some farmers manage their fields?

Answering the Guiding Questions
2. *Analyzing* Do you use a mental map or a paper map more often? Explain.
3. *Identifying* Give examples of three types of thematic maps.
4. *Identifying* If a museum map shows the location of each exhibit, is it a large-scale or a small-scale map? Explain.
5. *Describing* Describe the three main features of a map.
6. *Informative/Explanatory Writing* What tools could you use for a long trip to a national park? Explain.

What Do You Think?

Is Globalization Destroying Indigenous Cultures?

Globalization makes it easier for people, goods, and information to travel across borders. Customers have more choices when they shop. Costs of goods are sometimes lower. However, not everyone welcomes these changes. Resistance is particularly strong among indigenous peoples. They see the expansion of trade and outside influences as a threat to their way of life. Is globalization deadly for indigenous cultures?

Yes!

PRIMARY SOURCE

" Globalization ... is a multi-pronged attack on the very foundation of [indigenous people's] existence and livelihoods. ... Indigenous people throughout the world ... occupy the last pristine [pure and undeveloped] places on earth, where resources are still abundant [plentiful]: forests, minerals, water, and genetic diversity. All are ferociously sought by global corporations, trying to push traditional societies off their lands. ... Traditional sovereignty [control] over hunting and gathering rights has been thrown into question as national governments bind themselves to new global economic treaties. ... Big dams, mines, pipelines, roads, energy developments, military intrusions all threaten native lands. ... National governments making decisions on export development strategies or international trade and investment rules do not consult native communities. ... The reality remains that without rapid action, these native communities may be wiped out, taking with them vast indigenous knowledge, rich culture and traditions, and any hope of preserving the natural world, and a simpler ... way of life for future generations. "

—International Forum on Globalization (IFG), a research and educational organization

The Maasai strive to keep their rich cultural heritage alive by passing along their traditions from one generation to the next. In this ceremony, Maasai women decorate their faces with a red mixture. They wear flat, round necklaces, which are made up of rows of beads.

The Maasai at one time ranged throughout much of Kenya and Tanzania. Much of their land has been taken over for farms, ranches, and parks.

No!

PRIMARY SOURCE

" Indigenous people have struggled for centuries to maintain their identity and way of life against the tide of foreign economic investment and the new settlers that often come with it. . . . But indigenous groups are increasingly assertive. Globalization has made it easier for indigenous people to organize, raise funds and network with other groups around the world, with greater political reach and impact than before. The United Nations declared 1995–2004 the International Decade for the World's Indigenous People, and in 2000 the Permanent Forum on Indigenous Issues was created. . . . Many states have laws that explicitly recognize indigenous people's rights over their resources. . . . Respecting cultural identity [is] possible as long as decisions are made democratically—by states, by companies, by international institutions and by indigenous people. "

—Report by the United Nations Development Programme (UNDP)

What Do You Think? DBQ

1 Citing Text Evidence According to the IFG, why are indigenous people at risk?

2 Describing According to the United Nations report, how has globalization given indigenous people more power?

Critical Thinking

3 Identifying One effect of globalization is that more tourists are visiting remote places such as wildlife areas in Africa. How do you think indigenous people feel about the growth of tourism in their communities?

Chapter 1 **35**

Chapter 1 ACTIVITIES

Directions: Write your answers on a separate piece of paper.

1 Exploring the Essential Question
INFORMATIVE/EXPLANATORY WRITING Read the children's fable "The City Mouse and the Country Mouse." Write a synopsis of the story showing the differences between an urban and a country lifestyle.

2 21st Century Skills
INTEGRATING VISUAL INFORMATION Interview someone who has lived in your community for a number of years to find out how the community has changed in size and appearance. Create a scrapbook of your findings, including text and photos or drawings.

3 Thinking Like a Geographer
INTEGRATING VISUAL INFORMATION Locate your city or town on a globe. Find your lines of latitude and longitude. Next, locate the place in the Eastern Hemisphere with the same lines of latitude and longitude. Do the same for the Southern Hemisphere. Name the sites in the other two hemispheres.

4 GEOGRAPHY ACTIVITY

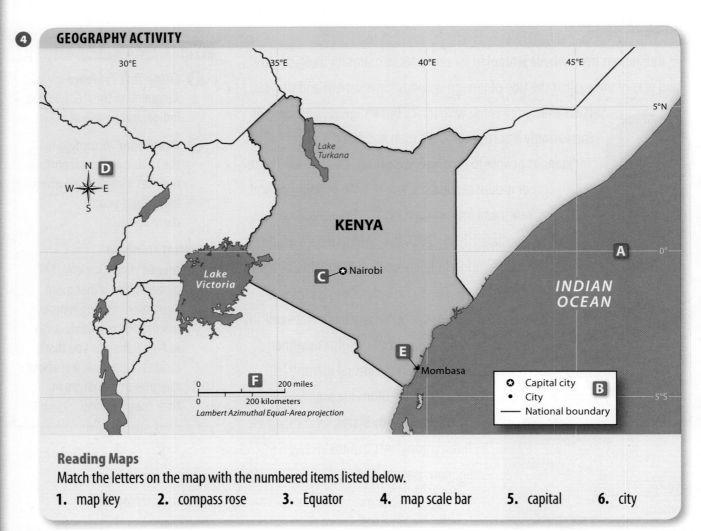

Reading Maps
Match the letters on the map with the numbered items listed below.

1. map key
2. compass rose
3. Equator
4. map scale bar
5. capital
6. city

Chapter 1 Assessment CCSS

REVIEW THE GUIDING QUESTIONS

Directions: Choose the best answer for each question.

1. The best way to represent abstract ideas is usually on a
   A. map.
   B. chart.
   C. diagram.
   D. graph.

2. Each hemisphere on Earth is equal to
   F. one-fourth of the planet.
   G. one-half of the planet.
   H. one continent plus one ocean.
   I. one pole and its surrounding land.

3. Because maps are flat representations of a sphere, they
   A. do not show as much detail as a globe.
   B. do not use a scale.
   C. must be very large.
   D. distort physical reality.

4. To understand the symbols on a map, you can use the
   F. scale bar.
   G. compass rose.
   H. title.
   I. key.

5. West of the river and north of the fire station describes
   A. relative location.
   B. absolute location.
   C. longitude.
   D. latitude.

6. Places that are close to one another and have some of the same characteristics are in the same
   F. landform.
   G. region.
   H. component.
   I. zone.

Chapter 1 ASSESSMENT (continued)

DBQ ANALYZING DOCUMENTS

7 CITING TEXT EVIDENCE In the excerpt below, two geographers summarize the content of their geography book.

> "In this book we... investigate the world's great geographic realms [areas]. We will find that each of these realms possesses a special combination of cultural... and environmental properties [characteristics]."
>
> —from H.J. de Blij and Peter O. Muller, *Geography*

What is one common way of dividing Earth into geographic realms?
A. by weather patterns
B. by distance to an ocean
C. by continent
D. by crops grown in an area

8 IDENTIFYING Which of the following is an example of a cultural characteristic of a realm?
F. the direction a river flows
G. people's actions and attitudes
H. the most common cloud formations of an area
I. the terrain of the land

SHORT RESPONSE

> "Hurricanes, wildfires, floods, earthquakes, and other natural events affect the Nation's economy,... property, and lives.... The USGS gathers and disseminates [gives out] real-time hazard data to relief workers, conducts long-term monitoring and forecasting to help minimize the impacts of future events, and evaluates conditions in the aftermath of disasters."
>
> —from United States Geological Service, *The National Map—Hazards and Disasters*

9 ANALYZING What is the benefit of evaluating conditions after a disaster occurs?

10 CITING TEXT EVIDENCE What is the primary focus of relief workers?

EXTENDED RESPONSE

Write your answer on a separate piece of paper.

11 INFORMATIVE/EXPLANATORY WRITING Large parts of Antarctica remain relatively unexplored. Research one of the bases that has been set up there. Write an essay explaining who uses the base and why. Also, describe some of the challenges researchers and scientists face in Antarctica and how they handle them.

Need Extra Help?

If You've Missed Question	1	2	3	4	5	6	7	8	9	10	11
Review Lesson	1	2	2	2	1	1	1	1	1	1	1

THE PHYSICAL WORLD

ESSENTIAL QUESTION • *How does geography influence the way people live?*

A geologist prepares to enter the crater of Ambrim Island volcano.

networks

There's More Online about The Physical World.

CHAPTER 2

Lesson 1
The Earth-Sun Relationship

Lesson 2
Forces Shaping Earth

Lesson 3
Earth's Land and Water

The Story Matters...

Earth is part of a larger physical system called the solar system. Earth's position in the solar system makes life on our planet possible. The planet Earth has air, land, and water that make it suitable for plant, animal, and human life. Major natural forces inside and outside of our planet shape its surface. Some of these forces can occur suddenly and violently, causing disasters that dramatically affect life on Earth.

FOLDABLES
Study Organizer

Go to the Foldables® library in the back of your book to make a Foldable® that will help you take notes while reading this chapter.

39

Chapter 2
THE PHYSICAL WORLD CCSS

The surface of Earth has changed greatly over time. Some changes come from internal forces associated with plate tectonics.

Step Into the Place

MAP FOCUS Use the map to answer the following questions.

1. **PHYSICAL GEOGRAPHY** Where are most of the world's volcanoes located?

2. **PHYSICAL GEOGRAPHY** Are earthquakes more common in Africa or South America?

3. **PHYSICAL GEOGRAPHY** Where do you think an earthquake is more likely to occur—along Australia's southern coast or Asia's east coast? Why?

4. **CRITICAL THINKING** **Integrating Visual Information** Are volcanoes and earthquakes more common in the Atlantic Ocean or the Pacific Ocean?

Tectonic Plate Boundaries

Step Into the Time

DRAWING EVIDENCE Choose one event from the time line and use it to explain how the natural forces that shape the physical geography of a particular place can have a worldwide impact.

1883 Krakatoa erupts

1931 Floods in China leave 80 million homeless

1500 — **1800** — **1900**

1556 Deadly earthquake strikes northern China

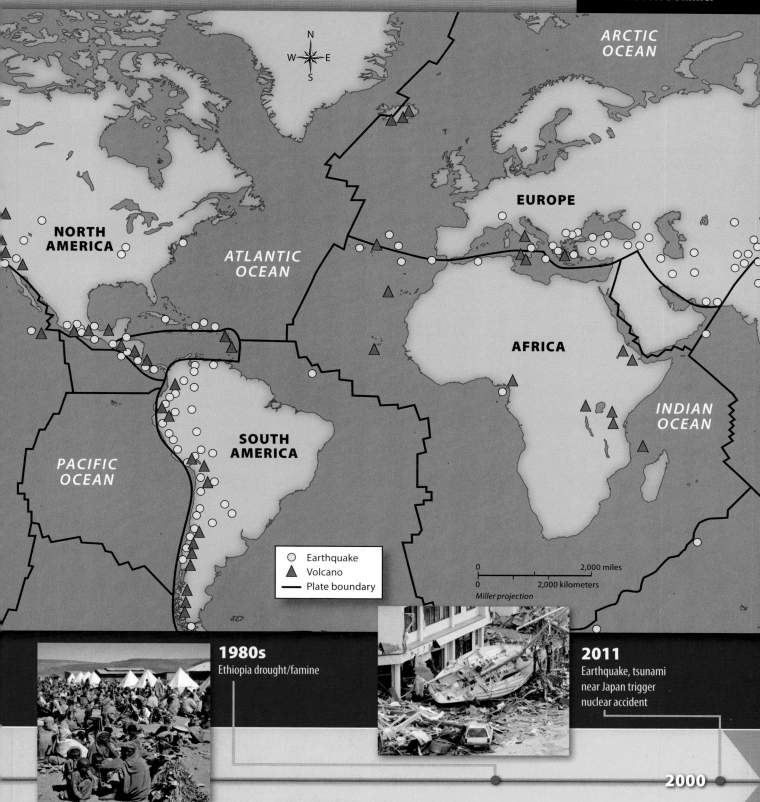

- ○ Earthquake
- ▲ Volcano
- — Plate boundary

1980s Ethiopia drought/famine

2011 Earthquake, tsunami near Japan trigger nuclear accident

2000

networks

There's More Online!

- ☑ **CHART** Climate Zones
- ☑ **ANIMATION** Earth's Daily Rotation
- ☑ **IMAGES** Seasons on Earth
- ☑ **MAP** Rain Shadow
- ☑ **SLIDE SHOW** Effects of Climate Change
- ☑ **VIDEO**

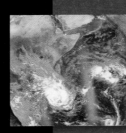

Lesson 1
The Earth-Sun Relationship

ESSENTIAL QUESTION • *How does geography influence the way people live?*

Reading HELPDESK

Academic Vocabulary
- accurate

Content Vocabulary
- orbit
- axis
- revolution
- atmosphere
- solstice
- equinox
- climate
- precipitation
- rain shadow

TAKING NOTES: *Key Ideas and Details*

Summarize As you read, complete a graphic organizer about Earth's physical system.

Element	Description
Hydrosphere	
Lithosphere	
Atmosphere	
Biosphere	

IT MATTERS BECAUSE
The processes that change Earth can act slowly or quickly with great fury.

Earth's Structure

GUIDING QUESTION *What is the structure of Earth?*

Earth is one of eight major planets that follow their own paths around the sun. The sun is just one of hundreds of millions of stars in our galaxy. Because the sun is so large, its gravity causes the planets to constantly **orbit**, or move around, it. The sun is the center of the solar system in which we live. Earth is a member of the solar system—planets and the other bodies that revolve around our sun.

Earth and the Sun

Life on Earth could not exist without heat and light from the sun. Earth's orbit holds it close enough to the sun—about 93 million miles (150 million km)—to receive a constant supply of light and heat energy. The sun, in fact, is the source of all energy on Earth. Every plant and animal on the planet needs the sun's energy to survive. Without the sun, Earth would be a cold, dark, lifeless rock floating in space.

As Earth orbits the sun, it rotates, or spins, on its axis. The **axis** is an imaginary line that runs through Earth's center from the North Pole to the South Pole. Earth completes one rotation every 24 hours. As Earth rotates, different areas are in sunlight and in darkness. The part facing toward the sun experiences daylight, while the part facing away has night. Earth makes one **revolution**, or complete trip around the sun,

42

in 365¼ days. This is what we define as one year. Every four years, the extra fourths of a day are combined and added to the calendar as February 29th. A year that contains one of these extra days is called a leap year.

Inside Earth

Thousands of miles beneath your feet, Earth's heat has turned metal into liquid. You do not feel these forces, but what lies inside affects what lies on top. Mountains, deserts, and other landscapes were formed over time by forces acting below Earth's surface—and those forces are still changing the landscape.

If you cut an onion in half, you will see that it is made up of many layers. Earth is also made up of layers. Earth's layers are made up of many different materials.

Layers of Earth

The inside of Earth is made up of three layers: the core, the mantle, and the crust. The center of Earth—the core—is divided into a solid inner core and an outer core of melted, liquid metal. Surrounding the outer core is a thick layer of hot, dense rock called the mantle. Scientists calculate that the mantle is about 1,800 miles (2,897 km) thick. The mantle also has two parts. When volcanoes erupt, the glowing-hot lava that flows from the mouth of the volcano is magma from Earth's outer mantle. Magma is melted rock. The inner mantle is solid, like the inner core. The outer layer is the crust, a rocky shell that forms the surface of Earth. The crust is thin, ranging from about 2 miles (3.2 km) thick under oceans to about 75 miles (121 km) thick under mountains.

DIAGRAM SKILLS >

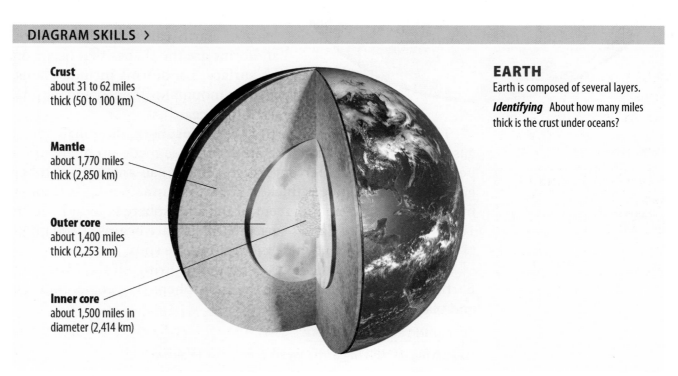

Crust
about 31 to 62 miles thick (50 to 100 km)

Mantle
about 1,770 miles thick (2,850 km)

Outer core
about 1,400 miles thick (2,253 km)

Inner core
about 1,500 miles in diameter (2,414 km)

EARTH
Earth is composed of several layers.

Identifying About how many miles thick is the crust under oceans?

Academic Vocabulary

accurate correct

The deepest hole ever drilled into Earth is about 8 miles (13 km) deep. That is still within Earth's crust. The farthest any human has traveled down into Earth's crust is about 2.5 miles (4 km). Still, scientists have developed an **accurate** picture of the layers in Earth's structure. One important way that scientists do this is by studying vibrations from deep within Earth. The vibrations are caused by earthquakes and explosions underground. From their observations, scientists have learned what materials are inside Earth and estimated the thickness and temperature of Earth's layers.

Earth's Physical Systems

There are powerful processes that operate below Earth's surface. Processes are also at work in the physical systems on the surface of Earth. Earth's physical systems consist of four major subsystems: the hydrosphere, the lithosphere, the atmosphere, and the biosphere.

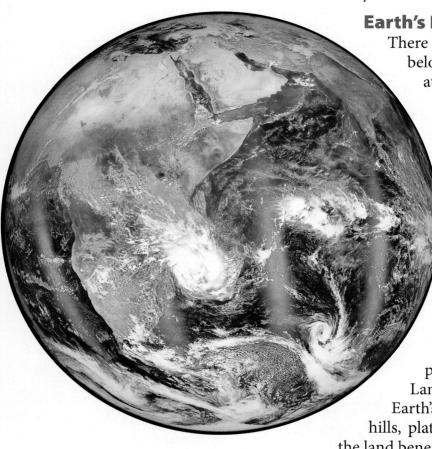

Earth's surface is a complex mix of landforms and water systems.
▶ **CRITICAL THINKING**
Analyzing How do Earth's landforms and water systems help support life on our planet?

About 70 percent of the earth's surface is water. The hydrosphere is the subsystem that consists of Earth's water. Water is found in many places: oceans, seas, lakes, ponds, rivers, groundwater, and ice.

Only about 30 percent of Earth's surface is land. Land makes up the part of Earth called the lithosphere. Landforms are the shapes that occur on Earth's surface. Landforms include plains, hills, plateaus, mountains, and ocean basins, the land beneath the ocean.

The air we breathe is part of the **atmosphere**, the thin layer of gases that envelop Earth. The atmosphere is made up of about 78 percent nitrogen, 21 percent oxygen, and small amounts of other gases. The atmosphere is thickest at Earth's surface and gets thinner higher up. Ninety-eight percent of the atmosphere is found within 16 miles (26 km) of Earth's surface. Outer space begins at 100 miles (161 km) above Earth, where the atmosphere ends.

The biosphere is made up of all that is living on the surface of Earth, close to the surface, or in the atmosphere. All people, animals, and plants live in the biosphere.

☑ **READING PROGRESS CHECK**

Identifying Which of Earth's layers contains the air we breathe?

Earth's Seasons

GUIDING QUESTION *How does Earth's orbit around the sun cause the seasons?*

Produce such as lettuce, grapes, and apples cannot grow when the weather is cold and icy. Yet grocery stores across America sell these ripe, colorful fruits all year, even in the middle of winter. Where is it warm enough to grow fruit in January? To find the answer, we start with the tilt of Earth.

Earth is tilted 23.5 degrees on its axis. If you look at a globe that is attached to a stand, you will see what the tilt looks like. Because of the tilt, not all places on Earth receive the same amount of direct sunlight at the same time.

As Earth orbits the sun, it stays in its tilted position. This means that one-half of the planet is always tilted toward the sun, while the other half is tilted away. As a result, Earth's Northern and Southern Hemispheres experience seasons at different times.

On about June 21, the North Pole is tilted toward the sun. The Northern Hemisphere is receiving the direct rays of the sun. The sun appears directly overhead at the line of latitude called the Tropic of Cancer. This day is the summer **solstice**, or beginning of summer, in the Northern Hemisphere. It is the day of the year that has the most hours of sunlight during Earth's 24-hour rotation.

Six months later—about December 22—the North Pole is tilted away from the sun. The sun's direct rays strike the line of latitude known as the Tropic of Capricorn. This is the winter solstice—when winter occurs in the Northern Hemisphere and summer begins in the Southern Hemisphere. The days are short in the Northern Hemisphere but long in the Southern Hemisphere.

DIAGRAM SKILLS >

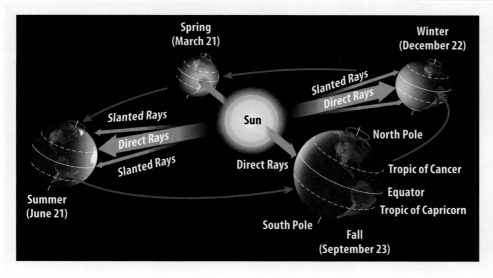

SEASONS
The tilt of Earth as it revolves around the sun causes the seasons to change.

▶ **CRITICAL THINKING**
Analyzing When does the Southern Hemisphere receive direct rays from the sun?

Midway between the two solstices, about September 23 and March 21, the sun's rays are directly overhead at the Equator. These are **equinoxes**, when day and night in both hemispheres are of equal length—12 hours of daylight and 12 hours of nighttime everywhere on Earth.

☑ **READING PROGRESS CHECK**

Identifying What is the difference between the solstices and the equinoxes?

Elements Affecting Climate

GUIDING QUESTION *How do elevation, wind and ocean currents, weather, and landforms influence climate?*

During the entire year, the sun's direct rays strike the low latitudes near the Equator. This area, known as the Tropics, lies mainly between the Tropic of Cancer and the Tropic of Capricorn. The Tropics circle the globe like a belt. If you lived in the Tropics, you would experience hot, sunny weather most of the year because of the direct sunlight. Outside the Tropics, the sun is never directly overhead. Even when these high-latitude areas are tilted toward the sun, the sun's rays hit Earth indirectly at a slant. This means that no direct sunlight shines on the high-latitude regions around the North and South Poles for as much as six months each year. Thus, the climate in these regions is always cool or cold.

Water swirls down the street of a small town in India after heavy rains.

▶ **CRITICAL THINKING**

Analyzing Why is the climate in low-latitude areas different from the climate in high-latitude areas?

Wind Patterns

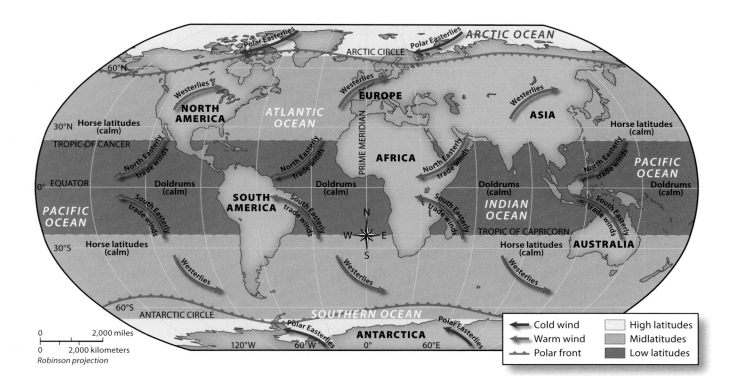

Elevation and Climate

At all latitudes, elevation influences climate. This is because Earth's atmosphere thins as altitude increases. Thinner air retains less heat. As elevation increases, temperatures decrease by about 3.5°F (1.9°C) for every 1,000 feet (305 m). For example, if the temperature averages 70°F (21.1°C) at sea level, the average temperature at 5,000 feet (1,524 m) is only 53°F (11.7°C). A high elevation will be colder than lower elevations at the same latitude.

Wind and Ocean Currents

In addition to latitude and elevation, the movement of air and water helps create Earth's climates. Moving air and water help circulate the sun's heat around the globe.

Movements of air are called winds. Winds are the result of changes in air pressure caused by uneven heating of Earth's surface. Winds follow prevailing, or typical, patterns. Warmer, low-pressure air rises higher into the atmosphere. Winds are created as air is drawn across the surface of Earth toward the low-pressure areas. The Equator is constantly warmed by the sun, so warm air masses tend to form near the Equator. This warm, low-pressure air rises, and then cooler, high-pressure air rushes in under the warm air, causing wind. This helps balance Earth's temperature.

MAP SKILLS

1. **PHYSICAL GEOGRAPHY** In what general direction does the wind pattern sometimes blow over Asia?

2. **PHYSICAL GEOGRAPHY** What air currents flow over the low latitudes?

Just as winds move in patterns, cold and warm streams of water, known as currents, circulate through the oceans. Warm water moves away from the Equator, transferring heat energy from the equatorial region to higher latitudes. Cold water from the polar regions moves toward the Equator, also helping to balance the temperature of the planet.

Weather and Climate

Weather is the state of the atmosphere at a given time, such as during a week, a day, or an afternoon. Weather refers to conditions such as hot or cold, wet or dry, calm or stormy, or cloudy or clear. Weather is what you can observe any time by going outside or looking out a window. **Climate** is the average weather conditions in a region or an area over a longer period. One useful measure for comparing climates is the average daily temperature. This is the average of the highest and lowest temperatures that occur in a 24-hour period. In addition to the average temperature, climate includes typical wind conditions and rainfall or snowfall that occur in an area year after year.

Rainfall and snowfall are types of precipitation. **Precipitation** is water deposited on the earth in the form of rain, snow, hail, sleet, or mist. Measuring the amount of precipitation in an area for one day provides data about the area's weather. Measuring the amount of precipitation for one full year provides data about the area's climate.

Landforms

It might seem strange to think that landforms such as mountains can affect weather and climate, but landforms and landmasses change the strength, speed, and direction of wind and ocean

GRAPH SKILLS >

RAIN SHADOW
A rain shadow affects the amount of rain a region receives.

▶ **CRITICAL THINKING**
1. **Determining Word Meanings** What do *windward* and *leeward* refer to?
2. **Describing** What happens to the winds after they release precipitation?

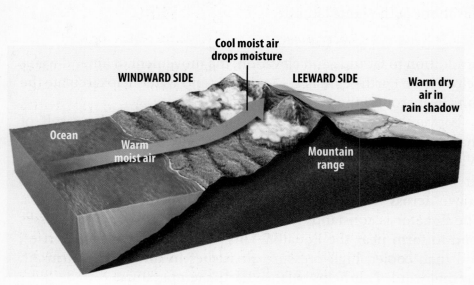

The climate in a zone affects how people live and work.

▶ **CRITICAL THINKING**
Identifying Point of View In which climate zone would people be more likely to live: polar or humid temperate? Why?

currents. Wind and ocean currents carry heat and precipitation, which shape weather and climate. The sun warms the land and the surface of the world's oceans at different rates, causing differences in air pressure. As winds blow inland from the oceans, they carry moist air with them. As the land rises in elevation, the atmosphere cools. When masses of moist air approach mountains, the air rises and cools, causing rain to fall on the side of the mountain facing the ocean. The other side of the mountain receives little rain because of the rain shadow effect. A **rain shadow** is a region of reduced rainfall on one side of a high mountain; the rain shadow occurs on the side of the mountain facing away from the ocean.

☑ **READING PROGRESS CHECK**

Identifying What measures are used to compare climates in different areas?

A Variety of Climate Zones

GUIDING QUESTION *What are the characteristics of Earth's climate zones?*

Why do so many people visit Florida and California? These places have cold or stormy weather at times, but their climates are generally warm, sunny, and mild. People in these states can enjoy activities outdoors all year long.

The Zones

In 1900 German scientist Wladimir Köppen invented a system that divides Earth into five basic climate zones. Climate zones are regions of Earth classified by temperature, precipitation, and distance from the Equator. Köppen used names and capital letters to label the climate zones as follows: Tropical (A); Desert (B); Humid Temperate (C); Cold Temperate (D); and Polar (E). Years later, a sixth climate zone was added: High Mountain (F).

Chapter 2 49

Each climate zone also can be divided into smaller subzones, but the areas within each zone have many similarities. Tropical areas are hot and rainy and often have dense forests. Desert areas are always dry, but they can be cold or hot, depending on their latitude. Humid temperate areas experience all types of weather with changing seasons. Cold temperate climates have a short summer season but are generally cold and windy. Polar climates are very cold, with ice and snow covering the ground most of the year. High mountain climates are found only at the tops of high mountain ranges such as the Rockies, the Alps, and the Himalaya. High mountain climates have variable conditions because the atmosphere cools with increasing elevation. Some of the highest mountaintops are cold and windy and stay white with snow all year.

Different types of plants grow best in different climates, so each climate zone has its own unique types of vegetation and animal life. These unique combinations form ecosystems of plants and animals that are adapted to environments within the climate zone. A biome is a type of large ecosystem with similar life-forms and climates. Earth's biomes include rain forest, desert, grassland, and tundra. All life is adapted to survive in its native climate zone and biome.

A bull moose stands alert on a tundra field in Alaska. Animals that live in that environment have unique adaptations that help them survive.

Changes to Climate

Many scientists believe climates are changing around the world. If this is true, the world could experience new weather patterns. These changes might mean more extreme weather in some places and milder weather in others. Human activities can affect weather and climate. For example, people have cut down millions of square miles of forests in Asia and Africa. As a result, there are fewer trees available to trap and release moisture into the air. This creates a drier climate in the region.

Metal, asphalt, brick, and concrete surfaces in cities absorb a huge amount of heat from the sun. An enormous mass of warmer air builds up in and around the city, affecting local weather.

In recent years, scientists have become aware of a problem called global warming. Global warming is an increase in the average temperature of Earth's atmosphere. Many scientists believe that a buildup of chemical pollution is contributing to the increasing temperature of Earth's atmosphere.

If the temperature of the atmosphere continues to rise, all of Earth's climates could be affected. Changes in climate might alter many natural ecosystems. Another consequence of climate change is that the survival of some plant and animal species will be threatened. It is also likely to be expensive and difficult for humans to adapt to these changes.

A thermal image shows heat escaping from the roofs of buildings.
▶ **CRITICAL THINKING**
Analyzing How is it possible for larger cities to change local weather?

☑ **READING PROGRESS CHECK**

Identifying Which climate zone experiences all types of weather with changing seasons?

 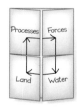

Include this lesson's information in your Foldable®.

LESSON 1 REVIEW CCSS

Reviewing Vocabulary
1. Explain the difference between the *revolution* and the rotation of Earth.

Answering the Guiding Questions
2. *Describing* Compare the layers of Earth to the layers of an orange.
3. *Analyzing* When South Africa is experiencing summer, what season is Central Asia having?
4. *Determining Word Meanings* Which side of a mountain experiences the rain shadow effect?
5. *Describing* What is the climate like in a large city like Hong Kong compared with the surrounding area?
6. *Informative/Explanatory Writing* Describe the climate of your area and the factors that contribute to it.

Chapter 2 51

networks

There's More Online!

- ☑ **MAP** Risk of Earthquakes in the United States
- ☑ **SLIDE SHOW** Human Impact on Earth
- ☑ **VIDEO**

Lesson 2
Forces Shaping Earth

ESSENTIAL QUESTION • *How does geography influence the way people live?*

Reading HELPDESK

Academic Vocabulary

- intense

Content Vocabulary

- continent
- tectonic plate
- fault
- earthquake
- Ring of Fire
- tsunami
- weathering
- erosion
- glacier

TAKING NOTES: *Key Ideas and Details*

Identify As you read, use a graphic organizer like this one to describe the external forces that have shaped Earth.

External Forces
_____
_____
_____

IT MATTERS BECAUSE
Internal and external forces change Earth, the setting for human life.

Constant Movement

GUIDING QUESTION *How was the surface of Earth formed?*

Since the formation of Earth, its surface has been moving continually. Landmasses have shifted and moved over time. Landforms have been created and destroyed. The way Earth looks from space has changed many times because of the movement of continents.

Earth's Surface

A **continent** is a large, continuous mass of land. Continents are part of Earth's crust. Earth has seven continents: Asia, Africa, North America, South America, Europe, Antarctica, and Australia. The region around the North Pole is not a continent because it is made of a huge mass of dense ice, not land. Greenland might seem as big as a continent, but it is classified as the world's largest island. Each of the seven continents has features that make it unique. Some of the most interesting features on the continents are landforms.

Even though you usually cannot feel it, the land beneath you is moving. This is because Earth's crust is not a solid sheet of rock. Earth's surface is like many massive puzzle pieces pushed close together and floating on a sea of boiling rock. The movement of these pieces is one of the major forces that create Earth's land features. Old mountains are worn down, while new mountains grow taller. Even the continents move.

52

Plate Movements

Earth's rigid crust is made up of 16 enormous pieces called **tectonic plates**. These plates vary in size and shape. They also vary in the amount they move over the more flexible layer of the mantle below them. Heat from deep within the planet causes plates to move. This movement happens so slowly that humans do not feel it. But some of Earth's plates move as much as a few inches each year. This might not seem like much, but over millions of years, it causes the plates to move thousands of miles.

Movement of surface plates changes Earth's surface features very slowly. It takes millions of years for plates to move enough to create landforms. Some land features form when plates are crushed together. At times, forces within Earth push the edge of one plate up over the edge of a plate beside it. This dramatic movement can create mountains, volcanoes, and deep trenches in the ocean floor.

At other times, plates are crushed together in a way that causes the edges of both plates to crumble and break. This event can form jagged mountain ranges. If plates on the ocean floor move apart, the space between them widens into a giant crack in Earth's crust. Magma rises through the crack and forms new crust as it hardens and cools. If enough cooled magma builds up that it reaches the surface of the ocean, an island will begin to form.

A path leads visitors to the Hakone Hot Springs (left) near Mount Hakone in Japan. Powerful forces within the earth heat groundwater, creating hot springs. Those forces also cause an eruption from Mount Lokon volcano (right) in Indonesia.

▶ **CRITICAL THINKING**
Describing What can happen if tectonic plates are pushed together?

Sudden Changes

Change to Earth's surface also can happen quickly. Events such as earthquakes and volcanoes can destroy entire areas within minutes. Earthquakes and volcanoes are caused by plate movement. When two plates grind against each other, faults form. A **fault** results when the rocks on one side or both sides of a crack in Earth's crust have been moved by forces within Earth. **Earthquakes** are caused by plate movement along fault lines. Earthquakes also can be caused by the force of erupting volcanoes.

Various plates lie at the bottom of the Pacific Ocean. These include the huge Pacific Plate along with several smaller plates. Over time, the edges of these plates were forced under the edges of the plates surrounding the Pacific Ocean. This plate movement created a long, narrow band of volcanoes called the **Ring of Fire**. The Ring of Fire stretches for more than 24,000 miles (38,624 km) around the Pacific Ocean.

The **intense** vibrations caused by earthquakes and erupting volcanoes can transfer energy to Earth's surface. When this energy travels through ocean waters, it can cause enormous waves to form on the water's surface. A **tsunami** is a giant ocean wave caused by volcanic eruptions or movement of the earth under the ocean floor. Tsunamis have caused terrible flooding and damage to coastal areas. The forces of these mighty waves can level entire coastlines.

People attempt to cross a collapsed bridge after a powerful earthquake in the Philippines.
▶ **CRITICAL THINKING**
Describing What is the relationship between a fault and an earthquake?

✓ **READING PROGRESS CHECK**

Determining Central Ideas How long does it take for moving plates to create landforms?

Academic Vocabulary

intense great or strong

Other Forces at Work

GUIDING QUESTION *How can wind, water, and human actions change Earth's surface?*

What happens when the tide comes in and washes over a sand castle on the beach? The water breaks down the sand castle. Similar changes take place on a larger scale across Earth's lithosphere. These changes happen much slower—over hundreds, thousands, or even millions of years.

Weathering

Some landforms are created when materials such as rocks and soil build up on Earth's surface. Other landforms take shape as rocks and soil break down and wear away over time. **Weathering** is a process by which Earth's surface is worn away by forces such as wind, rain, chemicals, and the movement of ice and flowing water. Even plants can cause weathering. Plant roots and small seeds can grow into tiny cracks in rock, gradually splitting the rock apart as the roots expand.

You may have seen the effects of weathering on an old building or statue. The edges become chipped and worn, and features such as raised lettering are smoothed down. Landforms such as mountains are affected by weathering, too. The Appalachian Mountains in the eastern United States have become rounded and crumbled after millions of years of weathering by natural forces.

Erosion

Erosion is a process that works with weathering to change the surface features of Earth. **Erosion** is a process by which weathered bits of rock are moved elsewhere by water, wind, or ice. Rain and moving water can erode even the hardest stone over time. When material is broken down by weathering, it can easily be carried away by the action of erosion. For example, the Grand Canyon was formed by weathering and erosion caused by flowing water and blowing winds. Water flowed over the region for millions of years, weakening the surface of the rock. The moving water carried away tiny bits of rock. Over time, weathering and erosion carved a deep canyon into the rock. Erosion by wind and chemicals caused the Grand Canyon to widen until it became the amazing landform we see today.

Weathering and erosion cause different materials to break down at different speeds. Soft, porous rocks, such as sandstone and limestone, wear away faster than dense rocks like granite. The spectacular rock formations in Utah's Bryce Canyon were formed as different types of minerals within the rocks were worn away by erosion, some more quickly than others. The result is landforms with jagged, rough surfaces and unusual shapes.

Wind erosion created these unusual rock formations in Algeria.
Identifying What are the greatest factors that cause erosion?

Water is released from the massive Three Gorges Dam in China.

▶ CRITICAL THINKING

Analyzing What are the dangers when humans change the natural course of land and waterways?

Changing Landforms

The buildup of materials creates landforms such as beaches, islands, and plains. Ocean waves pound coastal rocks into smaller and smaller pieces until they are tiny grains of sand. Over time, waves and ocean currents deposit sand along coastlines, forming sandy beaches. Sand and other materials carried by ocean currents build up on mounds of volcanic rock in the ocean, forming islands. Rivers deposit soil where they empty into larger bodies of water, creating coastal plains and wetland ecosystems.

Entire valleys and plains can be formed by the incredible force and weight of large masses of ice and snow. These masses are often classified by size as glaciers, polar ice caps, or ice sheets. A **glacier**, the smallest of the ice masses, moves slowly over time, sometimes spreading outward over the surface of the land. Although glaciers are usually thought of as existing during the Ice Age, glaciers can still be found on Earth today.

Ice caps are high-altitude ice masses. Ice sheets, extending more than 20,000 square miles (51,800 sq. km), are the largest ice masses. Ice sheets cover most of Greenland and Antarctica.

Human Actions

Natural forces have the awesome power to change the surface of Earth. Human actions, however, have also changed Earth in many ways. Activities such as coal mining have leveled entire mountains. Humans use explosives such as dynamite to blast tunnels through mountain ranges when building highways and railroads. Canals dug by humans change the natural course of waterways.

Humans have cut down so many millions of acres of forests that deadly landslides and terrible erosion occur on the deforested lands. When the tree roots are no longer there to hold the soil, wind and water can carry the soil away. This can result in a loss of nutrient rich topsoil. Rain forests and trees also help support the water cycle and help replace the oxygen in the atmosphere.

Pollution caused by humans can change Earth, as well. When people burn gasoline and other fossil fuels, toxic chemicals are released into the air. These chemicals settle onto the surfaces of mountains, buildings, oceans, rivers, grasslands, and forests. The chemicals may poison waterways, kill plants and animals, and cause erosion. The buildings in many cities show signs of being worn down by chemical erosion.

Studies show that humans have changed the environment of Earth faster and more broadly in the last 50 years than at any time in history. One major reason is that the demand for food and natural resources is greater than ever, and the demand continues to grow.

Changes to Earth's surface caused by natural weathering and erosion happen slowly. They create different kinds of landforms that make our planet unique. Erosion and other changes caused by humans, however, can damage Earth's surface quickly. Their effects threaten our safety and survival. We need to protect our environment to ensure that the quality of life improves for future generations.

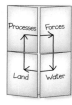

Include this lesson's information in your Foldable®.

✓ **READING PROGRESS CHECK**
Describing How does water erosion form valleys and canyons?

LESSON 2 REVIEW (CCSS)

Reviewing Vocabulary
1. How are *faults* and *earthquakes* related?

Answering the Guiding Questions
2. **Citing Text Evidence** Give three examples of forces that can lead to weathering.
3. **Describing** Describe one way that new islands can form.
4. **Determining Central Ideas** Why are Earth's plates in constant motion?
5. **Identifying** Place the following types of ice masses in order from smallest to largest: ice cap, ice sheet, glacier.
6. **Describing** What are some effects that might result from the development of manufacturing in China?
7. **Distinguishing Fact From Opinion** Is the following statement a fact or an opinion? Explain.

   Humans need to stop engaging in activities that lead to pollution so we can all enjoy a cleaner world.
8. **Informative/Explanatory Writing** Explain the difference between weathering and erosion, and give an example of each that you have seen personally.

networks

There's More Online!

- ☑ **IMAGE** The Ocean Floor
- ☑ **ANIMATION** How the Water Cycle Works
- ☑ **VIDEO**

Reading HELPDESK

Academic Vocabulary
- transform

Content Vocabulary
- plateau
- plain
- isthmus
- continental shelf
- trench
- desalinization
- groundwater
- delta
- water cycle
- evaporation
- condensation
- acid rain

TAKING NOTES: *Key Ideas and Details*

Describing Using a chart like this one, describe two kinds of landforms and two bodies of water.

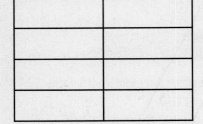

Lesson 3
Earth's Land and Water

ESSENTIAL QUESTION • *How does geography influence the way people live?*

IT MATTERS BECAUSE
Earth's landforms and bodies of water influence our ways of life.

Various Landforms

GUIDING QUESTION *What kinds of landforms cover Earth's surface?*

What is the land like near an ocean? Are unique landforms located in deserts or on mountains? Have you ever wondered how different kinds of landforms were created? The surface of Earth is covered with landforms and bodies of water. Our planet is filled with variety on land and under water.

Surface Features on Land

When scientists study landforms, they find it useful to group them by characteristics. One characteristic that is often used is elevation.

Elevation describes how far above sea level a landform or a location is. Low-lying areas, such as ocean coasts and deep valleys, may be just a few feet above sea level. Mountains and highland areas can be thousands of feet above sea level. Even flat areas of land can have high elevations, especially when they are located far inland from ocean shores.

Plateaus and plains are flat, but a **plateau** rises above the surrounding land. A steep cliff often forms at least one side of a plateau. **Plains** can be flat or have a gentle roll. They can be found along coastlines or far inland. Some plains are home to grazing animals, such as horses and antelope. Farmers and ranchers use plains areas to raise crops and livestock. A valley is a lowland area between two higher sides. Some valleys are small, level places surrounded by hills or mountains. Other valleys are huge expanses of land with highlands or mountain ranges on either side. Because they are often supplied with water runoff and topsoil from the higher lands around them, many valleys have rich soil and are used for farming and grazing livestock.

Another way to classify some landforms is to describe them in relation to bodies of water. Some types of landforms are surrounded by water. Continents are the largest of all landmasses. Most continents are bordered by land and water. Only Australia and Antarctica are completely surrounded by water. Islands are landmasses that are surrounded by water, but they are much smaller than continents.

A peninsula is a long, narrow area that extends into a river, a lake, or an ocean. Peninsulas at one end are connected to a larger landmass. An **isthmus** is a narrow strip of land connecting two larger land areas. One well-known isthmus is the Central American country of Panama. Panama connects two massive continents: North America and South America. Because it is the narrowest place in the Americas, the Isthmus of Panama is the location of the Panama Canal, a human-made canal connecting the Atlantic and Pacific oceans.

The Ocean Floor

The ocean floor is also covered by different landforms. The ocean floor, like the ground we walk on, is part of Earth's crust. In many ways, the ocean floor and land are similar. If you could see an ocean without its water, you would see a huge expanse of plains, valleys, mountains, hills, and plateaus. Some of the landforms were shaped by the same forces that created the features we see on land.

This map reveals ridges that are underwater mountain chains.

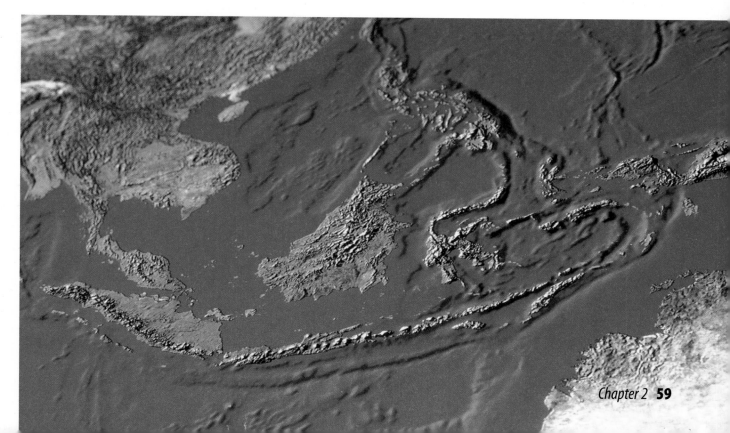

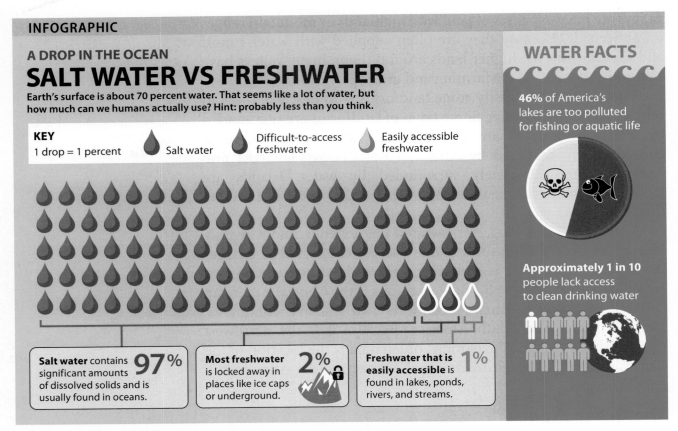

The surface of Earth is made up of water and land. Oceans, lakes, rivers, and other bodies of water make up a large part of Earth.

▶ **CRITICAL THINKING**

Analyzing Why is it important to keep freshwater sources clean and easily accessible?

One type of ocean landform is the continental shelf. A **continental shelf** is an underwater plain that borders a continent. Continental shelves usually end at cliffs or downward slopes to the ocean floor.

When divers explore oceans, they sometimes find enormous underwater cliffs that drop off into total darkness. These cliffs extend downward for hundreds or even thousands of feet. The water below is so deep it is beyond the reach of the sun's light. The deepest location on Earth is the Mariana Trench in the Pacific Ocean. A **trench** is a long, narrow, steep-sided cut in the ground or on the ocean floor. At its deepest point, the Mariana Trench is more than 35,000 feet (10,668 m) below the ocean surface.

Other landforms on the ocean floor include volcanoes and mountains. When underwater volcanoes erupt, islands can form because layers of lava build up until they reach the ocean's surface. Mountains on the ocean floor can be as tall as Mount Everest. Undersea mountains can also form ranges. The Mid-Atlantic Ridge, the longest underwater mountain range, is longer than any mountain range on land.

☑ **READING PROGRESS CHECK**

Determining Word Meanings What aboveground landform is a continental shelf similar to?

Locating Water

GUIDING QUESTION *What types of water are found on Earth's surface?*

Different forms of water are available all around you. Water in each of the three states of matter—solid, liquid, and gas—can be found all over the world. Glaciers, polar ice caps, and ice sheets are large masses of water in solid form. Rivers, lakes, and oceans contain water in liquid form. The atmosphere contains water vapor, which is water in the form of a gas.

Two Kinds of Water

Water at Earth's surface can be freshwater or salt water. Salt water is water that contains a large percentage of salt and other dissolved minerals. About 97 percent of the planet's water is salt water. Salt water makes up the world's oceans and also a few lakes and seas, such as the Great Salt Lake and the Dead Sea.

Salt water supports a huge variety of plant and animal life, such as whales, many kinds of fish, and other sea creatures. Because of its high concentration of minerals, humans and most animals cannot drink salt water. Humans have developed a way to remove minerals from salt water. **Desalinization** is a process that separates most of the dissolved chemical elements to produce water that is safe to drink. People who live in dry regions of the world use desalinization to process seawater into drinking water. But this process is expensive.

Freshwater makes up the remaining 3 percent of the water on Earth. Most freshwater stays frozen in the ice caps of the Arctic and Antarctic. Only about 1 percent of all water on Earth is the liquid freshwater that humans and other living organisms use. Liquid freshwater is found in lakes, rivers, ponds, swamps, and marshes, and in the rocks and soil underground.

The water contained inside Earth's crust is called **groundwater**. Groundwater is an important source of drinking water, and it is used to irrigate crops. Groundwater often gathers in aquifers. These are underground layers of rock through which water flows. When humans dig wells down into rocks and soil, groundwater flows from the surrounding area and fills the well. Groundwater also flows naturally into rivers, lakes, and oceans.

Bodies of Water

You are probably familiar with some of the different kinds of bodies of water. Some bodies of water contain salt water, and others hold freshwater. The world's largest bodies of water are its five vast, salt water oceans.

A young girl from Gambia in West Africa pumps water from a well.
▶ **CRITICAL THINKING**
Describing How does groundwater flow inside Earth's crust?

Humans use bodies of water for recreational activities (top). Bodies of water also help people earn a living like these women collecting seaweed to sell (bottom).

▶ **CRITICAL THINKING**
Describing Describe three ways water affects the lives of people who live near it.

From largest to smallest, the oceans are the Pacific, Atlantic, Indian, Southern, and Arctic. The Pacific Ocean covers more area than all of Earth's land combined. The Southern Ocean surrounds the continent of Antarctica. Although it is convenient to name the different oceans, it is important to remember that these bodies of water are connected and form one global ocean. Things that happen in one part of the ocean can affect the ocean all around the world.

When oceans meet landmasses, unique land features and bodies of water form. A coastal area where ocean waters are partially surrounded by land is called a bay. Bays are protected from rough ocean waves by the surrounding land, making them useful for docking ships, fishing, and boating. Larger areas of ocean waters partially surrounded by landmasses are called gulfs. The Gulf of Mexico is an example of ocean waters surrounded by continents and islands. Gulfs have many of the features of oceans, but they are smaller and they are affected by the landmasses around them.

Bodies of water such as lakes, rivers, streams, and ponds usually hold freshwater. Freshwater contains some dissolved minerals, but only a small percentage. The fish, plants, and other life-forms that live in freshwater cannot live in salty ocean water.

Freshwater rivers are found all over the world. Rivers begin at a source where water feeds into them. Some rivers begin where two other rivers meet; their waters flow together to form a larger river. Other rivers are fed by sources such as lakes, natural springs, and melting snow flowing down from higher ground.

A river's end point is called the mouth of the river. Rivers end where they empty into other bodies of water. A river can empty into a lake, another river, or an ocean. A **delta** is an area where sand, silt, clay, or gravel is deposited at the mouth of a river. Some deltas flow onto land, enriching the soil with the nutrients they deposit. River deltas can be huge areas with their own ecosystems.

Bodies of water of all kinds affect the lives of people who live near them. Water provides food, work, transportation, and recreation to people in many parts of the world. People get food by fishing in rivers, lakes, and oceans. The ocean floor is mined for minerals and drilled for oil. All types of waters have been used for transportation for thousands of years. People also use water for sports and recreation, such as swimming, sailing, fishing, and scuba diving. Water is vital to human culture and survival.

☑ **READING PROGRESS CHECK**

Describing How is water important in your life?

Recycling the Water Supply

GUIDING QUESTION *What is the water cycle?*

Water is necessary for all living things. Humans and other mammals, birds, reptiles, insects, fish, green plants, fungi, and bacteria must have water to survive. Water is essential for all life on Earth. To provide for the trillions of living organisms that use water every day, the planet needs a constant supply of fresh, clean water. Fortunately, water is recycled and renewed continually through Earth's natural systems of atmosphere, hydrosphere, lithosphere, and biosphere.

A Cycle of Balance

When it rains, puddles of water form on the ground. Have you noticed that after a day or two, puddles dry up and vanish? Where does the water go? It might seem as if water disappears and then new water is created, but this is not true. Water is not made or destroyed; it only changes form. When a puddle dries up, the liquid water has turned into gas vapor that we cannot see. In time, the vapor will become liquid again, and perhaps it will fill another puddle someday.

Scientists believe the total amount of water on Earth has not changed since our planet formed billions of years ago. How can this be true? It is possible because the same water is being recycled. At all times, water is moving over, under, and across Earth's surface and changing form as it is recycled. Earth's water-recycling system is called the **water cycle**. The water cycle keeps Earth's water supply in balance.

Water Changes Form

The sun's energy warms the surface of Earth, including the surface of oceans and lakes. Heat energy from the sun causes liquid water on Earth's surface to change into water vapor in a process called **evaporation**. Evaporation is happening all around us, at all times. Water in oceans, lakes, rivers, and swimming pools is constantly evaporating into the air. Even small amounts of water—in the soil, in the leaves of plants, and in the breath we exhale—evaporate to become part of the atmosphere.

Academic Vocabulary

transform to change

Air that contains water vapor is less dense than dry air. This means that moist air tends to rise. As water evaporates, tiny droplets of water vapor rise into the atmosphere. Water vapor gathers into clouds of varying shapes and sizes. Sometimes clouds continue to build until they are saturated with water vapor and can hold no more. A process called **condensation** occurs, in which water vapor **transforms** into a denser liquid or a solid state.

Condensation causes water to fall back to Earth's surface as rain, hail, or snow. Hail and snow either build up and stay solid or melt into liquid water. Snow stays solid when it falls in cold climates or on frozen mountaintops. When snow melts, it flows into rivers and lakes or melts directly into the ground.

Liquid rainwater returns water to rivers, lakes, and oceans. Rainwater also soaks into the ground, supplying moisture to plants and refilling underground water supplies to wells and natural springs. Much of the rainwater that soaks into the ground filters through soil and rocks and trickles back into rivers, lakes, and oceans. In this way, water taken from Earth's surface during evaporation returns in the form of precipitation. This cycle repeats all over the world, recycling the water every living organism needs to survive.

DIAGRAM SKILLS >

Water is constantly moving—from the oceans to the air to the ground and finally back to the oceans. The water cycle is the name given to this regular movement of water.

▶ **CRITICAL THINKING**
Analyzing How does the water on land evaporate?

Human actions have damaged the world's water supply. Waste from factories and runoff from toxic chemicals used on lawns and farm fields has polluted rivers, lakes, oceans, and groundwater. Chemicals such as pesticides and fertilizers seep into wells that hold drinking water, sometimes poisoning the water and causing diseases.

The fossil fuels we burn release poisonous gases into the atmosphere. These gases combine with water vapor in the air to create toxic acids. These acids then fall to Earth as a deadly mixture called **acid rain**. Acid rain damages the environment in several ways. It pollutes the water humans and animals drink. The acids damage trees and other plants. As acid rain flows over the land and into waterways, it may kill plant and animal life in bodies of water. This upsets the balance of the ecosystem.

Some human activities pollute our lakes and rivers.
▶ **CRITICAL THINKING**
Describing How does acid rain affect animal and plant life?

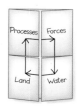

Include this lesson's information in your Foldable®.

LESSON 3 REVIEW

Reviewing Vocabulary
1. Why do people who live in extremely dry regions sometimes use the process of *desalinization*?

Answering the Guiding Questions
2. ***Determining Word Meanings*** How are plateaus and plains similar and different?
3. ***Analyzing*** Why do farmers in some locations want to farm near deltas?
4. ***Identifying*** How does driving a car contribute to acid rain?
5. ***Determining Central Ideas*** How do underwater features form and change?
6. ***Narrative Writing*** Imagine that you are a drop of water. Describe your day as you transform from one state to each of the others.

Chapter 2 ACTIVITIES CCSS

Directions: Write your answers on a separate piece of paper.

1 Use your **FOLDABLES** to explore the Essential Question.
INFORMATIVE/EXPLANATORY WRITING Select two of the following regions: Asia, Africa, Oceania, Australia, Antarctica. Describe three unique landforms that can be seen in each region. Include photos.

2 21st Century Skills
INTEGRATING VISUAL INFORMATION With a partner, research the water cycle. Create a slide show presentation illustrating why the water you drink today might have been consumed by a dinosaur millions of years ago.

3 Thinking Like a Geographer
DESCRIBING Select one volcano that has erupted in Asia. Write an essay explaining the immediate and long-term changes that resulted from this volcano.

4 **GEOGRAPHY ACTIVITY**

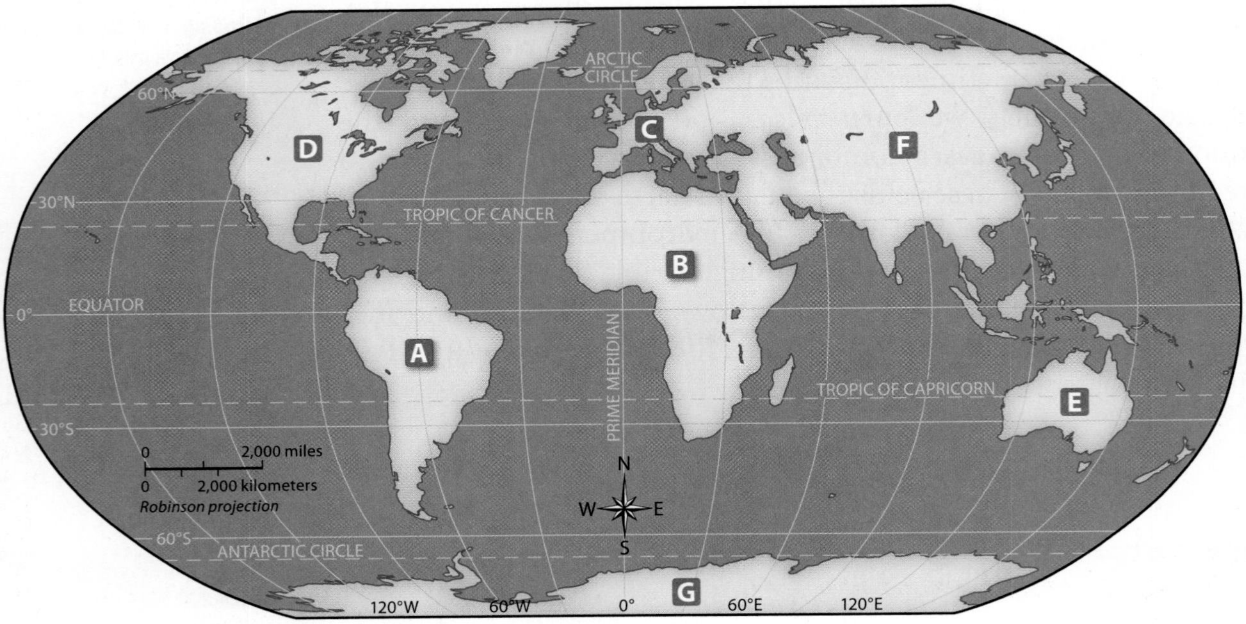

Locating Places
Match the letters on the map to the numbered continents below.

1. Africa
2. Australia
3. Europe
4. North America
5. Asia
6. South America
7. Antarctica

66 Chapter 2

Chapter 2 Assessment CCSS

REVIEW THE GUIDING QUESTIONS

Directions: Choose the best answer for each question.

1. The center of Earth is called its
   A. mantle.
   B. core.
   C. magma.
   D. crust.

2. An example of a biome is
   F. a rain forest.
   G. cold temperate.
   H. northerly.
   I. climate.

3. Earth's tectonic plates
   A. all move in the same direction.
   B. all move at the same speed.
   C. vary in size and shape.
   D. are about the same size and shape.

4. Antarctica is mostly covered by
   F. ice caps.
   G. glaciers.
   H. thick grasses.
   I. ice sheets.

5. When water has disappeared from a puddle on a sunny day, we say that it has
   A. evaporated.
   B. condensed.
   C. precipitated.
   D. cycled.

6. A long, narrow, steep-sided cut in the ground or on the ocean floor is called
   F. a ravine.
   G. a trench.
   H. a continental shelf.
   I. an isthmus.

Chapter 2 Assessment (continued)

DBQ ANALYZING DOCUMENTS

7 ANALYZING In this paragraph, a science writer describes the two types of planets in our solar system.

"*The planets can be divided quite easily into two categories—the inner planets, which are small and rocky, and the gas giants that circle through the outer reaches of the solar system. Within each class, the planets bear a striking resemblance to each other, but the two classes themselves are very different.*"

—from James S. Trefil, *Space, Time, Infinity*

What characteristic do Jupiter and Saturn share that the author would have taken into consideration?

A. presence of moons
B. presence of rings
C. their distance from the sun
D. lack of water

8 IDENTIFYING Which feature of Mars would the author consider to classify the planet?

F. temperature
G. size
H. inability to support life
I. lack of water

SHORT RESPONSE

"*There was a great rattle and jar. . . . [Then] there came a really terrific shock; the ground seemed to roll under me in waves, interrupted by a violent joggling up and down, and there was a heavy grinding noise as of brick houses rubbing together.*"

—from Mark Twain, *Roughing It*

9 DETERMINING WORD MEANINGS What do you think Twain meant by "violent joggling up and down"?

10 ANALYZING Based on Twain's description, do you think this was something he had experienced before? Explain.

EXTENDED RESPONSE

11 INFORMATIVE/EXPLANATORY WRITING The Pacific Ring of Fire is filled with more than 400 volcanoes, and it is the location of about 90 percent of the world's earthquakes. What have researchers and scientists done to help safeguard the residents of this vast area of the eastern coast of the Eastern Hemisphere? What precautions do the residents have in place in case of disaster? Research these questions and write a letter to the head of a government there, making further suggestions and recommendations.

Need Extra Help?

If You've Missed Question	1	2	3	4	5	6	7	8	9	10	11
Review Lesson	1	1	2	2	3	3	1	1	2	2	2

The Human World

ESSENTIAL QUESTIONS • How do people adapt to their environment?
• What makes a culture unique? • Why do people make economic choices?

Young girl from the town of Dori in Burkina Faso

networks
There's More Online about The Human World.

CHAPTER 3

Lesson 1
Earth's Population

Lesson 2
The World's Cultures

Lesson 3
Economies of the World

The Story Matters...

As part of our study of geography, we study culture, which is the way of life of people who share similar beliefs and customs. A particular culture can be understood by looking at the languages the people speak, what beliefs they hold, and what smaller groups form as parts of their society.

Study Organizer

Go to the Foldables® library in the back of your book to make a Foldable® that will help you take notes while reading this chapter.

69

Chapter 3
THE HUMAN WORLD CCSS

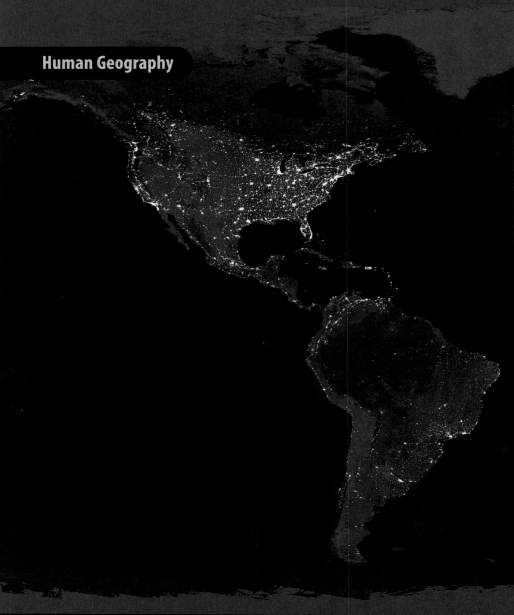

Human Geography

This image shows the world at night. Although the world's population is increasing, people still live on only a small part of Earth's surface. Some people live in highly urbanized areas, such as large cities. Others, however, live in areas where they may not have access to electricity or even running water.

Step Into the Place

MAP FOCUS Use the image to answer the following questions.

1. **HUMAN GEOGRAPHY** What do you think the brightly lit areas on the map represent?

2. **HUMAN GEOGRAPHY** Why might some areas be brighter than others?

3. **ENVIRONMENT AND SOCIETY** Are the lights evenly distributed across the land?

4. **CRITICAL THINKING ANALYZING** Do you think the darker areas have fewer people than brightly lit areas? Why or why not?

Step Into the Time

IDENTIFYING POINT OF VIEW Research one event from the time line. Write a journal entry describing the daily life of the time.

c. 4500–4000 B.C. Sumer civilization develops

c. 3500 B.C. Egypt builds trade network with West Asia

c. 1271 Marco Polo travels to China

1200s West African kingdom of Benin trade center

1845–1849 Ireland's Great Famine leads to mass emigration

1980 China is the first country to reach 1 billion population

2011 South Sudan secedes, forming new nation

networks

There's More Online!

- ☑ **CHART/GRAPH** Understanding Population
- ☑ **MAP** Landforms, Waterways, and Population
- ☑ **SLIDE SHOW** Places around the World
- ☑ **VIDEO**

Lesson 1
Earth's Population

ESSENTIAL QUESTION • How do people adapt to their environment?

Reading HELPDESK

Academic Vocabulary

- mature

Content Vocabulary

- **death rate**
- **birthrate**
- **doubling time**
- **population distribution**
- **population density**
- **urban**
- **rural**
- **emigrate**
- **immigrate**
- **refugee**
- **urbanization**
- **megalopolis**

TAKING NOTES: *Key Ideas and Details*

Determine Cause and Effect
As you read, use a graphic organizer like this one to take notes about the causes of population growth and migration.

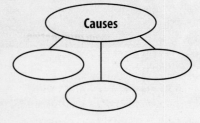

IT MATTERS BECAUSE
Billions of people share Earth. They have many different ways of life.

A Growing Population

GUIDING QUESTION *What factors contribute to Earth's constantly rising population?*

How fast has the world's population grown? In 1800 about 860 million people lived in the world. During the next 100 years, the population doubled to nearly 1.7 billion. By 2012, the total passed 7 billion.

What Causes Population Growth?

How has Earth's population become so large? What has caused our population to grow so quickly? Many factors cause populations to increase. One major cause of population growth is a falling death rate. The death rate is the number of deaths compared to the total number of individuals in a population at a given time. On average, about 154,080 people die every day worldwide. The **death rate** has decreased for many reasons. Better health care, more food, and cleaner water have helped more people—young and old—live longer, healthier lives.

Another major cause of population growth is the global birthrate. The **birthrate** is the number of babies born compared to the total number of individuals in a population at a given time. On average, about 215,120 babies are born each day worldwide. In time, the babies born today will

(l to r) MIXA/Getty Images; Lissa Harrison; ©George Hammerstein/Corbis; Carlos Spottorno/Getty Images News/Getty Images; Leung Cho Pan/Flickr/Getty Images

72

mature and have children and grandchildren of their own. This is how more and more people join the human population with each passing day.

However, during the past 60 years, the world's human birthrate has been decreasing slowly, although the global birthrate is still higher than the global death rate. This means that at any given time, such as a day or a year, more births than deaths occur. This results in population growth.

Growth Rates

In some countries, a high number of births has combined with a low death rate to greatly increase population growth. As a result, **doubling time**, or the number of years it takes a population to double in size based on its current growth rate, is relatively short. In some parts of Asia and Africa, for example, the doubling time is 25 years or less. In contrast, the average doubling time of countries with slow growth rates, such as Canada, can be more than 75 years.

Despite the fact that the global population is growing, the rate of growth is gradually slowing. The United Nations Department of Economic and Social Affairs predicts that the world's population will peak at 9 billion by the year 2050. After that, the population will begin to decrease. This means that for the next few decades, Earth's population will continue to grow. In time, however, this growth trend is expected to stop.

Academic Vocabulary

mature fully grown and developed as an adult; also refers to older adults

GRAPH SKILLS

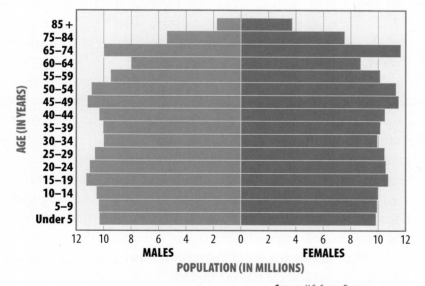

Source: U.S. Census Bureau, Statistical Abstract of the United States: 2012

POPULATION PYRAMID

Population pyramids use two bar graphs to show a country's population by age and gender. A growing population will be wide at the bottom. This represents a large number of young people. A declining population will be wide at the top. This means the country's population is mostly elderly. A stable population is represented by similar-length bars over several age ranges. This graph shows the population of the United States.

▶ **CRITICAL THINKING**

1. *Integrating Visual Information* Which age group has the most females?

2. *Integrating Visual Information* About how many males are in the 10–14 age group?

Chapter 3

Farmers in many parts of the world clear land by cutting and burning forests.

▶ **CRITICAL THINKING**
Analyzing What is the purpose of slash-and-burn agriculture?

Population Challenges

When human populations grow, the places people inhabit can become crowded. In many parts of the world, cities, towns, and villages have grown and expanded beyond a comfortable capacity.

When the population of an already-crowded area continues to grow, serious problems can arise. For example, diseases spread quickly in crowded environments. Sometimes there is not enough work for everyone, and many households live in ongoing poverty. Where many people share tight living spaces, crime can be a serious problem and pollution can increase.

Environmental Effects

On a global scale, rapid population growth can have a harmful effect on the environment. Each year, more people are sharing the same amount of space. People demand fuel for their cars and power for their homes. Miners drill and dig into the earth in a constant search for more energy resources. Forests are cut down to make farms for growing crops and raising livestock to feed hungry populations. Factory workers build cars, computers, and appliances. Some factories dump chemicals into waterways and vent poisonous smoke into the air.

Over time, and with many thousands of factories all over the world, chemical wastes have polluted Earth's atmosphere. Many groups and individuals are working to clean up the environment and restore polluted areas.

Humans have many methods of finding and using the resources we need for survival. Some of these methods are wasteful and destructive. However, humans are also creative in solving modern problems. People in all parts of the world have invented new ways to produce power and harvest resources. For example, in areas that receive enough sunshine, solar panels can be installed on the roofs of buildings. These panels collect energy from the sun, which can be used to produce heat and electric energy. Wind, solar, and geothermal energy are resources that do not pollute the environment. Humans are rising to the challenge of finding new ways to use these natural resources.

Population Growth Rates

Human populations grow at different rates in different areas of the world for many reasons. Often, the number of children each family will have is influenced by the family's culture and religion. In some cultures, families are encouraged to have as many children as they can. Although birthrates have fallen greatly in many parts of Asia, Africa, and South America in recent decades, the rates are still higher than in industrialized nations.

In locations with the largest and fastest-growing populations, the need for resources, jobs, health care, and education is great. When millions of people living in a small area need food, water, and housing, there is sometimes not enough for everyone. People in many parts of the world go hungry or die of starvation. Water supplies in crowded cities are often polluted with wastes that can cause diseases. Some areas do not have enough land resources and materials for people to build safe, sturdy homes.

Children in areas affected by extreme poverty often do not receive an education. In areas with job shortages, people are forced to live on low incomes. People living in poverty often live in crowded neighborhoods called *slums*. Slums surround many of the world's cities. These places are often dirty and unsafe. Governments and organizations such as the United Nations are working to make these areas safer, healthier places to live. Because populations grow at different rates, some areas experience more severe problems.

☑ **READING PROGRESS CHECK**

Determining Word Meanings What is the difference between the birthrate and the death rate?

Solar panels collect energy from the sun.

Some cities, like Seoul, South Korea, have a high population density.

▶ **CRITICAL THINKING**
Analyzing How is population density calculated?

Where People Live

GUIDING QUESTION *Why do more people live in some parts of the world than in others?*

Some families live in the same town or on the same land for generations. Other people move frequently from place to place.

Population Distribution

Population growth rates vary among Earth's regions. The **population distribution**, or the geographic pattern of where people live on Earth, is uneven as well. One reason people live in a certain place is work. During the industrial age, for example, people moved to places that had important resources such as coal or iron ore to make and operate machinery. People gather in other places because these areas hold religious significance or because they are government or transportation centers.

Population Density

One way to look at population is by measuring **population density**—the average number of people living within a square mile or a square kilometer. To say that an area is *densely populated* means the area has a large number of people living within it.

Keep in mind that a country's population density is the average for the entire country. Population is not distributed evenly throughout a country. As a result, some areas are more densely populated than their country's average indicates. In Egypt, for example, the population is concentrated along the Nile River; in China, along its eastern seaboard; and in Mexico, on the Central Plateau.

Anna Creek Station, located in a rural region in south-central Australia, is the world's largest working cattle station.

Where People Are Located

Urban areas are densely populated. **Rural** areas, in contrast, are sparsely populated. People inhabit only a small part of Earth. Remember that land covers about 30 percent of Earth's surface, and half of this land is not useful to humans. This means that only about 15 percent of Earth's surface is inhabitable. Large cities have dense populations, while deserts, oceans, and mountaintops are uninhabited.

The main reason people settle in some areas and not in others is the need for resources. People live where their basic needs can be met. People need shelter, food, water, and a way to earn a living. Some people live in cities, which have many places to live and work. Other people make their homes on open grasslands where they build their own shelters, grow their own food, and raise livestock.

☑ **READING PROGRESS CHECK**

Identifying Give one example of an urban area and one example of a rural area.

Changes in Population

GUIDING QUESTION *What are the causes and effects of human migration?*

The populations of different areas change as people move from one area to another. When many people leave an area, that area's population decreases. When large numbers of people move into a city, a state, or a country, the population of that area increases. Moving from one place to another is called *migration*.

Chapter 3 **77**

People in rural areas often obtain food in outdoor markets such as this Laotian market.

▶ **CRITICAL THINKING**
Analyzing What is the main reason people choose to settle in one area and not in another?

Think Again

Early immigrants came to the United States for both push and pull factors.

True. Some came to avoid war or to escape persecution. In addition to these "push" factors, others were "pulled" to gain freedom, education, and economic opportunity.

Migration is one of the main causes of population shifts in our world today. What causes people to leave their homelands and migrate to different parts of the world?

Causes of Migration

To **emigrate** means "to leave one's home to live in another place." Emigration can happen within the same nation, such as when people move from a village to a city inside the same country. Often, emigration happens when people move from one nation to another. For example, millions of people have emigrated from countries in Europe, Asia, and Africa to start new lives in the United States. The term *immigrate* is closely related to *emigrate*, but it does not mean the same thing. To **immigrate** means "to enter and live in a new country."

The reasons for leaving one area and going to another are called push-pull factors. *Push* factors drive people from an area. For example, when a war breaks out in a country or a region, people emigrate from that place to escape danger. People who flee a country because of violence, war, or persecution are called **refugees**. Sometimes people emigrate from an area after a natural disaster such as a flood, an earthquake, or a tsunami has destroyed their homes and land. If the economy of a place becomes so weak that little or no work is available, people emigrate to seek new opportunities.

Pull factors attract people to an area. Some people move to new places to be with friends or family members. Many young people move to cities or countries to attend universities or other schools. Some relocate in search of better jobs. Families sometimes move to places where their children will be able to attend good schools.

Effects of Migration

The movement of people to and from different parts of the world can affect the land, resources, culture, and economy of an area. Some of these effects are positive, but others can be harmful.

One positive effect of migration is cultural blending. As people from diverse cultures migrate to the same place and live close together, their cultures become mixed and blended. This blending creates new, unique cultures and ways of life. Artwork and music created in diverse urban areas is often an interesting mixture of styles and rhythms from around the world. Food, clothing styles, and languages spoken in urban areas change when people migrate into that area and bring new influences.

Some families and cultural groups work to preserve their original culture. These people want to keep their cultural traditions alive so they can be passed down to future generations. For example, the traditional Chinese New Year is an important celebration for many Chinese American families. Chinese Americans can be part of a blended American culture but still enjoy traditional Chinese foods, music, and arts, and celebrate Chinese holidays. It is possible to adapt to a local culture yet maintain strong ties to a home culture.

Some people migrate by choice. Others, such as the Libyan refugees shown here, are forced to flee to another country to live.

▶ **CRITICAL THINKING**
Citing Text Evidence What are examples of "pull" causes of migration?

Nearly half the world's people live in urban areas. Many live in large cities such as Hong Kong.
▶ **CRITICAL THINKING**
Determining Word Meanings What is urbanization?

Causes and Effects of Urbanization

Another effect of migration is the growth of urban areas. **Urbanization** happens when cities grow larger and spread into surrounding areas. Migration is a primary reason that urbanization occurs.

People move to cities for many reasons. The most common reason is to find jobs. Transportation and trade centers draw people primarily by creating new opportunities for business. As the businesses grow and people move into an area, the need for services also grows. Workers fill positions in medical services, education, entertainment, housing, and food sectors.

As more people migrate to cities, urban areas become increasingly crowded. When populations within urban areas increase, cities grow and expand. Farmland is bought by developers to build homes, apartment buildings, factories, offices, schools, and stores to provide for the growing number of people. The loss of farmland means that food must be grown farther from cities, resulting in additional shipping and related pollution.

Urbanization is happening in cities all over the world. In some places, cities have grown so vast that they have reached the outer edges of other cities. The result is massive clusters of urban areas that continue for miles. A huge city or cluster of cities with an extremely large population is called a **megalopolis**. These huge cities are growing larger every day, and they face the challenges that come with population growth and urbanization.

Delhi, one of India's largest cities, is a megalopolis. Its sprawling land area takes in a section called the Old City, which dates from the

mid-1600s. It also encompasses New Delhi, the modern capital city built by British colonial rulers in the early 1900s.

The largest megalopolis in the Americas is Mexico City. Because of its size and influence, Mexico City is a **primate city**, an urban area that dominates the economy and political affairs of its country. Primate cities include Cairo, Egypt, in Africa; Amman, Jordan, in Asia; and Paris, France, in Europe.

Examples of Urbanization

Urbanization takes place around the world, but for different reasons and at different rates.

Africa south of the Sahara, for example, is one of the least urbanized regions of the world. The region's urban areas, however, are growing so rapidly that Africa has a high rate of urbanization. Many Africans leave rural villages in order to find better job opportunities, health care, and public services in urban areas. At the same time, population growth has caused cities to spread into the countryside.

Europe is highly urbanized. Beginning in the late 1700s, the Industrial Revolution transformed Europe from a rural, agricultural society to an urban, industrial society. The growth of industries and cities began first in western Europe. Later, after World War II, the process spread to eastern Europe.

✔ **READING PROGRESS CHECK**

Describing In your own words, briefly summarize the main reasons people emigrate from their homelands.

Include this lesson's information in your Foldable®.

LESSON 1 REVIEW

Reviewing Vocabulary
1. Why might people *immigrate* to a new country?

Answering the Guiding Questions
2. ***Analyzing*** If parts of Asia and Africa see a doubling of their population every 25 years, how do their birthrate and death rate correlate?
3. ***Identifying*** What factors might explain why western China is less densely populated than eastern China?
4. ***Determining Word Meanings*** Explain the difference between push and pull factors as they pertain to human migration.
5. ***Informative/Explanatory Writing*** Imagine that a war has broken out in your homeland and you have decided to leave. Select a country in Asia, Australia, or Africa as your "home" and describe in a three-paragraph essay what features you will seek in a specific new location.

Chapter 3

networks

There's More Online!

- ☑ **MAP** Population Map
- ☑ **IMAGE** Cultural Change
- ☑ **VIDEO**

Reading HELPDESK

Academic Vocabulary
- behalf

Content Vocabulary
- culture
- ethnic group
- dialect
- cultural region
- democracy
- representative democracy
- monarchy
- dictatorship
- human rights
- globalization

TAKING NOTES: *Key Ideas and Details*

Organize On a graphic organizer like this one, take notes about the different forms of government.

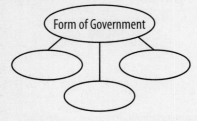

Lesson 2
The World's Cultures

ESSENTIAL QUESTION • *What makes a culture unique?*

IT MATTERS BECAUSE
Culture shapes the way people live and how they view the world.

What Is Culture?

GUIDING QUESTION *How is culture part of your life?*

What are some of your favorite foods? Do you like pizza, rice and beans, pasta, *samosas*, or corn on the cob? Have you ever thought about the people and cultures that invented the foods you enjoy eating? Millions of Americans eat foods created, grown, or developed by people of different cultures.

Culture is the set of beliefs, behaviors, and traits shared by a group of people. The term *culture* can also refer to the people of a certain culture. For example, saying "the Hindu culture" can mean the Hindu cultural traditions, the people who follow these traditions, or both.

You might be part of more than one culture. If your family has strong ties to a culture, such as that of a religion or a nation, you might follow this cultural tradition at home. You also might be part of a more mainstream American culture while at school and with friends.

If your family emigrated from Somalia to the United States, for example, you might speak the Somali language, wear traditional Somali clothing, and eat Somali foods. Your family might celebrate holidays observed in Somalia as well as American holidays, such as Thanksgiving and Independence Day. When you are with your friends, you might speak English, listen to American music, and watch American sports.

Ethnic Groups

We can look at members of a culture in terms of age, gender, or ethnic group. An **ethnic group** is a group of people with a common racial, national, tribal, religious, or cultural background. Members of the same Native American nation are an example of people of the same ethnic group. Other examples include the Maori of New Zealand and the Han Chinese. Large countries such as China can be home to hundreds of different ethnic groups. Some ethnic groups in a country are minority groups—people whose race or ethnic origin is different from that of the majority group. The largest ethnic minority groups in the United States are Hispanic Americans and African Americans.

Members of a culture might have special roles or positions as part of their cultural traditions. In some cultures, women are expected to care for and educate children. Most cultures expect men to earn money to support their families or to provide in other ways, such as by hunting and farming. Many cultures respect the elderly and value their wisdom. The leaders of older, traditional cultures are often elderly men or women who have leadership experience. Most cultures have clearly defined roles for their members. From an early age, young people learn what their culture expects of them. It is possible, too, to be part of more than one culture.

Language

Language serves as a powerful form of communication. Through language, people communicate information and experience and pass on cultural beliefs and traditions. Thousands of different languages are spoken in the world. Some languages have become world languages, or languages that are commonly spoken in many different parts of the world.

Some languages are spoken differently in different regions or by different ethnic groups. A **dialect** is a regional variety of a language with unique features, such as vocabulary, grammar, or pronunciation. People who speak the same language can sometimes understand other dialects, but at times, the pronunciation, or accent, of a dialect can be nearly impossible for others to understand.

Cities often have communities within them that share a distinct and common culture, language, and customs.
▶ **CRITICAL THINKING**
Describing What is an ethnic group?

CHART SKILLS

MAJOR WORLD RELIGIONS

Religion	Major Leader	Beliefs
Buddhism	Siddhārtha Gautama, the Buddha	Suffering comes from attachment to earthly things, which are not lasting. People become free by following the Eightfold Path, rules of right thought and conduct. People who follow the Path achieve nirvana—a state of endless peace and joy.
Christianity	Jesus Christ	The one God is Father, Son, and Holy Spirit. God the Son became human as Jesus Christ. Jesus died and rose again to bring God's forgiving love to sinful humanity. Those who trust in Jesus and follow his teachings of love for God and neighbor receive eternal life with God.
Hinduism	No one founder	One eternal spirit, Brahman, is represented as many deities. Every living thing has a soul that passes through many successive lives. Each soul's condition in a specific life is based on how the previous life was lived. When a soul reaches purity, it finally joins permanently with Brahman.
Islam	Muhammad	The one God sent a series of prophets, including the final prophet Muhammad, to teach humanity. Islam's laws are based on the Quran, the holy book, and the Sunnah, examples from Muhammad's life. Believers practice the five pillars—belief, prayer, charity, fasting, and pilgrimage—to go to an eternal paradise.
Judaism	Abraham	The one God made an agreement through Abraham and later Moses with the people of Israel. God would bless them, and they would follow God's laws, applying God's will in all parts of their lives. The main laws and practices of Judaism are stated in the Torah, the first five books of the Hebrew Bible.
Sikhism	Guru Nanak	The one God made truth known through 10 successive gurus, or teachers. God's will is that people should live honestly, work hard, and treat others fairly. The Sikh community, or Khalsa, bases its decisions on the principles of a sacred text, the Guru Granth Sahib.

Religion

Religion has a major influence on how people of a culture see the world. Religious beliefs are powerful. Some individuals see their religion as merely a tradition to follow during special occasions or holidays. Others view religion as the foundation and most important part of their life. Religious practices vary widely. Many cultures base their way of life on the spiritual teachings and laws of holy books. Religion is a central part of many of the world's cultures. Throughout history, religious stories and symbols have influenced painting, architecture, and music.

Customs

Customs are also an important outward display of culture. In many traditional cultures, a woman is not permitted to touch a man other than her husband, even for a handshake. In modern European cultures, polite greetings include kissing on the cheeks. People of many cultures bow to others as a sign of greeting, respect, and goodwill. The world's many cultures have countless fascinating customs. Some are used only formally, and others are viewed as good manners and respectful, professional behavior.

History

History shapes how we view the world. We often celebrate holidays to honor the heroes and heroines who brought about successes. Stories about heroes reveal the personal characteristics that people think are important. Groups also remember the dark periods of history when they met with disaster or defeat. These experiences too influence how groups of people see themselves. Cultural holidays mark important events and enable people to celebrate their heritage.

The Arts and Sports

Dance, music, visual arts, and literature are important elements of culture. Nearly all cultures have unique art forms that celebrate their history and enrich people's lives. Some art forms, such as singing and dancing, are serious parts of religious ceremonies or other cultural events. Aboriginal people of the Pacific Islands have songs, dances, and chants that are vital parts of their cultural traditions. Art can be forms of personal expression or worship, entertainment, or even ways of retelling and preserving a culture's history.

In sports, as in many other aspects of culture, activities are adopted, modified, and shared. Many sports that we play today originated with different culture groups in the past. Athletes in ancient Japan, China, Greece, and Rome played a game similar to soccer. Scholars believe that the Maya of Mexico and Central America developed "ballgame," the first organized team sport. Playing on a 40- to 50-foot long (12-m to 15-m) recessed court, the athletes' goal was to kick a rubber ball through a goal.

Government

Government is another element of culture. Despite differences, governments around the world share certain features. They maintain order within an area and provide protection from outside dangers. Governments also provide services to citizens. Such services typically include education and transportation infrastructure. Different cultures have different ways of distributing power and making rules.

Soccer is one of the most popular international sports.

Identifying Which culture first played the game of soccer?

Economy

Economies control the use of natural resources and define how goods are produced and distributed to meet human needs. Some cultures have their own type of economy, but most follow the economy of the country or area where they live. This allows people of different cultures living in an area to trade and conduct other types of business with one another. For example, many people in Benin, West Africa, sell goods in open-air markets. Some people bring items to the markets to trade for the goods they need, but others pay for goods using paper money and coins.

Cultural Regions

A **cultural region** is a geographic area in which people have certain traits in common. People in a cultural region often live close to one another to share resources, for social reasons, and to keep their cultures and communities strong. Cultural regions can be large or relatively small. For example, one of the world's largest cultural areas stretches across northern Africa and Southwest Asia. This cultural region is home to millions of people of the Islamic, or Muslim, culture. A much smaller cultural region is Spanish Harlem in New York City. This cultural region is home to a large and growing Hispanic culture.

✓ **READING PROGRESS CHECK**

Identifying Point of View What cultural traditions do you practice? Make a list of the beliefs, behaviors, languages, foods, art, music, clothing, and other elements of culture that are part of your daily life.

On July 4, 1776, the Second Continental Congress approved the Declaration of Independence, establishing the United States as an independent country.

▶ **CRITICAL THINKING**
Describing What form of government does the United States have? How does that form of government work?

Government

GUIDING QUESTION *How does government affect way of life?*

All nations need some type of formal leadership. What differs among countries is how leaders are chosen, who makes the rules, how much freedom people have, and how much control governments have over people's lives. Many different kinds of government systems operate in the world today. Three of the most common are democracy, monarchy, and dictatorship.

Democracy

In a democracy, the people hold the power. Citizens of a nation make the decisions themselves. A **democracy** is a system of government that is run by the people. In democratic systems of government, people are free to propose laws and policies. Citizens then vote to decide which laws and policies will be set in place. When people run the government, their rights and freedoms are protected.

In some democracies, the people elect leaders to make and carry out laws. A **representative democracy** is a form of democracy in which citizens elect government officials to represent the people; the government representatives make and carry out laws and policies on **behalf** of the people. The United States is an example of a representative democracy.

The queen is the symbolic head of the United Kingdom, but elected leaders hold the power to rule.

Monarchy

A **monarchy** is ruled by a king or a queen. In a monarchy, power and leadership are usually passed down from older to younger generations through heredity. The ruler of a monarchy, called a *monarch*, is usually a king, a queen, a prince, or a princess. In the past, monarchs had absolute power, or complete and unlimited power to rule the people. Today, most monarchs only represent, or stand for, a country's traditions and values, while elected officials run the government. The United Kingdom is an example of a monarchy.

Academic Vocabulary

behalf in the interest of; in support of; in defense of

Dictatorship

A **dictatorship** is a form of government in which one person has absolute power to rule and control the government, the people, and the economy. People who live under a dictatorship often have few rights. With absolute power, a dictator can make laws with no concern for how just, fair, or practical the laws are. North Korea is an example of a dictatorship.

Some dictators abuse their power for personal gain. One negative consequence of abuse of power is lack of personal freedoms and human rights for the general public. **Human rights** are the rights that belong to all individuals. Those rights are the same for every human in every culture. Some basic human rights are the right to life, liberty, security, privacy, freedom from slavery, and fair treatment before the law, as well as the right to marry and have children.

☑ READING PROGRESS CHECK

Describing In your own words, describe the system of government that is used in the United States.

Cargo containers are stockpiled and ready to be loaded onto ships in the port of Johor, Malaysia. International trade is the exchange of goods and services between countries. When people trade, they not only trade goods, they also trade customs and ideas.

Cultural Shifts

GUIDING QUESTION *How do cultures change over time?*

Cultures change over time for many reasons. When people relocate, they bring their cultural traditions with them. The traditions often influence or blend with the cultures of the places where they settle. Over time, as people of many cultures move to a location, the culture of that location takes on elements of all the cultures within it. Cities, such as London and New York, are examples of areas that have richly diverse cultures.

Cultural Change

Change can also occur as a result of trade, travel, war, and exchange of ideas. Trade brings people to new areas to sell and barter goods. Whenever people travel, they bring their language, customs, and ideas with them. They also bring elements of foreign cultures back with them when they return home. Throughout history, traders and explorers have brought home new foods, clothing, jewelry, and other goods. Some of these, such as gold, chocolate, gunpowder, and silk, became popular all over the world. Trade in these items changed the course of history.

Culture also can change as a result of technology. The telegraph, telephones, and e-mail have made communication increasingly faster and easier. Television and the Internet have given people in all parts of the world easy access to information and new ideas. Elements of culture such as language, clothing styles, customs, and behaviors spread quickly as people discover them by watching television and using the Internet.

Global Culture

Today's world is becoming more culturally blended every day. As cultures combine, new cultural elements and traditions are born. The spread of culture and ideas has caused our world to become globalized. **Globalization** is the process by which nations, cultures, and economies become integrated, or mixed. Globalization has had the positive effect of making people more understanding and accepting of other cultures. In addition, it has helped spread ideas and innovations. Technology has made communication faster and easier. Travel also has become faster and easier, allowing more people to visit more places in less time. This is resulting in cultural blending on a wider scale than ever before.

The process of cultural blending through globalization is not always smooth and easy. Sometimes it produces tension and conflict as people from different cultures come into contact with one another. Some people do not want their culture to change, or they want to control the amount of change. Sometimes the changes come too fast, and cultures can be damaged or destroyed.

Just as no one element defines a culture, no one culture can define the world. All cultures have value and add to the human experience. As the world becomes more globalized, people must continue to respect other ways of life. We have much to learn, and much to gain, from the many cultures that make our world a fascinating place.

✓ **READING PROGRESS CHECK**

Determining Word Meanings What is globalization?

Widespread use of technology, such as cell phones, allows us to share information with a larger audience.
▶ **CRITICAL THINKING**
Analyzing How does technology help spread new ideas?

Include this lesson's information in your Foldable®.

LESSON 2 REVIEW

Reviewing Vocabulary
1. Which *human rights* do you think are the hardest to safeguard? Why?

Answering the Guiding Questions
2. *Determining Central Ideas* Explain how a language becomes a world language, and predict whether any languages from Asia or Africa are likely to soon become a world language.

3. *Identifying Point of View* Why might a dictator want to be in power?

4. *Analyzing* Consider countries that are not yet developed. Write an essay explaining whether their cultures change more quickly or slowly than a culture of a developed country and explain your reasoning.

5. *Informative/Explanatory Writing* Think about the various cultures that you belong to. Write a short essay describing these cultures and your place in them.

GLOBAL CONNECTIONS

Rain Forest Resources

Many medicines that we use today come from plants found in rain forests. From these plants, we derive medicines to treat or cure diabetes, heart conditions, glaucoma, and many other illnesses and physical problems.

Largest Rain Forests The world's largest rain forests are located in the Amazon Basin in South America, the Congo Basin in Africa, and the Indonesian Archipelago in Southeast Asia. The Amazon rain forest makes up more than half of Earth's remaining rain forest.

The Planet's Lungs Rain forests are often called the "lungs of the planet" for their contribution in producing oxygen, which all animals need for survival. Rain forests also provide a home for many people, animals, and plants. They are also an important source of medicine and foods.

> " Every year, less and less of the rain forest remains. Human activity is the main cause of this deforestation. "

The People of the Congo Since ancient times, hunter-gatherers have lived in the Congo Basin Rain Forest. The groups, including the Mbuti, Baka, and Aka, are diverse; all have separate languages, religions, and customs.

Deforestation Every year, less and less of the rain forest remains. Human activity is the main cause of this deforestation. Humans cut rain forests for grazing land, agriculture, wood, and the land's minerals. Deforestation harms the native people who rely on the rain forest. The loss of rain forests also has an extreme impact on the environment because the rich biological diversity of the rain forest is lost as the trees are cut down.

Preserving Rain Forests More and more people realize that keeping the rain forests intact is critical. Groups plant trees on deforested land in the hope that forests will eventually recover. More companies are operating in ways that minimize damage to rain forests.

More Research Thirty years ago, very little research on the medicines of the rain forest was being done. Today, many drug companies and several branches of the U.S. government, including the National Cancer Institute, are taking part in research projects to find medicines and cures for viruses, infections, cancer, and AIDS.

▶ This unique plant is common in the rain forest of Malaysia. In the Malay language, the plant is called *Jangkang*, which means "stilt roots."

THERE'S MORE ONLINE

HEAR why the rain forest is important • *SEE* rain forest reduction • *WATCH* plants become drugs

GLOBAL CONNECTIONS

These numbers and statistics can help you learn about the resources of the rain forest.

1.4 Billion Acres

The Amazon rain forest covers 1.4 billion acres (5,665,599 sq km). If the rain forest were a nation, it would be the world's thirteenth-largest country.

OVER SEVEN PERCENT

Tropical rain forests make up about 7 percent of the world's total landmass. But found within the rain forest are half of all known varieties of plants.

40 Years

In 1950 rain forests covered about 14 percent of Earth's land. Rain forests cover about 7 percent today. Scientists estimate that, at the present rate, all rain forests could disappear from Earth within 40 years.

80%

About 80 percent of the diets of developed nations of the world originated in tropical rain forests. Included are such fruits as oranges and bananas; corn, potatoes, and other vegetables; and nuts and spices.

120

Today, 120 prescription drugs sold worldwide are derived from rain forest plants. About 65 percent of all cancer-fighting medicines also come from rain forest plants. An anticancer drug derived from a special kind of periwinkle plant has greatly increased the survival rate for children with leukemia.

50,000 Square Miles

When rain forests are cleared for land, animal and plant life disappear. Almost half of Earth's original tropical forests have been lost. Every year, about 32 million acres—50,000 square miles (129,499 sq km)—of tropical forest are destroyed. That's roughly the area of Nicaragua or the state of Alabama.

800,000 POPULATION

An Iban shaman, or healer, in traditional dress along a river side. For centuries, Iban people who lived in the rain forests of Malaysia relied on shamans to cure their ills. Today shamans still play an important role for some of the 800,000 Iban people.

ONE PERCENT

Although ingredients for many medicines come from rain forest plants, less than 1 percent of plants growing in rain forests have been tested by scientists for medicinal purposes.

92 Chapter 3

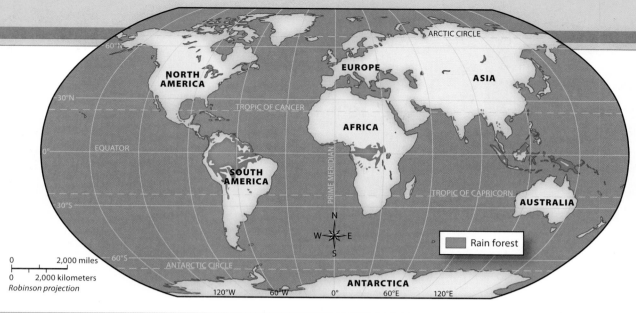

GLOBAL IMPACT

THE WORLD'S RAIN FORESTS Rain forests are located in a belt around Earth near the Equator. Abundant rain, relatively constant temperatures, and strong sunlight year-round are ideal conditions for the plants and animals of the rain forest.

Rain forests cover only a small part of Earth's surface. The Amazon Basin in South America is the world's largest rain forest area.

Rain Forest Research

Laboratories provide a research base for scientists to conduct environmental research. This laboratory in Mumbai attracts forest scientists from around the world.

Thinking Like a Geographer

1. **Environment and Society** Why do you think scientists only know about a small fraction of potential medicines from the rain forest?

2. **Environment and Society** How do you think native doctors in the Amazon rain forest discovered medical uses for plants?

3. **Human Geography** Write your government representatives and encourage them to support plans that help save rain forests. State at least three reasons why it is important to save rain forests.

Chapter 3   93

networks

There's More Online!

- ☑ **CHART/GRAPH** Economic Questions
- ☑ **GAME** Bartering and Trade
- ☑ **VIDEO**

Reading HELPDESK

Academic Vocabulary
- currency

Content Vocabulary
- renewable resource
- nonrenewable resource
- economic system
- traditional economy
- market economy
- command economy
- mixed economy
- gross domestic product
- standard of living
- productivity
- export
- import
- free trade
- sustainability

TAKING NOTES: *Key Ideas and Details*

Organize As you read, summarize the key ideas about each economic system.

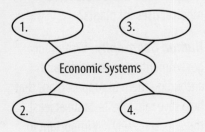

Lesson 3
Economies of the World

ESSENTIAL QUESTION • *Why do people make economic choices?*

IT MATTERS BECAUSE
People strive to meet their basic needs and their desire for a better life.

Resources

GUIDING QUESTION *How do people get the things they want and need?*

All human beings have wants and needs. How do you get the things you want and the things you need? To obtain these items, people use resources. Resources are the supplies that are used to meet our wants and needs. Some types of resources, such as water, soil, plants, and animals, come from the earth. These are called natural resources.

Other resources are supplied by humans. Human resources include the labor, skills, and talents people contribute. Countries also have wants and needs. Like individuals, nations must use resources to meet their needs.

Wants and Resources

What would happen if 14 students each wanted a glass of lemonade from a pitcher that contained only 12 glasses of lemonade? What if more students wanted a glass of lemonade? No matter how many people want lemonade, the pitcher still contains just 12 glasses. It does not hold enough for everyone to have a full glass. This is an example of a limited supply and unlimited demand. This situation is not uncommon. It happens to individuals and also to countries. You probably can think of many personal examples, as well as current and historical examples, of limited supply and unlimited demand.

One type of resource everyone needs is energy. Energy is the power to do work. Energy resources are the supplies that provide the power to do work. Many types of energy resources exist in our world. Energy resources can be renewable or nonrenewable. **Renewable resources** are resources that can be totally replaced or are always available naturally. They can be regenerated and replenished. Examples of renewable resources include water, trees, and energy from the wind and the sun.

In contrast are nonrenewable resources. **Nonrenewable resources** can not be totally replaced. Once nonrenewable resources are consumed, they are gone. Examples of nonrenewable resources include the fossil fuels oil, coal, and natural gas. These fuels received their name because they formed millions of years ago. Because of our increasing need for energy, supplies of nonrenewable resources are shrinking.

Making Choices

If the people of all nations have unlimited wants but face limited resources, what must happen? We must make choices. Do we continue to use nonrenewable resources? If so, at what rate should we be using them? Should we switch to renewable resources?

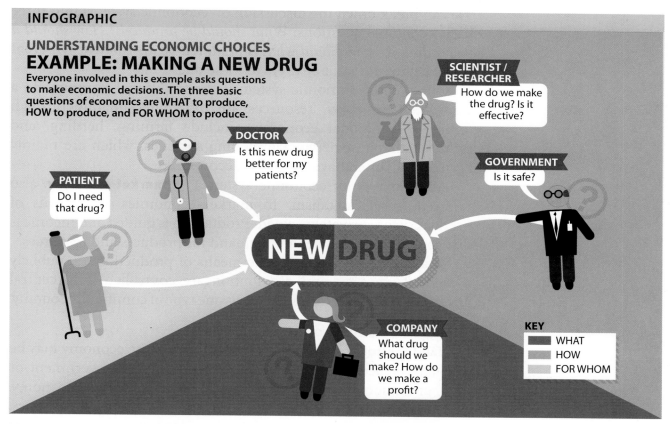

Every economic system must address three basic questions: What to produce? How to produce? For whom to produce?

▶ **CRITICAL THINKING**
Determining Central Ideas What is scarcity? How are the three basic economic questions related to the problem of limited supply?

> **Think Again**
>
> **No countries today rely on a command economic system.**
>
> **Not true.** Some countries still have planned economies, including Cuba, Saudi Arabia, Iran, and North Korea. These nations have an economic system in which supply and prices are regulated by the government, not by the market.

We must weigh the opportunity cost, or the value of what we must give up to acquire something else, of using renewable resources versus nonrenewable resources. We must take into account these and many more considerations as we make choices now and in the future.

☑ **READING PROGRESS CHECK**

Describing How are renewable and nonrenewable resources alike, and how are they different?

Economic Resources

GUIDING QUESTION *What kinds of economic systems are used in our world today?*

Economic resources are another important resource. Economic resources include the goods and services a society provides and how they are produced, distributed, and used. How a society decides on the ownership and distribution of its economic resources is its **economic system**. Do you ever stop to think about the goods and services you use in a single day? How do these goods and services become available to you?

Different Economic Systems

We can break down the discussion on economic systems into three basic economic questions: *What should be produced? How should it be produced? How should what is produced be distributed?* Different nations have different answers to these questions.

One type of economic system is the traditional economy. In a **traditional economy**, resources are distributed mainly through families. Traditional economies include farming, herding, and hunter-gatherer societies. Developing societies, which are mainly agricultural, often have traditional economies.

Another type of economic system is a **market economy**, also referred to as capitalism. In market economies, the means of production are privately owned. Production is guided and income is distributed through sales and demand for products and resources.

In a **command economy**, the means of production are publicly owned. Production and distribution are controlled by a central governing authority. Communism is one type of command economy.

What Is a Mixed Economy?

A **mixed economy** is just that—mixed. Parts of the economy may be privately owned, and parts may be owned by the government or another authority. The United States has a mixed market economy. Another economic system is socialism. In socialist societies, property and the distribution of goods and income are controlled by the community. How do individuals get the goods or services they need under each of these economic systems?

Buyers and sellers come together in an outdoor market in Kunduz, Afghanistan.

▶ **CRITICAL THINKING**
Describing Is the seller involved in a primary, a secondary, or a tertiary economic activity?

Parts of the Economy

We can break down the economy into parts. In economics, *land* is a factor of production that includes natural resources. Another factor of production is *labor*, which refers to all paid workers within a system. The other factor is *capital*, the human-made resources used to produce other goods. An *industry* is a branch of a business. For example, the *agriculture industry* grows crops and raises livestock. *Service industries* provide services rather than goods. Banking, retail, food service, transportation, and communications are examples of service industries.

Types of Economic Activities

Another way to view the parts of the economy is by the type of economic activity. Economists use the terms *primary sector*, *secondary sector*, and *tertiary sector* to group these activities. The primary sector includes activities that produce raw materials and basic goods. These activities include mining, fishing, agriculture, and logging.

The secondary sector makes finished goods. This sector includes home and building construction, food processing, and aerospace manufacturing. The tertiary sector of the economy is the service industry. Service sectors include sales, restaurants, banking, information technology, and health care.

Chapter 3

DIAGRAM SKILLS

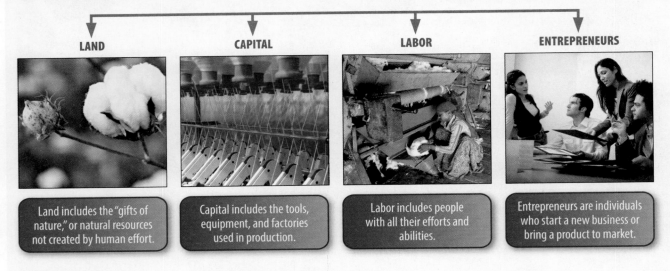

FACTORS OF PRODUCTION
The factors of production are broad categories of resources we need to produce the goods and services we want.

▶ **CRITICAL THINKING**
Determining Central Ideas Machines, tools, and equipment are examples of which factor of production?

Economic Performance

Economic performance measures how well an economy meets the needs of society. Economic performance can be determined by several factors that measure economic success. The **gross domestic product (GDP)** is the total dollar value of all final goods and services produced in a country during a single year. The **standard of living** is the level at which a person, a group, or a nation lives as measured by the extent to which it meets its needs. These needs include food, shelter, clothing, education, and health care. Per capita income is the total national income divided by the number of people in the nation.

When referring to economics, **productivity** is a measurement of what is produced and what is required to produce it. Sustainable growth is the growth rate a business can maintain without having to borrow money. The employment rate is the percentage of the labor force that is employed. These factors help determine a nation's economic strength and performance.

Types of National Economies

National economies also can be classified by types. Developed countries are industrialized countries. Developing countries are less industrialized, agricultural countries that are working to become more advanced economically. Developing countries often have weak economies, and most of their population lives in poverty. Newly industrialized countries (NICs) are in the process of becoming developed and economically secure. Their economies are growing and struggling to become fully developed, but they still face many economic and social challenges.

✓ **READING PROGRESS CHECK**

Describing How is standard of living a sign of economic performance?

A World Economy

GUIDING QUESTION *How do the world's economies interact and affect one another?*

You have read about different economic systems and different types of economies. All the world's nations can be classified into the different economic categories. All nations must find ways to interact with one another. Look at the labels in your clothes or on other products you buy. How many different country names can you find? How and why do we get goods from far across the world?

Trade

Trade is the business of buying, selling, or bartering. When you buy something at the store, you are trading money for a product. On a much bigger scale, nations trade with each other. Countries have different resources. Resources can include raw materials, such as iron ore. Labor also might be cheaper in another country where workers earn lower wages. As a result, goods can be produced in some countries more easily or efficiently than in other countries.

Trade can benefit countries. One country can **export**, or send to another country, a product that it is able to produce. Another country **imports**, or buys that product from the exporting country.

GRAPH SKILLS >

GDP COMPARISON
Gross Domestic Product (GDP) represents the total dollar value of all goods and services produced over a specific period of time. The graph compares the GDP of eight countries over one year's time.

▶ **CRITICAL THINKING**
1. ***Integrating Visual Information*** Name the countries whose GDP surpasses $6 trillion.

Country	GDP (U.S. Dollars in Trillions)
Mexico	1.15
Australia	1.48
India	1.67
United Kingdom	2.41
Brazil	2.49
Japan	5.86
China	7.29
United States	15.09

Source: CIA World Factbook, 2011

International trade involves preparing cargo for shipping (left) and transporting goods (right).

▶ **CRITICAL THINKING**
Describing What are the advantages of trade?

The country that imported the product can in turn export its products to another country. In global trade, extra fees are often added to the cost of importing products by a country's government. The extra money is a type of tax called a tariff. Governments often create tariffs to persuade their people to buy products made in their own country.

Sometimes a quota, a limit on the amount of one particular good that can be imported, is set. Quotas prevent countries from exporting goods at much lower prices than the domestic market can sell them for. A group of countries may decide to set minimal or no tariffs or quotas when trading among themselves. This is called **free trade**.

Advantages and Disadvantages of Trade

Trade has advantages and disadvantages. Trade can help build economic growth and increase a nation's income. On the other hand, jobs might be lost because of importing certain goods and services. With its benefits and its barriers, increasing trade leads to globalization. Economic globalization takes place when businesses move past national markets and begin to trade with other nations around the world.

Economic Organizations

In recent years, nations have become more interdependent, or reliant on one another. As they draw closer together, economic and political ties are formed. The World Trade Organization (WTO) helps regulate trade among nations. The World Bank provides financing, advice, and research to developing nations to help them grow their economies. The International Monetary Fund (IMF) is a group that monitors economic development. The IMF also lends money to nations in need and provides training and technical help. One well-known policy and organization that promotes global trade is the North American Free Trade Agreement (NAFTA). NAFTA encourages free trade among the United States, Canada, and Mexico.

The European Union (EU) is a group of European countries that operate under one economic unit and one **currency**, or type of money—the euro. The Mercado Camon del Sur (formerly called MERCOSUR) is a group of South American countries that promote free trade, economic development, and globalization. The Mercado Camon del Sur helps countries make better use of their resources while preserving the environment.

The Association of Southeast Asian Nations (ASEAN) is a group of countries in Southeast Asia that promote economic, cultural, and political development. The Dominican Republic–Central America Free Trade Agreement (CAFTA-DR) is an agreement among the United States, five developing Central American countries, and the Dominican Republic. The agreement promotes free trade.

Whether a nation produces its own goods or trades, one basic principle must be considered: sustainability. The principle of **sustainability** is central to the discussion of resources. When a country focuses on sustainability, it works to create conditions where all the natural resources for meeting the needs of society are available.

What can countries do to ensure sustainability now and into the future? What can you do to plan for your future and the future of your community? Just as every nation is part of a global system, you are part of your community. The choices you make affect you and those around you. What can you do now to plan for a bright economic future?

Academic Vocabulary

currency paper money and coins in circulation

Include this lesson's information in your Foldable®.

✅ **READING PROGRESS CHECK**

Analyzing What are some possible disadvantages of trade?

LESSON 3 REVIEW

Reviewing Vocabulary
1. Provide an example of a *renewable resource* and a *nonrenewable resource*.

Answering the Guiding Questions
2. *Identifying* Think of two examples of limited supply and unlimited demand.

3. *Identifying Point of View* Why might a government want to control certain parts of a mixed economy?

4. *Determining Central Ideas* Why is global trade necessary?

5. *Informative/Explanatory Writing* Write a short essay that addresses how people in developing versus developed countries likely affect sustainability.

Chapter 3 ACTIVITIES CCSS

Directions: Write your answers on a separate piece of paper.

1 Exploring the Essential Question
INFORMATIVE/EXPLANATORY WRITING Select a geographical region in Africa. Conduct research about the culture of the area. Write a letter to a student there explaining how your culture differs from his or hers.

2 21st Century Skills
INTEGRATING VISUAL INFORMATION Examine at least 25 items in your home, including your closet. Look at the labels to see where they were made. Make a graph that shows the number of items that were made in China, India, and Japan.

3 Thinking Like a Geographer
DETERMINING CENTRAL IDEAS Look up the gross domestic product of 10 countries in Asia and Africa. List them in order from greatest to least GDP. Include a paragraph exploring the success of the country with the largest GDP and another paragraph that explores the struggles for the country with the smallest GDP.

4 GEOGRAPHY ACTIVITY

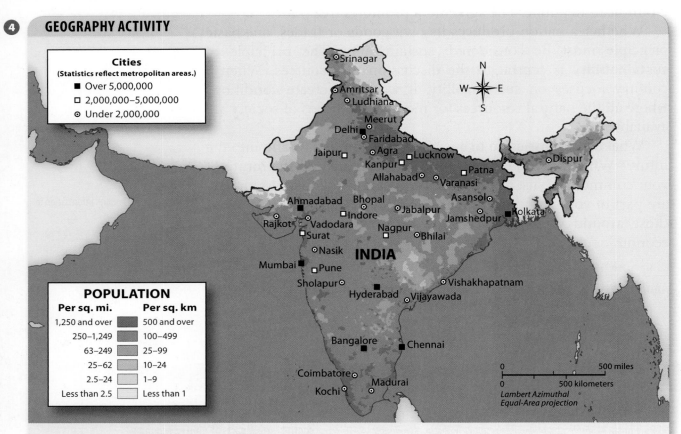

Study the map above and answer the following questions.
1. What is the population of Mumbai, India?
2. What area in India has the highest population per square mile?

Chapter 3 Assessment

REVIEW THE GUIDING QUESTIONS

Directions: Choose the best answer for each question.

1 Push and pull factors refer to
   A. rural and urban areas.
   B. the reasons people leave or go to an area.
   C. the increase and decrease in birthrate and death rate.
   D. changing forms of government.

2 A dialect is
   F. an ethnic group.
   G. a religious group.
   H. harmful to the culture group.
   I. a regional variety of language.

3 Under a dictatorship,
   A. the people have few rights.
   B. human rights are of central importance.
   C. people are all treated equally.
   D. power is passed on through heredity.

4 Why do governments create tariffs?
   F. to encourage free trade among nations
   G. to achieve a level of sustainability
   H. to prevent countries from exporting goods at lower prices
   I. to persuade their people to buy products made in their own country

5 If the economic performance of a country is improving, then
   A. its standard of living is dropping.
   B. its standard of living is improving.
   C. its gross domestic product is down.
   D. its per capita income is shrinking.

Chapter 3 ASSESSMENT (continued)

DBQ ANALYZING DOCUMENTS

6 DETERMINING WORD MEANINGS A news story reports on India's population growth.

"*India…will surpass China to become the world's most populous country in less than two decades. The population growth will mean a nation full of working-age youth, which economists say could allow the already booming economy to maintain momentum.*"

—from Anjana Pasricha, "India Challenged to Provide Jobs, Education to Young Population," October 2011

What does the author mean by "booming economy"?
- F. The economy will explode and shrink.
- G. The economy is growing rapidly.
- H. The economy will change forms.
- I. The economy will stay at its current level.

7 IDENTIFYING From the text, you can tell that the increase in India's population is largely the result of
- A. improvements in health care.
- B. a significant decline in the death rate.
- C. an increase in the birthrate.
- D. immigration.

SHORT RESPONSE

"*A school that is . . . easy for students, teachers, [and] parents . . . to reach on foot or by bicycle helps reduce the air pollution from automobile use, protecting children's health. Building schools . . . in the neighborhoods they serve minimizes the amount of paved surface . . . , which can help protect water quality by reducing polluted runoff.*"

—from Environmental Protection Agency, "Smart Growth and Schools"

8 IDENTIFYING In what other ways might building new schools help the environment?

9 ANALYZING What might be a problem with building new schools in neighborhood locations?

EXTENDED RESPONSE

10 INFORMATIVE/EXPLANATORY WRITING Although the increase in India's population will provide India with workers for decades, it will also place stresses on the environment and the people who live there. Research and discuss in an essay what stresses this might include, and also address what the government of India might be able to do to help with the problems that are surely coming. Also consider what steps the people of India should or should not take to curb their population growth. Include a future population projection for about a decade from now.

Need Extra Help?

If You've Missed Question	1	2	3	4	5	6	7	8	9	10
Review Lesson	1	2	2	3	3	3	1	1	1	1

ASIA

UNIT 2

Chapter 4 East Asia | **Chapter 5** Southeast Asia | **Chapter 6** South Asia | **Chapter 7** Central Asia, the Caucasus, and Siberian Russia | **Chapter 8** Southwest Asia

Explore the Continent

ASIA is the largest continent on Earth. It extends from the Mediterranean Sea in the west to the Pacific Ocean in the east. Asia is home to about 60 percent of the world's population. Yet, at least two-thirds of the land area is too cold or too dry to support a large population. Winds called monsoons affect climate in much of Asia. They bring cool, dry weather in winter and heavy rainfall and floods in summer.

① NATURAL RESOURCES Asia's numerous resources include petroleum, copper, rice, and fish. Growing populations and industries in cities such as Shanghai, China, demand more and more resources. Asian countries are quickly trying to meet this need.

② LANDFORMS Asia's landforms are varied. Mountain ranges, plateaus, and plains dominate western and central areas. Hundreds of islands dot coastlines in the east. Plate movements formed many of these islands. The island country of Indonesia has about 130 active volcanoes, more than any other country on Earth.

UNIT 2

③ BODIES OF WATER Water is plentiful in eastern and southern Asia, where many people depend on it for a living. In Sri Lanka, fishers use baitless hooks to snare mackerel and herring. In parts of western Asia, however, water is scarce. The lack of water is a major issue for countries in these areas.

Fast FACT

Asia comprises 30 percent of the world's land area.

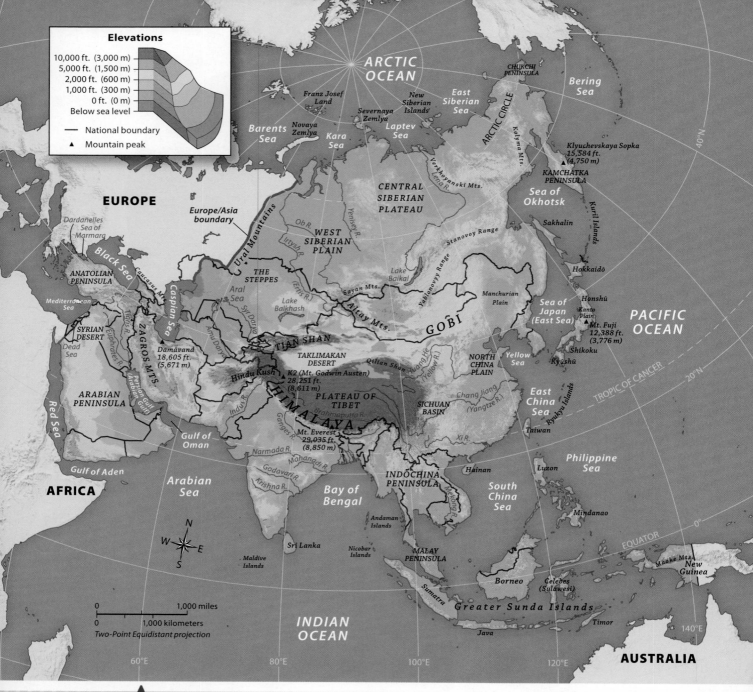

ASIA

PHYSICAL

MAP SKILLS

1. **PHYSICAL GEOGRAPHY** What part of Asia has the highest elevation?

2. **THE GEOGRAPHER'S WORLD** Which ocean borders the Central Siberian Plateau?

3. **PLACES AND REGIONS** How would you describe the region of Southeast Asia?

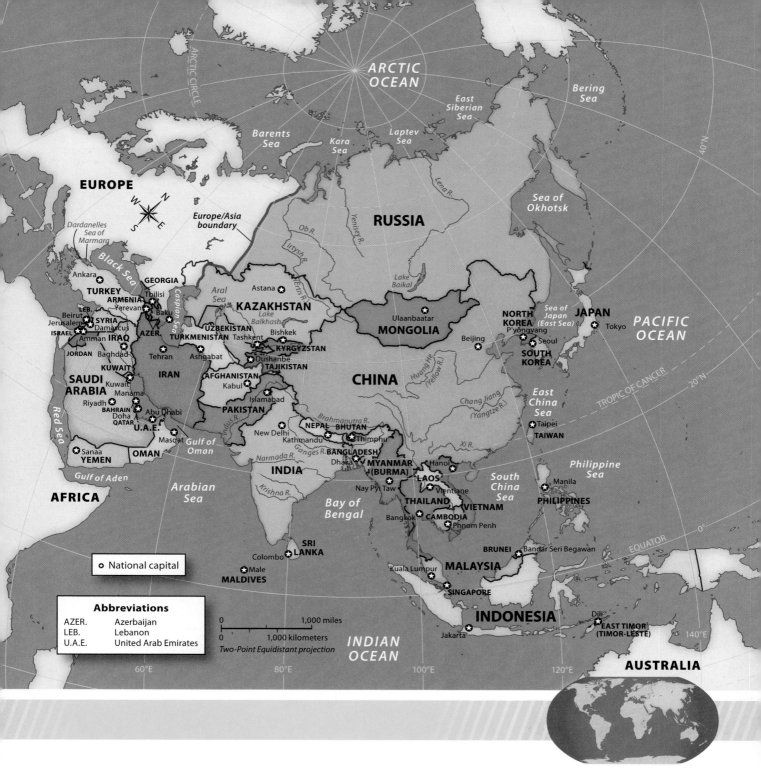

POLITICAL

MAP SKILLS

1. **PLACES AND REGIONS** Which country in the region is the largest?

2. **THE GEOGRAPHER'S WORLD** Which country is located between China and Russia?

3. **THE GEOGRAPHER'S WORLD** What body of water lies between Kazakhstan and Iran?

Unit 2 109

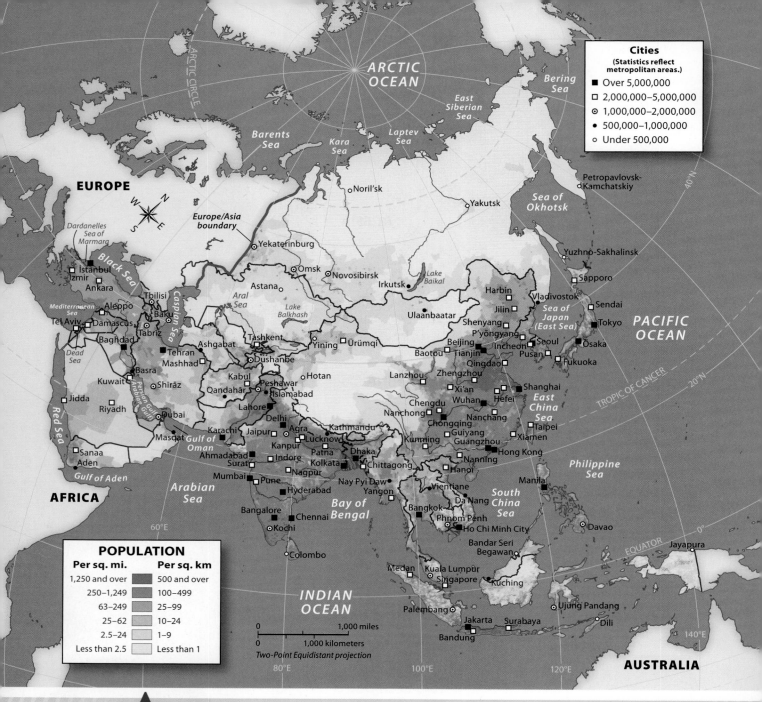

ASIA

POPULATION DENSITY

MAP SKILLS

1 THE GEOGRAPHER'S WORLD What parts of Asia are the most densely populated?

2 THE GEOGRAPHER'S WORLD Which part of the region has the lowest population density?

3 THE GEOGRAPHER'S WORLD In general, what population pattern do you see in Southwest Asia?

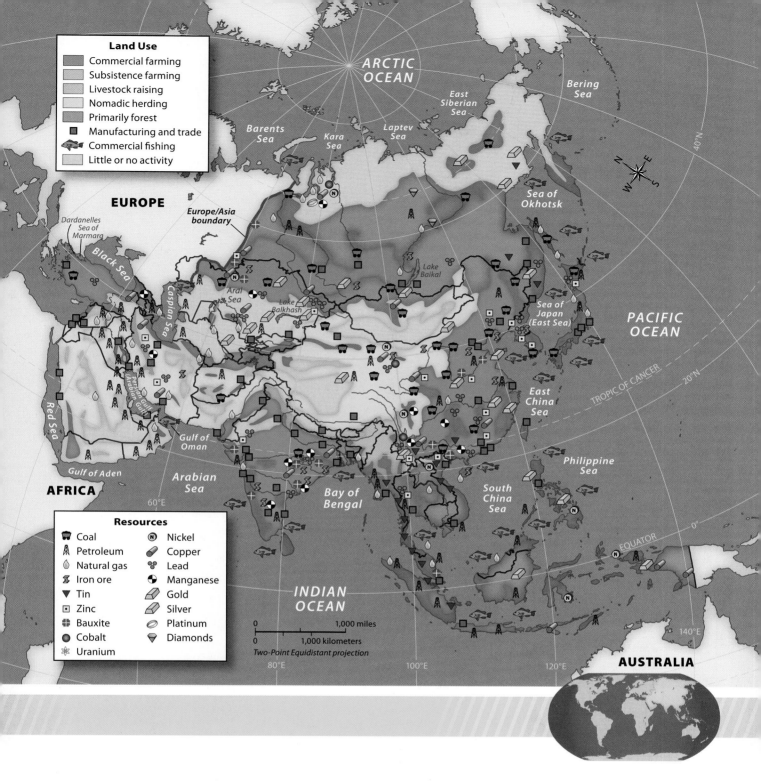

ECONOMIC RESOURCES

MAP SKILLS

1. **PLACES AND REGIONS** What mineral resources can be found around the Persian Gulf?

2. **HUMAN GEOGRAPHY** What economic activity takes place in the South China Sea?

3. **THE GEOGRAPHER'S WORLD** In what part of Russia are gold and silver deposits found?

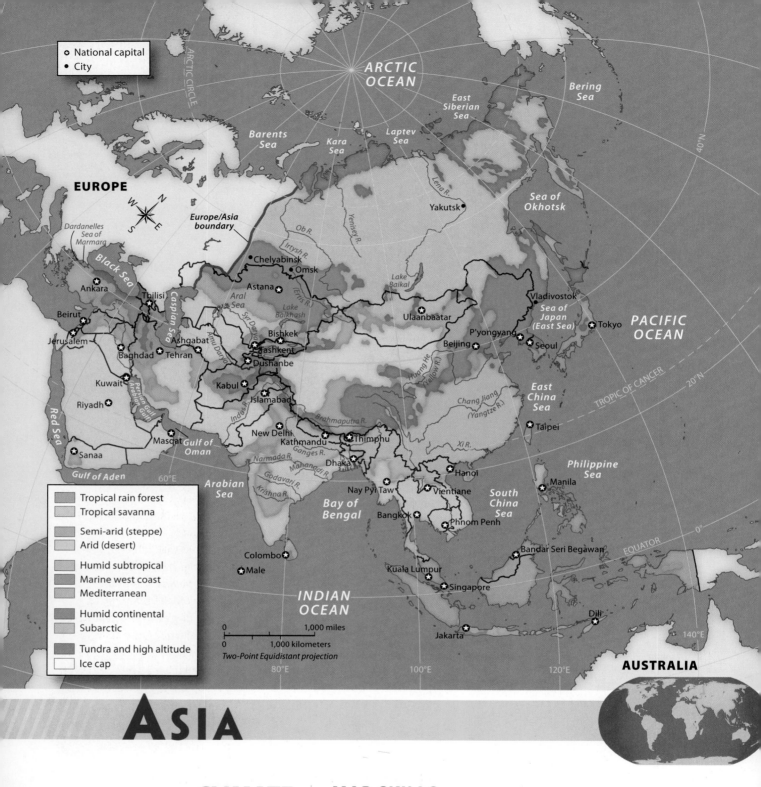

ASIA

CLIMATE

MAP SKILLS

1. **PHYSICAL GEOGRAPHY** How would you describe the climate zones found along the Equator?

2. **PLACES AND REGIONS** Which city do you think receives more rain each year—Beijing or Ulaanbaatar? Why?

3. **PHYSICAL GEOGRAPHY** In general, how does the climate of India's coastal areas differ from the climate of the Ganges River area?

EAST ASIA

ESSENTIAL QUESTIONS • How does geography influence the way people live? • What makes a culture unique? • How do cultures spread? • Why do people trade? • How does technology change the way people live?

Performers with the Beijing Opera wear facial paint and colorful costumes.

networks

There's More Online about East Asia.

CHAPTER 4

Lesson 1
Physical Geography of East Asia

Lesson 2
History of East Asia

Lesson 3
Life in East Asia

The Story Matters...

East Asia can trace many of its cultural features to an ancient civilization that arose in China thousands of years ago. In the centuries that followed, powerful dynasties ruled China, creating an enormous empire that influenced the entire region. Today, migration and trade have paved the way for an exchange of ideas and practices between East Asia and other parts of the world. Japan, Taiwan, and South Korea have become modern industrial nations.

FOLDABLES
Study Organizer

Go to the Foldables® library in the back of your book to make a Foldable® that will help you take notes while reading this chapter.

113

Chapter 4
EAST ASIA CCSS

East Asia borders Central Asia, South Asia, and Southeast Asia and extends to the Pacific Ocean. As you study the map of East Asia, look for the geographic features that make this region of Asia unique.

Step Into the Place

MAP FOCUS Use the map to answer the following questions.

1. **THE GEOGRAPHER'S WORLD** What country lies south of Mongolia?
2. **PLACES AND REGIONS** What seas border China?
3. **THE GEOGRAPHER'S WORLD** What bodies of water lie between Japan and the Koreas?
4. **CRITICAL THINKING**
   Analyzing If you traveled east along 40° N latitude, which East Asian countries would you travel through?

BUSY CITY MARKET One of the main attractions in Hong Kong, China, is the Temple Street Night Market. Each night, crowds visit the market to buy a variety of goods, such as clothes and household appliances.

REMOTE MYSTERIOUS HIGHLAND A nomadic sheep herder crosses the vast spaces of the Plateau of Tibet in western China.

Step Into the Time

TIME LINE Choose an event from the time line and write a paragraph predicting the effect of that event on the future of East Asia.

c. 550 B.C. Confucius is born

1766 B.C. Rule of China's Shang dynasty begins

1000 B.C.

networks

There's More Online!

- ☑ **IMAGE** Mt. Fuji in Art
- ☑ **MAP** Climates of East Asia
- ☑ **VIDEO**

Reading HELPDESK

Academic Vocabulary
- dominate

Content Vocabulary
- de facto
- archipelago
- tsunami
- loess

TAKING NOTES: *Key Ideas and Details*

Summarize As you study the lesson, use a graphic organizer like the one below to write one summary sentence for each topic.

Topic	Summary
Landforms	
Waterways	
Resources	

Lesson 1
Physical Geography of East Asia

ESSENTIAL QUESTION • *How does geography influence the way people live?*

IT MATTERS BECAUSE

East Asia is where some of the most densely populated areas in the world are found. It also has vast areas where few people live. To understand why these extreme differences exist, it is necessary to understand the region's physical geography.

Landforms and Waterways

GUIDING QUESTION *What are the main physical features and physical processes in East Asia?*

The landscapes and physical features of East Asia are varied and sometimes unusual. East Asia is home to the world's highest mountain range, as well as a vast plateau that sits miles high. It is also the site of islands created by volcanoes and fertile plains where hundreds of millions of people live.

A Regional Overview

East Asia covers much of the eastern half of the Asian continent. Its eastern boundary stretches along the Pacific Ocean. Bordering East Asia to the north are Russia and Central Asia. To the south are the regions of South Asia and Southeast Asia.

East Asia is made up of six countries: China, Japan, Mongolia, North Korea, South Korea, and the de facto country of Taiwan. A **de facto** country is one not legally recognized by other countries. China is the largest country in the region. It has more than four-fifths of the region's total land area. Slightly smaller than the United States, China is the world's fourth-largest country in land area.

At the opposite extreme is Taiwan, the region's smallest country. It is roughly the size of Massachusetts and Connecticut combined. China and Mongolia sit on the Asian mainland, while the other four countries occupy islands or a peninsula.

The Mainland

Geographers often divide mainland East Asia, which consists of China and Mongolia, into three broad geographic subregions. Because of differences in elevation, the subregions are like steps in a staircase. The highest step is the Plateau of Tibet, a vast area of mountains and uplands in southwestern China. The plateau is sometimes called "the roof of the world" because of its extremely high elevation. Much of the land sits more than 2.5 miles (4 km) above sea level.

Lofty mountains circle the Plateau of Tibet. The Kunlun Shan range runs along the northern edge, and the Himalaya—the tallest mountains in the world—rise along the southern edge. The Himalaya reach their highest point at Mount Everest, which soars higher than any other mountain in the world. When mountain climbers reach Everest's peak, they stand nearly 5.5 miles (8.9 km) above sea level.

North and east of Tibet, land elevation drops sharply to the second step in the staircase. Mountains and plateaus **dominate** this subregion, too, but they are generally much lower than those of Tibet. Some of the mountain ranges lie along the edges of enormous basins, or natural land depressions. Much of the land in the northern part of the subregion is desert or near desert, with little or no vegetation.

Land along the southern part of the subregion is forested. Some of the steepest and deepest canyons in the world lie where the land descends from Tibet.

The third and lowest step in the staircase covers most of the eastern third of China. The main landforms in this subregion are low hills and plains. Most Chinese live on these plains.

Academic Vocabulary

dominate to have the greatest importance

A train crosses a bridge spanning a deep valley in western China. The Qingzang, the world's highest rail system, links newly developing areas in the west to densely populated plains areas in eastern China.

▶ **CRITICAL THINKING**

Describing Why is western China's Plateau of Tibet called "the roof of the world?"

A Peninsula and Many Islands

In addition to the mainland, East Asia includes a large peninsula and, running parallel to the eastern coast, a long string of islands. The Korean Peninsula is a thumb of land that juts southward from the mainland between the Yellow Sea and the Sea of Japan (East Sea). It is home to two countries: North Korea and South Korea. The peninsula is mountainous, especially in the northeast. In the south and west, broad plains stretch between the mountains and the coast.

Along the eastern edge of the Sea of Japan, an arc of islands stretches for roughly 1,500 miles (2,414 km). The islands—four large ones and thousands of much smaller ones—form the **archipelago** (ahr kuh PEH luh goh) of Japan. An archipelago is a group or chain of islands.

The islands of Japan are part of the Ring of Fire, which nearly encircles the Pacific Ocean. In this area, huge sections of Earth's crust grind against each other and cause earthquakes and volcanic eruptions. The islands of Japan were formed by volcanic eruptions millions of years ago.

Forested mountains cover nearly three-fourths of the land on the islands. Plains are generally small and isolated. On Honshū, Japan's largest island, a beautiful, cone-shaped volcano called Mount Fujiyama or Mount Fuji rises about 12,400 feet (3,780 m) above the Kanto plain. Snow-covered Mount Fuji is a well-known symbol of Japan. Although it has not erupted in nearly 300 years, scientists believe it could.

Japan is one of the most earthquake-prone countries in the world. More than 1,000 small earthquakes shake the country every year. Major quakes occur less often, but they can cause tremendous damage and loss of life. When an earthquake occurs below or close to the ocean, it can trigger a **tsunami** (soo NAH mee). This is a huge wave that gets higher as it approaches the coast. Tsunamis can wipe out coastal cities and towns. In 2011 the most powerful earthquake in Japan's history produced massive tsunamis that devastated areas along the northeastern coast.

Hundreds of miles southwest of Japan's main islands lies another large island, Taiwan. Like Japan, Taiwan was formed as a result of volcanic activity. Mountains form a rugged spine that stretches the length of this sweet-potato-shaped

Mount Fuji, on Japan's Honshū island, soars above blooming cherry trees. Each spring, people come to the five lakes around Mount Fuji to view the cherry blossoms with the beautiful mountain in the distance.

▶ **CRITICAL THINKING**
Describing What physical features make up the islands of Japan?

island. The spine is actually the edge of a huge, tilting block of Earth's crust. The western face of the block slopes much more gradually than the steep eastern face. Broad plains spread across the western part of the island.

Bodies of Water

Bodies of water in East Asia provide food for its people, give them a place to live, move their goods, and power factories to light homes. Fish and other seafood caught in the seas and ocean make up an important part of the diet of many East Asians.

Four large seas sit along East Asia's eastern edge. The largest is the South China Sea. It is partly enclosed by Taiwan, southeastern China, and islands of Southeast Asia. Because the South China Sea lies between important ports in the Pacific and Indian oceans, it has some of the busiest shipping lanes in the world.

The East China Sea lies between China and Japan's Ryukyu Islands, a long archipelago extending southwestward toward Taiwan. In the north, this sea meets the Yellow Sea, which is shaped by the Korean Peninsula and the northeastern coast of China. Farther north, Japan, the Korean Peninsula, and the Asian mainland together are shaped like a corral that almost entirely encircles the Sea of Japan (East Sea).

Rivers in China

The water of East Asia's two most important rivers, the Huang He (Yellow River) and the Chang Jiang (Yangtze River), flow across China. Both of these rivers begin high on the Plateau of Tibet in southwestern China and flow down the eastern slope of the plateau. The Huang He flows by twists and turns to the Yellow Sea far to the east. Along the way, it picks up a tremendous amount of yellow-brown silt called **loess** (LEHS). This silt gives the river and the sea their name and color.

In eastern China, silt deposited by floods over millions of years has created a broad, fertile plain called the North China Plain. This is one of China's most productive farming areas. Throughout history, however, floods have regularly destroyed homes and crops and have drowned many people. For this reason, the Huang He is sometimes called "China's sorrow."

Like the Huang He, the Chang Jiang begins on the Plateau of Tibet. From its headwaters, it flows about 3,450 miles (5,552 km) to its mouth at the port city of Shanghai on the East China Sea. It is the longest river in Asia and the third longest in the world. Only the Nile and the Amazon are longer.

The Chang Jiang is China's principal waterway. It also provides water for a fertile farming region where more than two-thirds of the country's rice is grown. Nearly one-third of China's people live in the river's basin.

Visual Vocabulary

Peninsula A peninsula is an area of land that juts into a lake or an ocean and is surrounded on three sides by water. The term comes from the Latin words *paene*, meaning "almost," and *insula*, meaning "island."

The Huang He flows peacefully past irrigated farmland in eastern China on its way to the Yellow Sea.
▶ **CRITICAL THINKING**
Describing Why is the Huang He sometimes called "China's sorrow?"

Rivers in Japan and Korea

Japan's major rivers are relatively short, steep, and swift. They flow down from the mountains in the interior of the islands to low plains along the coast. Most of the rivers, including the two longest, the Shinano and the Tone, generate hydroelectric power.

The main rivers of the Korean Peninsula flow from inland mountains westward toward the Yellow Sea. The Han River flows through South Korea's capital, Seoul. North Korea's longest river, the Yalu (or Amnok), forms the country's border with China.

☑ **READING PROGRESS CHECK**

Identifying What are some ways the people of East Asia depend on rivers?

Climate

GUIDING QUESTION *What are the main factors that affect climate in different parts of East Asia?*

A traveler to East Asia would encounter a great range of climates, from hot and rainy to cold and dry. This is partly due to the area's vast size and partly because of its range of elevation.

Climate Factors

East Asia spans a tremendous distance north to south. The region's southernmost lands lie within Earth's hot tropical zone, while the northernmost lands sit closer to the frigid North Pole than to the Equator. Much of the climate variation results from these differences in latitude.

Another important factor is land elevation. Higher areas are generally colder than lower areas. Two areas at the same latitude can have very different climates if one area is much higher than the other. This situation is found in many parts of East Asia.

Thinking Like a Geographer

Rice as a Staple Food

Rice is the main source of food for over half the world's people. Most East Asians depend on it as a staple food. Rice is such an important part of Asian cultures that the word for *rice* in some languages is the same as the word for *food*. What foods are considered staples of the American diet?

120 Chapter 4

Air masses also play a major role in shaping East Asia's climate zones. A cold, dry, polar air mass spreads southward from northern Asia during the colder months of the year. A warm, moist, tropical air mass spreads northward and eastward from the Pacific Ocean during the warmer months.

Climates in the East

Southeastern China stays hot and rainy through much of the year. Vegetation is lush, and conditions are ideal for growing grain, especially rice. Farther north in eastern China, more seasonal variation occurs. Summers are hot and rainy, but winters are cold and fairly dry. China's capital, Beijing, lies in this part of the country. In the summer, temperatures in Beijing often soar above 90°F (32°C). In the winter, icy winds from the northwest whip through the city, and temperatures can drop below 0°F (−18°C).

Being surrounded or nearly surrounded by water affects the climates of East Asia's island and peninsula countries: Taiwan, Japan, and North and South Korea. These countries are generally wetter and experience milder temperatures than mainland areas at the same latitudes.

Climates in the North, Northwest, and Southwest

Across north-central and northwestern China and neighboring Mongolia, the climate ranges from semiarid (somewhat dry) to arid (very dry). Winters are bitterly cold. This sparsely populated area includes the vast, rocky Gobi Desert and the sandy Taklimakan Desert. It also has great expanses of treeless grasslands.

The Plateau of Tibet in southwestern China also has a dry climate. This is mainly because the towering Himalaya block moist air flowing northward from the Indian Ocean. Because the plateau sits at such a high elevation, the weather is cold and windy throughout the year.

✅ **READING PROGRESS CHECK**

Identifying How do the Himalaya affect the climate of the Plateau of Tibet?

Natural Resources

GUIDING QUESTION *What mineral resources are most abundant in East Asia?*

East Asia is rich in minerals, forests, and other natural resources. But the resources are not evenly distributed.

The nomadic Tsaatan herd reindeer and hunt for gold in mountainous areas of northern Mongolia.
▶ **CRITICAL THINKING**
Describing What is the climate like in Mongolia?

Minerals

China, which covers a large portion of East Asia, holds by far the greatest share of the region's resources. China is a world leader in the mining of tin, lead, zinc, iron ore, tungsten, and other minerals. Manufacturers use tungsten to make high-quality steel, lightbulbs, rockets, and electrical equipment.

Japan is one of the world's leading industrial countries, but the islands of Japan have few mineral resources. Japan has coal, copper, some iron ore, and a few other minerals, but it must import a variety of raw materials. Taiwan, like Japan, is a major industrial country, but it has limited mineral reserves. As a result, Taiwan also has to import various minerals to meet increasing demand.

Cultured pearls are harvested in the seas surrounding the Japanese islands. They range in color from white to silver and pink. They are known for their beautiful, smooth, shiny appearance.

Energy Resources

East Asia has a variety of energy resources, including coal, oil, and natural gas. The largest deposits of these fossil fuels are in China. China is the world's largest producer of coal. Its deposits are larger

Workers haul coal to barges on the Chang Jiang River in China. The barges transport the coal downstream to power plants.

than those of any other country except the United States and Russia. China also has substantial oil and natural gas reserves under the South China Sea and in the Taklimakan Desert in the far west.

Despite these fossil fuel resources, China still cannot meet all of its energy needs. The country's economy is growing rapidly. As a result, China is turning to energy-rich countries in Central and Southwest Asia for supplies of oil and natural gas. China also imports coal from Southeast Asia and Australia.

Several East Asian countries use hydroelectric power to help meet their energy needs. China produces electricity from the Three Gorges Dam on the Chang Jiang. This massive dam also helps control floods and provides water for irrigation. Hydroelectric dams in Japan harness the power of the country's swift-flowing rivers.

Forests

Much of western and northwestern China is so dry that trees cannot grow. Forestland once blanketed the eastern part of the country, but over the centuries, forests became smaller as people cut down trees for heating, building, and to create farmland. Today, forested areas cover less than one-sixth of the country.

Other East Asian countries, however, still have extensive forested areas. More than half of Taiwan's rugged landscape is covered by forests. The country used to be a major exporter of timber and wood products. Now much of the forested land is protected, so the country relies on imports to meet its need for wood products.

Thick forests cover the steep hillsides of Japan's inland areas. Almost two-thirds of the country is forested. Logging is limited in parts of the country, however, because the Japanese consider many forest areas to be sacred.

In the Korean Peninsula, many trees have been cleared for farmland, but forests still cover mountainous areas. About three-fourths of North Korea's rugged landscape is forested.

✓ **READING PROGRESS CHECK**

Analyzing Why is it necessary for people in Taiwan and Japan to import wood products?

Include this lesson's information in your Foldable®.

LESSON 1 REVIEW

Reviewing Vocabulary
1. How did Japan's *archipelago* form?

Answering the Guiding Questions
2. *Citing Text Evidence* How do the climates of East Asia's island and peninsula areas differ from climates of the mainland areas?

3. *Analyzing* How has the Huang He helped and hurt the people of China?

4. *Identifying* How does Japan make use of its short, swift rivers?

5. *Describing* Why must China import petroleum, natural gas, and coal?

6. *Informative/Explanatory Writing* Think about how elevation affects climate. Write a paragraph that explains how mountains and plateaus affect East Asia's climate.

networks

There's More Online!
- ☑ **IMAGES** China's Great Wall
- ☑ **MAP** The Silk Road and Trade
- ☑ **VIDEO**

Reading HELPDESK

Academic Vocabulary
- intertwine

Content Vocabulary
- dynasty
- shogun
- samurai
- sphere of influence
- communism

TAKING NOTES: *Key Ideas and Details*

Describe Use a chart like the one below to describe two key events in the histories of China, Japan, and Korea.

	Key Events
China	1. 2.
Japan	1. 2.
Korea	1. 2.

Lesson 2
History of East Asia

ESSENTIAL QUESTIONS • *What makes a culture unique?* • *How do cultures spread?*

IT MATTERS BECAUSE
East Asia has a long, rich, and fascinating history. Learning about the events, innovations, and ideas that shaped the region will give you a better understanding and appreciation of its countries, cultures, and peoples.

Early East Asia

GUIDING QUESTION *What important inventions from East Asia spread across the rest of the world?*

China's civilization is more than 4,000 years old. Throughout its history, Chinese civilization influenced the development of other East Asian countries.

Early China

For many centuries until the early 1900s, rulers known as emperors or empresses governed China. A **dynasty**, or line of rulers from a single family, would hold power until it was overthrown. Then a new leader would start a new dynasty. Under the dynasties, China built a highly developed culture and conquered neighboring lands.

As their civilization developed, the Chinese tried to keep out foreign invaders. In many ways, this was easy. On most of China's borders, natural barriers such as seas, mountains, and deserts already provided protection. Still, invaders threatened from the north. To defend this area, the Chinese began building the Great Wall of China about 2,200 years ago. Over the centuries, the wall was continually rebuilt and lengthened. In time, it snaked more than 4,000 miles (6,437 km) from the Yellow Sea in the east to the deserts of the west. It remains in place today.

124

China's Dynasties

The Shang was the first dynasty to leave written records. From the writings, we learned that the Shang took power about 1600 B.C. in the North China Plain. Like all succeeding dynasties, the Shang faced rebellions by local lords, attacks by Central Asian nomads, and natural disasters such as floods. When the government was stable, it could defend its people against some of these problems. Eventually, however, the dynasty weakened and fell.

After the Shang, the Zhou dynasty ruled for about 800 years, beginning around 1045 B.C. Under Zhou rule, Chinese culture spread, trade grew, and the Chinese began making iron tools.

After the Zhou, powerful dynasties expanded China's territory. In the 200s B.C., the Han united all of China and started building the Great Wall that exists today. Under the Han and Tang dynasties, traders and missionaries spread Chinese culture to all of East Asia.

In the early 1400s, under the Ming dynasty, the naval explorer Zheng He reached as far as the coast of East Africa. The last dynasty of China, the Qing, ruled from the mid-1600s to the early 1900s.

Achievements and Ideas

China underwent many changes during the Zhou dynasty. During this period, laws were recorded for the first time, the first coins were created, and farmers began to use plows pulled by oxen. It was also an age of great thinkers. One of these thinkers was a man named Confucius. He thought people should be morally good and loyal to their families. He also believed that a ruler should lead his people as though he were the head of a family.

Another of the thinkers was Laozi, who founded an important belief system called Daoism. Laozi thought that people should live in harmony with nature.

Think Again

The Great Wall of China is a single, continuous wall.

Not true. The Great Wall, which runs for about 5,500 miles (8,851 km) across northern China, is actually made up of many different walls. Some of the walls run parallel to each other. The best-preserved sections of wall were constructed of bricks and stones during the Ming dynasty (1368–1644). Some walls are made of dirt, gravel, stone, and wood. Long stretches of wall have been destroyed by erosion or buried by blowing sand.

To build the Great Wall, Chinese rulers used teams of workers that included border guards, peasant farmers, and prison convicts.

▶ **CRITICAL THINKING**

Describing Why was the Great Wall built?

CHART SKILLS >

DYNASTIES OF CHINA

Dynasty	Time Span
Xia	2200–1700 B.C.
Shang	1766–1080 B.C.
Zhou	1045–221 B.C.
Qin	221–206 B.C.
Han	206 B.C.–A.D. 221
Sui	A.D. 581–618
Tang	A.D. 618–907
Song	A.D. 960–1279
Yuan (Mongol)	1279–1368
Ming	1368–1644
Qing	1644–1911

For centuries, rulers of various dynasties used their power to expand and unite China.

Identifying Which dynasties lasted 500 years or more?

Later, another important belief system called Buddhism was introduced to China from India. Confucianism, Daoism, and Buddhism have been major influences on China and the rest of East Asia for many centuries.

The Han dynasty (202 B.C.–A.D. 220) had such an impact that many of China's people today call themselves the People of Han. With Han rule came unity and stability. The arts and sciences flourished. Papermaking was invented, and government officials began using paper for keeping records. Han rulers encouraged trade along the Silk Road. This was a caravan route that stretched 4,000 miles (6,437 km) between China and Southwest Asia, and then extended into Europe and South Asia. The Chinese sent goods such as silk, tea, spices, paper, and fine porcelain (POHR•suh•luhn) as far west as ports along the Mediterranean Sea in exchange for wool, gold, and silver.

The Han dynasty was followed by several centuries of decline and disunity as individual states fought to gain power. When China was eventually reunified, the stage was set for a long period of stability and cultural advancement under the Tang dynasty (A.D. 618–907) and Song dynasty (A.D. 960–1279).

During the Tang dynasty, probably in the A.D. 800s, a new type of printing emerged. The Chinese developed a way to use blocks of wood and clay to print characters on paper. This invention, known as woodblock printing, made it possible to print large numbers of books quickly, which allowed ideas to spread more rapidly. Another important invention was gunpowder, which was used in explosives and fireworks. The Chinese also invented the magnetic compass, which helped sailors find their direction at sea.

Early Korea and Japan

Korea was first populated thousands of years ago by people who migrated from northern Asia. By the first century A.D., the peninsula was divided among three rival kingdoms. Six centuries later, one of these kingdoms—the Silla—conquered the other two and unified the peninsula. After enduring for three centuries, the Silla kingdom was succeeded in A.D. 935 by the Koryo.

In the 1200s, the Mongols, a people from the steppes of Central Asia, had conquered North China, parts of Asia, and the northern half of the Korean peninsula. The vast territory became part of the Mongol's Yuan dynasty. At the end of the 1300s, the Mongols were driven out of Korea and a new Korean dynasty called the Choson came to power. It would stay in power until modern times.

Korea went through religious changes during this period. Buddhism had spread from India to China to Korea in the A.D. 300s and became popular during the Koryo dynasty. Later, during the Chosun dynasty, Confucianism became Korea's dominant religion. Chinese characters came to be used for Korean writing. Korean artists and writers were inspired by the art and literature of China. Korean rulers also adopted Confucianism as a basis for government. In some periods, China provided Korea with military protection. In other periods, Koreans lived in fear of Chinese invasion.

The history of Japan is **intertwined** with that of its neighbors to the west. The Japanese islands were settled by people from Korea and China. By A.D. 500, the clans and tribal kingdoms of Japan had close ties to Korea. Eventually, the ties extended to China, beginning a flow of ideas and culture that transformed Japan. Japanese people began using the Chinese calendar and the Chinese system of writing.

Academic Vocabulary

intertwine to twist or twine together

MAP SKILLS

1. **THE GEOGRAPHER'S WORLD** What was the Silk Road?

2. **PLACES AND REGIONS** Which areas of the world were linked by the Silk Road?

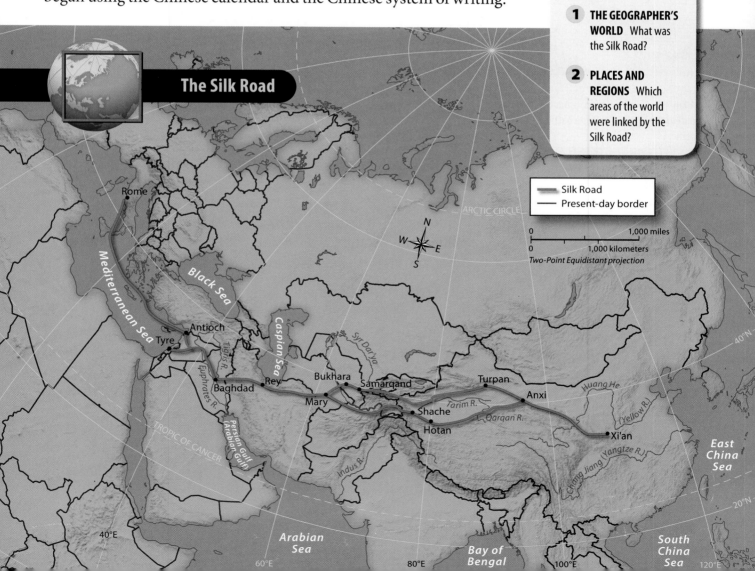

A Japanese folding screen, produced during the 1700s, displays a battle scene.

▶ CRITICAL THINKING
Citing Text Evidence How was early Japan ruled?

In addition, the Japanese adopted Chinese technology and Buddhism spread to the islands from Korea and mixed with Shinto, a Japanese religion. Shinto, or "Sacred Way," stressed that all parts of nature—humans, animals, plants, and rivers—have spirits.

Japan was ruled by emperors, but over time, they lost power. Eventually landowning families set up a feudal system. Under the system, high nobles gave land to lesser nobles in return for their loyalty and military service. At the bottom of the social ladder, peasants farmed nobles' estates in exchange for protection. By the 1100s, armies of local nobles had begun fighting for control of Japan. Minamoto Yoritomo (mee•nah•moh•toh yho•ree•toh•moh) became Japan's first **shogun**, or military leader. Landowning warriors, who were called **samurai**, (SA•muh•RY) supported the shogun. Although the emperor kept his title, the shoguns held the real power.

☑ READING PROGRESS CHECK

Describing What are some ways in which China influenced Japan?

Change in East Asia

GUIDING QUESTION *How did increased contact with the West influence the region?*

Throughout much of its history, East Asia was mostly isolated from the rest of the world. High mountains, harsh deserts, and vast distances limited the flow of ideas and goods between the region and other parts of Eurasia. From the 1500s onward, however, increasing trade brought East Asian countries into greater contact with other cultures, especially Europe.

Spheres of Influence

By the early 1800s, internal problems had weakened China. Meanwhile, European countries were growing more powerful and making stronger claims. By the 1890s, European governments and Japan had claimed large areas of China as spheres of influence. A **sphere of influence** is an area of a country where a single foreign power has been granted exclusive trading rights.

The foreign intrusion in their country made many Chinese people angry. Their anger fueled a revolution in 1911, and the new government could not control the country. By 1927, a military leader named Chiang Kai-shek had formed the Nationalist government.

Two Chinas

Meanwhile, Chiang's rival, Mao Zedong (MOW dzuh•DUNG), gained support from Chinese farmers. Mao believed in **communism**, a system in which the government controls all economic goods and services. After years of civil war, the Communists won power in 1949. They set up the People's Republic of China on the Chinese mainland. The Nationalists fled to the island of Taiwan. There, they set up a government called the Republic of China.

Rise of Japan

Around 1542, a Portuguese ship heading to China was blown off its course and landed in Japan. The traders on the ship became the first Europeans to visit Japan. Soon, more traders began arriving, along with Christian missionaries. By the early 1600s, Japan's rulers had begun to fear that European powers were planning a military conquest of the islands. They decided to isolate Japan by forcing all foreigners to leave, banning European books, and blocking nearly all relations with the outside world.

Japan's isolation lasted for roughly two centuries. In 1854 U.S. naval officer Matthew C. Perry sailed to Japan with four warships. He pressured the Japanese to end their isolation and open their country to foreign trade. Not long afterward, rebel samurai forced the shoguns to return full power to the emperor.

Recognizing that European countries were far more advanced and powerful, Japan set out to transform itself by learning everything it could about the West. The country soon became an industrial and military power, and it began developing an empire.

By 1940, Japanese forces had gained control of Taiwan, Korea, parts of mainland Asia, and some Pacific islands. This expansion was one factor that led Japan to fight the United States and its allies in World War II.

✓ **READING PROGRESS CHECK**

Analyzing How was Korea affected by Japanese expansion?

U.S. naval officer Matthew Perry met representatives of the Japanese shogun at the port of Yokohama.
▶ CRITICAL THINKING
Describing How did Perry's visit change Japan?

Modern East Asia

GUIDING QUESTION *What conflicts divided East Asian countries?*

After World War II, East Asia saw substantial changes in its governments and economies. Some of the region's countries developed into important economic powers.

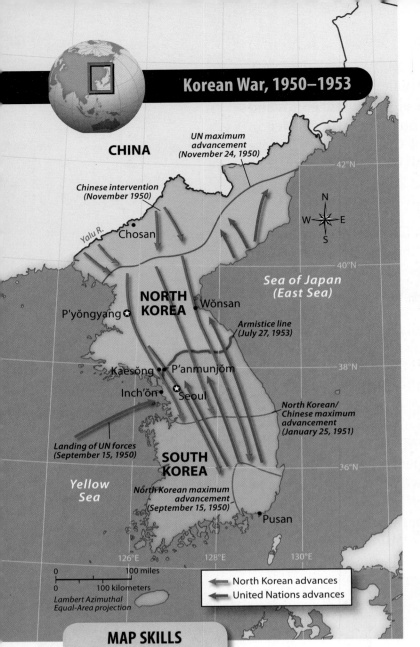

Korean War, 1950–1953

MAP SKILLS

1. **THE GEOGRAPHER'S WORLD** When and where did the North Koreans make their maximum advance?

2. **HUMAN GEOGRAPHY** What is the armistice line?

Modern China

After 1949, China became "two Chinas"— one was mainland China ruled by the Communists, and the other was the Nationalist government on the island of Taiwan. In the 1950s, the Communist mainland government took control of businesses and industry. It also took land and created state-owned farms.

In the late 1950s, Mao Zedong introduced the Great Leap Forward. This program's goal was to increase China's industrial output. Many peasants left the fields and began working in factories. Cities grew rapidly. The program failed, however. Poor planning, natural disasters, and a drop in food production led to widespread famine.

During China's Cultural Revolution in the late 1960s, intellectuals such as doctors and teachers were ordered to work on farms. Students also were taken from school and sent to the countryside to work. In this way, Mao hoped to get rid of any cultural elements that did not support his idea of communism.

Taiwan, on the other hand, pursued a goal of "one China"— two parts of one nation moving toward reunification. Taiwan wanted the mainland Communist government to negotiate with Taiwan as an equal, but Communist leaders said no.

At first, the Taiwanese government limited the freedom of its people. By 1970, however, Taiwan's leaders had instituted democratic reform and developed an economy based on capitalism. Prosperity transformed the island into an economic powerhouse.

By comparison, China's Communist government was stagnant. After Mao's death in 1976, though, Chinese leaders started to open China to the West. Economic reforms helped China become a rising global power. Chinese leaders gradually allowed a market economy to develop alongside the communist economy by letting people own businesses and sell products and services freely.

A Divided Korea

After World War II ended, Korea was divided into two countries: South Korea and North Korea. South Korea was supported by the United States, and Communist North Korea had strong ties to China and to the Soviet Union, a powerful communist country that stretched across northern Europe and Asia.

Both Koreas claimed the entire peninsula. In 1950 North Korea invaded South Korea. United Nations forces and the United States rushed to support South Korea, and China helped North Korea. The Korean War ended in 1953 without a peace treaty or a victory for either side. A buffer zone, called the demilitarized zone, was established to separate the two countries.

Stark differences developed between the two Koreas after the war. Over several decades, South Korea followed the path of capitalism and the country's economy began to grow rapidly. In contrast, North Korea's economy, which is strictly controlled by its Communist government, has struggled. North Koreans face many hardships because most resources go to the military.

Modern Japan

After being defeated by the United States and its allies in World War II, Japan was stripped of its overseas territories and military might. The country adopted a democratic constitution, and women and workers gained more rights. During the Korean War, the United States needed Japanese factories to provide supplies for its war effort. Japanese shipbuilders, manufacturers, and electronics industries benefited from giving this assistance.

The Japanese government worked closely with businesses to plan the country's economic growth. Both invested in research and development of electronics products for the home. A highly skilled workforce and the latest technology helped Japan develop its industries. Within a few decades, the Japanese were leading producers of ships, cars, cameras, and computers. By the 1990s, world demand for Japanese-made goods had turned it into a global economic power. Despite some economic ups and downs in the 2000s, Japan's economy remains one of the world's strongest. Based on gross national product, Japan ranks third in the world, trailing only the United States and China.

Include this lesson's information in your Foldable®.

✅ **READING PROGRESS CHECK**
Determining Central Ideas What led to the growth of China's economy beginning in the 1970s?

LESSON 2 REVIEW

Reviewing Vocabulary
1. How can a new *dynasty* form?

Answering the Guiding Questions
2. *Describing* How did Buddhism spread across East Asia?
3. *Identifying* Who ruled during the Yuan dynasty?

4. *Determining Central Ideas* Why did Europeans want access to China and Japan?
5. *Analyzing* What led to the creation of "two Chinas"?
6. *Argument Writing* Write a one-page essay explaining which early Chinese invention had the greatest effect on the rest of the world and why.

networks

There's More Online!

- ☑ **IMAGE** Baseball in Japan
- ☑ **ANIMATION** Population Pyramid of East Asia
- ☑ **SLIDE SHOW** Influence of Japanese Anime
- ☑ **VIDEO**

Reading HELPDESK CCSS

Academic Vocabulary
- structure

Content Vocabulary
- urbanization
- megalopolis
- trade deficit
- trade surplus

TAKING NOTES: *Key Ideas and Details*

Organize Information As you read the lesson, use a diagram like the one below to record four key facts about the people of East Asia.

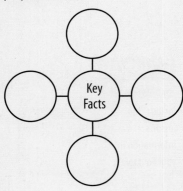

Lesson 3
Life in East Asia

ESSENTIAL QUESTIONS • Why do people trade? • How does technology change the way people live?

IT MATTERS BECAUSE
East Asia is one of the largest population centers on Earth. China alone holds about one-fifth of the world's people. The powerful economies of the East Asian countries affect global trade and manufacturing.

The People

GUIDING QUESTIONS Which areas in East Asia have the highest population densities?

Most people in East Asia live crowded together in river valleys, basins, and deltas, or on coastal plains. These lands and climates are favorable to agriculture and industry and are among the most densely populated places on Earth. In contrast, vast areas in the west and northwest are sparsely populated.

Population Patterns

Throughout much of history, China has had a large population. Two thousand years ago, the number of inhabitants had already reached an astonishing 59 million. Even today, only about two dozen countries have this many people.

For many centuries, China's population growth was slowed by epidemics, famines, warfare, and other factors. After the 1300s, however, growth began to increase steadily, and in the middle of the 1900s, it became explosive. Because the uncontrolled growth was causing many problems, the government enacted a policy in 1979 that required families to have no more than one child. This "one-child" policy helped to slow China's growth. China's 2010 Census showed a population of 1.37 billion people on the mainland and in

132

Taiwan and Hong Kong. Far more densely populated in the east than in the west and northwest, China has an average population density of about 140 persons per square mile (54 per sq. km). In contrast, its northern neighbor, Mongolia, has a population density of less than 4 persons per square mile (1.6 persons per sq. km).

Population growth in some other parts of East Asia slowed at the end of the 1900s. Japan's low birthrate means that the average age of its population is increasing. Nearly one-quarter of the population is age 65 or older. Since the mid-1990s, Japan has encouraged more births by providing programs such as child care. Soon the country could face a shortage of workers. A shortage might encourage leaders to allow more foreign workers into the country.

Where People Live

Throughout China's history, most of its people lived off the land as farmers. Economic reforms in the late 1970s, however, caused a surge of **urbanization**. Millions of peasants left their farms and moved to booming cities in eastern China. Nearly half of the country's people now live in cities. Shanghai has about 11 million people, making it the largest city in China. Just over 7 million people live in Beijing, the capital. Hong Kong, which China regained from Britain in 1997, has more than 5 million people. Dozens of other Chinese cities have populations greater than 1 million.

In other East Asian countries, urbanization began earlier and is farther advanced than in China. In Japan, two-thirds of the people live in cities. The cities of Tokyo, Osaka, Nagoya, and Yokohama form a **megalopolis**, or supersized urban area, along the coast. Greater Tokyo, Japan's capital and its largest city, is home to about 32 million people.

As South Korea industrialized, more people moved to cities. Now, 83 percent of South Koreans live in urban areas. The country's capital city, Seoul, has more than 10 million people. Across East Asia, the standard of living for people in cities is generally higher than that of people in rural areas.

✓ **READING PROGRESS CHECK**

Determining Central Ideas How can a country's growth rate influence its economy?

Because of the lack of building space, Hong Kong's urban skyline is dominated by many high-rise office and apartment buildings. As a result, Hong Kong is described as "one of the most vertical places on Earth."

▶ **CRITICAL THINKING**
Describing What major change has affected East Asia's cities in recent decades?

Culture in East Asia

GUIDING QUESTION *What are some of the cultural differences among East Asian countries?*

The people of East Asia have a rich cultural heritage. Traditions, beliefs, art, literature, and other elements that make up their heritage have been shaped by contributions from many different groups. The rise of a global culture has also brought change.

Ethnic and Language Groups

In each East Asian country, people tend to be ethnically similar. In Japan, about 99 percent of the population is ethnic Japanese and speaks the Japanese language. Nearly all the people in North and South Korea are ethnic Koreans who speak the Korean language. About 95 percent of Mongolia's people are ethnic Mongolian, and almost all of them speak the Khalkha Mongolian language.

In China, the Han ethnic group makes up about 92 percent of the population. The other 8 percent belong to more than 50 different ethnic groups. The official language in China is Mandarin Chinese, but many dialects are spoken. In Taiwan, Mandarin Chinese is the official language. Most of the people of Taiwan are Taiwanese, and many also speak the Taiwanese, or Min, language. The official languages of Hong Kong are Cantonese and English.

Worshipers often leave flowers and light candles at street-corner religious shrines in Macau. The territory was under Portuguese rule until becoming part of China in 1999.

▶ **CRITICAL THINKING**
Describing What role does China's government play in setting religious policy?

Religion and the Arts

The people of East Asia have many religions and belief systems. Many Chinese practice a mix of Buddhism, Daoism, and Confucianism. The Communist government that took over in 1949 believed that religion had no place in a communist country, so it began limiting religious practice. In recent decades, however, antireligious policies have been relaxed somewhat.

Buddhism has a large following in Korea and Japan, although North Korea's government also limits religious practice. In Japan, many people combine Buddhism with Shinto, the country's traditional religion. In South Korea, Christianity has a strong presence.

A number of art forms have long been popular in East Asia. In China, Korea, and Japan, many artists have painted the rugged landscapes of their countries. Their works reflect a reverence for nature that is part of Daoism and Shinto. Ceramics and pottery have been important parts of East Asian art since ancient—even prehistoric—times. Craftspeople in East Asia are also skilled at weaving, carving, and lacquerwork.

Calligraphy, the art of turning the written word into beautiful, expressive images, is considered one of the highest forms of art in China and Japan. Chinese characters are visually interesting and complex, so they lend themselves well to calligraphy. It is common for East Asians to display works of calligraphers in their homes.

East Asians also have strong literary and theatrical traditions. Japanese poets often write haiku, brief poems that follow a specific **structure**. Japan is famous for its traditional forms of theater. Today, Japan is also known for anime, a type of animation. Comic books and cartoons using this style have become popular all over the world. Jingxi, known to English-speakers as Peking opera, is a popular type of musical theater in China. In Jingxi performances, actors and actresses wear colorful costumes, sing in high-pitched voices, and use gestures, postures, and steps to reveal the attitudes of their characters.

Along with art and literature, East Asians have developed new forms of expression. South Korean "K-pop" is a popular type of music that young people in Japan and other East Asian countries enjoy. It has its roots in dance and electronic music from the West. Communication technology, the Internet, and travel increase the reach of new art forms. At the same time, different countries adopt and change them in unique ways.

For centuries, East Asian artisans have used lacquer, a clear or colored wood finish, to protect and add luster to jewelry boxes, furniture, and other art objects.

Academic Vocabulary

structure organization

Daily Life

Traditionally, the family is the center of social life in East Asia. In rural areas of East Asia, for example, different generations of one family may share the same home. In crowded cities, tall apartment buildings provide housing for many families. As more people have moved into urban areas, some traditional attitudes have begun to change. For example, many women in the region now work outside the home.

East Asian cultures place a high value on education. Teachers are greatly respected, and children are expected to work hard. At a young age, children begin taking important exams that can determine whether they will get into top colleges.

Rice and noodles are staples in the diets of most East Asians, but otherwise cuisine varies widely across the region. For example, China's Sichuan (or Szechuan) Province is known for its bold, spicy dishes. Cantonese cuisine from the Guangdong Province is generally milder, with a careful balance of flavors. In Japan, meals are often built around seafood and soybean-based foods such as tofu. In Mongolia, where many people raise livestock, the cuisine features meat and dairy products.

People in East Asia enjoy many different pastimes, some traditional and some modern. Millions of people young and old practice martial arts such as tai chi (TY CHEE) and tae kwon do (TY KWAHN DOH), both of which originated in the region centuries ago. Baseball, introduced from North America, is widely

Billboards of anime characters form a backdrop to young fans lining up to attend an exhibition of animation, comics, toys, and games in Hong Kong.

Identifying What other new forms of expression have recently developed in East Asia?

popular in Japan, Taiwan, and South Korea. Many children play on Little League baseball teams, and Japan has its own major baseball leagues. Basketball, also imported from America, has become one of the top sports in China.

Holidays are also important in East Asia. The biggest holiday of the year in China is Spring Festival, also known as Chinese New Year. It is celebrated with dances, fireworks, traditional foods, and religious ceremonies to honor ancestors. In Japan, people observe New Year's Day by visiting temples and shrines. South Korea's most popular holiday is the Harvest Moon Festival, which is something like Thanksgiving Day in the United States. During this festival, South Koreans celebrate the fall harvest and hold special ceremonies in honor of their ancestors.

☑ **READING PROGRESS CHECK**

Analyzing Why is religious activity limited in China?

Members of a Little League team in Tokyo, Japan, prepare for a baseball game. In 2012 a Japanese team won the Little League World Series Championship held in South Williamsport, Pennsylvania.

Identifying What other American sport has become popular in East Asia?

Current Issues in East Asia

GUIDING QUESTION *How do East Asian economies affect economies around the world?*

In the last half-century, rapid economic growth has transformed East Asia. China and Japan now have larger economies than any other country in the world except the United States. With this growth, however, have come problems, including environmental damage.

Economies and Environments

In China, factories, coal-burning power plants, and the growing numbers of cars and trucks have led to dramatic increases in air pollution. Factory waste, sewage, and farm chemicals have poisoned water. Rapidly growing urban areas have eaten up valuable farmland. Many cities face a constant shortage of water.

Japan has similar issues. Polluted air from power plants has produced acid rain and other problems. But as a more developed country with a longer history of industrialization, Japan has engaged in stronger environmental protection efforts. In addition, Japan faces a constant threat of earthquakes.

Chapter 4   **137**

A boy runs across a rubbish-covered beach in Hainan, a tropical island in the South China Sea. Hainan, a part of China, has seen its economy grow rapidly in recent years.

▶ CRITICAL THINKING
Determining Central Ideas What have been the effects of rapid economic growth in East Asia?

In 2011 the strongest earthquake ever recorded in Japan killed thousands of people and damaged several nuclear power plants. It also disrupted trade and manufacturing around the world.

Trade

Many of the goods manufactured in East Asia are shipped to the United States and Europe. China's exports to the United States include electronic goods, toys and games, clothing, and shoes. From the United States, China imports soybeans, cotton, automobiles, and many other goods. Trade between the two countries, however, is not balanced. In 2010 the U.S. trade deficit with China rose to more than $273 billion. The **trade deficit** means the United States imports more goods from China than it sells to China. A **trade surplus** occurs when a country exports more products than it imports.

Challenges Facing East Asia

Japan and China are dealing with the challenges of population growth. Ever-growing populations put a strain on limited resources and services. In 1979 China began a policy that allowed each family to have no more than one child. The policy did slow population growth. However, with fewer children and better living standards, the percentage of elderly has grown substantially. Economic growth and productivity depend on a labor force of young adult workers. To increase the population of young people, some Chinese are demanding that the government change or repeal the law that penalizes families that have more than one child.

GRAPH SKILLS

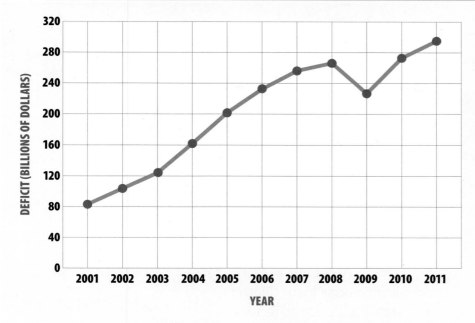

U.S. TRADE DEFICIT WITH CHINA, 2001–2011

The United States depends on imports from China. Except for the 2009 recession, the U.S. trade deficit with Chinas has risen dramatically.

▶ **CRITICAL THINKING**

1. **Integrating Visual Information** In what years since 2001 has the United States had a trade surplus with China?

2. **Identifying** In what years was the trade deficit greater than $280 billion?

Political differences in East Asia are another challenge. Japan is in dispute with Russia over a long chain of islands known as the Kuril Islands, which lie north of Japan. Russia claims the entire archipelago, but Japan claims the southernmost islands. North Korea's efforts to develop nuclear weapons have drawn harsh criticism from several countries.

Both China and North Korea face questions about human rights in their countries, and China continues to receive international pressure for its views on Tibet and Taiwan. China's economic boom has drawn many people out of poverty, but the country faces a growing income gap between people in its cities and the countryside.

Include this lesson's information in your Foldable®.

☑ **READING PROGRESS CHECK**

Drawing Conclusions How might an earthquake in Japan affect the economies of other parts of the world?

LESSON 3 REVIEW

Reviewing Vocabulary
1. What parts of Japan make up its *megalopolis*?

Answering the Guiding Questions
2. **Describing** How is Japan trying to compensate for the rise in the average age of its people?

3. **Determining Central Ideas** Today, many East Asian families are becoming scattered as people move to cities. Also, more women now work outside the home. How might these changes affect the structure and role of the family?

4. **Analyzing** In recent decades, hobbies and sports have become more important in the lives of people in China and other East Asian countries. What is the connection between this change and economic growth?

5. **Determining Central Ideas** What are some of the negative results of economic growth in East Asia?

6. **Argument Writing** Would you rather live in urban or rural areas in East Asia? Write a paragraph explaining your answer.

GLOBAL CONNECTIONS

The Fury of a TSUNAMI

A tsunami is a series of ocean waves generated by earthquakes, landslides, volcanic activity, or other disturbances below or on the seafloor. Scientists say that earthquakes cause about 90 percent of all tsunamis.

Warning Signs A tsunami can strike quickly. There are a few warning signals that typically occur. If an earthquake strikes, do not stay in low areas or near water. Earthquakes often trigger a tsunami. Sometimes, the ocean appears to drain away before an approaching tsunami hits. This is a warning sign that a tsunami is approaching.

4 of every 5 About four in every five tsunamis occur in the "Ring of Fire" in the Pacific Ocean. Tectonic shifts make volcanoes and earthquakes common there.

500 MPH Some tsunamis are 100 miles (161 km) long or longer. They can travel as fast as 500 miles (805 km) per hour—as fast as a commercial jet plane. At that speed, they can cross the Pacific Ocean in less than a day.

Read about two significant tsunamis in recent times:

July 9, 1958 The largest recorded tsunami in modern times luckily struck the isolated region around Lituya Bay, Alaska. Waves rose 1,700 feet (518 m). That is taller than the Empire State Building.

> They can travel as fast as 500 miles (805 km) per hour—as fast as a commercial jet plane.

Dec. 26, 2004 The deadliest tsunami in recorded history, the Indian Ocean tsunami killed an estimated 230,000 people. The tsunami swept through a wide area, including Indonesia, India, Madagascar, and Ethiopia. Many people died in the weeks after the tsunami hit because of lack of water and medical treatment.

Waters flood the city of Miyako shortly after the March 2011 earthquake struck northern Japan. ▶

Did You Know?

The Richter scale is used to rate the magnitude—or amount of energy released—of an earthquake. Most earthquakes that occur today rate less than 3 on the Richter scale and do not produce much damage. On average, at least one earthquake with a magnitude of 8.0 occurs each year.

THERE'S MORE ONLINE

HEAR stories from tsunami survivors • *SEE* before and after images • *WATCH* an animation of a tsunami

GLOBAL CONNECTIONS

These numbers and statistics can help you understand the full effects of the earthquake and tsunami that struck Japan in March 2011.

8.9
MAGNITUDE

Japan was hit by an 8.9 magnitude earthquake on March 11, 2011, that triggered a deadly, 23-foot tsunami in the northern part of the country. In comparison, the 1998 earthquake that caused massive damage in California was only a magnitude 6.9 on the Richter scale.

40 Years
The tsunami in Japan recalled the 2004 disaster in the Indian Ocean. On December 26, a 9.0 magnitude earthquake—the largest earthquake in 40 years—ruptured in the Indian Ocean, off the northwestern coast of the Indonesian island of Sumatra. The earthquake stirred up the deadliest tsunami in world history. It was so powerful that the waves caused loss of life on the coast of Africa and were even detected on the East Coast of the United States.

100,000
At a news conference on March 13, Prime Minister Naoto Kan, who later gave the disaster the name "Great East Japan Earthquake," emphasized the gravity of the situation: "I think that the earthquake, tsunami, and the situation at our nuclear reactors makes up the worst crisis in the 65 years since the war. If the nation works together, we will overcome." The government called in 100,000 troops to aid in the relief effort.

200,000 Evacuated

Cooling systems in one of the reactors at the Fukushima Daiichi nuclear power station in the Fukushima prefecture on the east coast of Japan failed shortly after the earthquake, causing a nuclear crisis. This initial reactor failure was followed by an explosion and eventual partial meltdowns in two reactors, then by a fire in another reactor that released radioactivity directly into the atmosphere. The nuclear troubles were not limited to the Daiichi plant; three other nuclear facilities also reported problems. More than 200,000 residents were evacuated from affected areas.

15,839

According to the official toll, the disasters left 15,839 dead and 3,647 missing. One year after the tsunami, about 160,000 people had not returned to their homes.

LEVEL 7
On April 12, 2011, Japan increased its assessment of the situation at the Fukushima Daiichi nuclear power plant to Level 7, the worst rating on the international scale, putting the disaster on par with the 1986 Chernobyl explosion.

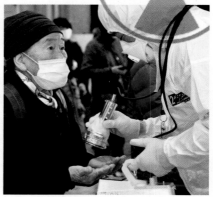

450 MPH

The waves travel in all directions from the epicenter of the disturbance. The waves may travel in the open sea as fast as 450 miles (724 km) per hour. As they travel in the open ocean, tsunami waves are generally not particularly large—hence the difficulty in detecting the approach of a tsunami. But as these powerful waves approach shallow waters along the coast, their velocity is slowed and they consequently grow to a great height before smashing into the shore.

December 26, 2004, Sumatra, Indonesia: A 9.1–9.3 earthquake triggered a devastating tsunami in the Indian Ocean. More than 200,000 people were estimated to have been killed.

October 25, 2010, Sumatra, Indonesia: A 7.7 earthquake and resulting tsunami caused the loss of 509 lives when it struck off the coast of Sumatra.

July 17, 2006, Java, Indonesia: 668 lives were lost when a 7.7 earthquake caused a tsunami on the island of Java.

March 11, 2011, Japan: A 9.0 earthquake caused a tsunami on Japan's northeast coast. More than 20,000 people were confirmed killed or missing.

GLOBAL IMPACT

DESTRUCTIVE FORCES Hundreds of small earthquakes shake Japan every year. Major quakes occur less often, but they may cause disastrous damage and loss of life. When an undersea earthquake generates a tsunami—a huge tidal wave that gets higher and higher as it approaches the coast—many lives may be lost. Because earthquakes and tsunamis are difficult to predict, some parts of the region, especially in high-population areas, rely on special building methods and emergency preparedness to help reduce casualties.

Natori, Japan

Natori (left) is shown before the March 2011 earthquake and tsunami. A photograph of the city (right) shows the effects afterwards.

BEFORE | AFTER

Thinking Like a Geographer

1. **Physical Geography** How are tsunamis created?

2. **Physical Geography** Research to find information about the 2004 Indian Ocean tsunami. Create a map that identifies the nations that suffered deaths and great destruction.

3. **Environment and Society** Research online to find out what happened when the Fukushima Daiichi nuclear power plant was hit by the 2011 tsunami. Create a PowerPoint® presentation that includes diagrams and photos to show the chain of events that occurred when the nuclear plant was damaged. Present your slides to

Chapter 4 ACTIVITIES

Directions: Write your answers on a separate piece of paper.

1 Use your **FOLDABLES** to explore the Essential Question.
INFORMATIVE/EXPLANATORY WRITING Why do people in China live in densely populated areas in river valleys and along the coastal plains when the country has so much sparsely populated land in its western regions?

2 21st Century Skills
DESCRIBING Working in small groups, choose an East Asian country and research how its schools operate. Find out what kinds of classes students take, how they are graded, what sports and activities are offered, and what a typical school day is like. Prepare a report or a slide show and share results with another class in your school.

3 Thinking Like a Geographer
IDENTIFYING The United States and China are about the same size and lie within the Northern Hemisphere at similar latitudes. Make a T-chart on a sheet of paper. On one side list similarities between the two countries, and on the other side list differences.

4 GEOGRAPHY ACTIVITY

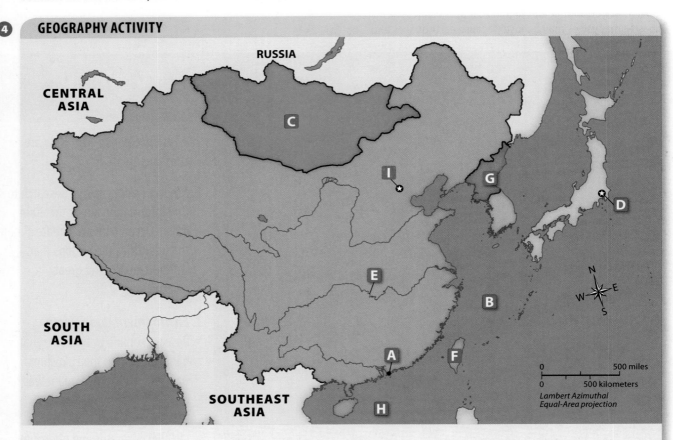

Locating Places
Match the letters on the map with the numbered places below.
1. Taiwan
2. East China Sea
3. South China Sea
4. Chang Jiang (Yangtze River)
5. Mongolia
6. Hong Kong
7. Beijing
8. North Korea
9. Tokyo

Chapter 4 Assessment

REVIEW THE GUIDING QUESTIONS

Directions: Choose the best answer for each question.

1. Why do most Japanese live within a few miles of the coast?
   A. The interior area of the country consists of steep, heavily forested mountains.
   B. The weather is better.
   C. Huge corporate farms take up most of the interior space.
   D. People like living near the beach.

2. China's major river systems
   F. flow through Beijing.
   G. begin on the Plateau of Tibet.
   H. empty into the South China Sea.
   I. are not navigable.

3. What product invented during China's Han dynasty is still in use around the world?
   A. fireworks
   B. paper
   C. gunpowder
   D. soap

4. Why did the United States send warships to Japan in 1854?
   F. to conquer the island
   G. to establish a naval base
   H. to pressure the Japanese to open their country to foreign trade
   I. to protect Japan from a Chinese invasion

5. During the Korean War, which country sent its troops to help North Korea and fight against American and United Nations forces?
   A. Taiwan
   B. Japan
   C. China
   D. South Korea

6. Which two East Asian countries have the largest economies?
   F. China and North Korea
   G. China and Taiwan
   H. China and Japan
   I. China and South Korea

Chapter 4 ASSESSMENT (continued)

DBQ ANALYZING DOCUMENTS

7 ANALYZING Read this passage about the earthquake and tsunami that struck Japan in 2011:

"*Very little of the devastation resulting from this earthquake was from the initial shaking. This is partly because of Japan's stringent [tough] building codes. But mainly because any damage from the seismic waves that sent skyscrapers in Tokyo swaying was dwarfed by the impact of the 10 metre tsunami that hit the Japanese coast less than an hour later.*"

—from Chris Rowan, "Japan Earthquake" (2011)

What does Rowan mean when he calls the tsunami a 10-meter (33-foot) tsunami?
  A. how far the wave traveled from the point of the earthquake
  B. how far along the coast the tsunami hit
  C. how far inland the wave traveled
  D. how high the tsunami was above sea level

8 CITING TEXT EVIDENCE What detail in the passage explains that the Japanese are aware of the dangers of living along the Ring of Fire?
  F. strict building codes that limit earthquake damage
  G. the height of the tsunami
  H. the short duration of the initial shaking caused by the quake
  I. heavy damage from the tsunami

SHORT RESPONSE

"*But for many [Chinese], the growing gap between rich and poor is the most pressing issue, especially in Beijing's slums. . . . Li Yulan, 78, runs a small shop. . . . She says the rich are too rich. The poor are too poor. Of eight people in her family, just two have income, she says.*"

—from Shannon Van Sant, "China Struggles to Bridge Gap Between Rich, Poor" (2012)

9 IDENTIFYING POINT OF VIEW Would China's middle class agree that the income gap is the country's most pressing issue? Why or why not?

10 DISTINGUISHING FACT FROM OPINION Is the example of Li Yulan convincing evidence of the income gap problem? Why or why not?

EXTENDED RESPONSE

11 INFORMATIVE/EXPLANATORY WRITING Every year America's trade deficit with China increases by billions of dollars. How do trade deficits affect the American economy? How might they affect you and your family personally?

Need Extra Help?

If You've Missed Question	1	2	3	4	5	6	7	8	9	10	11
Review Lesson	1	1	2	2	2	3	1	1	3	3	3

SOUTHEAST ASIA

ESSENTIAL QUESTIONS • How does geography influence the way people live? • What makes a culture unique? • Why does conflict develop?

networks
There's More Online about Southeast Asia.

CHAPTER 5

Lesson 1
Physical Geography of Southeast Asia

Lesson 2
History of Southeast Asia

Lesson 3
Life in Southeast Asia

The Story Matters...

Southeast Asia is made up of a mainland and hundreds of islands. The mainland has several mountain ranges, coastal plains, and major rivers. The islands also have varied landforms, but their rivers are short and steep. Shaped by tectonic plate movements, these islands have many active volcanoes. Today, plate movements cause dangerous earthquakes, sometimes generating tsunamis. Recovering from such natural disasters presents many challenges to the people of the region.

FOLDABLES
Study Organizer

Go to the Foldables® library in the back of your book to make a Foldable® that will help you take notes while reading this chapter.

Royal guard at the Grand Palace in Bangkok, Thailand

147

Chapter 5
SOUTHEAST ASIA

Two peninsulas form Southeast Asia's mainland—the Indochina and the Malay Peninsulas. Like the mainland, most of the region's islands are mountainous.

Step Into the Place

MAP FOCUS Use the map to answer the following questions.

1. **THE GEOGRAPHER'S WORLD** What is the capital of Cambodia?

2. **THE GEOGRAPHER'S WORLD** In what country is the city of Manila located?

3. **THE GEOGRAPHER'S WORLD** In which direction would you travel to go from Hanoi to Mandalay?

4. **CRITICAL THINKING** Describing What does it mean to say a country is "landlocked"? Does Southeast Asia have any landlocked countries?

URBAN SKYLINE Skyscrapers dominate Jakarta, the capital and leading economic center of Indonesia. With more than 11 million people, Jakarta has the largest population of any city in Southeast Asia.

FLOATING MARKET Fruit and vegetable dealers sell their wares from boats on Inya Lake in Myanmar (Burma). The artificial lake was created in the 1880s to provide water to Yangon (Rangoon), the capital.

Step Into the Time

TIME LINE Choose an event from the time line and write a paragraph predicting the social, political, or economic causes and effects of that event.

802 A.D. Khmer empire, based in what is now Cambodia, rules much of region

939 Vietnam becomes semi-independent from China after 1,000 years of domination

c. 1150 King Suryavarman II has the Angkor Wat temple built in Cambodia

1565 The Spanish establish their first settlement in the Philippines

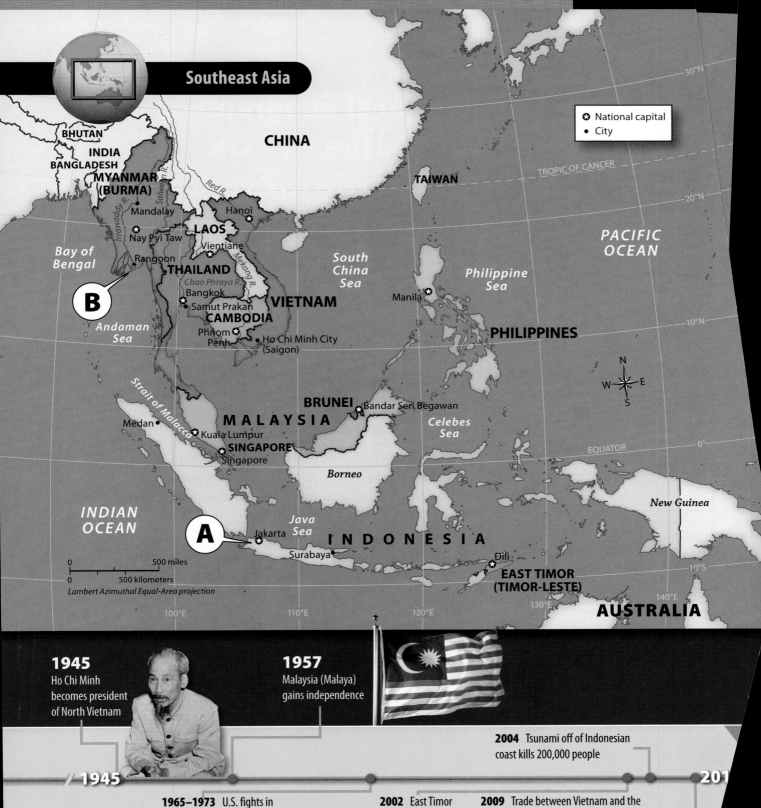

Southeast Asia

- ◉ National capital
- • City

1945 Ho Chi Minh becomes president of North Vietnam

1957 Malaysia (Malaya) gains independence

1965–1973 U.S. fights in Vietnam War

2002 East Timor gains independence

2004 Tsunami off of Indonesian coast kills 200,000 people

2009 Trade between Vietnam and the United States exceeds $15 billion annually

networks

There's More Online!

- **IMAGE** Indonesian Tsunami
- **MAP** Malaysia/Indonesian Islands
- **SLIDE SHOW** Wildlife of Southeast Asia
- **VIDEO**

Reading HELPDESK

Academic Vocabulary
- commodity

Content Vocabulary
- insular
- flora
- fauna
- endemic

TAKING NOTES: *Key Ideas and Details*

Identify As you read the lesson, use a graphic organizer like this one to identify the countries that occupy or share the region's main peninsulas and islands.

Peninsula or Island	Country or Countries
Indochinese and Malay Peninsulas	
Borneo	
New Guinea	
Java	
Timor	
Luzon	

Lesson 1
Physical Geography of Southeast Asia

ESSENTIAL QUESTION • *How does geography influence the way people live?*

IT MATTERS BECAUSE
Southeast Asia is like a challenging puzzle. Some of its main pieces, its countries, are divided into smaller pieces. Learning how the region's pieces fit together will help you answer a question geographers ask: What makes it a region?

Landforms and Resources

GUIDING QUESTION *How are the landforms of Southeast Asia's mainland different from the landforms of its islands?*

Southeast Asia can be divided into two main parts: a mainland area and an **insular** area, an area comprised of islands. (*Insular* comes from the Latin word *insula,* meaning "island." Another term that comes from *insula* is *peninsula,* which means "almost an island.")

The mainland sits at the southeastern corner of the Asian continent, bordering the world's two most populous countries: China and India. In this area, where the Indian Ocean meets the Pacific, thousands of islands stretch across miles of tropical waters.

Peninsulas and Islands

Most of the mainland occupies a large peninsula that juts southward from the Asian continent. Located between India and China, this extension of land is known as the Indochinese Peninsula, or simply Indochina.

Of Southeast Asia's 11 countries, 6 are located at least partly on the mainland peninsulas. These countries are Cambodia, Laos, Malaysia (the western region of this divided country), Myanmar (also known as Burma), Thailand, and Vietnam. The tiny island country of Singapore sits just off the southern tip of the Malay Peninsula. A bridge connects Singapore's main island to the peninsula.

The region's other four countries, along with the eastern part of Malaysia, occupy islands in the vast Malay Archipelago. Larger in area than any other island group in the world, it contains more than 24,000 islands, stretching from mainland Southeast Asia to Australia. Among the islands are 7 that rank among the 20 largest in the world: New Guinea, Borneo, Sumatra, Sulawesi, Java, Luzon, and Mindanao.

More than 17,000 of the islands in the Malay Archipelago are part of Indonesia, which is by far the largest country in Southeast Asia. Indonesia shares the islands of Timor, New Guinea, and Borneo with other countries. East Timor, one of the world's newest countries, occupies the eastern half of Timor. Malaysia's eastern region spreads across northern Borneo and surrounds the small country of Brunei.

The northernmost islands in the Malay Archipelago are the Philippines, which form their own archipelago east of the mainland. The Philippines are a country of more than 7,000 islands and islets.

Mountains and Volcanoes

Much of the land in Southeast Asia is rugged and mountainous. On the mainland, mountain ranges generally have a north-south orientation. They include a group of ranges along the western edge of Myanmar that follow the border between Myanmar and Thailand and form the backbone of the Malay Peninsula. The Annamese Cordillera, a range that stretches through Laos and Vietnam, runs parallel to the mainland's eastern coast.

Many of the mountains on Southeast Asia's islands have volcanic origins. The islands lie along the Ring of Fire, a seismically active zone that encircles much of the Pacific Ocean. Most of the world's earthquakes and volcanic eruptions occur within this belt.

This aerial view of Komodo National Park in Indonesia reveals that mountainous islands are a major part of the landscape in Southeast Asia.

Identifying What Southeast Asian island group has a larger area than any other island group in the world?

A farmer tends to his rice crop near Mount Mayon, the most active volcano in the Philippines.

▶ **CRITICAL THINKING**
Analyzing Despite the dangers, how do farmers in the region benefit from volcanic eruptions?

Academic Vocabulary

commodity a material, resource, or product that is bought and sold

In Southeast Asia, four major plates meet: the Eurasian Plate, the Indo-Australian Plate, the Pacific Plate, and the Indian Plate. The pressures and tensions produced by the meeting of these plates have fractured Earth's crust into many smaller plates across the region. This action has also produced the many islands and the fractured geography, as well as the volcanoes.

Indonesia has more than 100 active volcanoes—more than any other country in the world. Most are in a long arc along the country's southern edge. One of the most famous volcanoes, Krakatoa, lies between the islands of Sumatra and Java. In 1883 Krakatoa erupted and collapsed into the sea, triggering tsunamis that claimed about 36,000 lives.

An even deadlier tsunami occurred in Indonesia in 2004. A powerful undersea earthquake off the coast of Sumatra produced huge waves that slammed into coastal areas of Southeast and South Asia, causing more than 230,000 deaths.

Natural Resources

Southeast Asia possesses a rich variety of mineral resources, including tin, copper, lead, zinc, gold, and gemstones such as rubies and sapphires. Indonesia, Malaysia, and Thailand rank among the world's top tin producers, with Indonesia accounting for roughly a fourth of the total world production of this **commodity**.

Teak, mahogany, ebony, and other hardwood trees that grow in Southeast Asia's tropical forests have long been in high demand. Many of the region's countries export wood and wood products. To combat deforestation, some countries have restricted logging.

Southeast Asia is also rich in fossil fuels. Indonesia and Malaysia rank among the top 30 countries in the world in oil reserves and production. They rank among the top 15 in natural gas reserves and production.

☑ **READING PROGRESS CHECK**

Analyzing What effect has the Ring of Fire had on the formation of the region?

Bodies of Water

GUIDING QUESTION *Why does Southeast Asia have so many different seas?*

Bodies of water are key parts of Southeast Asia's geography and identity. The region encompasses about 5 million square miles (13 million sq. km), but only a third of the area is land.

Oceans and Seas

The Malay Peninsula and Indonesia's Sunda Isles represent the boundary between two oceans. To the west and south lies the Indian Ocean, and to the north and east lies the Pacific Ocean. The region's largest seas are the South China Sea and the Philippine Sea. West of the Malaysian Peninsula lies the Andaman Sea.

Some of the busiest lanes in the world pass through Southeast Asia's seas and their waterways. About a quarter of the world's trade, including half of all sea shipments of oil, goes through the Strait of Malacca, which links the South China and Andaman Seas. As a result, Singapore, which controls the Strait of Malacca, has become one of the most important ports in the world.

River Systems

Southeast Asia's longest and most important rivers are on the mainland. Most rain flows into one of five major rivers. From west to east, the rivers are the Irrawaddy, Salween, Chao Phraya, Mekong, and Red. Each river generally flows from highlands in the north to lowlands in the south before emptying into the sea.

The Irrawaddy, Salween, and Mekong begin high on the Plateau of Tibet. The Irrawaddy flows almost straight south through a valley in Myanmar's center. The river plays a major role in the region's transportation system. Its vast fertile delta is important for farming.

The Salween runs southward through the eastern part of Myanmar, forming part of that country's border with Thailand. Like the Irawaddy, it drains into the Andaman Sea. The Mekong is Southeast Asia's longest river. It twists and turns for about 2,700 miles (4,345 km) through or near Myanmar, Thailand, Laos, Cambodia, and Vietnam. The river's drainage basin is twice the size of California, and its enormous delta is one of the world's most productive agricultural regions.

✓ **READING PROGRESS CHECK**

Analyzing Why do you think Southeast Asia's longest rivers are found on the mainland and not on islands?

The port city of Banda Aceh in Indonesia suffered great damage from a tsunami that struck on December 26, 2004.

A plantation worker in Indonesia gathers latex, a milky liquid tapped from rubber trees. The latex is eventually refined into useable rubber.

▶ **CRITICAL THINKING**
Analyzing Why do rubber trees and plants grow so well in Indonesia and other parts of Southeast Asia?

Climate, Vegetation, and Wildlife

GUIDING QUESTION *In what ways does Southeast Asia's location shape its climate?*

Climates in Southeast Asia are generally hot and humid. Much of the region receives more than 60 inches (152 cm) of rain fall each year. Because of these weather conditions and the variety of habitats in Southeast Asia, the region is home to an astonishing abundance of plant and animal life.

A Tropical Region

Latitude and air currents play major roles in shaping Southeast Asia's climates. Nearly the entire region lies within the Tropics—the zone that receives the hottest, most direct rays of the sun. From November to March, the direct rays of the sun are south of the Equator in the Southern Hemisphere. This produces areas of low pressure that draw monsoon winds that blow across the region from the northeast to the southwest. These winds bring cooler, drier air to much of the mainland but deliver heavy rains to the southern Malay Peninsula and to the islands. From May to September, the pattern reverses. The direct rays of the sun and the associated low pressure are north of the Equator. This causes monsoon winds to blow from the southwest to the northeast, bringing warm air and rain to the islands and the mainland.

Most of the region's land is surrounded or nearly surrounded by the sea, which has a moderating effect on air temperatures. Elevation also affects weather conditions. Temperatures in the highland areas are generally cooler than in the lowland areas. At the top of Indonesia's highest peak, Puncak Jaya (PUHN-chock JAH-yuh), which is close to the Equator, glaciers are visible.

The sea, the elevation, and air currents combine to create four climate zones. The southern Malay Peninsula, the southern Philippines, and most of Indonesia have a tropical rain forest climate. A tropical monsoon climate, with rainy and dry seasons, prevails in the northern Philippines, the northern Malay Peninsula, and coastal areas of the mainland. Most inland areas fall within a tropical savanna climate zone with less distinct rainy and dry seasons. The northernmost mainland has a humid subtropical climate in which summers are hot and wet and winters are mild and dry.

Weather in Southeast Asia sometimes turns deadly. Intense tropical storms called typhoons form over warm Pacific Ocean waters and slowly spin westward, often targeting the Philippines. With winds that can exceed 150 miles (241 km) per hour and torrential rains that sometimes continue for days, typhoons can wipe out homes and buildings, cause devastating floods, destroy crops, and kill large numbers of people.

> **Think Again**
>
> *Hurricane, typhoon,* and *cyclone* refer to different types of storms.
>
> **Not true.** All three terms refer to a large, powerful, circular storm that originates over warm tropical ocean waters and is characterized by strong winds and heavy rain. These storms have different names in different parts of the world.

Plants and Animals

Few places in the world rival Southeast Asia in diversity of **flora**, or plant life. Indonesia, for example, has more than 40,000 species of flowering plants, including about 5,000 species of orchids and more than 3,000 species of trees.

Blanketing much of the region are either tropical rain forests or forests with a mix of evergreen and deciduous trees. In coastal areas, forests of mangrove trees, which have aboveground roots, form a border between land and sea.

Southeast Asia's **fauna**, or animal life, consists of a wide variety of animals. Many species, including mammals, birds, fish, and insects, are **endemic**, or found nowhere else in the world. Fires, logging, mining, agriculture, and poaching have reduced the habitat of many animals.

Include this lesson's information in your Foldable®.

☑ **READING PROGRESS CHECK**

Identifying What are three factors that affect climate in Southeast Asia?

LESSON 1 REVIEW

Reviewing Vocabulary
1. What peninsula forms a long bridge between mainland and *insular* Southeast Asia?

Answering the Guiding Questions
2. *Identifying* What are some of Southeast Asia's most important mineral resources?
3. *Analyzing* Why are most of Indonesia's volcanoes located along its southern edge?
4. *Analyzing* What is the connection between Singapore's status as one of the most important ports in the world and its location at the southern end of the Strait of Malacca?
5. *Describing* How do monsoon winds cause wet and dry seasons on Southeast Asia's mainland?
6. *Argument Writing* Write a persuasive letter urging the government leaders of Southeast Asia to take urgent action to protect the region's tropical forests from uncontrolled logging. Discuss the importance of the forests and the consequences of deforestation.

networks

There's More Online!

- ✓ **IMAGE** Angkor Wat
- ✓ **MAP** History of Southeast Asian Civilizations
- ✓ **VIDEO**

Reading HELPDESK

Academic Vocabulary
- ultimate

Content Vocabulary
- sultan
- plantation
- absolute monarchy
- constitutional monarchy

TAKING NOTES: *Key Ideas and Details*

Describe Select one of the countries from the lesson. As you read, describe three important events in that country's history on a web diagram like the one below.

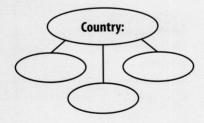

Lesson 2
History of Southeast Asia

ESSENTIAL QUESTION • *How does geography influence the way people live?*

IT MATTERS BECAUSE
Southeast Asia sometimes gets less attention than its two giant neighbors, China and India. The region has played an important role in world history in recent centuries, however, and it is likely to play an even larger role in the future.

Kingdoms and Empires

GUIDING QUESTION *What role has trade played in Southeast Asia's history?*

Southeast Asia is known as "the Crossroads of the World" because it is located along important maritime trade routes. Trade has exposed the region to many different cultural influences. It also has made the region prey for foreign powers seeking to increase wealth and power by controlling the routes.

Prehistoric Cultures

Humans have lived in Southeast Asia for at least 40,000 years. For much of this period, the region looked quite different from today. Earth was in the grip of an ice age, so sea levels were lower. Much of the continental shelf lay above water as dry land. This land connected the mainland to many areas that are now islands, including Borneo, Sumatra, and Java. The mainland area, therefore, was much larger than it is now, and the island area was much smaller. As the ice age waned, the seas began to rise. They reached their present levels about 8,000 years ago.

Throughout the prehistoric period, Southeast Asia's inhabitants survived by hunting and gathering. They used stone tools and weapons. Around 6,000 years ago, people had begun practicing agriculture, growing rice in fertile river valleys and deltas. Agriculture allowed for a more settled existence and led to the development of complex societies.

Powerful Societies Emerge

Several thousand years ago, early metalworking societies arose in Southeast Asia. Using sophisticated techniques, these cultures produced bronze tools, weapons, ornaments, and ceremonial drums. The most famous of these cultures, the Dong Son, was centered in northern Vietnam.

In the middle of the 100s B.C., China and India began to exert a powerful influence. China conquered the Red River delta in the northeast and made Vietnam a province of the Han empire. For the next 1,000 years, Vietnam remained under Chinese control. Chinese culture became embedded in the cultures of Vietnam and its neighbors.

People from India came to Southeast Asia as traders and missionaries. India's culture and its religions, Hinduism and Buddhism, spread along trade routes from India into Southeast Asia. Societies with Indian roots began to develop throughout the region.

Trading Societies

Funan, one of the first important trade-based states in the region, was established in the A.D. 100s. It covered parts of what are now Cambodia, Thailand, and Vietnam. Its people traded with China and India, but its cultural influences were mostly Indian.

Around the A.D. 600s, a kingdom called Srivijaya arose on the island of Sumatra and gained control of the Strait of Malacca. This waterway connecting the Pacific and Indian oceans was important because of the many trade goods that had passed through it by boat.

The main Hindu temple at Angkor Wat displays ancient building styles of India and Southeast Asia. Angkor Wat forms the largest single religious site in the world. Today, it is the national symbol of Cambodia, and the main temple appears on that country's flag.

Identifying How did Hinduism spread to Southeast Asia?

This modern mosque built on stilts is located near the port city of Malacca in Malaysia. When the tide is high and the water level rises, the building appears to be floating.
▶ **CRITICAL THINKING**
Describing How did the religion of Islam reach Southeast Asia?

This included much of the spice trade, which connected China and Southeast Asia with India, Southwest Asia, Africa, and Europe. By controlling trade, Srivijaya grew into an empire that dominated commerce in the region for 600 years. Its capital, Palembang, became an important center of Buddhism.

Agricultural Societies

Southeast Asia's major agricultural societies developed where rice could be grown. The Pagan kingdom sprang up in Myanmar's Irrawaddy delta; Vietnamese society took root in the Red River delta; and the Khmer empire was centered near a large lake called Tonle Sap in Cambodia. The Khmer invented a water-management system to improve crops. The empire is also known for architecture, especially the temple complexes Angkor Wat and Angkor Thom. Built 800 years ago, the complexes still stand, drawing millions of visitors annually.

Islamic States

Islam, the religion of Muslims, could have reached Southeast Asia as early as the A.D. 800s or A.D. 900s when traders from the Middle East or western India traveled the sea route to China. By the 1200s, two Islamic kingdoms had been established in northern Sumatra. From there, Islam gradually spread eastward. By the 1400s, other Islamic kingdoms had risen near ports located along the main trade route through the region's waterways. By the 1600s, Islam had become the dominant religion across the Malay Archipelago.

The most important Islamic kingdom was centered on the port of Malacca on the Malay Peninsula. Founded in about 1400, Malacca capitalized on its location along the Strait of Malacca and developed into a powerful trading empire. Its **sultans**, or kings, ruled over much of the peninsula and the island of Sumatra across the strait.

☑ **READING PROGRESS CHECK**

Identifying What are some ways in which India influenced Southeast Asia?

Western Colonization

GUIDING QUESTION *How did European colonization change Southeast Asia?*

During the period known as the Age of Discovery, European explorers made long voyages in search of gold, silver, spices, and other sources of wealth. They also sought to spread Christianity and to map the world. The arrival of European ships in Southeast Asia marked the beginning of conquest and colonization that would leave the region dramatically changed.

European Traders

From ancient times through the Middle Ages, spices from Southeast Asia had reached Europe through Chinese, Indian, and Arab traders. Europeans used spices such as ginger, cinnamon, cloves, nutmeg, and mace to flavor food, to preserve meat, and to make perfumes and medicines. The demand was great and supplies were limited, so traders were able to charge high prices. Some spices were worth more than gold. The fabulous wealth of the spice trade led European powers to seek ocean routes to the source of the spices, and to gain control of the trade.

These spices on display on the Indonesian island of Bali include nutmeg, lemon grass, and cinnamon.

▶ **CRITICAL THINKING**
Describing Why did Europeans value Southeast Asian spices?

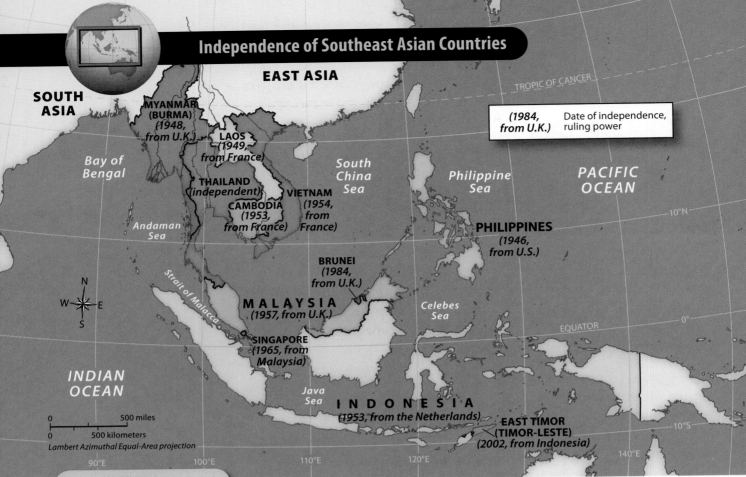

Independence of Southeast Asian Countries

MAP SKILLS

1. **PLACES AND REGIONS** Which Southeast Asian country does not have an independence date? Why?

2. **PHYSICAL GEOGRAPHY** Why might Indonesia find it difficult to achieve a sense of national unity?

In the early 1500s, Portuguese navigators discovered they could reach India and Southeast Asia by sailing around the southern tip of Africa. In 1511 the Portuguese conquered Malacca. That same year, they discovered the sources of cloves, nutmeg, and mace: the Moluccas and the Banda Islands, which became known as the Spice Islands. Through much of the 1500s, wealth generated by spices enriched Portugal's monarchy.

Meanwhile, other European powers sought alternate routes that would allow them to profit from the spice trade. The explorer Ferdinand Magellan commanded five Spanish ships that reached the Philippines by sailing across the Pacific Ocean from Mexico. Soon the Philippines became a Spanish colony.

At the beginning of the 1600s, Holland (known today as the Netherlands) jumped into the spice trade. Because the Portuguese controlled the Strait of Malacca, the Dutch charted a new route to the Spice Islands. By the middle of the century, they replaced the Portuguese as the dominant trading power.

Colonial Rule

During the 1800s and early 1900s, European countries gained control over other parts of Southeast Asia. Burma and Malaysia became colonies of Great Britain, and Vietnam, Laos, and Cambodia became colonies of France. In addition to the spice trade, European countries developed tin and coal mines and built factories to process the raw materials. They also established **plantations**, or large farms on which a single crop is grown for export. Plantation crops included tea, coffee, tobacco, and rubber trees. Thousands of laborers were brought in from China and India to work in the mines and on plantations. Many of them stayed on as permanent residents of the colonies.

Thailand (Siam)

Thailand, then known as Siam, was the only state in Southeast Asia that Europeans did not colonize. It was ruled by an **absolute monarchy** from the mid-1300s until 1932. In an absolute monarchy, one ruler has **ultimate** governing power over the country. The rulers of Siam had resisted threats from Burma since the 1500s. When the British declared war there, Siam's leaders agreed to allow free trade with Great Britain and other Western countries. This admission allowed the country to remain independent, acting as a buffer state between British and French possessions.

Academic Vocabulary

ultimate most extreme or greatest

☑ **READING PROGRESS CHECK**

Identifying In what ways did European powers gain wealth from their colonies in Southeast Asia?

Independent Countries

GUIDING QUESTION *What events ended the colonial era in Southeast Asia?*

In the early 1900s, nearly all of Southeast Asia was under the control of foreign powers. Colonial rule was often harsh, unjust, and exploitive, and sometimes it was met with violent resistance by the region's people. It was not this resistance, however, that ended colonialism. Instead, it was dramatic events that unfolded elsewhere in the world.

Dawn of Freedom

The first Southeast Asian colony to glimpse independence was the Philippines, which had been ruled by Spain since the 1500s. The United States took control of the colony in 1898 after defeating Spain in the Spanish-American War. During World War II, Japan sent its military forces to take control of Southeast Asian lands.

A Buddhist monk (right) in Cambodia casts his vote in a local election held in 2012. Today, Cambodia is struggling to build a democracy after years of civil war.

▶ **CRITICAL THINKING**
Describing Why were the 1970s a difficult decade for Cambodia's people?

Chapter 5 **161**

Viet cong fighters patrol a waterway in the Mekong Delta during the Vietnam War. The Viet cong carried out armed attacks on U.S. and South Vietnamese forces. Their goal was to unite all of Vietnam under Communist rule.

▶ CRITICAL THINKING
Predicting What was the outcome of the Vietnam War?

Japan's nearness to Southeast Asia was a great advantage. The Japanese hoped to use the region's oil, rubber, and timber to help its economy and enable it to continue the war. Japan's rule lasted four years and caused the region's people great hardship.

The war ended in 1945, and the United States granted independence to the Philippines in 1946. Soon, other Southeast Asian colonies began to break free of European rulers. Myanmar negotiated its independence from Britain in 1948, and Indonesia was freed from the Netherlands in 1949. In the 1950s, Vietnam, Laos, and Cambodia freed themselves from French rule, and Malaysia and Singapore became independent of Britain.

The last country to gain its freedom was East Timor, sometimes called by its Portuguese name, Timor-Leste (TEE-mor LESS-tay). After being ruled by Portugal since the 1500s, East Timor declared independence in 1975 but was invaded by Indonesia. Indonesia's harsh, violent rule lasted there until 2002.

Regional Conflicts

The newly independent countries of Southeast Asia faced many challenges. Wars, revolutions, and dictatorships tested the new leaders. In Vietnam, Communist forces defeated the French in 1954 and ruled the northern part of the country. The United States supported leaders in the south. Fighting led to the Vietnam War, which lasted until 1975 and took more than 2 million lives. After the war ended, North Vietnam united the country under Communist rule.

The Vietnam War affected other parts of Southeast Asia. In Laos, a conflict pitted the government of Laos against nationalists allied with Vietnamese communists. In 1975 the government of Laos collapsed, and many people fled to other countries.

In 1970, Cambodia's military took over the government. Five years later, a rural communist movement, called the Khmer Rouge, overthrew this government. The country's new leaders began a brutal campaign of terror and destruction. Between 1975 and 1979, at least 1.5 million people died or were killed. In 1978 Vietnamese forces invaded Cambodia and installed a new government controlled by Vietnam, sparking almost 13 years of civil war.

Modern Southeast Asia

The economic growth of East Asian countries such as China and Taiwan in the late 1900s helped some countries in Southeast Asia. Thailand, Malaysia, Indonesia, the Philippines, and Vietnam became centers of manufacturing in the region. Textiles and tourism are important parts of Cambodia's economy. Singapore, which gained independence from Great Britain in 1959 and then from Malaysia in 1965, is one of the world's most prosperous countries.

Thailand has been a **constitutional monarchy** since 1932. A constitutional monarchy's ruler is legally bound by a constitution and government. Thailand's government has often been threatened by the military. Numerous military coups took place in the 1900s. By the 1980s, Thailand had combined a tradition of monarchy with democratic reforms, and it enjoyed high rates of economic growth through the 1990s.

Myanmar has struggled since gaining independence. In 1962 the military seized power and established a socialist government. Since then, leaders have closed off the country to outside influences. Even a cyclone in 2008 that killed more than 100,000 people did not persuade the government to open its borders to allow aid from abroad.

Include this lesson's information in your Foldable®.

☑ **READING PROGRESS CHECK**

Identifying What colonial power ruled Vietnam before 1954?

LESSON 2 REVIEW

Reviewing Vocabulary
1. What crops were grown on *plantations* in Southeast Asian colonies?

Answering the Guiding Questions
2. *Describing* How did the geography of Southeast Asia change as a result of a rise in sea level?
3. *Analyzing* What made the Strait of Malacca such an important waterway?
4. *Determining Central Ideas* Why did European countries want to control the spice trade?
5. *Analyzing* How could the resources of Southeast Asia help Japan's economy and its ability to continue fighting in World War II?
6. *Informative/Explanatory Writing* Knowing about the history of colonization in Southeast Asia, write a paragraph in which you answer these questions: What are the dangers of resisting colonization? What are the consequences of not resisting?

There's More Online!

- ☑ **IMAGE** Minority Groups in Southeast Asia
- ☑ **MAP** Urban Southeast Asia
- ☑ **VIDEO**

Reading HELPDESK

Academic Vocabulary
- exploit

Content Vocabulary
- primate city
- minority
- Pacific Rim
- subsistence farming
- ecotourism

TAKING NOTES: *Key Ideas and Details*

Identify As you read, use a graphic organizer like this one to note key facts about the people of Southeast Asia.

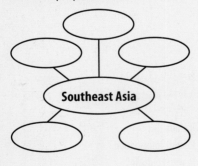

Lesson 3
Life in Southeast Asia

ESSENTIAL QUESTIONS • *What makes a culture unique?* • *Why does conflict develop?*

IT MATTERS BECAUSE
Southeast Asia is one of the most culturally diverse regions in the world. Learning about this diversity will lead to a better understanding of the region.

People and Places

GUIDING QUESTION *How is Southeast Asia's population shifting?*

In population size, Southeast Asia has fewer people than its neighbors, China and India. Compared to other areas, however, Southeast Asia's population is high. This lesson describes where the region's people live. The lesson also discusses how the population is expanding and changing.

Population Profile

Southeast Asia is home to about 625 million people. Indonesia has a population of approximately 250 million, which is 40 percent of the regional total.

Many Southeast Asian countries experienced rapid population growth in the 1900s, but today the region's growth rate is only slightly above the world average. Population is not evenly distributed. The highest densities generally are found in areas where good soil and abundant water allow agriculture to thrive. These areas include coastal plains, river valleys, and deltas. One of the greatest population densities is on the island of Java, whose volcanic soils are exceptionally fertile.

People on the Move

Historically, Southeast Asian societies were mostly rural. Since World War II, however, people have moved steadily from rural areas to cities. This movement is called urbanization. Urbanization is occurring most rapidly in countries that recently were or are currently the most rural.

Urbanization has produced explosive population growth in some cities. Manila, the capital of the Philippines, is home to more than 11 million people. Manila is a **primate city**—a city so large and influential that it dominates the rest of the country. Like many large cities, Manila suffers from overcrowding. More than a third of its inhabitants live in slums, where people are desperately poor and most dwellings are crudely built shacks.

Rapid population growth in and around Jakarta, the capital of Indonesia, has created a continuous urban chain of cities that were once separated. This megalopolis, as such areas are called, is one of the largest in the world, with more than 26 million people.

☑ READING PROGRESS CHECK

Identifying What types of geographical areas in Southeast Asia have the highest population densities?

People and Cultures

GUIDING QUESTION *How have China and India influenced Southeast Asian cultures?*

Peoples from other regions have migrated to Southeast Asia for more than 2,000 years for many different reasons. As a result, Southeast Asia has a rich mix of peoples and cultures.

Southeast Asia's cities provide contrasts in the ways people live. In Manila, the Philippines (left), residents crowd into slum areas not far from the city's modern business center. Singapore's Chinatown district (right) is known for its restored historic buildings and numerous shops and restaurants.

▶ CRITICAL THINKING

Describing How have Southeast Asia's cities changed since World War II?

Chapter 5 165

Ethnic and Language Groups

Southeast Asia's population consists of many ethnic groups. Five main groups dominate the mainland: the Burmese in Myanmar, the Siamese in Thailand, the Malay in Malaysia, the Mon-Khmer in Laos and Cambodia, and the Vietnamese in Vietnam. The largest group in Indonesia is the Javanese, and the largest group in the Philippines is the Tagalog. Many smaller groups, sometimes called **minorities**, also live in each country.

The people of Southeast Asia speak many languages. Traveling throughout Indonesia, one could encounter several hundred different languages. Most of the region's languages are indigenous, or native to the region. Others, including English, Spanish, and French, arrived with trade and colonialism.

A woman of the Padaung ethnic group drives a car in Myanmar near Yangon (Rangoon). Many Southeast Asians today blend traditional and modern ways.

▶ **CRITICAL THINKING**
Describing What is a major characteristic of peoples and cultures in Southeast Asia?

Religion and the Arts

Buddhism is the most widely practiced religion across most of the mainland. Islam is dominant on the southern Malay Peninsula and across the Indonesian islands. In the Philippines and East Timor, most people are Roman Catholic. In some remote areas, people practice animist religions. Animism is based on the belief that all natural objects, such as trees, rivers, and mountains, have spirits.

When people from India brought Buddhism and Hinduism to the region, they also brought rich traditions in architecture, sculpture, and literature. Some of the greatest architecture and sculpture are Hindu and Buddhist temples built centuries ago.

The wide variety of art forms that flourish in Southeast Asia reflect the region's great diversity of cultures. In Thailand, plays known as *likay* feature singing, dancing, and vibrantly colored costumes. Actors improvise song lyrics, dialogue, and plots. In the popular shadow-puppet theater in Indonesia, Malaysia, and Cambodia, one puppeteer sings, chants, and controls the puppets behind a screen that is lit from the back. An orchestra of metal instruments provides music.

Daily Life

Although Southeast Asian cities are growing rapidly, three-fourths of the region's people live in rural areas. Many rural people move to the cities, but some leave the region to work in other countries. The money they send home helps their families survive.

Education and literacy rates vary across the region. The highest literacy rate is in Vietnam, where 94 percent of the people can read and write. At the other extreme are East Timor, Laos, and Cambodia, all of which have literacy rates below 75 percent. Roughly 95 percent of the schools in East Timor were destroyed during the country's fight for independence from Indonesia. The government of the new country is working to build new schools.

Many sports are popular in Southeast Asia. Some are traditional and generally unknown outside the region, while others—including soccer, badminton, martial arts, and volleyball—are international. All 11 countries in the region participate in the Southeast Asian Games that take place every two years. The countries take turns hosting the games, which include Olympic sports and traditional regional sports like *sepaktakraw*, a kind of volleyball played with the feet, knees, head, and chest.

✔ **READING PROGRESS CHECK**

Identifying Which religions are most widespread in Southeast Asia?

Issues in Southeast Asia

GUIDING QUESTION *In what ways have human activities affected the environment in Southeast Asia?*

Southeast Asia is a region in transition. In recent decades, the rapid economic growth along much of the **Pacific Rim**—the area bordering the Pacific Ocean—has brought great changes to some of the region's countries. With these changes, however, have come new challenges and problems.

Residents of Manila, the Philippines, celebrate a religious festival that has its roots in the culture of Spain. The Philippines was part of the Spanish Empire from the 1500s to the late 1800s.

A young American visitor poses with a produce seller on a busy street in Hanoi, Vietnam. Tourism is a growing industry in Vietnam and many other Southeast Asian countries.

Identifying What other new industries have developed in Southeast Asia in recent decades?

Earning a Living

Farming remains the most common livelihood in most countries. Many rural villages depend on rice, the most important food staple in the region, as a cash crop. Rice is by far the most widely grown crop. Two of the region's countries, Thailand and Vietnam, lead the world in rice exports. Plantations located in Malaysia, Indonesia, and southern Thailand produce natural rubber and palm oil, while coconuts and sugar are important crops in the Philippines. Southeast Asia's other top export crops include cacao, coffee, and spices.

Many farmers grow food only to feed themselves and their families. This is called **subsistence farming**. Some practice subsistence farming but also work seasonally to earn money.

Since the end of the colonial period, many countries have focused on industry. The tiny country of Singapore has developed into a major industrial center. Mining contributes to the economies of some countries in the region. Indonesia, Malaysia, and Thailand together produce more than half of the world's tin. Large deposits of copper and gold are mined on the Indonesian portion of the island of New Guinea.

Fishing is an important livelihood in Thailand, Indonesia, Malaysia, and the Philippines. Tourism is a growing industry in countries such as Cambodia, Thailand, and Vietnam. An important part of this industry is **ecotourism**, or touring natural environments such as rain forests and coral reefs.

Finance, communication, and information technology have also improved greatly in the region. Many people in the cities use cell phones and have access to the Internet. Due to its large and generally inexpensive labor force, many U.S., Japanese, and European industries employ workers in the region.

Making Connections

In 1968, Indonesia, Malaysia, the Philippines, Singapore, and Thailand joined together to form the Association of Southeast Asian Nations (ASEAN). The organization seeks to increase economic development, social progress, and cultural development in the region. All the countries in the region except East Timor are members. Although many countries have experienced economic growth since the late 1960s, Cambodia, Laos, Myanmar, and Vietnam rank among the poorest countries in the world.

Economic and Environmental Challenges

Southeast Asia faces many challenges. The gap between rich and poor has widened in recent decades. Economic growth has helped some people out of poverty, but many still struggle to meet daily needs. In Manila, Jakarta, and other big cities, rapid urbanization has led to overcrowding and water shortages.

Protecting the environment is another challenge. Industries provide an economic boost, but they sometimes **exploit** and harm the environment. Tin mining has created huge wastelands in Malaysia, Thailand, and Indonesia. Commercial logging and farming destroy the tropical forests at an alarming rate. Indonesia and Malaysia recently introduced laws to slow deforestation while still promoting economic growth.

Dams along the Mekong River create hydroelectric power, but that also has created problems for the fishing industry. Vietnam, Cambodia, Laos, and Thailand established the Mekong River Commission to encourage safe management and usage of the river and its resources. Conflict over control of oil and gas resources in the South China Sea is another challenge. Vietnam, the Philippines, and Malaysia surround the sea. The countries' claims on the resources are in dispute.

Academic Vocabulary

exploit to use a person, a resource, or a situation unfairly and selfishly

Political Challenges

Since the end of the colonial era, economic and social progress have been slowed by political instability. After the military took over the government of Myanmar in 1962, the country slid into isolation, poverty, and brutality. The government has refused democratic elections and limits any protest by its people. Some people in Myanmar have tried to bring democracy to the country. In 2011 the military government began to allow more freedom for its people. In 2012 the United States reestablished diplomatic ties with Myanmar and assigned an ambassador to the country for the first time in more than 20 years.

Include this lesson's information in your Foldable®.

✓ **READING PROGRESS CHECK**

Determining Central Ideas What is the mission of the Association of Southeast Asian Nations (ASEAN)?

LESSON 3 REVIEW CCSS

Reviewing Vocabulary

1. ***Identifying*** What are some of the problems that a *primate city* might experience?

Answering the Guiding Questions

2. ***Analyzing*** Why might the literacy rate in East Timor soon begin to rise?

3. ***Determining Central Ideas*** What are some of the environmental problems facing Southeast Asia?

4. ***Analyzing*** What might happen to agricultural production in Southeast Asia if the urbanization trend continues?

5. ***Determining Word Meanings*** What is *animism*, and where is it practiced in Southeast Asia?

6. ***Argument Writing*** Imagine that your class is trying to pick a destination for an overseas field trip. Write a letter to your classmates encouraging them to select Southeast Asia. Give specific reasons why this region would make a good destination.

Chapter 5 ACTIVITIES

Directions: Write your answers on a separate piece of paper.

1 Use your **FOLDABLES** to explore the Essential Question.
INFORMATIVE/EXPLANATORY WRITING Explain in a paragraph where Southeast Asia's agricultural societies began and why they originated in those locations.

2 21st Century Skills
ANALYZING Research information to write about the economies of Vietnam and Cambodia. Discuss which country you think will undergo stronger economic development in the next decade.

3 Thinking Like a Geographer
IDENTIFYING The population of the Greater Jakarta, Indonesia, metropolitan area reached 26 million people. How many large U.S. cities, including the largest city in your state, would have to be combined to reach a population of 26 million? Use the chart to list several American cities and their respective populations. Add the totals until you reach approximately 26 million.

City	Population

4 GEOGRAPHY ACTIVITY

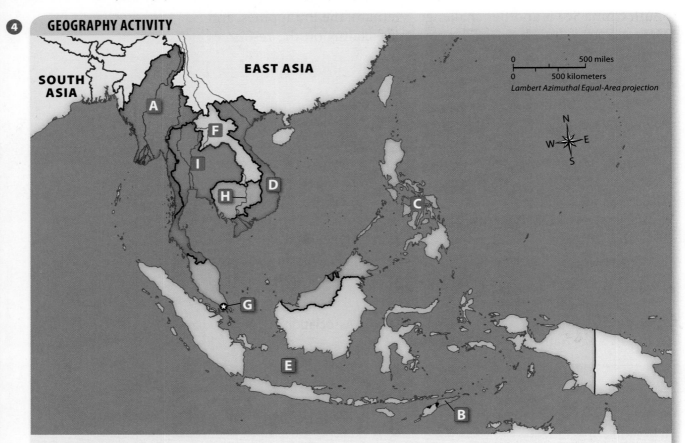

Locating Places
Match the letters on the map with the numbered places below.

1. Philippines
2. Thailand
3. Laos
4. Java Sea
5. Myanmar
6. Singapore
7. Mekong River
8. East Timor (Timor-Leste)
9. Vietnam

Chapter 5 Assessment

REVIEW THE GUIDING QUESTIONS

Directions: Choose the best answer for each question.

1. The region of Southeast Asia consists mainly of
   A. constitutional monarchies.
   B. Singapore.
   C. peninsulas and islands.
   D. war-ravaged countries.

2. Which country has the greatest number of active volcanoes in the world?
   F. Indonesia
   G. Vietnam
   H. Krakatoa
   I. Thailand

3. Why did foreign powers begin to colonize Southeast Asian countries?
   A. to defeat local uprisings
   B. to control the profitable spice trade
   C. to test new oceangoing navigation instruments
   D. to enslave the Southeast Asian peoples

4. In which country did the Khmer Rouge arise?
   F. the Philippines
   G. Laos
   H. Myanmar
   I. Cambodia

5. Forty percent of the population of Southeast Asia live in
   A. Manila.
   B. Brunei.
   C. Indonesia.
   D. Malaysia.

6. What is one of the greatest challenges Southeast Asian countries face?
   F. loss of trade
   G. environmental damage
   H. loss of cottage industries
   I. devaluation of their monetary systems

Chapter 5 ASSESSMENT (continued)

DBQ ANALYZING DOCUMENTS

7 IDENTIFYING Read the following passage:

"*The opening of the Suez Canal in 1869 and the advent of steamships launched an era of prosperity for Singapore as . . . trade expanded. . . . In the 20th century, the automobile industry's demand for rubber from Southeast Asia and the packaging industry's need for tin helped make Singapore one of the world's major ports.*"

—from U.S. State Department, "Singapore: History" (2011)

What kind of change was the initial cause of Singapore's growth?
- A. political
- B. technological
- C. cultural
- D. geographical

8 ANALYZING Where was the tin mentioned in the passage most likely mined?
- F. southern Africa
- G. Southeast Asia
- H. East Asia
- I. Oceania

SHORT RESPONSE

"*The Chinese conquerors referred to Vietnam as Annam, the 'pacified south.' But it was not peaceful. . . . Revolts recurred chronically [regularly], and . . . [leaders] stressed that Vietnam's customs, practices, and interests differed from those of China.*"

—from Stanley Karnow, *Vietnam: A History* (1983)

9 IDENTIFYING POINT OF VIEW Why would leaders of Vietnamese revolts emphasize that Vietnam's culture was different from China's?

10 ANALYZING How were the events that took place early in Vietnam's history, described here, similar to its later history?

EXTENDED RESPONSE

11 INFORMATIVE/EXPLANATORY WRITING Vietnam has one of the highest literacy rates in Southeast Asia and one of the highest poverty rates. With this in mind, research online and at the library to write about this seeming contradiction. In your essay, explore the effects of poverty on child labor. What impact does the purchase of high-priced consumer goods made in factories staffed by children have on the practice of child labor?

Need Extra Help?

If You've Missed Question	1	2	3	4	5	6	7	8	9	10	11
Review Lesson	1	1	2	2	3	3	3	1	2	2	3

SOUTH ASIA

ESSENTIAL QUESTIONS • *How does geography influence the way people live?*
• *How do governments change?* • *What makes a culture unique?*

A man in traditional clothes celebrates at a Hindu festival in northern India.

There's More Online about South Asia.

CHAPTER 6

Lesson 1
Physical Geography of South Asia

Lesson 2
History of South Asia

Lesson 3
Life in South Asia

The Story Matters...

The landscape of South Asia is one of contrasts—from lowlands only a few feet above sea level to the Himalaya, the highest mountains in the world. Many of the rivers of South Asia begin in the Himalaya and flow down onto valleys and plains, providing the fertile soils needed for farming. Many of the countries of South Asia have their roots in ancient civilizations and religions that developed in these river valleys.

Go to the Foldables® library in the back of your book to make a Foldable® that will help you take notes while reading this chapter.

173

Chapter 6
SOUTH ASIA

The northern part of South Asia is separated from the rest of Asia by mountains. The southern part of South Asia juts out of the Asian continent and into the waters of the Arabian Sea, Indian Ocean, and Bay of Bengal.

Step Into the Place

MAP FOCUS Use the map to answer the following questions.

1. **PLACES AND REGIONS** What is the capital of Bangladesh?
2. **THE GEOGRAPHER'S WORLD** What are the two island countries of South Asia?
3. **PHYSICAL GEOGRAPHY** Which South Asian countries are landlocked?
4. **CRITICAL THINKING**
   Analyzing Use the scale bar to estimate the distance between Mumbai and New Delhi.

MOUNTAIN GATEWAY A road winds through the Khyber Pass at the border between Pakistan and Afghanistan. Invaders and traders of long ago used the pass to enter South Asia from the northwest.

HISTORIC MONASTERY Built in the 1600s, this Buddhist monastery overlooks Thimphu, the capital of Bhutan. Today, the monastery holds government offices and the throne room of Bhutan's king.

Step Into the Time

TIME LINE Choose two events from the time line and explain the connection between those events and their effects on the people of South Asia today.

c. 3000 B.C. Indus River Valley civilization emerges

1500 B.C. Aryans migrate south into India

c. 500s B.C. Buddhism emerges in South Asia

A.D. 1526 Mughal Empire begins reign

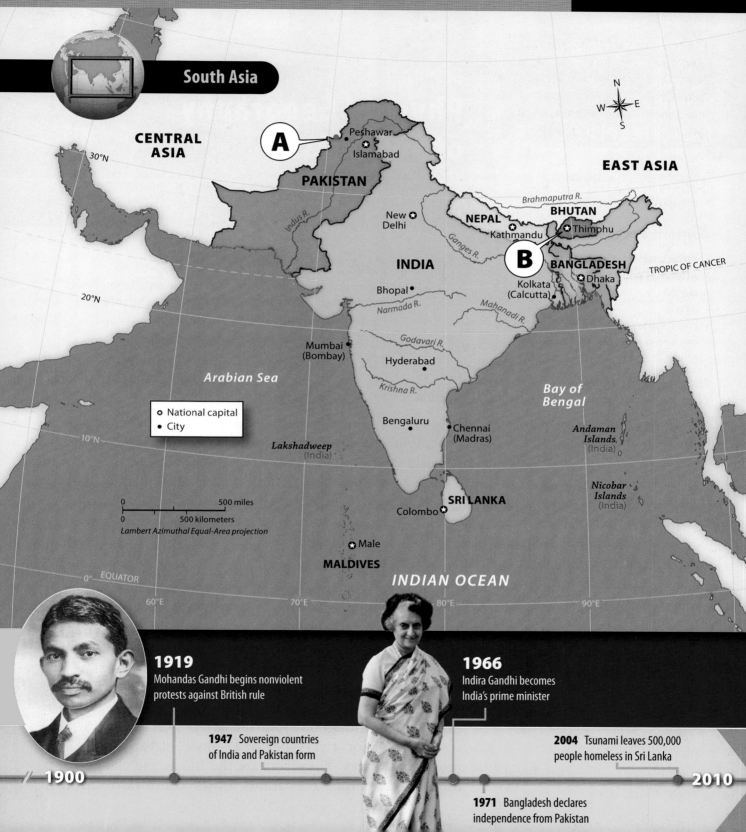

networks

There's More Online!

- ☑ **CHART/GRAPH** Water Wells
- ☑ **MAP** Monsoons
- ☑ **ANIMATION** Rivers in South Asia
- ☑ **VIDEO**

Reading HELPDESK

Academic Vocabulary
- annual

Content Vocabulary
- subcontinent
- alluvial plain
- delta
- atoll
- monsoon
- cyclone

TAKING NOTES: *Key Ideas and Details*

Organize As you read the lesson, identify at least three facts about the physical geography of South Asia and write them in a graphic organizer like the one below.

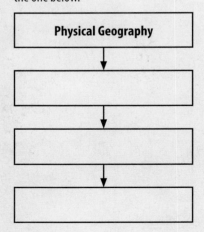

Lesson 1
Physical Geography of South Asia

ESSENTIAL QUESTION • *How does geography influence the way people live?*

IT MATTERS BECAUSE
South Asia is a land of contrasts, with snowcapped mountains towering over parched deserts. More than one-sixth of the world's people live in this region.

South Asia's Physical Features

GUIDING QUESTION *What physical features make South Asia unique?*

South Asia forms a **subcontinent**. A subcontinent is a geographically or politically unique part of a larger continent. Seven countries make up the region of South Asia. Of these, India is by far the largest. The other six countries are Pakistan, Nepal, Bhutan, Bangladesh, Maldives, and Sri Lanka.

Northern Mountains and Plains

Three mighty mountain ranges form South Asia's northern border. They are the Hindu Kush, the Karakoram, and the Himalaya. The Himalaya range includes the highest mountain in the world: Mount Everest, at 29,028 feet (8,848 m). The mountain ranges form a physical barrier. Invaders and traders could enter South Asia through only a few openings, such as the Khyber Pass between Afghanistan and Pakistan. Plate tectonics created the ranges millions of years ago. Today, the mountains are still rising. Plate movements also cause earthquakes throughout South Asia.

Three major rivers begin as small streams from the mountain ranges. The rivers are the Indus, the Ganges, and the Brahmaputra. The Indus flows southward through

Pakistan to the Arabian Sea. The Ganges and the Brahmaputra flow east and southeast to the Bay of Bengal.

The three rivers cross a vast plains area, where their annual flooding has deposited rich soil for growing crops. One-tenth of the world's people now live in the **alluvial plain** created by the Ganges River. This alluvial plain, or area of fertile soil deposited by floodwaters, is the world's longest.

The Brahmaputra and the Ganges come together and form the largest **delta** on Earth. Deltas are places where rivers deposit soil at the mouth of a river. The Brahmaputra/Ganges delta has some of the world's richest farmland.

Central and Southern Highlands

Mountains and rivers also dominate central and southern parts of South Asia. They physically and culturally separate India into northern and southern parts. Much of southern India is a high, flat area called the Deccan Plateau. Two low mountain ranges, the Western and Eastern Ghats, form the plateau's edges. A narrow coastal plain lies between each mountain range and the seacoast. The soils of the plain are rich and fertile.

Islands of South Asia

Sri Lanka and Maldives are the two island countries of South Asia. Sri Lanka, shaped like a teardrop, lies off the southeastern tip of India. Maldives lies southwest of India's tip and is made up of numerous islands. Many of the islands are small, ring-shaped islands called **atolls**.

✓ **READING PROGRESS CHECK**

Describing Describe the main physical regions of South Asia.

People wash clothes along the banks of the Ganges River in Varanasi, India. The Ganges is important for daily activities, but to Hindus it is also sacred. The city of Varanasi draws religious pilgrims from all over India.

▶ **CRITICAL THINKING**

Describing How does the Ganges help India's economy?

Chapter 6 177

Visual Vocabulary

Atoll An atoll is a small, ring-shaped island formed when coral builds up around the edges of an ancient, underwater volcano.

Academic Vocabulary

annual yearly or each year

South Asia's Climates

GUIDING QUESTION *How does climate affect people's lives in South Asia?*

Climate is closely related to the physical features of the region. Because the physical features of South Asia are so diverse, the region's climate is diverse, too. About half of South Asia has a tropical climate, with different kinds of plant life. Much of the northern half of the region enjoys a warm temperate climate. You can also find cool highlands in the north and scorching deserts to the west.

Monsoons

Much of South Asia's climate is a result of seasonal wind patterns called **monsoons**. Most of the region has little or no rainfall for eight months of the year. Then, beginning in May and early June, temperatures begin to rise sharply. Heated air causes a change in wind direction. Winds from the Indian Ocean carry moisture inland, bringing heavy rains and flooding. Most areas along the coast get at least 90 inches (229 cm) of rain per year. Millions of family farms depend on rain for survival.

The annual monsoon rains support South Asia's large population. Without the rains, the region could not grow enough food for its people. The floods come at a cost, though—they damage property and can cause loss of life.

Other natural hazards include tropical **cyclones**. These large, swirling storms often slam into the coast along the Bay of Bengal. Their violent winds and heavy rains can cause devastation. The strong winds push water from the Bay of Bengal to the shore, flooding low-lying areas far inland. One cyclone can kill tens of thousands of people. The delta lands of the Brahmaputra and Ganges rivers are especially vulnerable to flooding.

Tropical and Dry Areas

Much of South Asia has a tropical wet/dry climate with just three seasons—hot, wet, and cool. The hot and cool seasons are dry. The three seasons are a result of the monsoon wind patterns.

Tropical wet climates are found along the western coast of India, southern Sri Lanka, and the Ganges Delta in Bangladesh. These areas get plenty of rain year-round and have thick, green vegetation.

Not all of South Asia gets drenched by seasonal monsoons. Some places are dry. Parts of the Deccan Plateau, for example, get little rain, because the Western Ghats block the winds and rains of the wet-season monsoons.

Northwestern South Asia has the region's driest climate. The Thar Desert straddles the border between Pakistan and India. The area gets relatively little rain; **annual** rainfall is less than 20 inches (51 cm). The vegetation is mostly low, thorny trees and parched grasses.

Highland and Temperate Climates

The tops of the huge mountain ranges on South Asia's northern border are covered year-round in snow. The mountains affect the climate of lower-lying areas. In winter, the Himalaya block the cold winds sweeping down from Central Asia. This forms a large temperate zone that stretches across Nepal, Bhutan, northern Bangladesh, and northeastern India. Farther south, the elevation of the Deccan Plateau combined with the wind-blocking effect of the Western and Eastern Ghats creates another temperate climate area.

✓ **READING PROGRESS CHECK**

Analyzing What positive and negative effects do the monsoons have on the lives of people in South Asia?

South Asia's Natural Resources

GUIDING QUESTION *Which natural resources are most important to South Asia's large population?*

South Asia has many natural resources, but they are not evenly distributed. As South Asia's largest country, India has the most productive land, as well as water and mineral resources.

South Asia: Seasonal Rains

Rainfall, May to September
- More than 60 in. (150 cm)
- 20–60 in. (50–150 cm)
- Less than 20 in. (50 cm)
- ← Winds

Rainfall, November to March
- More than 60 in. (150 cm)
- 20–60 in. (50–150 cm)
- Less than 20 in. (50 cm)
- ← Winds

MAP SKILLS

1. **PHYSICAL GEOGRAPHY** What time of year brings wet weather to much of South Asia? What time of year brings drier weather? Why?

2. **PLACES AND REGIONS** Which parts of South Asia receive the least amounts of rainfall during the summer period?

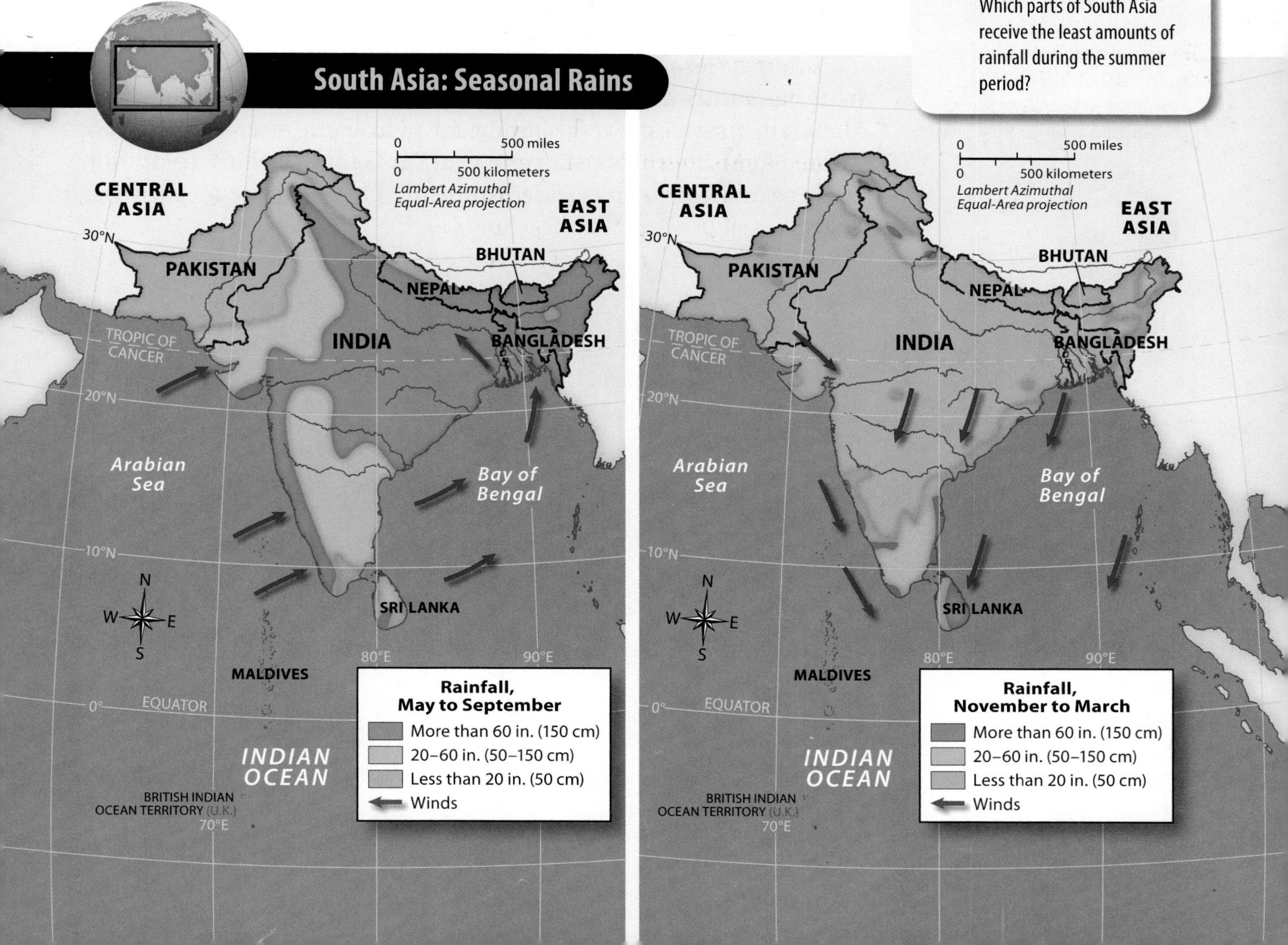

Water Resources

South Asians depend on rivers for irrigation, drinking and household water, and transportation. Water in rivers is also considered sacred in Hinduism, the principal religion in India. Hindus revere the Ganges, named for the goddess Ganga.

Today, water is an important source of energy for South Asia. Mountains provide swift-flowing rivers that can be used to generate electricity. Several dams, such as the Narmada River project, are being built, but hundreds more are planned. The Indian government argues that the projects will provide water for drinking, irrigation, and electricity. Opponents point out that areas must be flooded to build dams. This will displace millions of people and destroy ecosystems. They favor smaller-scale projects and traditional ways to manage water needs.

Mineral and Energy Resources

India has most of South Asia's mineral resources. These include iron ore, manganese, and chromite, all used in making steel. India also has large quantities of mica, a rock used to manufacture electrical equipment.

Nepal's natural resources include mica and copper. To the south, Sri Lanka boasts some of the world's finest gemstones, including sapphires and rubies. Sri Lanka also has large quantities of graphite. This is the "lead" that is used in pencils. Graphite is also used in batteries and as a lubricant.

South Asia has several important petroleum reserves. They are located in northern Pakistan and near the Ganges Delta. Exploration in the Arabian Sea has yielded some oil. One offshore oil field was

Villagers in northwestern India try to get water from a large well. When this region is affected by drought, dams, wells, and ponds often go dry.
▶ **CRITICAL THINKING**
Describing What are the physical characteristics of northwestern India?

discovered in the mid-1970s about 100 miles (161 km) west of the Mumbai (Bombay) coast. The field accounted for a large portion of India's domestic oil production. Overall, though, South Asia depends on imported oil. Natural gas fields are found in southern Pakistan and in Bangladesh. India also has an important deposit of uranium north of the Eastern Ghats. The uranium is used in the country's nuclear power plants.

Forests and Wildlife

Like rivers, forests have greatly influenced the history of South Asia. In colonial times, when the British ruled much of the subcontinent, forests were admired for their beauty but exploited for their commercial value. Important timber resources then included teak, sal, and sandalwood. The woods are still valuable today. There is much debate about how they should be used or whether they should be conserved.

Each kind of wood has qualities that make it valuable. Teak is a strong, attractive wood used to make high-quality furniture. Sal is a hardwood used for construction. Sandalwood, with its sweet scent, is often used to make decorative objects.

Forests are, however, more than resources for wood products. They take in carbon dioxide—a greenhouse gas—and release oxygen. Tree roots hold soil in place, reducing erosion. People live in the forests and depend on leaves and fruits for food. Forests also provide habitat for much of South Asia's unique wildlife. Indian forests, for example, are home to three of Earth's most endangered mammals: the tiger, the Asian elephant, and the one-horned rhinoceros. South Asians are working to reverse some of the region's wildlife losses. The creation of wildlife reserves and laws controlling hunting and logging have started to make a difference.

✓ **READING PROGRESS CHECK**

Analyzing Think about how people use rivers in South Asia. How is it similar to how rivers are used in other parts of the world?

A porter in Nepal uses a headband to carry a heavy load in the snowy, high altitudes of the Himalaya.
▶ **CRITICAL THINKING**
Explaining Why are Himalaya snows important to people living in the plains areas of South Asia?

Include this lesson's information in your Foldable®.

LESSON 1 REVIEW

Reviewing Vocabulary
1. Why is a *delta* often used as an agricultural area?

Answering the Guiding Questions
2. ***Analyzing*** Why was the Khyber Pass so important to South Asia for centuries?
3. ***Determining Central Ideas*** What might happen in South Asia if there were no monsoons?
4. ***Analyzing*** What might be the consequences of cutting down a forest in South Asia?
5. ***Informative/Explanatory Writing*** Take notes about the physical features of the countries of South Asia. Use your notes to write about the features. Use descriptive terms to contrast the mountains, deserts, plains, and rivers of the region.

networks

There's More Online!

- ☑ **IMAGE** Taj Mahal
- ☑ **MAP** The British in India
- ☑ **VIDEO**

Reading HELPDESK (CCSS)

Academic Vocabulary
- policy

Content Vocabulary
- caste
- reincarnation
- Raj
- boycott
- civil disobedience
- nuclear proliferation

TAKING NOTES: *Integration of Knowledge and Ideas*

Sequence As you read about the history of South Asia, use a time line like this one to order events from the beginning of the Indus Valley civilization to independence.

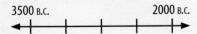
3500 B.C.   2000 B.C.

Lesson 2
History of South Asia

ESSENTIAL QUESTION • *How do governments change?*

IT MATTERS BECAUSE
South Asia is home to one of the oldest known civilizations. You will learn about the birth of a civilization and the struggle for independence by studying its rich history.

Early South Asia

GUIDING QUESTION *How did South Asia's early history lay the foundation for modern life in the region?*

The history of South Asia stretches back thousands of years. The region has experienced much change, yet many social structures and religious beliefs are steeped in tradition.

Early Civilizations

In the early 1920s, archaeologists in South Asia discovered the remains of one of the oldest known civilizations. Because of its location near the Indus River, this culture is known as the Indus Valley civilization. Its origins date back to 3500 B.C. The Indus Valley civilization formed about the same time as other river-valley civilizations around the world.

Two major centers of Indus Valley culture were located at Harappa and Mohenjo-Daro in modern-day Pakistan. These were large cities. Archaeologists have also discovered dozens of smaller settlements.

Written records by the Indus Valley people—mainly strings of symbols—have never been fully deciphered, but archaeological excavations indicate that the civilization was advanced. City streets were laid out in a grid-like pattern. Most houses were made of brick and had their own wells. Houses even had bathrooms and drains. Although most residents were farmers, craftsmanship flourished, as is evident from toys and other artifacts. Some of the same artifacts show that people traded over long distances.

The Indus Valley culture lasted for about 1,000 years. What brought it to an end? No one knows for sure. A natural disaster such as an earthquake may have occurred. Disease or enemy invasions might have played a part in bringing down the civilization.

Equally mysterious are the beginnings of the Aryans, a group that swept into what is now India about 1500 B.C. They came from the northwest, probably from southern Russia and central Asia. Were they invaders, migrants, or wandering nomads? There are different theories about the Aryans. They likely herded sheep, cows, and other livestock in their homeland. Once in India, they settled down to become farmers.

The Aryan civilization lasted for about 1,000 years. It left behind two important legacies for South Asians. The first legacy was social. Aryans believed that society could be successful only if people followed strict roles and tasks. So, they established a system of *varnas,* or castes. **Castes** were social classes. The top class was made up of Brahmans, or priests. Next came the warriors. Third were the merchants. The bottom class consisted of laborers and peasants.

The caste system had a deep impact on South Asia for thousands of years, causing much inequality. If they were born into a lower caste, people could not move up in society, regardless of their talents. After India won its independence in 1947, the new country outlawed the caste system. Some effects of the system are still present though.

The second major legacy of the Aryans was literary. Recall that scholars have not been able to fully decipher the Indus Valley writing. In contrast, the Aryans composed long poetic texts, called the Vedas, in the ancient Sanskrit language. Sanskrit is the parent language of Hindi, one of the most important languages in modern India. Sanskrit also greatly influenced the development of ancient Greek and Latin.

A worker replaces broken bricks with new ones at the ruins of the ancient city of Mohenjo-Daro in Pakistan.
▶ **CRITICAL THINKING**
Describing What have archaeological digs revealed about the Indus Valley civilization and its large cities?

The Vedas were religious hymns handed down orally for many centuries before they were written down. The *Rig Veda* [rihg vay•duh] likely took shape around 1200 B.C. This poem is a series of hymns in honor of Aryan deities. The hymns are full of vivid imagery and philosophical ideas. This poem laid the foundation for the growth of Hinduism.

Religious Traditions

South Asia is the birthplace of several major religions. The first of these is Hinduism. Often described as a way of life, Hinduism has no founder, no holy book, and no central set of core beliefs. Hindus usually pay respect to the Vedas and take part in religious rituals, either at home or in a local temple.

Hindus believe in **reincarnation**, or the rebirth of a soul in another body. Related to this idea is karma—the belief that actions in this life can affect your next life. After many lifetimes, an enlightened soul can be released from the reincarnation cycle of birth, death, and rebirth. Then the soul enters nirvana, a state of eternal bliss.

Around 500 B.C., two new religions arose in South Asia in response to Hinduism and its emphasis on the caste system. The first was Jainism. This religion was based on the Hindu principle of ahimsa, or noninjury. Followers of Jainism turned from farming to commerce so they would not have to kill or injure any living creature.

The second new religion began in northeastern India. It was founded by a noble prince named Siddhārtha Gautama. When he was 29 years old, Siddhārtha gave up his wealthy lifestyle and traveled in poverty, searching for spiritual truth. He reached his goal at the age of 35 and became known as "the Buddha," or "the enlightened one." He passed on to his followers what he believed to be the Four Noble Truths:

A steplike Hindu temple (top) and a wall statue of the Buddha (bottom) show the effects of Hinduism and Buddhism on South Asia's architecture and art.

- Life is full of suffering.
- The cause of suffering is selfish desire.
- Suffering can be stopped by conquering desire.
- Desire can be conquered by following the Eightfold Path: right view, right intention, right speech, right action, right way of living, right effort, right mindfulness, and right concentration.

Buddhists, like Jains, largely rejected the caste system. Hinduism remains the major religion throughout India today. Buddhism has spread to other Asian countries, while Jainism remains a small but vibrant religion in India.

Three Indian Empires

The Aryan civilization faded by 500 B.C. Around 200 years later, another power arose: the Mauryas. The Mauryas conquered much of South Asia. Their most famous ruler was named Ashoka. A highly successful warrior, he converted to Buddhism around 260 B.C. and adopted a life of nonviolence. His conversion influenced many people throughout the subcontinent. Trade and culture thrived under his rule.

Hundreds of years later, another empire, the Guptas, managed to unify much of northern India. Under the ruler Chandragupta I, science, medicine, mathematics, and the arts flourished. Gupta scholars developed the decimal system in mathematics that we still use today.

Finally, during the 1500s and 1600s, India witnessed the flowering of a third great empire: the Mughals. In contrast to earlier emperors, Mughal rulers were Muslim rather than Hindu. They were the first Indian emperors to be members of a minority religion. During this era, many South Asians converted to Islam.

Some of the Mughals were tolerant. Akbar the Great, who ruled from 1556 to 1605, was a devoted Muslim, but he encouraged freedom of religion. He regularly held discussions with religious scholars. As with the Mauryas and the Guptas, culture, science, and the arts flourished under the Mughals. The architectural monument known as the Taj Mahal was constructed by the fifth Mughal emperor, Shah Jahan, in memory of his beloved wife.

A reflecting pool leads to the magnificent Taj Mahal in Agra, northern India. Built of white marble, the Taj Mahal is considered the greatest example of Muslim architecture in South Asia.

▶ **CRITICAL THINKING**
Describing What was the purpose of the Taj Mahal?

✅ **READING PROGRESS CHECK**

Identifying What two especially important legacies did the Aryans leave behind?

Modern South Asia

GUIDING QUESTION *How has conflict in South Asia led to change?*

Since their early history, South Asians have been no strangers to conflict. Then, beginning in the late 1800s, a combination of internal and external factors led to great changes in the region.

The British in South Asia

Beginning in the 1500s, European countries used improved ships and maps in exploration. In the 1600s, British traders established settlements in India. With the decline of the Mughal Empire, the traders became a powerful presence in South Asia. The British were especially interested in textiles, timber, and tea.

After a bloody rebellion in northern India in 1857, the British government took direct control of most of South Asia. They ruled over what is today India, Pakistan, and Bangladesh. The British used the name India to refer to the entire area. By this time, India was a British colony rather than a trading partner. Although the British built railways, schools, and ports, Indians resented a foreign presence in their land. In the late 1800s, an independence movement began.

Achieving Independence

In 1885, less than 30 years after the rebellion of 1857, Indian supporters of independence formed the Indian National Congress. The British, however, were reluctant to give up the **Raj**, as their imperialist rule of India was called. The Congress responded by endorsing a **boycott**. A boycott means refusing to buy or use certain goods—in this case, Indians refused to buy imported British goods.

In the early 1900s, two members of the Congress became leaders. The first was Mohandas K. Gandhi. Often called "Mahatma," which

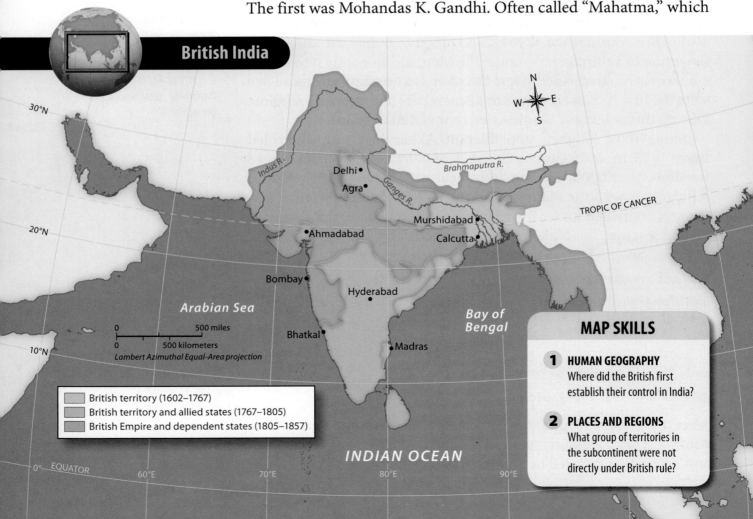

British India

MAP SKILLS

1 HUMAN GEOGRAPHY
Where did the British first establish their control in India?

2 PLACES AND REGIONS
What group of territories in the subcontinent were not directly under British rule?

means "great soul," Gandhi studied law. He practiced law in South Africa, where the racist **policies** shocked and angered him. He returned to India determined to fight for independence from the British.

Gandhi was deeply opposed to violence. His most powerful weapon against British rule was **civil disobedience**, or nonviolent resistance. He was joined by a younger leader, Jawaharlal Nehru. Also trained as a lawyer, Nehru was the son of one of the original leaders of the Congress. Gandhi and Nehru finally succeeded in persuading the British to leave South Asia and surrender colonial rule. India became independent in August 1947. Conflicts between Hindus and Muslims, however, divided the subcontinent.

India and Pakistan

Since Mughal times, South Asia has been troubled by religious and cultural divisions between Hindus and Muslims. In 1947, as part of the independence settlement, the British negotiated a division of the subcontinent. Two countries were created: India (mainly Hindu) and Pakistan (mainly Muslim). To complicate matters, Pakistan was divided into western and eastern sectors. In the 1970s, East Pakistan achieved independence after a civil war and became the country of Bangladesh.

Tensions between India and Pakistan did not die down. The two countries have fought several wars and are involved in a dispute over the region of Kashmir, located in the Himalaya and Karakoram mountain ranges. In the late 1990s, both countries developed nuclear weapons. Because of this **nuclear proliferation**, or the spread of enormously powerful atomic weapons, conflict between the two countries could prove dangerous.

✓ **READING PROGRESS CHECK**

Describing What countries were created out of the South Asia subcontinent? What religions do the people of these countries follow?

Jawaharlal Nehru (left) and Mohandas K. Gandhi (right) were the main leaders of India's independence movement during the 1930s and 1940s.

▶ **CRITICAL THINKING**
Describing What method did Gandhi and Nehru use to get the British to leave India?

Academic Vocabulary

policy plan or course of action

Include this lesson's information in your Foldable®.

LESSON 2 REVIEW

Reviewing Vocabulary
1. How has the *caste* system influenced life in South Asia?

Answering the Guiding Questions
2. *Describing* How are the religions of Jainism and Buddhism alike? How are they different?

3. *Determining Central Ideas* Why are Hindu-Muslim conflicts in South Asia so significant to the history of the region?

4. *Narrative Writing* You are a time traveler whose machine can transport you to any one of the following: the Indus Valley cities of Harappa or Mohenjo-Daro; the empires of Ashoka or Akbar the Great; or the struggle for independence by the Indian National Congress under the leadership of Mohandas Gandhi. Write a brief story describing what you see and hear in the time and setting of your choice.

networks

There's More Online!

- ☑ **CHART/GRAPH** Population of India
- ☑ **SLIDE SHOW** Culture Groups in South Asia
- ☑ **VIDEO**

Reading HELPDESK CCSS

Academic Vocabulary
- establish

Content Vocabulary
- sitar
- green revolution
- cottage industry
- outsourcing
- *dalit*

TAKING NOTES: *Key Ideas and Details*

Describe As you read the lesson, create a chart like the one below and write one key fact about each topic.

Urban growth	
Languages	
Religion	

Lesson 3
Life in South Asia

ESSENTIAL QUESTION • *What makes a culture unique?*

IT MATTERS BECAUSE
By 2030, India will become the most populous country on the planet. Among the challenges India faces are raising the standard of living of its people.

People and Places

GUIDING QUESTION *What are the population patterns of South Asia?*

South Asia is about half the size of the lower 48 states of the United States, but more than 1.5 billion people live in the region. That is about five times the size of the American population.

Population Profile
The largest countries in South Asia—in geographical area and in population—are India, Pakistan, and Bangladesh. The estimated population of India is 1.22 billion. This makes it the world's second-most-populous country, after China. India's birth rate, though, is higher than China's. By 2030, India will be the world's most populous country. Today, India alone has about 17.3 percent of all the world's people. The median age of Indians is 25 years old.

Pakistan's population is about 190 million. It is the world's sixth-largest country. The population of Bangladesh is about 160 million. By 2025, that number could grow to 205 million.

Where People Live
South Asians live mainly in areas that are good for farming. This means, for example, that a high percentage of Indians live in the fertile Ganges Plain in the north-central and northeastern parts of the country.

188

Agriculture plays a major role in India's economy. This is also true of other countries in South Asia. Although India has many large cities, about 7 of every 10 people live in the country's small villages. In contrast, the United States was already more urban than rural as early as 1920. Today, more than 7 of every 10 U.S. residents live in urban areas.

Population trends are changing, though, as more and more Indians leave their villages every year. The 2011 census showed more growth in urban areas than in rural areas—a first for India. People migrate to big cities, called "metros," in hopes of finding better jobs and a higher standard of living. Major cities in India include Mumbai, New Delhi (the capital city), Chennai (Madras), Kolkata (Calcutta), Bengaluru (Bangalore), and Hyderabad.

Census data show that Indian cities are experiencing rapid growth. For example, Mumbai is the business center of the country. The population of Mumbai and its surrounding urban area has grown from a few million around the time of independence to about 20 million. That makes Mumbai India's largest city and the fourth-largest city in the world. Bengaluru, India's third-largest city and a center for technology industries, is growing rapidly, as well.

GRAPH SKILLS >

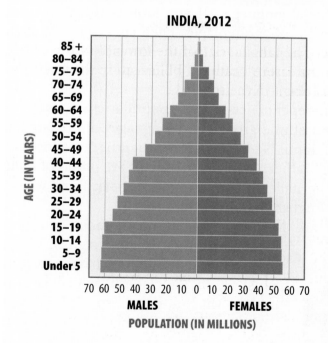
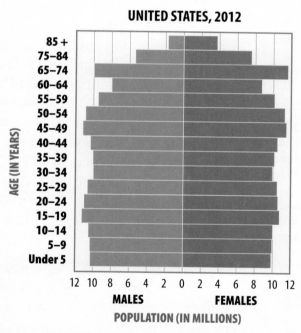

Source: U.S. Census Bureau, Statistical Abstract of the United States: 2012

POPULATION PYRAMIDS
Population pyramids use bar graphs to show a country's population by age and gender. A population pyramid shaped like India's indicates high birthrates and short life expectancy. A population pyramid shaped like the United States's indicates low birthrates and long life expectancy.

▶ **CRITICAL THINKING**
1. *Analyzing* How many males and females in India are between 10 and 14 years of age? How many in the United States?
2. *Analyzing* Compare the *Population in Millions* numbers at the bottom of each pyramid. What do the numbers tell you about the total population of India and the United States?

Population *density* is the number of people who live in a given unit of area. In Mumbai, the population density is 80,100 per square mile (30,900 per sq. km)—seven times the world's average.

The rapid growth of South Asia has put a strain on its resources. Cities struggle to provide essential services to all people, some of whom live in slums surrounding the city centers. Air and water pollution have increased with overcrowded conditions.

☑ **READING PROGRESS CHECK**

Citing Text Evidence Why is India projected to overtake China as the most populous country by 2030?

People and Cultures

GUIDING QUESTION *How are the diverse cultures in South Asia rooted in ethnic and religious traditions?*

The cultures of South Asia are highly diverse. We think of the United States as a "melting pot" or a "nation of immigrants." The same might be said of South Asia.

Ethnic and Language Groups

If you look at an Indian currency note, you will see that all the important information appears in Hindi and in English. These are the country's two official languages. If you look at the fine print, though, you will find 15 other languages. Each language is spoken by millions of people. The languages include Punjabi in the north; Bengali and Oriya in the east; Marathi and Gujarati in the west; and Kannada, Tamil, and Telugu in the south.

After India won independence in 1947, boundaries for its states were based mainly on ethnic groups and languages. For example, the southern state of Tamil Nadu was **established** because most of its people spoke Tamil rather than Hindi. Indians are proud of their ethnic and language heritage. In fact, many who speak Tamil would rather talk to you in English than in Hindi.

Most Pakistanis are Muslim. The country's most common languages are Urdu and English. Urdu, like Hindi, developed from ancient Sanskrit. In Bangladesh, which is also mainly Muslim, people speak Bangla, a variation of Indian Bengali. In Sri Lanka, most people speak Sinhalese. This language has its own unique alphabet. An important minority in Sri Lanka speaks Tamil, mainly because Sri Lanka is so close to the southern state of India, called Tamil Nadu.

Academic Vocabulary

establish to set up

Indian currency note with portrait of Mohandas K. Gandhi

Identifying What two official languages appear on Indian currency notes?

Religion and the Arts

South Asia's diversity is also apparent in religion and the arts. Six main religions are practiced in the region: Hinduism, Islam, Buddhism, Jainism, Sikhism, and Christianity. You have already learned about the first four religions. Sikhism developed in the 1500s, nearly 2,000 years after Buddhism and Jainism. Yet it, too, was a reaction to Hinduism—by and large, Sikhs reject the Hindu caste system. Sikhs deeply respect their original guru, or teacher. Like Muslims and Christians, they are *monotheists*, meaning they believe in only one God.

A South Asian family enjoys a meal of dishes of their region.

Identifying What foods make up a typical meal in South Asian homes?

Artistic expression is rich in South Asia. Two great epic poems of ancient India, the *Ramayana* and the *Mahabharata*, embody Hindu social and religious values. Their impact extends far beyond the Hindu community, though. South Asian children of all religions can tell you about the plots and characters of these epics. They are part of the region's heritage.

South Asia also is a center for classical dance. Music is a thriving art, and many of the region's musicians, singers, and composers are popular. Ravi Shankar was probably the best-known Indian musician and composer. He performed on the **sitar**, a stringed instrument. Motion pictures first arrived in India in 1896. The country now has the largest film industry in the world. The Hindi film industry, nicknamed "Bollywood," is based in Mumbai. Kolkata is also a filmmaking center.

Daily Life

Life in South Asia centers on family. Several generations of family members often live together. They share household chores and finances. Within the family, age and gender play important roles. Elders are respected, and females are often subordinate to males.

Arranged marriages are still common for many South Asians, but less so than in the past. Parents often introduce couples to each other but allow them to decide whether to marry. When couples meet independently of their parents, it is still important to get approval to marry from their parents.

A typical, South Asian family meal would include rice, legumes, and flatbreads. Curry is a combination of spices that is part of many dishes. South Asians do not eat a great deal of meat, in part because of religious guidelines.

☑ **READING PROGRESS CHECK**

Describing How do South Asian culture and American culture compare?

Thinking Like a Geographer

What's In a Name?

Why have the names of many Indian cities been changed? National pride sparked Indians to discard names linked to colonial history. For example, the name Bombay came from the Portuguese *bom baia*, meaning "good bay." The city was renamed Mumbai in 1995 in honor of an ancient Hindu goddess.

Issues in South Asia

GUIDING QUESTION *What impact do economic and environmental issues have on life in South Asia?*

South Asia is a region of rapid growth. As the population grows, it will be ever more important for South Asians to work together to resolve their differences and jointly address the issues they face.

Earning a Living

One of the biggest challenges for South Asians is making a living. Many people are farmers, but good cropland is scarce. Yet South Asians have managed to grow enough food to feed their huge population. How do they do it?

Agricultural advances known as the **green revolution** have helped increase crop yields. The green revolution involves the use of irrigation, fertilizers, and high-yielding crops. Because of these improvements, India has not had to import food to feed its people since the 1970s. This situation might change, however. Green revolution methods are no longer yielding increases in crop productivity to meet the growing need.

In addition to farming, some South Asians make a living from mining or fishing. Others own or work for **cottage industries**—small businesses that employ people in their homes. Cottage industries include textile weaving, making jewelry and furniture, and wood carving. These small businesses help traditional crafts survive.

Advanced technology is a fast-growing part of the South Asian economy. Entire cities have grown up around the high-tech field. Indian computer specialists, engineers, and software designers are in high demand throughout the world, including the United States.

Another growing part of South Asia's economy is ecotourism. Ecotourism combines recreational travel and environmental awareness.

Making Connections

The efficiency of communication systems varies. Newspapers are thriving, cheap, and widely read, but landline phones are often unreliable. Cell phones have helped ease communication problems.

Because of its vast numbers of English speakers, India is well suited for **outsourcing**. Outsourcing occurs when a company hires an outside company or individual to do work. U.S. companies often outsource to foreign countries where workers may be more flexible or willing to work for lower wages. A U.S. resident who calls customer service with a laptop problem might speak to an agent in India.

Customers can apply for visas, check e-mail and Web sites, and receive computer instruction at a cybercafe in Bengaluru, India.

▶ **CRITICAL THINKING**

Describing How has South Asia been affected by the global revolution in communications?

Meeting Challenges

Despite South Asia's rich resources, the region faces a number of challenges. One challenge is the long-standing conflict between India and Pakistan over the largely Muslim territory of Kashmir. The area containing the headwaters of the Indus River is one source of conflict. Both countries want to control the source of the river. Today, control of Kashmir is divided between the two countries.

Sometimes, political conflicts have resulted in tragedy. Two of India's prime ministers have been assassinated in recent decades: Indira Gandhi (the daughter of Jawaharlal Nehru) in 1984, and her son Rajiv Gandhi in 1991. Both prime ministers were killed by extremists who opposed their views. Another issue in India involves oppression of the **dalits**, or so-called "untouchables." For centuries, these people have been discriminated against as outcasts, barely clinging to the lowest rung on the social ladder.

Other South Asian countries also face challenges. In Sri Lanka, Buddhists and Hindus engaged in a civil war that lasted several decades and cost more than 60,000 lives. In Nepal, a rebel group has been struggling since 1996 to overthrow democracy and establish a communist government. Tribal people control large parts of the Pakistan-Afghanistan border.

Other challenges involve health and environmental issues. Rapid population growth and lack of infrastructure have led to water pollution and air pollution in many parts of South Asia. Deforestation is also a major problem. Most of India's forests have been cut down for timber or development. Water quality, wildlife habitats, and climate have been affected as a result.

South Asia has achieved some success in meeting the challenges. The civil war in Sri Lanka has ended, restoring peace to that island country. India has implemented stricter controls on air and water pollution. Government expenditures for wildlife conservation have increased many times over.

Include this lesson's information in your Foldable®.

☑ **READING PROGRESS CHECK**

Analyzing Why might a U.S. company outsource its customer-service operations to a company in South Asia?

LESSON 3 REVIEW

Reviewing Vocabulary
1. Why has the *green revolution* been important for South Asians?

Answering the Guiding Questions
2. ***Determining Central Ideas*** Why might population growth in South Asia be a major problem in the years ahead?
3. ***Determining Word Meanings*** Why is South Asia sometimes called a religious and ethnic "melting pot"?
4. ***Identifying*** Give two examples of environmental issues confronting South Asia today.
5. ***Informative/Explanatory Writing*** Assume you are a seventh-grader in India, Pakistan, Bangladesh, Sri Lanka, or another South Asian country. Write a journal entry describing your typical day. Include some information about your family, as well as your friends and your favorite hobbies.

Chapter 6 ACTIVITIES

Directions: Write your answers on a separate piece of paper.

1 Use your **FOLDABLES** to explore the Essential Questions.
INFORMATIVE/EXPLANATORY WRITING Choose one of the countries that make up South Asia. Compare the physical and population maps of that country found in the front of the chapter. Explain in two or more paragraphs how the geographical features of that country influence where people live.

2 **21st Century Skills**
ANALYZING Use the Internet and print resources to find three or more facts about the economy of one of the South Asian countries. Present your facts in a chart. Then write at least two paragraphs explaining the effects physical geography has on the country's economy.

3 **Thinking Like a Geographer**
INTEGRATING VISUAL INFORMATION Use a web graphic organizer like the one shown here to list the six main religions practiced by people in South Asia, along with one fact you think is important to remember about each one.

4 **GEOGRAPHY ACTIVITY**

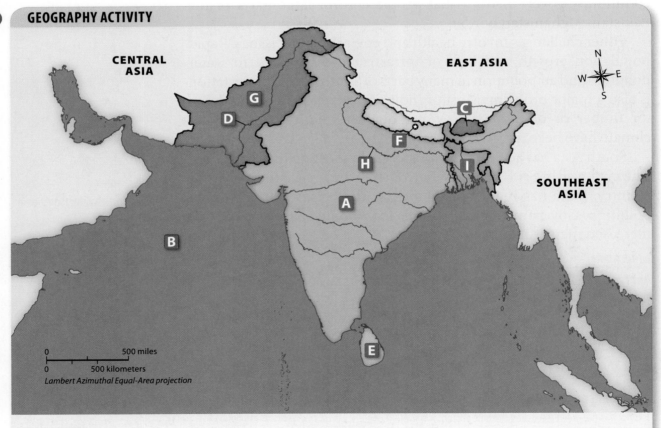

Locating Places
Match the letters on the map with the numbered places listed below.

1. Ganges River
2. Bhutan
3. Sri Lanka
4. Arabian Sea
5. Kathmandu
6. Pakistan
7. Indus River
8. Bangladesh
9. India

Chapter 6 Assessment

REVIEW THE GUIDING QUESTIONS

Directions: Choose the best answer for each question.

1. The subcontinent of South Asia is separated from the rest of Asia by the
   A. Bay of Bengal.
   B. Ganges River.
   C. Hindu Kush, the Karakoram Range, and the Himalaya.
   D. Thar Desert.

2. Much of South Asia's weather is directly influenced by wind patterns called
   F. cyclones.
   G. northerlies.
   H. doldrums.
   I. monsoons.

3. Which ancient Indian civilization was responsible for establishing the caste system?
   A. Mauryans
   B. Aryans
   C. Guptas
   D. Mughals

4. In what year did India finally obtain its independence from Great Britain?
   F. 1885
   G. 1918
   H. 1947
   I. 1960

5. Which South Asian nation is expected to become the most heavily populated country on Earth within the next 15 to 20 years?
   A. Pakistan
   B. India
   C. Bangladesh
   D. Sri Lanka

6. Which area has been the source of political conflict and violence between India and Pakistan?
   F. Nepal
   G. Kashmir
   H. Tibet
   I. Bhutan

Chapter 6 ASSESSMENT (continued)

DBQ ANALYZING DOCUMENTS

7 ANALYZING Read the following passage:

"*Because it cannot produce enough electricity to meet demand, [Pakistan's] government shuts off power for extended periods of time. These chronic blackouts, called load-shedding, sometimes last up to 18 hours a day and hamper economic activity, particularly affecting the country's textile industry, and leave people across a wide socio-economic spectrum in sweltering heat.*"

—from Azmat Khan, "You Aren't Hearing About Pakistan's Biggest Problems" (2011)

How is Pakistan's population related to this problem?
- A. The large number of young people creates high demand for electricity.
- B. The aging population relies on energy-demanding health care.
- C. A growing population causes growing demand for electricity.
- D. Consumers' demand for electric cars means an increase in energy needs.

8 IDENTIFYING Which of the following is an example of an economic problem that can result from a loss of electricity?
- F. lack of power for cooking meals
- G. people being unable to recharge electronics
- H. lack of hot water for cleaning
- I. factory machinery that cannot run

SHORT RESPONSE

"*Buddhism, like other faiths of India, believes in a cycle of rebirth. Humans are born many times on earth, each time with the opportunity to perfect themselves further. And it is their own karma—the sum total of deeds [actions in life], good and bad—that determines the circumstances of a future birth.*"

—from Vidya Dehejia, "Buddhism and Buddhist Art"

9 ANALYZING How does the Buddhist belief in rebirth fit in the broader culture of India?

10 CITING TEXT EVIDENCE In Buddhism, is a person always reborn into better circumstances in a new life? Why or why not?

EXTENDED RESPONSE

11 INFORMATIVE/EXPLANATORY WRITING What do you think is the biggest challenge facing South Asia in the twenty-first century? In several paragraphs, explain what you would do to meet that challenge if you were a leader in that region of the world.

Need Extra Help?

If You've Missed Question	❶	❷	❸	❹	❺	❻	❼	❽	❾	❿	⓫
Review Lesson	1	1	2	2	3	3	3	3	2	2	3

CENTRAL ASIA, THE CAUCASUS, AND SIBERIAN RUSSIA

ESSENTIAL QUESTIONS • How does geography influence the way people live? • How do governments change?

networks

There's More Online about Central Asia, the Caucasus, and Siberian Russia.

CHAPTER 7

Lesson 1
Physical Geography of the Regions

Lesson 2
History of the Regions

Lesson 3
Life in the Regions

The Story Matters...

Central Asia, the Caucasus, and Siberian Russia share a harsh land. Siberia is an enormous and cold land, but it contains vast and economically important oil and natural gas resources. The mountainous terrain, semiarid steppes, and vast deserts of Central Asia are home to many ethnic groups, some with roots in ancient cultures. Throughout history, people living in Central Asia have been affected by conflicts in the region, as well as expansion of the Russian Empire and the Soviet Union.

FOLDABLES
Study Organizer

Go to the Foldables® library in the back of your book to make a Foldable® that will help you take notes while reading this chapter.

A woman of northwestern Siberia in traditional dress

197

Chapter 7: Central Asia, the Caucasus, and Siberian Russia

Siberian Russia, the Central Asian countries, and the Caucasus region make up a large area. The regions were once part of the Soviet Union.

Step Into the Place

MAP FOCUS Use the map to answer the following questions.

1. **PHYSICAL GEOGRAPHY** The Aral Sea extends into which two countries?

2. **THE GEOGRAPHER'S WORLD** Which country borders Kyrgyzstan to the north?

3. **PLACES AND REGIONS** Yerevan is the capital city of which country?

4. **CRITICAL THINKING Analyzing** In which direction do most Siberian rivers flow? Why do you think this is so?

Step Into the Time

TIME LINE Choose one event on the time line and describe what effects that event might have on the people of the region.

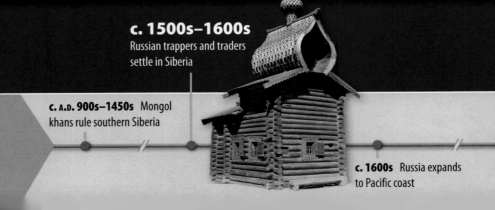

- **c. A.D. 900s–1450s** Mongol khans rule southern Siberia
- **c. 1500s–1600s** Russian trappers and traders settle in Siberia
- **c. 1600s** Russia expands to Pacific coast

networks

There's More Online!

- ☑ IMAGE Breaking Ice
- ☑ MAP Understanding the Region
- ☑ VIDEO

Reading HELPDESK

Academic Vocabulary
- define

Content Vocabulary
- tundra
- taiga
- permafrost
- steppe
- deciduous

TAKING NOTES: *Key Ideas and Details*

Summarize As you read the lesson, list important details for each topic of the lesson on a graphic organizer like the one below.

Topic	Details
Regions	
Landforms and Climates	
Waterways	
Natural Resources	

Lesson 1
Physical Geography of the Regions

ESSENTIAL QUESTION • *How does geography influence the way people live?*

IT MATTERS BECAUSE
Russian Siberia, the Central Asian countries, and the Caucasus make up a large area. The regions cover about one-ninth of the world's total land area.

Regions

GUIDING QUESTION *Which countries make up the regions?*

Siberian Russia, Central Asia, and the Caucasus are separate regions. They share certain characteristics, such as rugged terrain, harsh climate, and sparse population.

Siberian Russia

The eastern part of Russia is known as Siberia. It stretches from the Ural Mountains in the west to the Pacific Ocean in the east. Siberia is 25 percent larger than Canada, the world's second-largest country. The region is so vast that people living on Siberia's Pacific coast are farther from Moscow, Russia's capital, than they are from Maine in the United States.

Central Asia

Central Asia is made up of five countries: Kazakhstan, Turkmenistan, Uzbekistan, Kyrgyzstan, and Tajikistan. Kazakhstan is the northernmost and largest country in Central Asia and the ninth-largest country in the world. It is also the most sparsely settled country in Central Asia.

Kazakhstan is bordered by China on the east, the Caspian Sea on the west, and Russia to its north. The Central Asian countries of Turkmenistan, Uzbekistan, and Kyrgyzstan lie along Kazakhstan's southern border.

Turkmenistan and Uzbekistan are about the same size. Both are about twice the size of Minnesota. Turkmenistan has the smallest population in Central Asia, and Uzbekistan has the largest.

Kyrgyzstan and Tajikistan form Uzbekistan's eastern border. Both are small countries with small populations. Kyrgyzstan is about the size of South Dakota, and the smaller Tajikistan is about the size of Iowa.

Turkmenistan, Kyrgyzstan, and Tajikistan are Central Asia's southernmost countries. South of the countries lie China, Pakistan, Afghanistan, and Iran.

The Caucasus Region

Between the Caspian Sea and the Black Sea and south of the Caucasus Mountains is a region called the Caucasus. Its three small countries—Georgia, Azerbaijan, and Armenia—together total about the size of North Dakota. To their north lies Russia. Turkey and Iran border the Caucasus region to the south.

☑ **READING PROGRESS CHECK**

Analyzing Why is Kazakhstan, which has a larger population than Turkmenistan, more sparsely settled than Turkmenistan?

Landforms and Climates

GUIDING QUESTION *What are the major landforms and climates of Siberia, Central Asia, and the Caucasus region?*

The variety of landforms and climates that characterize the Caucasus, Central Asia, and Siberia make for one of the most geographically interesting regions of the world.

Mountains

The Caucasus Mountains **define** the Caucasus region. These mountains generally mark the border between Europe and Asia. Many of the mountains in this range are volcanoes. Although they have not erupted in thousands of years, the forces that formed them make the region prone to frequent earthquakes.

To the east, two high mountain ranges lie along Central Asia's border with China. The rugged Tian Shan cover most of Kyrgyzstan and extend into eastern Kazakhstan and Uzbekistan. On their southwestern edge is a smaller range called the Pamirs. They cover most of Tajikistan. Like the Caucasus, the Tian Shan and the Pamirs are an earthquake zone.

Academic Vocabulary

define to describe the nature or extent of something

Musicians perform in a mountain village in the Caucasus of southern Russia.
Identifying Where is the Caucasus located?

The Ural Mountains, which run north and south through Russia, are yet another border range. These heavily forested mountains are much lower in elevation than the other ranges. Geographers consider the Ural Mountains to be a boundary between Europe and Asia. The Urals also mark the western border of Russia's Siberia region. The eastern third of Siberia is another mountainous region that extends south along Russia's border with Mongolia.

Plains and Deserts

Plains and deserts also characterize the regions. In Siberia, the west Siberian plain extends from the Ural Mountains east to the Yenisey River. Covering an area of almost 1 million square miles (2.6 million sq. km), the west Siberian plain is one of the world's largest and flattest plains. Its lowland areas are poorly drained, with many swamps and marshes. East of the Yenisey River, the land rises to form the central Siberian plateau, a rugged region of hills cut by deep river gorges.

Siberia's geography changes from north to south as well as from west to east. Extreme northern Siberia is mostly **tundra**—a treeless zone found near the Arctic Circle or at high mountain elevations. Northern Siberia is a harsh environment of bare, rocky ground with patches of small shrubs, mosses, and lichens.

South of the tundra lies a vast area of **taiga**—a zone of coniferous forest. Siberia's taiga is swampy, because, like the tundra, the region is covered in **permafrost**, a layer of permanently frozen ground that lies beneath the surface soil and rocks. Permafrost covers about two-thirds of Siberia. Southern portions of the western plains and central plateau contain dry grasslands called **steppe;** the steppe extends into Kazakhstan in Central Asia.

Central Asia is made up of lowland mountains and dune-covered deserts. Most of Kazakhstan consists of dry grassland plains and plateaus, with lowlands on its coast along the Caspian Sea. In eastern Kazakhstan, the Kyzyl Kum desert stretches south into Uzbekistan. Another desert, the Kara–Kum, extends over most of nearby Turkmenistan. Together, they form a harsh region, nearly the size of Texas, consisting of sand ridges, scattered grasses, and desert plants.

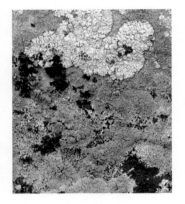

Visual Vocabulary

Lichen a plantlike organism that consists of algae and fungus

Nomadic people live in Siberia and Central Asia. A family, including reindeer and a dog, gather outside their home (left). A shepherd and his flock (right) cross the steppe near a Central Asian mountain range in Kazakhstan.

▶ **CRITICAL THINKING**
Describing What landscapes are found in much of Central Asia?

DIAGRAM SKILLS

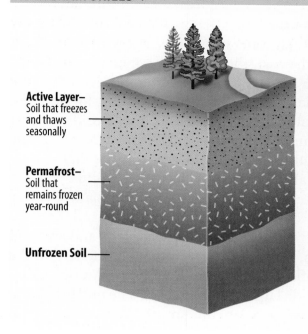

BELOW GROUND IN SIBERIA
This cutaway diagram shows how permafrost penetrates below ground in Siberia.

Active Layer— Soil that freezes and thaws seasonally

Permafrost— Soil that remains frozen year-round

Unfrozen Soil

▶ **CRITICAL THINKING**
1. *Identifying* What is the active layer?
2. *Describing* How might climate change affect permafrost?

A Variety of Climates

The regions have many climates, ranging from arctic climates in Siberia to desert climates in Central Asia. Dry conditions prevail across much of Central Asia, characterizing what is called an arid climate. The steppe of southwest Siberia and northern Kazakhstan receives a little more rain and has a semiarid climate. Moving north and east, summers become shorter and cooler, and the winters become colder. Summer in the Siberian tundra, for example, lasts only two months. Temperatures rarely rise above 50°F (10°C). Summers in the mountain valleys of eastern Siberia are longer and milder. Eastern and central Siberia have some of the coldest winters on Earth. There, the temperature has reached as low as −96°F (−71°C). The west Siberian plain experiences heavy snows, but elsewhere snowfall is light.

The mountain areas of Central Asia have a humid continental climate. Temperatures and precipitation in these areas vary according to location and elevation. Mountain valleys have hot, dry summers and cold winters. Rainfall varies from 4 inches to 20 inches (10 cm to 51 cm) per year. Mountain foothills are cooler and get more rain. The high elevations are even colder, with more rain and snow.

The Caucasus Mountains give Georgia, Azerbaijan, and Armenia a climate much like the mountains of eastern Central Asia. The presence, however, of two large bodies of water—the Caspian Sea and the Black Sea—makes the region's summers cooler and winters warmer. The Black Sea gives Georgia's coastal lowlands a humid subtropical climate, with up to 100 inches (254 cm) of rainfall in a year. Farther east, in the mountains and mountain valleys, the climate is much drier.

Waterways

The Caspian Sea separates the Caucasus and Central Asia. At nearly the size of California, this saltwater lake is the largest inland body of water in the world. To the east, the Aral Sea, a much smaller saltwater lake, straddles the Kazakhstan-Uzbekistan border. Once the world's fourth-largest lake, the Aral has been shrinking for decades.

Lake Baikal, in southeastern Siberia, is the world's largest freshwater lake by volume. It holds about 20 percent of all the freshwater on Earth. With a maximum depth of more than 1 mile (0.6 km), Lake Baikal is also the deepest lake in the world.

The dry climate forces many people in Central Asia to depend heavily on the region's two major rivers, the Syr Dar'ya and the Amu Dar'ya. The rivers flow from mountains across deserts and are used for irrigation. In Siberia, four great rivers—the Ob', the Irtysh, the Yenisey, and the Lena—flow north to empty into the Arctic Ocean. They are among the world's largest river systems. The north-flowing Siberian rivers flood vast areas in the spring. Temperatures are warmer where the rivers begin in the south than at their mouths in the north. Ice in the north blocks the rivers from emptying into the Arctic Ocean, resulting in floods.

Unlike other main Siberian rivers, the Amur River drains eastward. Some of it forms the border between Russia and China. Affected by summer winds, the Amur river valley is warmer than the rest of Siberia. It is Siberia's main food-producing area.

The nuclear-powered ship *Rossiya* breaks through Arctic Sea ice to open passageways for other ships. Nuclear-powered icebreakers like *Rossiya* can travel only in cold water because their reactors need to be kept cool.

▶ CRITICAL THINKING
Describing How are Siberia's rivers affected by the ice in the far north?

✓ READING PROGRESS CHECK

Citing Text Evidence What type of terrain is common in Siberia, Central Asia, and the Caucasus region?

Natural Resources

GUIDING QUESTION *Which natural resources of Siberia, Central Asia, and the Caucasus are economically important?*

Siberia holds some of the greatest wealth in natural resources on Earth. The Caucasus is also rich in natural resources. Central Asia is rich in natural resources too, but the area has little water.

Vast Forests

Siberia's vast taiga contains about 20 percent of all the world's trees. The economic value of this resource is limited, however. One reason is that a lack of roads makes logging in the region difficult. Another is that most of the taiga's trees are larch, one of the few species that will grow on permafrost. Because the quality of larch wood is poor, it is not in demand. Central Asia has few trees because of the arid climate. Parts of the Caucasus are forested. Oak and other **deciduous** trees are found in some lowlands and at lower mountain elevations. Higher in the mountains, pine, fir, and other coniferous trees grow.

Energy and Mineral Resources

All three regions are important producers of oil and natural gas. Fields in the tundra and taiga of Siberia's Ob' River basin make Russia a major provider of these fuels. The Central Siberian Plateau supplies most of Russia's coal. Other huge coal deposits exist in the Lena River valley in southeast Siberia. These areas are so remote, however, that most of their resources remain untapped. Eastern Siberia also holds most of Russia's gold, lead, and iron ore. The tundra region near the mouth of the Yenisey River is one of the world's leading producers of nickel and platinum.

Important oil and gas resources are found in Kazakhstan, Uzbekistan, and Turkmenistan in Central Asia. Large coal deposits are present in Kazakhstan, Tajikistan, and Uzbekistan. Kazakhstan is a major producer of uranium. The mountains of Tajikistan and Kyrgyzstan contain rich mineral resources, as do the eastern mountain areas of Kazakhstan and Uzbekistan. Gold, mercury, copper, iron, tin, lead, zinc, and other metals are mined there.

Most of the Central Asian countries have harnessed their rivers, especially in mountain areas, to produce electricity. The same is true of the Caucasus region, which does not have the rich energy resources of Siberia or Central Asia. Only Azerbaijan is a major oil and gas producer, mainly from fields near or in the Caspian Sea.

✔ **READING PROGRESS CHECK**

Identifying Which three resources are economically important in all three regions?

A worker welds a pipeline at the Urengoy gas field in northwestern Siberia. The Urengoy is one of the largest gas fields in the world. Its natural gas is sent by pipeline to customers as far away as Western Europe.

Identifying What minerals other than natural gas are found in Siberia?

Include this lesson's information in your Foldable®.

LESSON 1 REVIEW

Reviewing Vocabulary
1. How does the *permafrost* affect the geography of Siberian Russia?

Answering the Guiding Questions
2. ***Identifying*** What countries make up the Caucasus region and Central Asia?

3. ***Identifying*** Where are Siberia's lowlands and mountains?

4. ***Identifying*** What geographic feature separates the Caucasus and Central Asia?

5. ***Analyzing*** Why are so many of Siberia's vast resources undeveloped?

6. ***Informative/Explanatory Writing*** Write a paragraph to answer the question: How do you think climate affects the way people of the regions make a living?

networks

There's More Online!

- ☑ **IMAGE** Samarkand
- ☑ **MAP** The Soviet Union
- ☑ **ANIMATION** Trans-Siberian Railroad
- ☑ **VIDEO**

Reading HELPDESK CCSS

Academic Vocabulary
- inhibit
- complex

Content Vocabulary
- pastoral
- collective
- irrigate

TAKING NOTES: *Key Ideas and Details*

Categorize As you read about the histories of Siberia, Central Asia, and the Caucasus, use a graphic organizer like this one to keep track of the invaders who conquered each region.

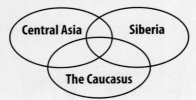

Lesson 2
History of the Regions

ESSENTIAL QUESTION • *How do governments change?*

IT MATTERS BECAUSE
Under communism, the Soviet Union was the main force in the regions. Today, the republics of Russia, the Caucasus, and Central Asia are running their own affairs.

Siberia

GUIDING QUESTION *How did the settlement of Siberia progress over time?*

Scientists are uncertain exactly when the first humans arrived in Siberia and whether they came from Europe or from central and eastern Asia. What is more certain is that thousands of years ago, some of them crossed from eastern Siberia to what is now Alaska to become the first people in the Americas.

Settlement, Invasion, and Conquest

After about 1000 B.C., Turkic, Iranian, Mongol, and Chinese people began migrating into southern Siberia from central and eastern Asia. These early people were nomads who lived in small groups. Some were hunter-gatherers. Others were **pastoral**; their lives were based on herding animals.

In the 200s B.C., invaders from Manchuria in northeast China drove many groups north onto the central Siberian plateau. The invaders from Manchuria were the first of several peoples to conquer parts of Siberia and add it to their empire. Other invaders included the Huns, Mongols, and Tartars. The Tartars were defeated by the Russians, who had been coming into Siberia to trade for furs.

Russian Siberia

As they pushed east, Russian traders built small forts and trading posts. Some of these sites eventually became towns.

206 Chapter 7

Some Siberian groups welcomed the Russians. Groups that resisted eventually were subdued. By 1700, Russian control extended to the Pacific Ocean, and some 230,000 Russians were living in Siberia. The czars who were the rulers of Russia also used Siberia as a place of exile for political prisoners.

Lack of roads and other transportation links **inhibited** Siberia's development until the Trans-Siberian Railroad was built across the region between 1891 and 1905. To encourage settlement, the czar began offering settlers free land. By 1914, more than 3 million people had settled in Siberia.

The railroad allowed easier export of products from the region. Coal mines were opened in several locations. Modern farming methods were introduced, and the farmers began producing dairy products and large amounts of grain.

Revolution and Development

When the czar was overthrown and the Communists took control in 1917, some Siberian leaders resisted the new government. In 1922 the Communists brought Siberia under control. Siberia became part of the Union of Soviet Socialist Republics (also called the Soviet Union and USSR)—the country they formed from the Russian Empire.

Academic Vocabulary

inhibit to restrict or prevent a process or an action

A richly decorated train station stands in the city of Irkutsk, a major stopping point on the Trans-Siberian Railroad.

Trans-Siberian Railroad

MAP SKILLS

1 THE GEOGRAPHER'S WORLD
What areas are linked by the Trans-Siberian Railroad?

2 ENVIRONMENT AND SOCIETY
How did the Trans-Siberian Railroad affect settlement?

Soviet leaders increased mining and industry in the region, especially along the Trans-Siberian Railroad. Much of this expansion came through the use of forced labor. Labor camps called gulags spread across Siberia in the 1930s. Millions of people who disagreed with Communist government policies were imprisoned in gulags. The Communists also combined the lands of small farmers into large, government-run farms called **collectives**. Farmers who resisted giving up their land were sent to the gulags.

When Germany invaded Russia during World War II, Soviet leaders relocated some factories to Siberia to keep them from being captured by German forces. Siberian industry continued to grow after the war. In the late 1990s, oil production increased but has not reached the levels of the late 1980s. Yet oil prices have risen, giving the Russian government a steady source of income. China's increased need for oil has led to calls to develop eastern Siberia's untapped reserves.

☑ **READING PROGRESS CHECK**

Determining Central Ideas How did control of Siberia change over time?

Central Asia

GUIDING QUESTION *How have invaders affected Central Asia's history and culture?*

For most of their history, the countries of Central Asia have been part of outside empires. In the 300s B.C., the region was the eastern border of the Greek empire created by Alexander the Great. In the centuries that followed, the Chinese, Huns, and other groups took turns ruling Central Asia. For many conquerors, the greatest prize was control of the Silk Road, the network of trade routes that linked China and Europe. Nomad groups that could not defeat the invaders often retreated into the region's deserts and dry plains, where they continued to live in freedom.

The Influence of Islam

The Arab conquest of Central Asia in the early A.D. 700s brought the religion of Islam to the region, where it remains the main religion today. The Arabs were soon replaced by the Persians, who ruled until the Mongols conquered the region in the early 1200s. In the mid-1300s, a Central Asian conqueror, Timur, overthrew the Mongol rulers. His armies soon conquered a huge region that stretched from the Caucasus to India.

The Silk Road city of Samarqand was the capital of Timur's empire. He made the city a center of culture by bringing in artists and

The city of Samarqand, in present-day Uzbekistan, was Central Asia's major political and cultural center during the 1300s and 1400s. Many buildings from this period, such as this mosque and burial site, still stand today.

▶ **CRITICAL THINKING**
Describing Why did Samarqand become an important center of the Islamic world during the 1300s and 1400s?

The Sayano-Shushenskaya Dam, in south central Siberia, was built during the Soviet era. For many years, it was Russia's largest hydroelectric project. In 2009 an accident at the dam's power plant caused flooding that killed 78 people and destroyed much property.

▶ CRITICAL THINKING

Describing What economic changes did Soviet rule bring to Central Asia?

scholars from other parts of the empire. His successors continued his support of the arts and sciences by establishing madrassas, Islamic centers of learning. Samarqand became a major center for the study of astronomy and mathematics.

Russian Rule and After

Czar Peter the Great began expanding the Russian Empire into Central Asia in the 1700s. By the mid-1800s, most of the region was under Russian control. The Russians' main interest was to grow cotton in Kazakhstan and Uzbekistan. To increase production, they began to **irrigate** the land. They also expanded the railroads to link the region with Russia.

When the Russian Revolution began in 1917, parts of Central Asia took advantage of the unrest to declare their independence. By 1920, Communist forces from Russia had regained control. Under a treaty of union, five of the Central Asian countries became part of the Soviet Union.

Soviet rule brought great change. The Soviets built dams on rivers to generate electric power and collect water for irrigation. They built one of the world's longest canals, the Kara-Kum Canal, to carry water from the Amu Dar'ya River more than 500 miles (805 km) to irrigate desert land near the Caspian Sea. They also increased industry by opening mines and factories to develop the region's mineral wealth.

The changes also had negative impacts. Soviet leaders resettled large numbers of Russians and other Europeans in the region. Local farmers and nomadic herders were forced onto collectives. The Soviets tried to uproot local cultures. They closed mosques to end the practice of Islam in the region.

When the Soviet Union collapsed in 1991, the republics declared their independence. The region's present-day countries were formed. Most of the countries are still struggling with problems that arose after Soviet rule. Among the problems are modernizing their economies and dealing with ethnic tensions and dictatorial rulers.

✓ **READING PROGRESS CHECK**

Analyzing Why was Central Asia difficult to conquer and control completely?

After independence, Georgia ended government economic controls and allowed free enterprise. As a result, many inefficient, Soviet-built factories were abandoned and left to decay.

▶ **CRITICAL THINKING**
Determining Central Ideas Why did the Soviet economic system fail?

Academic Vocabulary

complex having interrelated parts that are difficult to understand or separate

The Caucasus

GUIDING QUESTION *How did the countries of Georgia, Armenia, and Azerbaijan develop?*

The Caucasus are a crossroads between Europe and Asia. This region has drawn conquerors since ancient times. It was part of the ancient Persian, Greek, and Roman empires. Later, Persians, Turks, and Russians dominated the region. This history of conquests and migrations makes the Caucasus one of the most ethnically **complex** places in the world today.

Early History

Georgia and Armenia have each had periods of self-rule during their long histories. Both existed briefly as powerful kingdoms in ancient times but came under the control of powerful neighbors. Around the year A.D. 300, each threw off Persian rule and established kingdoms again. They became two of the earliest countries to convert to the Christian religion. Beginning in the A.D. 600s, Muslim conquerors brought Islam to the Caucasus.

Later, much of the region was conquered by the Mongols in the 1200s and became part of Timur's Central Asian empire in the late 1300s. The Byzantines—Christians who controlled the eastern part of the old Roman Empire—and Persians also sought control of the region. By the late 1400s, the struggle had shifted to the Persians and a new power in Southwest Asia—the Ottoman Turks. The Ottomans and Persians competed for the Caucasus for the next 300 years.

Russian and Soviet Rule

The Christian Armenians and Georgians suffered under Persian and Turkish control. In the early 1700s, they turned to Russia for protection from their Muslim rulers. By the early 1800s, Russia had annexed most of the region. Only part of Armenia remained in the Ottoman Empire.

During World War I, hundreds of thousands of Armenians died at the hands of Ottoman troops, and Russia took the rest of Armenia. Then, the Russian Revolution gave the entire Caucasus the chance to break free of Russian control. Georgia, Armenia, and Azerbaijan briefly formed independent states. Following the Communist victory in Russia's civil war, the Soviet Union annexed the Caucasus in 1922 and eventually created three Soviet socialist republics from the region—the present-day countries of Georgia, Armenia, and Azerbaijan.

The Caucasus benefited but also suffered under Soviet rule. Communists transformed the Caucasus from a largely agricultural area to an urban and industrial one. Soviet power finally ended centuries of invasion and instability, but crushed all opposition. Then during World War II, Germany invaded the Caucasus.

Some of the region's ethnic groups were accused of helping the German invaders. After the war, these groups were broken apart and resettled in various other republics of the USSR. The Soviets also punished other Caucasus people for showing pride in and loyalty to their ethnic identity and culture. Persecution eased after the death of the brutal Soviet dictator Joseph Stalin in 1953. Caucasus leaders took steps toward freedom when the Soviet Union began falling apart in the late 1980s. Non-Communists were already in power in Armenia and Georgia when the Soviet Union dissolved in 1991.

Like the Central Asian republics, Georgia, Armenia, and Azerbaijan declared their independence in 1991. As in Central Asia, though, their 70-some years under Soviet control continue to affect these countries. Since independence, these countries have struggled with economic changes, ethnic tensions, and border conflicts.

Include this lesson's information in your Foldable®.

☑ **READING PROGRESS CHECK**

Citing Text Evidence Which Caucasus country did not have a history of independence before the end of Soviet rule?

LESSON 2 REVIEW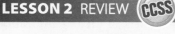

Reviewing Vocabulary
1. How did Soviet *collectives* change agriculture in Siberia and Central Asia?

Answering the Guiding Questions
2. ***Describing*** What factors contributed to Siberia's economic growth and development?
3. ***Analyzing*** How might Central Asia's culture and history have been different if Arabs had not been among the region's many conquerors?
4. ***Determining Central Ideas*** Why are the countries of the Caucasus so culturally complex?
5. ***Argument Writing*** Do you believe that the Soviet Union's persecution of groups for their ethnic loyalty and pride was a good policy? Why or why not?

networks

There's More Online!

- ☑ **IMAGES** Living in a Yurt
- ☑ **MAP** Understanding Time Zones
- ☑ **VIDEO**

Reading HELPDESK

Academic Vocabulary
- significant

Content Vocabulary
- oasis
- homogeneous
- yurt

TAKING NOTES: Key Ideas and Details

Summarize As you read the lesson, list important details for each region using a graphic organizer like the one below.

Siberian Russia	1. 2.
The Caucasus	1. 2.
Central Asia	1. 2.

212

Lesson 3
Life in the Regions

ESSENTIAL QUESTION • *How does geography influence the way people live?*

IT MATTERS BECAUSE
To live in these regions, people must adjust to a harsh and mountainous landscape.

People and Places

GUIDING QUESTION *Who are the peoples of Siberia, Central Asia, and the Caucasus, and where do they live?*

Siberia and most of Central Asia share much the same settlement pattern—the regions have a few large cities and sparsely settled remote areas. The Caucasus is a more urban region, and its rural areas are more densely populated.

Siberian Russia

More than half of Siberians live in the hundreds of cities and towns in the region. Most are alongside or south of the route of the Trans-Siberian Railroad, especially on the west Siberian plain. Novosibirsk, a city of nearly 1.5 million, is located there. It is Siberia's largest city and the third largest in Russia.

A few small cities and many towns also exist along the Ob', the Yenisey, and the Lena rivers and their tributaries, as well as along Siberia's eastern coast. Vladivostok, a city of 500,000 on the Sea of Japan in extreme southeastern Siberia, is located at the eastern end of the Trans-Siberian Railroad. It is a major port for shipping goods into and out of all of Russia.

The resettlement programs of the czars and Soviets have made Siberia's population overwhelmingly Russian. There are still populations of Mongol and Turkic groups, as well as other indigenous peoples. Most people have settled on farms or in cities and towns. A few in the north still live as pastoral nomads.

Central Asia

Kazakhstan is the only country in Central Asia with a largely urban population. About two-thirds of the people are ethnic Kazakhs. Another 25 percent are Russians. Most of the Russians and many Kazakhs live in Alma-Ata. With 1.4 million people, it is the country's largest city and industrial center. About 40 percent of Kazakhstan's people live in rural areas. Most live in the small towns and farm villages that are scattered across the lowlands and plateaus of the steppe.

Turkmenistan is about evenly divided between people who live in cities and large towns and those who live in rural settlements. The greatest number of people live along the Amu Dar'ya and in **oasis** areas in the south. An oasis is a green area by a water source in a dry region. Ethnic Turkomans make up about three-fourths of the population. Russians, at about 10 percent, form the second-largest ethnic group. Most Turkomans live in villages. Most Russians live in the capital, Ashkhabad, which has a population of about 850,000, and in Turkmenistan's two smaller cities.

Most people in Uzbekistan live in the eastern half of the country, almost two-thirds of them in rural areas. Heavily populated oases are covered with orchards, farm fields, and irrigation canals. Most rural people are Uzbeks. Most of the city dwellers are Russian and Kazakh. Uzbekistan's capital city, Tashkent, is home to more than 2 million people. It is Central Asia's largest city and its economic and cultural center.

Thinking Like a Geographer

World Time Zones

The world has 24 official time zones, each 15° of longitude apart. The Prime Meridian is the reference for measuring time at 0°. Traveling east from 0°, the time is one hour later in each time zone. Traveling west from 0°, the time is one hour earlier. When crossing the International Date Line at 180° longitude from west to east, one day is lost; when crossing from east to west, one day is gained.

MAP SKILLS

PLACES AND REGIONS How many time zones are in the area from the Caucasus to Siberia's easternmost point at the International Date Line?

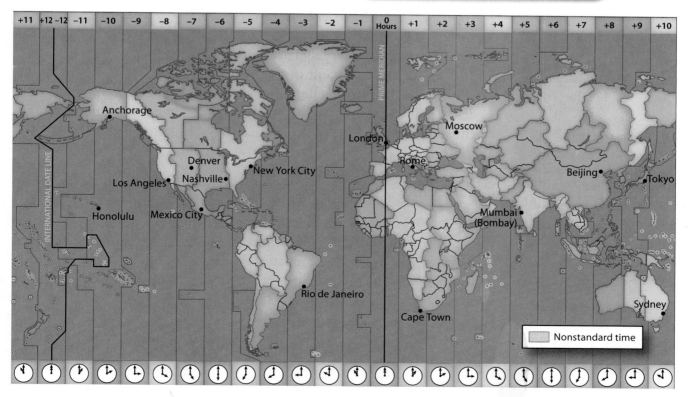

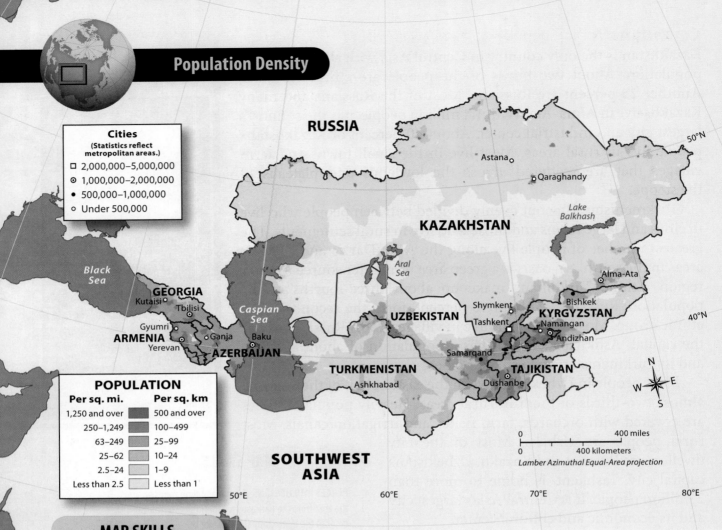

Population Density

Cities (Statistics reflect metropolitan areas.)
- ☐ 2,000,000–5,000,000
- ◉ 1,000,000–2,000,000
- ● 500,000–1,000,000
- ○ Under 500,000

POPULATION

Per sq. mi.	Per sq. km
1,250 and over	500 and over
250–1,249	100–499
63–249	25–99
25–62	10–24
2.5–24	1–9
Less than 2.5	Less than 1

MAP SKILLS

This map shows the population density and major cities of Central Asia and the Caucasus.

1. **HUMAN GEOGRAPHY** What city on the map has the largest population?

2. **PLACES AND REGIONS** How does population density in Armenia compare with Turkmenistan's?

Academic Vocabulary

significant a noticeably large amount; important

About two-thirds of Kyrgyzstan's people are Kirghiz. The Kirghiz were once nomads who were settled onto collectives during Soviet rule. They remain largely rural today.

Less than 40 percent of the population live in cities and towns. Most of the town dwellers are Uzbeks and Russians. Tajikistan is also mainly rural, but the country's largely mountainous terrain causes its settled areas to be densely populated. Villages dot the foothills and mountain valleys. In dry regions, irrigation has created densely populated oases. About 80 percent of the people are ethnic Tajiks. Most of the rest are Uzbeks.

The Caucasus

Nearly all of Armenia's people are ethnic Armenians. Most of Azerbaijan's people are ethnic Azeris. Georgia is largely ethnic Georgian but has a population that contains **significant** minorities—mainly Armenians and Azeris. The Caucasus is also home to about 50 smaller ethnic groups.

About half the region's people are urban, and half are rural. Armenia, however, has a larger urban population overall. Two-thirds of Armenians live in cities and towns.

The region's rural population is not evenly distributed. Mountain valleys and foothills and some parts of the mountains are heavily settled. Few people live above 6,000 feet (1,829 m). The area along Georgia's Black Sea coast is also densely populated. At the same time, one of every four Georgians lives in Tbilisi, its capital city of 1.1 million.

Armenia's Ararat plain, near the country's border with Turkey and Iran, is its most heavily populated region. The plain and the surrounding foothills and mountains are Armenia's economic and cultural center. Yerevan, its capital city of 1.1 million people, is located there.

Azerbaijan's most densely populated area is around its capital, Baku, on a peninsula in the Caspian Sea. Baku, with about 2 million people, is the largest city and most important industrial center in the Caucasus. Azerbaijan's most heavily settled rural region is in the extreme southeast, between the Caspian Sea and its border with Iran.

☑ READING PROGRESS CHECK

Identifying Where does most of Central Asia's Russian population live?

People and Cultures

GUIDING QUESTION *How are the cultures of Central Asia and the Caucasus alike and different?*

Siberia, Central Asia, and the Caucasus not only have different population distributions, their cultures are also very different. In parts of Central Asia, Soviet rule left lasting Russian influences. In parts of the Caucasus, Russian influences are strong but less widespread.

Language and Religion

Russian is the main language in Siberia, and Christianity, mainly Russian Orthodox, is the dominant religion, along with Buddhist and Muslim minorities. Russian is widely spoken in Central Asia. In Kazakhstan and Kyrgyzstan, Russian is an official language, in addition to the language of the country's ethnic majority. Russian is often used in business. In Kazakhstan, 95 percent of the population speaks it, mostly as a second language.

About half of Kazakhstan's people are Christians, and half are Muslims. In the rest of Central Asia, between 75 percent and 90 percent of the people practice Islam, and most of the rest are Christians.

In the Caucasus, most Armenians and Georgians are Christians, and most Azerbaijanis are Muslims. Georgia has a significant Muslim minority as well. In each country, the language of its ethnic majority is the official language and is widely spoken.

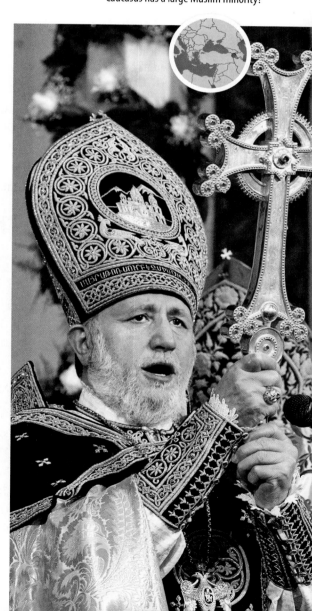

Armenia was the first country in the world to adopt Christianity as its official religion—in A.D. 301. Today, most Armenians belong to the Armenian Apostolic Church.

Identifying Which country in the Caucasus has a large Muslim minority?

Visual Vocabulary

Yurt a round tent with a domed roof covered with heavy, waterproof wool felt or animal skins

Village women in Uzbekistan and other parts of Central Asia have traditionally woven carpets and rugs. After years of decline under the Soviets, the art of carpet weaving is now being taught at workshops in some urban areas.

▶ **CRITICAL THINKING**

Analyzing What economic change has affected the making of traditional goods by hand?

Ethnic Unrest

Compared to the rest of the Caucasus, Armenians are **homogeneous**, or of the same kind. The Turks and Azeris, who live in the countries bordering Armenia, historically have been the Armenians' enemies. In the 1890s and again during World War I, the Turks massacred millions of Armenians who were living under Turkish rule. Meanwhile, Christian Armenians were fighting the influence of Islam being spread by Azeris living in the Caucasus and Iran.

More recently, Armenians and Azeris have been fighting over Nagorno-Karabakh, a region with an Armenian majority that is part of Azerbaijan. In 1988 Armenia seized Nagorno-Karabakh and the territory that connected it to Azerbaijan. As tensions and violence increased, many Azeris fled the region. The dispute remains unsettled.

Georgia also has experienced much ethnic unrest. Its Armenian minority has long sought greater self-rule. In 1992 ethnic minority groups in Georgia's Abkhaz Republic region revolted for independence. In 2008 other minorities in its South Ossetia region did the same. Russia sent troops to each region to help the rebels and increase its influence over Georgia. Pressure from other countries caused the Russians to withdraw from South Ossetia, but tensions over both regions continue.

In Central Asia, Tajikistan was torn by civil war after Soviet rule ended. Tajik Communists, backed by Russian troops, fought Islamic groups and their allies. Some 10,000 people were killed and 500,000 were forced to flee before a peace agreement was reached in 1997.

Culture and Daily Life

Most urban Kazakhs live and dress like Europeans. Rural Kazakhs, women especially, continue to wear the heavily embroidered and colorful traditional Kazakh clothing. Many rural Kazakhs also live in **yurts**. Colorful embroidered rugs decorate traditional Kazakh homes. Such rugs are typical of most Central Asian cultures.

Some Kirghiz and other rural Central Asians also live in sturdy yurts. Rural Kirghiz women stay at home caring for their children while the men tend to the family's crops and livestock. Raising and racing horses are important in most Central Asian cultures.

Uzbek families are often large. Family relationships are close, and parents have great authority. Children help with tasks around the home. In cities and towns, most families live in

In a nomad camp, Kirghiz herders tend to their horses. Equestrian sports (sports with horses), such as racing and a form of polo, are important parts of the culture of Central Asia's nomadic people.

houses with a courtyard, where they spend much of their leisure time. Uzbeks also decorate their homes with colorful rugs, but most wear Western-style clothing.

Most Tajiks live in the western half of the country in the valleys that lie between the steep mountains. They live in villages of 200 to 700 flat-roofed houses along an irrigation canal or a river. A mud fence surrounds each house and the orchard or vineyard next to it. In mountain communities, the villages are smaller. On steep slopes, the houses' flat roofs are the yards for the houses above them.

✓ **READING PROGRESS CHECK**

Citing Text Evidence What ethnic unrest exists in the Caucasus today?

Relationships and Challenges

GUIDING QUESTION *What challenges lie ahead for the regions?*

Most Caucasus and Central Asian countries have struggled to establish stable, democratic governments since the end of Soviet rule. They are also trying to develop free-enterprise economies after years of Communist government controls.

The Soviet Legacy

The countries of the Caucasus and Central Asia adopted new constitutions after they gained independence. The constitutions created democratic governments, gave the people rights and freedoms, and set the stage for free market economies. In practice, however, the promises were hard to keep.

Chapter 7 **217**

Old rusting ships lie in the sand where the Aral Sea used to be. The Aral Sea supported a major fishing industry before its water level dropped.

▶ **CRITICAL THINKING**

Analyzing How have Soviet policies of the past affected Central Asia's environment today?

In Azerbaijan and Kyrgyzstan, former Communists came to power and were reluctant to change their authoritarian ways. Uzbekistan's Communist party stayed in control by changing its name. In Tajikistan and Georgia, civil war and government corruption brought hardships to citizens.

Turkmenistan's first leader was not a Communist. He seized all power, however, declared himself president for life, and showed little concern for the people. The country's first real elections were not held until a year after his death in 2006. Most other countries in the two regions are now also moving slowly toward democracy.

Russians have been leaving Central Asia gradually since the end of Soviet rule. This emigration has deprived some countries of workers with the skills needed to run the mines, factories, and other businesses created during the Soviet era. Students who are graduating from the many universities, scientific and technical institutes, trade schools, and public schools the Soviets created are closing the gap. Another benefit of schooling is that nearly all adults in Central Asia and the Caucasus can now read and write.

Public health in Central Asia has suffered. The Soviets' carelessness in developing agriculture and industry has resulted in high levels of pollution and related health risks. The environmental disaster they created in the Aral Sea is a good example.

The Shrinking Aral Sea

The Aral Sea was once the fourth-largest inland lake in the world. Today it has lost about 90 percent of its water. Towns that were once important fishing ports are now dozens of miles from the sea.

The problem began in the 1960s, when the Soviets built huge farms in the desert and steppes; they dug long canals to irrigate them. Not enough water from the Amu Dar'ya and Syr Dar'ya reached the sea to replace the water lost to evaporation and irrigation. The sea's water level dropped, and its shoreline receded.

What was once the seafloor became dry land filled with salt and farm chemicals from decades of runoff from the irrigated fields. Winds whipping across this polluted wasteland carried salt and chemicals hundreds of miles away into people's lungs. Lung diseases and cancer rates rose as a result.

Kazakhstan is now trying to raise the sea's water level again; some improvement has resulted. Uzbekistan and Turkmenistan have not been able, however, to agree on ways to reduce their use of the Amu Dar'ya's water. Without a regional solution, restoring the sea might not be possible.

Territorial Issues

The dispute over the Amu Dar'ya is just one of several regional issues caused by Soviet rule. When the Soviets created the borders of their Central Asian and Caucasus republics, they paid little attention to where each region's ethnic groups lived. Ethnic-related disputes exist along every shared boundary in Central Asia.

Azerbaijan, Iran, Russia, Turkmenistan, and Kazakhstan also cannot agree on how to share the surface and seabed of the Caspian Sea. At stake is access to minerals such as oil and natural gas, fishing rights, and transportation for the landlocked countries of Azerbaijan and Turkmenistan. Russia and Japan have a long-running dispute over the Kuril Islands, an island chain that extends south from Siberia's Kamchatka peninsula. Japan still claims rights to the southernmost islands, which it lost to the Soviet Union during World War II.

Include this lesson's information in your Foldable®.

☑ **READING PROGRESS CHECK**

Determining Central Ideas Why were the Caucasus and Central Asia slow to establish democratic governments when the Soviet Union collapsed?

LESSON 3 REVIEW CCSS

Reviewing Vocabulary
1. Why are *oases* important in Central Asia?

Answering the Guiding Questions
2. **Describing** Why does most of Siberia's population live along the Trans-Siberian Railroad?
3. **Identifying** What cultural characteristics do many of Central Asia's main ethnic groups share?
4. **Analyzing** How have ethnic tensions affected relationships in the Caucasus and Central Asia?
5. **Narrative Writing** Create a journal entry describing a day on the Trans-Siberian Railroad or in a rural community in Central Asia.

Chapter 7 ACTIVITIES CCSS

Directions: Write your answers on a separate piece of paper.

1 Use your **FOLDABLES** to explore the Essential Questions.
INFORMATIVE/EXPLANATORY WRITING Review the population map at the beginning of this unit. Explain why some areas are heavily populated. Describe factors that influenced this pattern of settlement in three or more paragraphs.

2 21st Century Skills
INTEGRATING VISUAL INFORMATION Research to find a a primary source and a secondary source on the collapse of the Soviet Union. Read and discuss the two documents. Answer the questions: What kind of information is provided by the primary source? Is it different from information provided by the secondary source? How? Which source provides information that is more accurate? Why?

3 Thinking Like a Geographer
INTEGRATING VISUAL INFORMATION Create a time line. List five key events in the history of the region. Include the entries on your time line.

4 GEOGRAPHY ACTIVITY

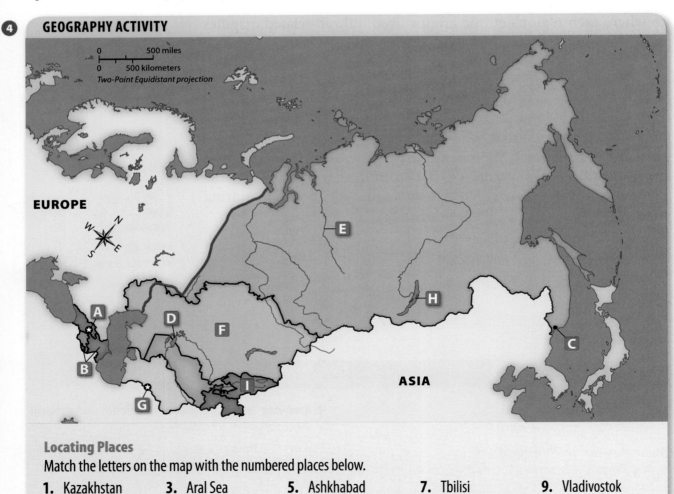

Locating Places
Match the letters on the map with the numbered places below.
1. Kazakhstan
2. Azerbaijan
3. Aral Sea
4. Kyrgyzstan
5. Ashkhabad
6. Yenisey River
7. Tbilisi
8. Lake Baikal
9. Vladivostok

Chapter 7 Assessment

REVIEW THE GUIDING QUESTIONS

Directions: Choose the best answer for each question.

1. Natural resources found in Siberian Russia and Central Asia include huge deposits of
   A. oil and natural gas.
   B. lead and antimony.
   C. copper and tungsten.
   D. bauxite and iron.

2. What is the name of the mountain chain that divides Europe from Asia?
   F. Ural Mountains
   G. Caucasus Mountains
   H. Syr Dar'ya
   I. Tien Shan

3. What technological development opened up Siberia for settlement and aided economic expansion?
   A. the telephone
   B. the steam engine
   C. the Trans-Siberian Railroad
   D. the Moscow to Novosibirsk Canal

4. What benefits did the Caucasus countries of Armenia, Georgia, and Azerbaijan gain from the years they were part of the Soviet Union?
   F. Roads and airports were built.
   G. They profited greatly from the Cold War arms race.
   H. Soviet military power reduced ethnic tensions and brought peace to the region.
   I. The Soviets encouraged the development of the area's natural resources.

5. Which large inland lake nearly dried up because of mismanagement and too much irrigation?
   A. Caspian Sea
   B. Aral Sea
   C. Black Sea
   D. Red Sea

6. The two most popular religions practiced by the people who live in Central Asia, the Caucasus, and Siberia are
   F. Protestantism and Catholicism.
   G. Islam and Hinduism.
   H. Hinduism and Christianity.
   I. Christianity and Islam.

Chapter 7 ASSESSMENT (continued)

DBQ ANALYZING DOCUMENTS

7 IDENTIFYING Read the following passage about the Sogdians, a group of people who were important during the early history of Central Asia:

> "*The Sogdians were the major participants in the Silk Road caravans, ... their alphabet the source of later alphabets to the east, they carried with them such religions as Zoroastrianism . . . and Nestorian Christianity.*"
>
> —from Albert Dien, "The Glories of Sogdiana"

What was the purpose of caravans along the Silk Road in ancient times?

A. to carry out trade of valuable goods
B. to achieve military conquest
C. to spread Sogdian culture to new areas
D. to enrich the Sogdian rulers

8 ANALYZING The Sogdians were important because they

F. built the silk road.
G. organized the caravans.
H. had a strong cultural impact.
I. ruled the region for many centuries.

SHORT RESPONSE

The following passage discusses the work of USAID, a U.S. government agency that provides humanitarian aid and assistance for development to other countries:

> "*When USAID arrived in Kazakhstan in 1992, . . . the country's centrally controlled [economic] system lay in ruins. No one could have imagined that, less than 15 years later, this once-underdeveloped . . . republic would be co-financing USAID's economic development work in the country.*"
>
> —from Geoff Minott and Leanne McDougall, "One Part U.S., Two Parts Kazakhstan" (2012)

9 DETERMINING WORD MEANINGS What do the authors mean when they say that now Kazakhstan is "co-financing" USAID projects?

10 ANALYZING What can you infer about Kazakhstan's economy from this report?

EXTENDED RESPONSE

11 INFORMATIVE/EXPLANATORY WRITING Research to find out about the construction of America's transcontinental railroad. In an essay, compare the building of that railroad with the building of the Trans-Siberian Railroad in Russia. In your writing, describe the effects each railroad had on settlement patterns and economic growth.

Need Extra Help?

If You've Missed Question	1	2	3	4	5	6	7	8	9	10	11
Review Lesson	1	1	2	2	3	3	2	2	3	3	2

SOUTHWEST ASIA

ESSENTIAL QUESTIONS • How does geography influence the way people live? • Why do civilizations rise and fall? • How does religion shape society?

A woman in Baghdad, Iraq, displays her inked fingers, showing that she has just voted in the election.

networks
There's More Online about Southwest Asia.

CHAPTER 8

Lesson 1
Physical Geography of Southwest Asia

Lesson 2
History of Southwest Asia

Lesson 3
Life in Southwest Asia

The Story Matters...

The area between the Tigris and Euphrates rivers in Southwest Asia is known as the Fertile Crescent. Here, fertile soil and water for irrigation supported the growth of an early civilization—Mesopotamia. Three major world religions—Judaism, Christianity, and Islam—began in this region. From ancient to modern times, the region's physical features, resources, and cultures have greatly influenced how people live.

FOLDABLES
Study Organizer

Go to the Foldables® library in the back of your book to make a Foldable® that will help you take notes while reading this chapter.

Chapter 8
SOUTHWEST ASIA

Southwest Asia lies where the continents of Asia, Africa, and Europe meet. Some of the world's earliest civilizations started here.

Step Into the Place

MAP FOCUS Use the map to answer the following questions.

1. **PLACES AND REGIONS**
   What is the capital city of Syria?

2. **THE GEOGRAPHER'S WORLD**
   What countries border Yemen?

3. **THE GEOGRAPHER'S WORLD**
   What body of water lies west of Saudi Arabia?

4. **CRITICAL THINKING**
   Describing In which direction would you travel from the Persian Gulf to the city of Ankara?

BEACH RESORT Vacationers gather at a beach along the Dead Sea. Because of the high salt content of the water, people who bathe in the Dead Sea can float on its surface without effort.

PORT CITY Modern buildings tower over the city of Jaffa, along Israel's Mediterranean coast. One of the world's oldest cities, Jaffa is known for its historic areas, fragrant gardens, and sprawling flea market.

Step Into the Time

TIME LINE Choose an event from the time line and write a paragraph explaining its effect on the region and the world.

c. 3000 B.C. The Sumerians develop early form of writing

c. A.D. 30 Jesus preaches in Jerusalem

c. 4000 B.C. People settle along the Tigris and Euphrates rivers

1792–1750 B.C. King Hammurabi develops written laws, known as the Code of Hammurabi

B.C. | A.D.

2003 The Iraq War begins

1290 The Ottoman Empire begins; lasts until 1922

1900

1948 Israel is founded

2010

A.D. 600s Muhammad begins preaching the teachings of Islam

1938 Oil is discovered in Saudi Arabia

1990 Persian Gulf War begins

2010 Arab Spring revolts take place in the Middle East

networks

There's More Online!

- ☑ **CHART/GRAPH** Oases
- ☑ **MAP** Bodies of Water in Southwest Asia
- ☑ **ANIMATION** Why Much of the World's Oil Supply is in Southwest Asia
- ☑ **VIDEO**

Reading HELPDESK

Academic Vocabulary
- vary

Content Vocabulary
- alluvial plain
- oasis
- wadi
- semiarid

TAKING NOTES: *Key Ideas and Details*

Identify As you read the lesson, complete the graphic organizer by listing the physical features found in this region.

Southwest Asia	Examples
Mountains	
Deserts	
Natural Resources	

Lesson 1
Physical Geography of Southwest Asia

ESSENTIAL QUESTION • *How does geography influence the way people live?*

IT MATTERS BECAUSE
Southwest Asia is characterized by a complex physical geography that influences its people, history, and importance in the world today.

Southwest Asia's Physical Features

GUIDING QUESTION *What are the main landforms and resources in Southwest Asia?*

Southwest Asia comprises 15 countries that lie in the area where Asia meets Europe and Africa. Similarities in physical geography help unite these countries into a single region. Mountains and plateaus formed by active plate tectonics can be seen throughout the region. Dry, desert climates are also widespread.

Mountains and Plateaus

Mountains and plateaus dominate the landscape of Southwest Asia. They have been created over the past 100 million years by collisions between four tectonic plates. This movement also caused earthquakes.

Southwest Asia's loftiest mountains rise in the Hindu Kush range, which stretches across much of Afghanistan and along Afghanistan's border with the South Asian country of Pakistan.

The Hindu Kush and neighboring ranges form natural barriers to travel and trade. As a result, mountain passes have been important in this area. One of the world's most famous

226

passes is the Khyber Pass. It links the cities of Kabul, Afghanistan, and Peshawar, Pakistan. The pass has served as a route for trade and invading armies for thousands of years.

A vast plateau, covering much of Iran, is encircled by high mountain ranges. The mountains of western Iran merge with those of eastern Turkey. Close to the border rises Turkey's highest peak, Ararat, a massive, snowcapped volcano that last erupted in 1840. An elevated area known as the Anatolian Plateau spreads across central and western Turkey.

The Arabian Peninsula consists of Saudi Arabia, Yemen, Oman, and several other countries. It is a single, vast plateau that slopes gently from the southwest to the northwest. Long mountain ranges that parallel the peninsula's southwestern, northwestern, and southeastern coasts are actually the deeply eroded edges of the plateau.

Bodies of Water

The region of Southwest Asia has thousands of miles of coastline. Turkey has coasts on the Mediterranean and Black seas. Syria, Lebanon, Jordan, and Israel have coasts on the Mediterranean Sea. Jordan, Saudi Arabia, and Yemen border the long, narrow Red Sea. The Red Sea has been one of the world's busiest waterways since Egypt's Suez Canal, connecting the Red Sea and the Mediterranean, was completed in 1869. To the southeast of the Arabian Peninsula lies a part of the Indian Ocean called the Arabian Sea. Yemen and its neighbor Oman have coasts along this sea.

A deadly earthquake in 1999 left widespread destruction in Turkey.
▶ **CRITICAL THINKING**
Describing What causes earthquakes throughout Southwest Asia?

Oil tankers pass through the Strait of Hormuz. The strait is the only sea passage from the Persian Gulf to the open ocean.

▶ CRITICAL THINKING

Analyzing Why is the Strait of Hormuz considered a strategic waterway?

Thinking Like a Geographer

Characteristics of Seas

The word *sea* is most often used to describe a large body of salt water that is part of, or connected to, an ocean. The Black Sea and the Mediterranean Sea, both of which are connected to the Atlantic Ocean, meet this description, as does the Red Sea, a narrow extension of the Indian Ocean. The landlocked Caspian and Dead seas, however, do not. They best fit the description of *lake*. Large lakes with salty water, however, are sometimes called seas.

In the northeast, the Arabian Peninsula is shaped by the Persian Gulf, which is connected to the ocean by a strategic waterway called the Strait of Hormuz. The Persian Gulf has become tremendously important in world affairs since the middle of the 1900s.

Eight of Southwest Asia's 15 countries border the Persian Gulf: Oman, the United Arab Emirates, Saudi Arabia, Qatar, Bahrain, Kuwait, Iraq, and Iran. In the north, Iran also borders the landlocked Caspian Sea.

The Dead Sea, which lies between Israel and Jordan, is also landlocked. It is far smaller than the region's other seas. At 1,300 feet (396 m) below sea level, it ranks as the world's lowest body of water, and its shore represents the lowest land elevation.

Southwest Asia's two longest and most important rivers are the Tigris and the Euphrates, which are often considered parts of the same river system. The rivers begin within 50 miles (80 km) of each other in the mountains of eastern Turkey. In their lower courses, they flow parallel to one another across a broad **alluvial plain**, a plain created by sediment deposited during floods. The plain covers most of Iraq as well as eastern Syria and southeastern Turkey. This area has been known since ancient times as Mesopotamia, which is Greek for "land between the rivers." Thousands of years ago, one of the world's earliest civilizations took root in the fertile lands of Mesopotamia.

Deserts

Desert landscapes spread across most of Southwest Asia. The Arabian Desert, which covers nearly the entire Arabian Peninsula, is the largest in the region and one of the largest in the world. It is made up of rocky plateaus, gravel-covered plains, salt-crusted flats, flows of black lava, and sand seas, which are unbroken expanses of sand.

In the southern part of the peninsula lies the largest sand sea in the world: the Rub' al-Khali, or Empty Quarter. Winds have sculpted its reddish-orange sands into towering dunes and long, winding ridges. The climate is so dry and hot that this starkly beautiful wilderness cannot support permanent human settlements. In some areas, nomadic people known as the Bedouin keep herds of camels, horses, and sheep.

The Arabian Desert is a harsh environment, but plants thrive in oases. An **oasis** is an area in a desert where underground water allows plants to grow throughout the year.

✅ **READING PROGRESS CHECK**

Analyzing How has tectonic activity—that is, movement of Earth's crustal plates—helped shape landforms in Southwest Asia?

Southwest Asia's Climates

GUIDING QUESTION *What are some ways that mountains, seas, and other physical features affect climate in Southwest Asia?*

A single type of climate dominates most of Southwest Asia. The only parts of the region with greater climatic variety lie in the northwest and northeast.

An Arid Region

Although this region is surrounded by seas and gulfs, water is a scarce resource here. Most of the region falls within an arid, or very dry, climate zone. Deserts—areas that receive less than 10 inches (25 cm) of annual rainfall—cover nearly the entire Arabian Peninsula as well as large parts of Iran. These deserts are part of a broad band of arid lands that stretch from western North Africa to East Asia. Southwest Asia's arid lands can be brutally hot in the summer. Temperatures in the Arabian Desert can soar as high as 129°F (54°C).

Mountain springs provide a constant flow of water through the Bani Wadi Khalid riverbed in northern Oman. Along the stream's course are large rock formations and shimmering pools of turquoise water.

▶ **CRITICAL THINKING**
Analyzing What feature makes the Bani Wadi Khalid area different from most other landscapes in the Arabian Peninsula?

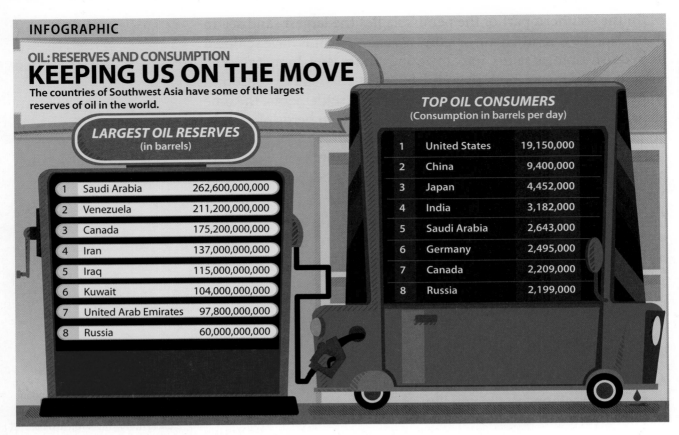

INFOGRAPHIC

OIL: RESERVES AND CONSUMPTION
KEEPING US ON THE MOVE
The countries of Southwest Asia have some of the largest reserves of oil in the world.

LARGEST OIL RESERVES (in barrels)

1	Saudi Arabia	262,600,000,000
2	Venezuela	211,200,000,000
3	Canada	175,200,000,000
4	Iran	137,000,000,000
5	Iraq	115,000,000,000
6	Kuwait	104,000,000,000
7	United Arab Emirates	97,800,000,000
8	Russia	60,000,000,000

TOP OIL CONSUMERS (Consumption in barrels per day)

1	United States	19,150,000
2	China	9,400,000
3	Japan	4,452,000
4	India	3,182,000
5	Saudi Arabia	2,643,000
6	Germany	2,495,000
7	Canada	2,209,000
8	Russia	2,199,000

Oil reserves are estimates of the amount of crude oil located in a particular economic region. Oil consumption is the amount of oil an economic region uses.

▶ CRITICAL THINKING

Describing Which Southwest Asian countries are among the countries with the largest oil reserves?

Academic Vocabulary

vary to show differences between things

Although rain is scarce in this region, rainfall can quickly transform the desert landscapes. Torrents of water race through **wadis** (WAH-deez), or streambeds that are dry. Buried seeds sprout within hours, carpeting barren gravel plains in green.

At the margins of Southwest Asia's dry zones lie areas that are considered **semiarid** (seh-mee-AIR-id), or somewhat dry. These areas are found in the highlands and mountain ranges of the region.

A Mediterranean climate prevails along Southwest Asia's Mediterranean and Aegean coasts and across much of western Turkey. Winds blowing off the seas bring mild temperatures and moderate amounts of rainfall during the winter months. The summer months are warm and dry.

Mountainous areas of eastern Turkey, western Iran, and central Afghanistan have continental climates in which temperatures **vary** greatly between summer and winter. The mountains of the Hindu Kush range in far eastern Afghanistan fall within a highland climate zone, and glaciers are found among the soaring peaks.

☑ READING PROGRESS CHECK

Identifying In what parts of Southwest Asia could farmers grow crops without irrigation?

Natural Resources

GUIDING QUESTION *How do natural resources influence the lives of people in Southwest Asia?*

Scarcity of water has shaped this region's human history and settlement patterns. Other natural resources, however, are found in abundance. The most important resources are two fossil fuels for which the world has a seemingly unquenchable thirst: oil and natural gas.

The gaseous form of petroleum is called natural gas, and the liquid form is called crude oil, or simply oil. Crude oil is refined to produce energy sources such as gasoline, diesel fuel, heating oil, and industrial fuel oil. Petroleum is also the basic raw material used to make many other products, such as plastics, bicycle tires, and cloth fibers.

The world's largest known deposits of petroleum are in Southwest Asia. Most are concentrated around and under the Persian Gulf. Together, five countries that border the gulf—Saudi Arabia, Iran, Iraq, Kuwait, and United Arab Emirates—hold more than half the oil that has been discovered in the world.

Most of the petroleum produced by these countries is exported to industrialized countries. Petroleum revenues have brought tremendous wealth to a few people in the exporting countries. But only in a relatively few areas has the wealth been used to improve the lives of the people or bring about modernization.

Southwest Asia also has a great variety of mineral resources. Large coal deposits are found in Turkey and Iran. Phosphates, used to make fertilizers, are mined in Iraq, Israel, and Syria. Between 2006 and 2010, American geologists conducting a survey of Afghanistan discovered enormous deposits of iron, copper, gold, cobalt, lithium, and other minerals such as rare earth elements used to make electronic devices.

Include this lesson's information in your Foldable®.

☑ **READING PROGRESS CHECK**

Identifying Five countries that border the Persian Gulf hold more than half the oil that has been discovered in the world. Name three of the countries.

LESSON 1 REVIEW

Reviewing Vocabulary
1. Describe the difference between a *wadi* and an *oasis*.

Answering the Guiding Questions
2. ***Identifying*** What makes the Dead Sea distinct?
3. ***Describing*** What are the major physical geography features of the Arabian Peninsula?

4. ***Describing*** If you were to travel across the Arabian Desert, what are two types of landscapes or landforms you might see?
5. ***Citing Text Evidence*** The United Nations ranks Afghanistan as one of the world's poorest countries. How might recent discoveries change that situation?
6. ***Narrative Writing*** Imagine that you are spending a few days exploring the area of the Arabian Desert called the Rub' al-Khali. Write a one-paragraph journal entry describing the experience.

networks

There's More Online!

- ☑ **CHART/GRAPH** Ziggurats
- ☑ **IMAGES** The Kurds
- ☑ **MAP** Islamic Expansion
- ☑ **VIDEO**

Reading HELPDESK

Academic Vocabulary
- expand
- collapse

Content Vocabulary
- polytheism
- millennium
- monotheism

TAKING NOTES: *Key Ideas and Details*

Sequence As you read about Southwest Asia's history, use a time line to put key events and developments in order.

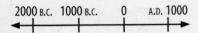

Lesson 2
History of Southwest Asia

ESSENTIAL QUESTION • *Why do civilizations rise and fall?*

IT MATTERS BECAUSE
Southwest Asia has played a large role in human history. The world's earliest civilization developed here, and three major religions were born. Great empires that arose in the region grew to cover parts of three continents.

Early Southwest Asia

GUIDING QUESTION *What are some of the most important advancements that occurred in Southwest Asia in ancient times?*

Mesopotamia

Throughout most of human history, people lived as hunter-gatherers. In small groups, they hunted wild animals and searched for wild fruits, nuts, and vegetables. They were nomadic, frequently moving from place to place. About 10,000 years ago, though, a dramatic change began to occur: People started practicing agriculture—raising animals and growing crops. One of the first places this agricultural revolution unfolded was in Mesopotamia. Mesopotamia was a fertile plain between the Tigris and Euphrates rivers in present-day Iraq.

With the shift to agriculture came a shift to a more settled lifestyle. Villages began to appear in Mesopotamia. Because food was plentiful, some villagers were freed up from farming and could undertake toolmaking, basket weaving, or record keeping. Over time, some villages grew into large, powerful cities that had their own governments and military forces. These cities represent the world's first civilizations.

Over thousands of years, Mesopotamian societies such as the Sumerians and the Babylonians invented sophisticated irrigation and farming methods. They built huge, pyramid-shaped temple towers, and made advances in mathematics, astronomy, government, and law. Using a writing system called cuneiform (kew-NAY-ih-form), they produced great works of literature, including a poem known as the *Epic of Gilgamesh*. Mesopotamia's achievements helped shape later civilizations in Greece, Rome, and Western Europe.

Birthplace of World Religions

Southwest Asia is also a cradle of religion. Three of the world's major religions originated there. In ancient times, most people in the region worshiped many gods. This practice is known as **polytheism**. During the second millennium B.C., a new religion arose. A **millennium** is a period of a thousand years. This new religion was based on **monotheism**—the belief in just one God—and developed among a people called the Israelites. This religion came to be known as Judaism and its followers as Jews.

Jews believe the father of the Israelite people was Abraham. According to the Hebrew Bible, God called Abraham to leave his home in Mesopotamia and found a new nation in a land called Canaan, between the Jordan River and the Mediterranean Sea. The area is shared today by Israel, the Palestinian territories, and Lebanon.

About A.D. 30, a Jewish teacher named Jesus began preaching in this area. The teachings of Jesus led to the rise of Christianity. This new religion spread rapidly throughout the Mediterranean world and into Europe. It also spread across Southwest Asia.

Then, in the A.D. 600s, Islam—the religion of Muslims—arose in the Arabian Peninsula. Muhammad, regarded by Muslims as the last and greatest of the prophets, announced his message in the desert city of Makkah (Mecca). Many of the teachings of Islam are similar to those of Judaism and Christianity. For example, all three religions are monotheistic and regard Abraham as the messenger of God who first taught this belief.

The Ziggurat of Ur was a temple built in Mesopotamia during the 2000s B.C. It served as the political and religious center of the city of Ur.
▶ **CRITICAL THINKING**
Determining Word Meanings
Why is early Mesopotamia called a civilization?

Southwest Asia is the birthplace of Judaism, Christianity, and Islam. The three religions still influence the region today. (left) A Jewish teenager carries a scroll of the Hebrew Bible at the Western Wall in Jerusalem. (center) Muslims circle the Kaaba, a cube-shaped shrine, at the Grand Mosque in Makkah, Saudi Arabia. (right) Christian clergy lead a procession held in Jerusalem shortly before Easter.

▶ **CRITICAL THINKING**

Describing How is the influence of Judaism, Christianity, and Islam reflected in Southwest Asia today?

Academic Vocabulary

expand to increase or enlarge

Islamic Expansion

The religion that Muhammad preached was relatively simple and direct. It focused on the need to obey the will of Allah, the Arab word for God. It obligated followers to perform five duties, which became known as the Pillars of Islam: promising faith to God and accepting Muhammad as God's prophet, praying five times daily, fasting during the month of Ramadan (RAHM-uh-don), aiding the poor and unfortunate, and making a pilgrimage to the holy city of Makkah.

In its first several years, Islam attracted few converts. By the time of Muhammad's death, however, in A.D. 632, it had **expanded** across the Arabian Peninsula. Under Muhammad's successors, known as caliphs (KAY-lifs), Arab armies began spreading the religion through military conquests. It was also spread by scholars, by religious pilgrims, and by Arab traders.

By about A.D. 800, Islam had spread across nearly all of Southwest Asia, including Persia (present-day Iran) and part of Turkey. It also extended into most of Spain and Portugal and across northernmost Africa. It later expanded to northern and eastern Africa, Central Asia, and South and Southeast Asia.

Islamic society was enriched by knowledge, skills, ideas, and cultural influences from many different peoples and areas. The influences contributed to a flowering of Islamic culture that lasted for centuries. During this period, great works of architecture were built, and centers of learning arose. Arab scholars made advances in math and science. This golden age was to have a lasting impact on every place it touched.

During the 1100s and 1200s, crusaders from Western Europe set up Christian states along Southwest Asia's Mediterranean coast. The Muslims fought back and gained control of these territories by 1300. However, in other areas, Muslim military power weakened.

In the middle of the 1200s, a Central Asian people known as the Mongols, led by the grandson of the famous leader Genghis Khan, conquered Persia and Mesopotamia. These areas became part of a vast Mongol empire that stretched across much of Eurasia.

As a result of the Mongol attacks, the Islamic world was fragmented and fell into decay. Soon, however, a new era of Islamic expansion began. At its heart were the Ottomans, a group of Muslim tribes who began building an empire on the Anatolian Peninsula in the early 1200s. By the mid-1300s, the Ottoman Empire had grown to include much of western Southwest Asia and parts of southeastern Europe and northern Africa. At its height, it was one of the world's most powerful states. It endured for six centuries before finally **collapsing** in the early 1900s.

Academic Vocabulary

collapse to break down completely

☑ **READING PROGRESS CHECK**

Determining Central Ideas What are some ways in which Islam was spread?

MAP SKILLS

1. **THE GEOGRAPHER'S WORLD** How far did Islam spread by A.D. 750?

2. **PLACES AND REGIONS** Why did Muslims from Arabia conquer lands such as Syria, Persia, and Egypt?

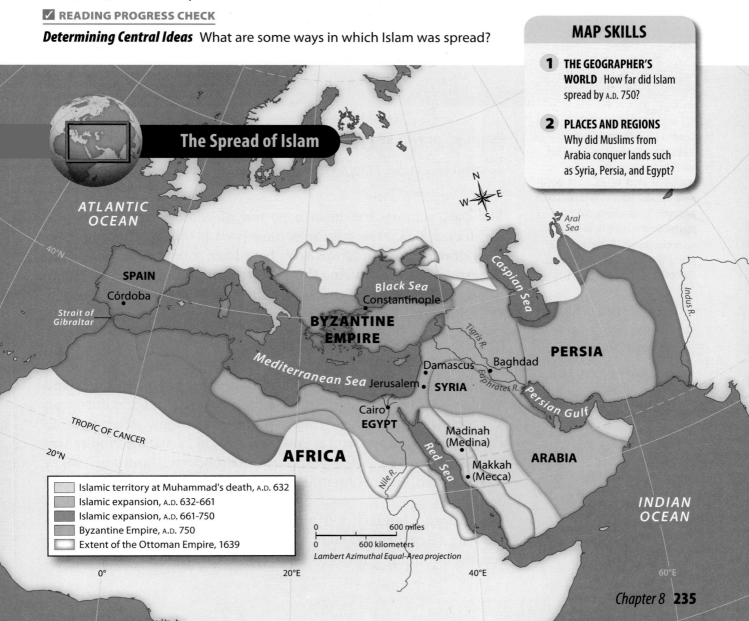

The Spread of Islam

- Islamic territory at Muhammad's death, A.D. 632
- Islamic expansion, A.D. 632–661
- Islamic expansion, A.D. 661–750
- Byzantine Empire, A.D. 750
- Extent of the Ottoman Empire, 1639

Chapter 8 235

General Mustafa Kemal reviews Turkish troops during the war that led to the creation of a Turkish republic in 1923. Kemal, a military hero, became Turkey's first president and introduced reforms to modernize the country. In honor of his achievements, Kemal was later called Atatürk, meaning "Father of the Turks."

Modern Southwest Asia

GUIDING QUESTION *What present-day issues facing Southwest Asia have their roots in ancient times?*

The past century has been a period of change and conflict for Southwest Asia. New countries have been born, new borders have been drawn, and numerous wars have been fought. Vast petroleum reserves discovered during this period have brought great wealth to some of the region's countries but have also created new tensions and conflicts.

Independent Countries

After reaching the peak of its power in the 1500s, the Ottoman Empire began to decline. The decline worsened in the 1800s and early 1900s. During that time, the empire lost African and European territories through wars, treaties, and revolutions. After fighting alongside the losing Central Powers in World War I, the empire was formally dissolved. A few years later, the modern country of Turkey was founded on the Anatolian Peninsula, where the empire had been born.

European interest and influence in Southwest Asia had been growing since the 1869 completion of the Suez Canal, which quickly became an important world waterway. In the peace settlement that ended World War I, Britain and France gained control over the Ottoman Empire's former territories under a mandate system. In this arrangement, the people of these territories were to be prepared for eventual independence.

Dividing up their territories, the British and French created new political boundaries that showed little regard for existing ethnic, religious, political, or historical divisions. These boundaries would take on deep importance when the territorial units became independent countries and when new discoveries of petroleum deposits were made.

Long-simmering resentment toward the European colonial powers soon grew into strong nationalist movements among Arabs, Persians, Turks, and other groups. Between 1930 and 1971, one country after another won its independence, and the map of Southwest Asia began to take its present form.

Arab-Israeli Conflict

One of the mandates received by Britain after World War I was the territory called Palestine. It roughly corresponded to the Land of Israel, which was the area inhabited by the Jewish people in ancient times. Most of the people living in Palestine at the time of the mandate were Muslim Arabs. During the same period, growing numbers of Jewish immigrants seeking to escape persecution had been arriving from Europe and other parts of the world. As the Jewish population increased, tensions between Palestinian Arabs and Jews deepened.

Jewish nationalists called for the reestablishment of their historic homeland in Palestine. This movement gained support as a result of the Holocaust—the systematic murder of 6 million European Jews by Nazi Germany during World War II. Hundreds of thousands of Jews who survived the Holocaust were now refugees in search of a place to live.

In 1947 the United Nations decided on the issue of Palestine. The United Nations voted to divide the territory into two states, one Arab and one Jewish. The proposal was rejected by the Arabs.

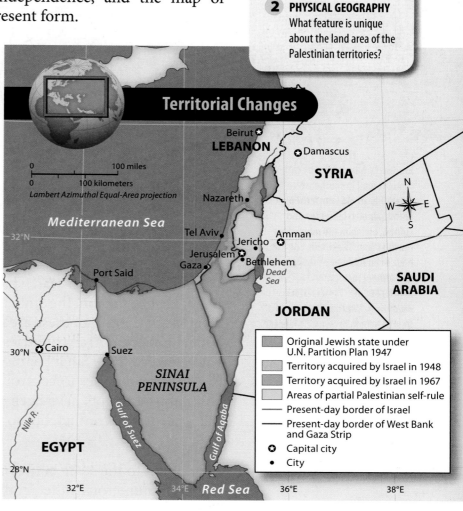

MAP SKILLS

1. **HUMAN GEOGRAPHY** How does Israel's territory today compare with Israel's territory in 1967?

2. **PHYSICAL GEOGRAPHY** What feature is unique about the land area of the Palestinian territories?

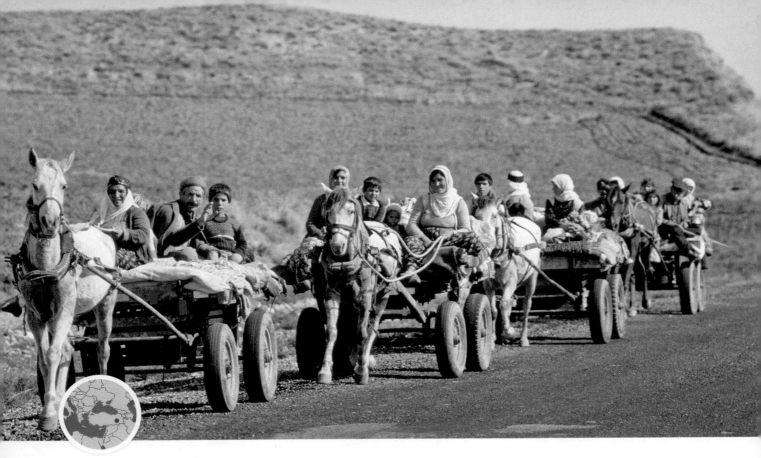

Kurdish families travel by cart over a modern road in southeastern Turkey. The Kurds are a Sunni Muslim people with their own language and culture. Living in the mountains north of Southwest Asia, the Kurds have been ruled by other people throughout history.

▶ **CRITICAL THINKING**
Analyzing Why has the demand of the Kurds for their own independent country been difficult to achieve?

The Arabs did not want to give up land. On the day in 1948 that Israel, the Jewish state, declared its independence, armies from neighboring Arab countries invaded. Hundreds of thousands of Palestinian Arabs became refugees after fleeing the violence. That war ended with a truce in 1948. Other major Arab-Israeli wars were fought, however, in the 1950s, 1960s, and 1970s.

During a brief 1967 war, Israel captured areas known as the West Bank, East Jerusalem, the Sinai Peninsula, the Gaza Strip, and Golan Heights. Its control of these areas was opposed by Palestinian Arabs and neighboring Arab countries, which led to further conflict. Israel withdrew from the Sinai Peninsula in 1982 and from the Gaza Strip in 2005. It continues to control the West Bank and East Jerusalem. Numerous attempts have been made to find a peaceful solution to the Arab-Israeli conflict, but so far none have been successful.

Civil Wars

In addition to the strife between Arabs and Israelis, Southwest Asia has seen numerous other conflicts since World War II. Ethnic, religious, and political differences have fueled many conflicts. So has the rise of Islamist movements that consider Islam to be a political system as well as a religion. The desire to control large oil fields has also caused, or contributed to, many of the conflicts.

Civil wars have torn apart Lebanon, Afghanistan, and Yemen. The Kurds, a fiercely independent people living in eastern Turkey, northern Iraq, and western Iran, have fought to gain their own country. A revolution in Iran in the late 1970s resulted in the overthrow of that country's monarchy and the establishment of an Islamic republic. Iraq invaded Iran in 1980, touching off an eight-year-long war. A decade later, Iraq invaded and annexed its small but oil-rich neighbor, Kuwait. This invasion triggered the Persian Gulf War, in which a coalition led by the United States quickly liberated Kuwait.

Conflict and Terrorism

On September 11, 2001, an Islamist organization called al-Qaeda carried out terrorist attacks on U.S. soil that killed nearly 3,000 people. The United States determined that Afghanistan's Islamist ruling group, the Taliban, was supporting al-Qaeda and sheltering its leaders. In October, forces led by the United States and the United Kingdom invaded Afghanistan and removed the Taliban from power.

Two years later, the Second Persian Gulf War began when forces from the United States and the United Kingdom invaded Iraq and overthrew the government of Saddam Hussein. Hussein was accused of possessing weapons of mass destruction, a suspicion that eventually was proved to be untrue.

Looking to the Future

Despite the many conflicts, there is hope for a more peaceful and brighter future in Southwest Asia. Revenue from petroleum has brought prosperity and modernization to oil-rich countries of the Persian Gulf. In 2010 and 2011, a popular uprising in Tunisia inspired democratic movements in Yemen, Bahrain, and Syria.

On the other hand, militant Islamic political movements limit the growth of democracy and civil rights. In addition, throughout the region, major gaps still exist in standards of living between the oil-rich countries and poorer countries.

Think Again

The Middle East and Southwest Asia are two names for the same region.

Not true. According to most authorities, the Middle East includes all or part of North Africa as well as all or most of Southwest Asia. Some authorities also consider the countries of Central Asia to be part of the Middle East.

Include this lesson's information in your Foldable®.

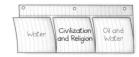

☑ READING PROGRESS CHECK

Determining Central Ideas What event in Europe helped spur the creation of a Jewish state in Southwest Asia?

LESSON 2 REVIEW

Reviewing Vocabulary
1. What is the difference between *monotheism* and *polytheism*?

Answering the Guiding Questions
2. **Identifying** What is one of the Pillars of Islam?

3. **Describing** What change in Mesopotamia around 10,000 years ago resulted in a less nomadic lifestyle?

4. **Identifying** What are two developments that occurred during Islam's golden age?

5. **Citing Text Evidence** What empire represented a second period of Islamic expansion, and where did that empire begin?

6. **Determining Central Ideas** How did the 2001 terrorist attacks on the United States lead to a U.S. invasion of Afghanistan?

7. **Informative/Explanatory Writing** Some conflicts in Southwest Asia relate to the struggle for a homeland by groups such as the Jews, the Palestinians, and the Kurds. Write a short essay discussing what a homeland is and why groups are willing to fight for one.

networks

There's More Online!
- ☑ **IMAGES** Foods of Ramadan
- ☑ **VIDEO**

Reading HELPDESK CCSS

Academic Vocabulary
- widespread

Content Vocabulary
- hydropolitics
- fossil water

TAKING NOTES: *Key Ideas and Details*

Determine the Main Idea As you read the lesson, write the main idea for each section on a graphic organizer like the one below.

Main Ideas

People and Places
People and Cultures
Issues

Lesson 3
Life in Southwest Asia

ESSENTIAL QUESTION • *How does religion shape society?*

IT MATTERS BECAUSE
Because of its strategic location at the convergence of three continents, its huge petroleum reserves, and the deep-rooted conflicts that divide its people, Southwest Asia occupies a central place in world affairs.

People and Places

GUIDING QUESTION *In what parts of Southwest Asia do most people live?*

Southwest Asia's population is slightly greater than that of the United States, although the region is only about three-fourths as large as the United States in area. Throughout history, population patterns in Southwest Asia have been shaped largely by the availability of water. In recent times, another resource—petroleum—has also played an important role.

Population Profile

Southwest Asia is home to about 330 million people. Iran and Turkey, its most populous countries, each have about 80 million people. Some oil-rich countries around the Persian Gulf are experiencing population booms as their fast-growing economies attract foreign workers. Qatar has had one of the world's highest population growth rates in recent years.

Today, many countries of Southwest Asia are highly urbanized. In Israel, Saudi Arabia, and Kuwait, for example, more than four of every five people live in cities. In Afghanistan and Yemen, however, more than two-thirds of the people live in rural areas. However, these countries have the region's highest annual urbanization rate as people move to the cities. As a whole, the region has a rapidly growing population and a high percentage of people below 15 years of age.

240

Where People Live

Population is not evenly distributed across Southwest Asia. The highest densities are in the region's northern and western parts and in its southern tip. These areas include parts of Turkey, Iraq, Iran, and Afghanistan; the countries along the coast of the Mediterranean Sea; and the highlands of southern Saudi Arabia and southwestern Yemen. Most of these areas have relatively higher rainfall.

Areas with dry or somewhat dry climates are more sparsely populated. These areas include the Arabian Desert and the desert lands that spread across central and eastern Iran. Some desert areas are almost completely uninhabited. One exception is Mesopotamia, the land between the Tigris and Euphrates rivers in Iraq. Although its climate is relatively dry, the area supports high population density because the rivers provide abundant water for irrigating crops.

Southwest Asia has metropolises, such as Istanbul, Damascus, Tehran, and Baghdad, that are home to millions of people. Gleaming modern cities, such as Dubai, Abu Dhabi, and Riyadh, rise from the sands of oil-rich Persian Gulf countries. Tel Aviv, Israel's largest city after Jerusalem, is a thriving urban center. These cities stand in sharp contrast to ancient rural villages that seem untouched by the passing of time. In some of the region's desert areas, nomads, known as Bedouins, sleep in tents and raise herds of camels, sheep, goats, and cattle.

✓ READING PROGRESS CHECK

Citing Text Evidence Why do some countries around the Persian Gulf have rapidly growing populations?

At a height of more than 2,700 feet (823 m), the Burj Khalifa (left) is the tallest building in the world. The building is located in Dubai, United Arab Emirates. (right) Adobe storage buildings stand along Al-Assad Lake, a reservoir on the Euphrates River in Syria. A network of canals carries water from the lake to irrigate land on both sides of the Euphrates.

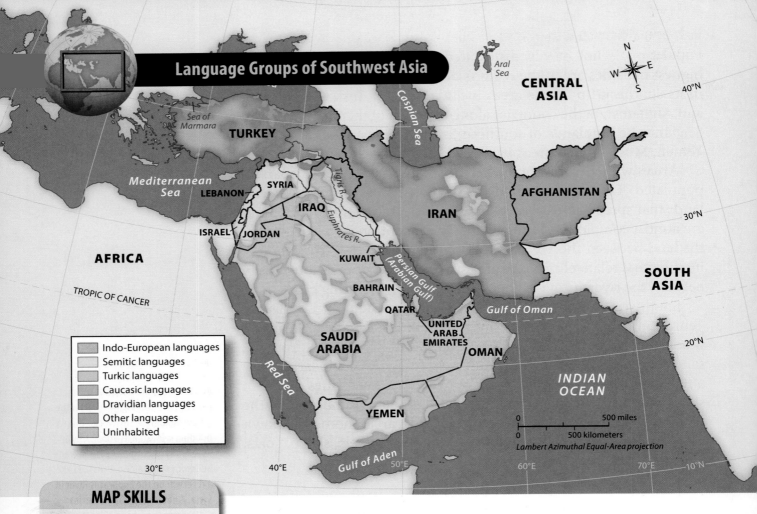

Language Groups of Southwest Asia

MAP SKILLS

1. **THE GEOGRAPHER'S WORLD** What language group is most common in Syria?

2. **THE GEOGRAPHER'S WORLD** What two language groups are most prevalent in Iran?

Academic Vocabulary

widespread spread out

People and Cultures

GUIDING QUESTION What cultural differences are found across Southwest Asia?

Southwest Asia is often thought of as an Arab or an Islamic realm. The reality, however, is more complex. The region, which has always been a crossroads of humanity, is home to many different people.

Ethnic and Language Groups

Arabs represent the largest group in Southwest Asia. In Saudi Arabia, Syria, Jordan, and other countries, 9 out of 10 people are Arab. The two most populous countries in the region, however, have only small Arab populations. In Turkey, Turks form the majority. In Iran, which once was the historical region called Persia, most people are Persian.

In Israel, which was founded as a Jewish state, Jews account for about three-fourths of the population. Kurds, who have no country of their own, represent significant minorities in Turkey, Iran, and Iraq. The region they inhabit is traditionally known as Kurdistan.

Arabic, spoken by Arabs, is the most **widespread** language in Southwest Asia. Other important languages include Turkish and Farsi, the language of Persians. Hebrew is the official language of Israel, and Kurdish is spoken by Kurds.

242 Chapter 8

Some of the region's countries have complex ethnic and linguistic makeups. Afghanistan, for example, is home to Pashtuns, Tajiks, Hazaras, Uzbeks, Aimaks, Turkmen, and Balochs. In addition to the official languages of Afghanistan—Pashto and Afghan Persian—the Afghani people speak Uzbek and more than 30 other languages.

The presence of so many ethnic and language groups in one country presents a challenge to national unity. Many people in Southwest Asia identify with their ethnic group more strongly than with the country they live in. This is clearly evident in countries such as Afghanistan, where people identify themselves as Pashtun or Hazari rather than as Afghani. Even in countries that are mostly Arab, such as Syria and Iraq, people identify with tribes that are based on family relationships. Tribal identity is often stronger than national identity.

Religion and the Arts

From its birthplace in the cities of the Arabian Peninsula, Islam spread across Southwest Asia some 1,300 years ago. It remains the region's dominant religion, helping to unite people of different ethnicity and languages. It is practiced by Arabs, Turks, Persians, Kurds, and many other groups.

Islam has two main branches, Sunni and Shia. Most of Southwest Asia's Muslims are Sunnis. In Iran, however, Shias—Muslims of the Shia branch—outnumber Sunnis nine to one.

Judaism is practiced by about three-fourths of the people in Israel. Christians represent about 40 percent of the population in Lebanon and 10 percent in Syria.

An Iranian woman copies the tile design on a wall at a mosque in Eşfahān, Iran. Islam discourages showing living figures in religious art, so Muslim artists often work in colorful geometric patterns, floral designs, and calligraphy. Passages from the Quran decorate the walls of many mosques.

▶ **CRITICAL THINKING**
Analyzing Why would Muslim artists use passages from the Quran in calligraphy?

A family in Iraq eats a pre-dawn meal before fasting on the second day of the Muslim holy month of Ramadan. Muslims believe that fasting will help people focus on God and on living better lives. According to the Quran, Muhammad first received teachings from God during the month of Ramadan.

▶ **CRITICAL THINKING**
Describing How does religion affect daily life in Southwest Asia?

Religion and art have been closely tied in Southwest Asia throughout history. Some of the region's most magnificent works of architecture are mosques, temples, and other religious structures. Sacred texts such as the Hebrew and Christian Bibles and Islam's Quran stand as works of literature as well as guides to their followers.

The region also has other rich artistic traditions, including calligraphy, mosaics, weaving, storytelling, and poetry. Colorful, handwoven carpets from Persia, or present-day Iran, have been famous for centuries, as has the collection of folktales known as *The Thousand and One Nights*.

Daily Life

Across Southwest Asia, daily life varies greatly. Some people live in cities, some live in villages, and a few live as nomads. Throughout the region's history, most people practiced traditional livelihoods such as farming, raising livestock, or fishing. In recent times, more people have been leaving the land to work in petroleum production, food processing, auto manufacturing, textiles, and construction.

Religion plays a central role in the daily lives of many people in Southwest Asia. Islam is a complete way of life, with rules regarding diet, hygiene, relationships, business, law, and more. To Muslims, families are the foundation of a healthy society; maintaining family ties is an important duty.

Ramadan, the ninth month of the Muslim calendar, is a holy month of fasting. Between dawn and dusk, Muslims are obligated to refrain from eating and drinking. After ending their fast with prayer each evening, people enjoy festive meals. The end of Ramadan is marked by a three-day celebration called *Eid al-Fitr*, which translates as Festival of Breaking Fast.

✓ **READING PROGRESS CHECK**

Identifying What is the major ethnic group in Iran, and what language does that group speak?

Issues

GUIDING QUESTION *How have oil wealth and availability of natural resources created challenges for countries of Southwest Asia?*

The period since World War II has brought a great deal of change and conflict to Southwest Asia. Looking to the future, the region faces many difficult issues. Some of them relate to resources and others to ethnic, religious, and cultural divisions. Some are new, and others are rooted in the distant past.

Oil Dependency and Control

The discovery in the mid-1900s of vast petroleum deposits in Southwest Asia had a strong impact on the region. Exports of petroleum products have brought great wealth to countries around the Persian Gulf, where the largest deposits are found. With this wealth came modernization in some countries. In other countries, little has changed, especially for the average person.

Petroleum has brought new challenges. Many people living in modern cities in oil-producing countries grew up living in tents and practicing traditional farming and herding. Some Muslims believe that increased exposure to Western ways is corrupting the region's people. Another issue is the growing gap between rich and poor countries. Qatar and Kuwait, for example, rank among the wealthiest countries in the world; Afghanistan ranks among the poorest. The struggle to control oil has led to tension and wars. It has also resulted in increased intervention in Southwest Asia by foreign powers.

In 1991 U.S.-led forces pushed Iraqi invaders out of Kuwait. As Iraqi troops left, they set fire to more than 600 oil wells. Tons of thick oil smoke filled the air, and unburned oil spilled into the Persian Gulf. This environmental disaster took place in the area around Kuwait. The air and the soil were polluted, and animal and sea life were destroyed.

Because of irrigation, farming is possible in some dry areas of Saudi Arabia. This wheat farm in Saudi Arabia's Nejd region is supported by center-pivot irrigation. This method involves a long pipe of sprinklers that moves in a circle around a deep well that supplies the water.

▶ **CRITICAL THINKING**
Determining Central Ideas Why is fossil water valuable to Saudi Arabia?

Oil dependency is also an issue. Exporting countries thrive when oil commands high prices, but they suffer when worldwide prices drop. Further, oil is not a renewable resource, and the countries have already depleted some of their reserves. To lessen their dependency on oil, exporting countries have invested money in other industries. Countries that import oil are investigating alternatives to oil.

Changing Governments

More than six decades after it began, the Arab-Israeli conflict continues as one of the biggest issues facing Southwest Asia. At the heart of the conflict are the Gaza Strip and West Bank territories, which Israel captured in 1967. Eruptions of violence have hampered progress toward a peaceful solution.

The years 2010 and 2011 marked the beginning of the Arab Spring, a wave of pro-democracy protests and uprisings in North Africa and Southwest Asia. Protests against authoritarian rulers broke out in Tunisia, Egypt, Libya, Yemen, Bahrain, Syria, Jordan, and Oman in early 2011.

By the end of the year, leaders in Tunisia, Egypt, Libya, and Yemen were overthrown. Protests in Bahrain were quashed by security forces, but the government later agreed to implement reforms. Syria fell into upheaval when the government used armed force to stop protests. More peaceful reform efforts are underway in Jordan and Oman.

Water Concerns

Scarcity of freshwater has plagued Southwest Asia throughout history. Dramatic population growth has produced greater demand for this precious resource, making the situation more dire and increasing the importance of **hydropolitics**, or politics related to water usage and access.

Water from the saltwater seas that surround Southwest Asia can be made into freshwater through desalination, or the removal of salt. Unfortunately, this process is expensive and therefore not practical for meeting the region's water needs.

Saudi Arabia, which has no rivers that flow year-round, has tapped into **fossil water**. This term refers to water that fell as rain thousands of years ago, when the region's climate was wetter, and is now trapped between rock layers deep below ground. By pumping the water to the surface for irrigation, Saudi Arabia has transformed areas of barren desert into productive farmland. Fossil water is not a renewable resource, however, and the underground reservoirs could soon run dry.

The region's greatest source of freshwater is the Tigris-Euphrates river system. From their sources in the mountains of eastern Turkey, the Tigris and Euphrates rivers flow southeastward through the desert plains of Syria and Iraq.

The three countries depend heavily on the rivers and their tributaries. In recent decades, all have built dams to control flooding, to generate electricity, and to capture water for irrigation. Syria and Iraq, which are downstream from Turkey, have bitterly opposed an ambitious, decades-long dam-building project in Turkey that threatens to reduce river flow.

Include this lesson's information in your Foldable®.

☑ **READING PROGRESS CHECK**

Describing What was the Arab Spring? What countries in Southwest Asia were involved?

LESSON 3 REVIEW

Reviewing Vocabulary

1. What is *fossil water*?

Answering the Guiding Questions

2. **Identifying** What are Southwest Asia's two most populous countries, and approximately how many people live in each country?

3. **Determining Word Meanings** What is hydropolitics?

4. **Analyzing** How might dams built on the Tigris and Euphrates rivers in Turkey affect agriculture in Syria and Iraq?

5. **Identifying** What are the two main branches of Islam, and to which branch do most Muslims in Southwest Asia belong?

6. **Identifying Point of View** How might Persian Gulf countries be affected if oil-importing countries begin turning to alternate energy sources?

7. **Describing** Who are the Bedouin?

8. **Narrative Writing** Imagine that have you spent your whole life in a poor village somewhere in Southwest Asia. Then, one day you visit Dubai, a bustling, modern city of skyscrapers and shopping malls. Write a letter to a friend or a family member back in your village describing your experience in Dubai.

What Do You Think?

Are Trade Restrictions Effective at Changing a Government's Policies?

Sometimes, one country restricts trade with another country as a way to force it to change its policies. For example, if the U.S. government wants a country to give its citizens more democratic rights, it might not allow that country to sell goods in the United States. Although the U.S. government often applies trade restrictions on countries, opinions differ about their effectiveness.

Yes!

PRIMARY SOURCE

" Sanctions aimed at achieving major policy objectives have the strongest chance of success if applied by a multilateral coalition [many countries]; it helps that the multilateral approach also shares the cost.... Smart sanctions have their use in tightening the screws on a recalcitrant [stubborn] or defiant [openly disobedient] regime [government] without inflicting collateral [additional] damage on the population.... And even when sanctions don't achieve stated objectives, they nonetheless signal resolve [determination], and may in fact be essential to prepare the political ground at home and abroad for military action against the target."

—Gordon Kaplan, "Making Economic Sanctions Work"

Rows of cargo containers await shipment at Singapore, a city-state in Southeast Asia. Singapore depends on international trade for its survival. Its port is one of the busiest cargo shipping centers in the world.

No!
PRIMARY SOURCE

"Imposing sanctions...are not only an act of war according to international law, they are most often the first step toward a real war starting with a bombing campaign. We should be using diplomacy rather than threats and hostility. Nothing promotes peace better than free trade. Countries that trade with each other generally do not make war on each other, as both countries gain economic benefits they do not want to jeopardize. Also, trade and friendship applies much more effective persuasion to encourage better behavior, as does leading by example."

—Ron Paul, "The Folly of Sanctions"

A port security worker checks cargo at a warehouse in London, England. Ships from all over the world pass through major international ports, such as London.

What Do You Think? DBQ

1. **Identifying Point of View** According to Kaplan, why are sanctions effective even if they do not achieve their objectives?

2. **Identifying Point of View** Why does Paul think that free trade is better than economic sanctions for promoting peace?

3. **Analyzing** Who do you think makes the stronger argument, Kaplan or Paul?

Chapter 8 ACTIVITIES CCSS

Directions: Write your answers on a separate piece of paper.

1 Use your FOLDABLES to explore the Essential Question.
INFORMATIVE/EXPLANATORY WRITING Choose one of the countries in this region and learn about its natural resources. Then write an essay to answer: Is the country using its resources wisely?

2 21st Century Skills
INTEGRATING VISUAL INFORMATION Working in small groups, identify one problem facing the countries of Southwest Asia. Research its effects on the people of the region and offer possible solutions. Produce a slide show or build a visual display to share your findings.

3 Thinking Like a Geographer
IDENTIFYING Identify five countries of Southwest Asia. List the countries and their capital cities. Then write one interesting fact about each country.

4 GEOGRAPHY ACTIVITY

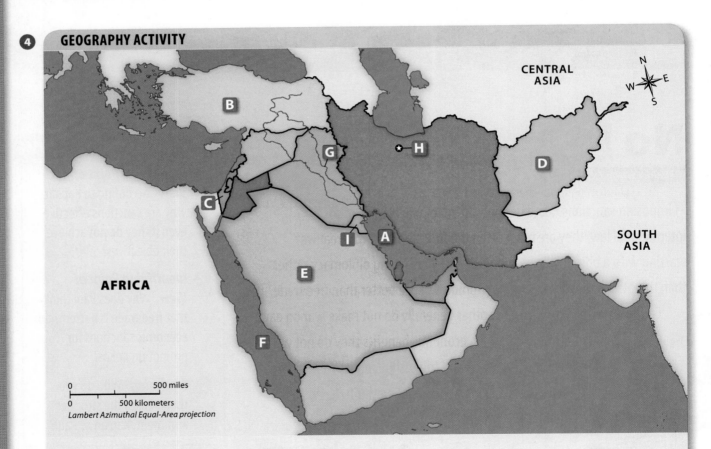

Locating Places
Match the letters on the map with the numbered places below.
1. Saudi Arabia
2. Tehran
3. Afghanistan
4. Israel
5. Red Sea
6. Kuwait
7. Persian Gulf
8. Tigris River
9. Turkey

Chapter 8 ASSESSMENT

REVIEW THE GUIDING QUESTIONS

Directions: Choose the best answer for each question.

1. Which scarce resource has most directly shaped Southwest Asia's history and settlement patterns?
    A. rich soil
    B. natural gas
    C. water
    D. forests

2. Southwest Asia's physical geography can be described as one of
    F. high mountains.
    G. extremes.
    H. little variety.
    I. sandy deserts.

3. How long ago did humans convert from living as hunter-gatherers to living in settlements and practicing agriculture?
    A. a million years ago
    B. 10,000 years ago
    C. 50,000 years ago
    D. 100,000 years ago

4. Which three major world religions originated in Southwest Asia?
    F. Sikhism, Hinduism, Judaism
    G. Christianity, Judaism, Hinduism
    H. Islam, Judaism, Christianity
    I. Judaism, Islam, Hinduism

5. What twentieth-century discovery brought great wealth to some countries in Southwest Asia?
    A. the cell phone
    B. vast petroleum deposits
    C. gold
    D. rubies and emeralds

6. The population of Southwest Asia is
    F. growing rapidly.
    G. declining.
    H. aging.
    I. leaving to find work in India.

Chapter 8 ASSESSMENT (continued)

DBQ ANALYZING DOCUMENTS

7 CITING TEXT EVIDENCE Read the following passage about Israel's economy:

"*Israel has a diversified, technologically advanced economy. . . . The major industrial sectors include high-technology electronic and biomedical equipment, metal products, processed foods, chemicals, and transport equipment. . . . Prior to the violence that began in September 2000, [Israel] was a major tourist destination.*"

—from U.S. State Department Background Notes, "Israel"

Which detail supports the idea that Israel's economy is technologically advanced?
- A. biomedical equipment industry
- B. processed food industry
- C. transport equipment industry
- D. tourism industry

8 ANALYZING What inference can you draw about the decline of tourism to Israel after 2000?
- F. Economic hard times reduced tourism to all locations.
- G. Other places became more fashionable as tourist destinations.
- H. Tourism declined because people were worried about their safety.
- I. Lower prices for air travel would revive tourism to Israel.

SHORT RESPONSE

"*One challenge the Saudis face in achieving their strategic vision to add production capacity is that their existing fields experience 6 to 8 percent annual 'decline rates' on average . . . , meaning that the country needs around 700,000 [billion barrels per day] in additional capacity each year just to [make up] for natural decline.*"

—from Energy Information Administration, *Saudi Arabia*

9 DETERMINING WORD MEANINGS Based on this passage, what do the "decline rates" refer to?

10 ANALYZING What could cause these decline rates?

EXTENDED RESPONSE

11 INFORMATIVE/EXPLANATORY WRITING Write an essay explaining why you think conflict in this part of the world since the end of World War II has increased so dramatically. Consider how conflict and wars in this part of the world affect you and your family.

Need Extra Help?

If You've Missed Question	1	2	3	4	5	6	7	8	9	10	11
Review Lesson	1	1	2	2	3	3	3	3	1	1	2

AFRICA

UNIT 3

Chapter 9	Chapter 10	Chapter 11	Chapter 12	Chapter 13
North Africa	East Africa	Central Africa	West Africa	Southern Africa

EXPLORE the CONTINENT

AFRICA
The continent of Africa covers about one-fifth of Earth's total land area. Africa is the world's second-largest continent, after Asia. For the 54 independent countries located in Africa, physical features such as the Sahara, the Congo River, and the Great Rift Valley have greatly influenced how people live in this environment.

1 NATURAL RESOURCES The region has an abundance of minerals. South Africa, Botswana, and the Democratic Republic of the Congo are among the countries that have valuable deposits of gold, uranium, diamonds, and other minerals. These miners work in one of the world's deepest gold mines in South Africa.

2 BODIES OF WATER Many of Africa's lakes lie in huge basins formed millions of years ago by the uplifting of the land. The Nile, the world's longest river, flows from the East African highlands through Egypt to the Mediterranean Sea.

UNIT 3

3 LANDFORMS Africa's geography is made up of a variety of landforms. Step-like plateaus rise from the coasts to inland mountains. Tropical grasslands with scattered trees cover almost one half of the continent. These giraffes are among the millions of animals—zebras, gazelles, lions, and cheetahs—that roam East Africa's Serengeti, one of the world's largest plains.

FAST FACT

Africa is three times the size of the United States.

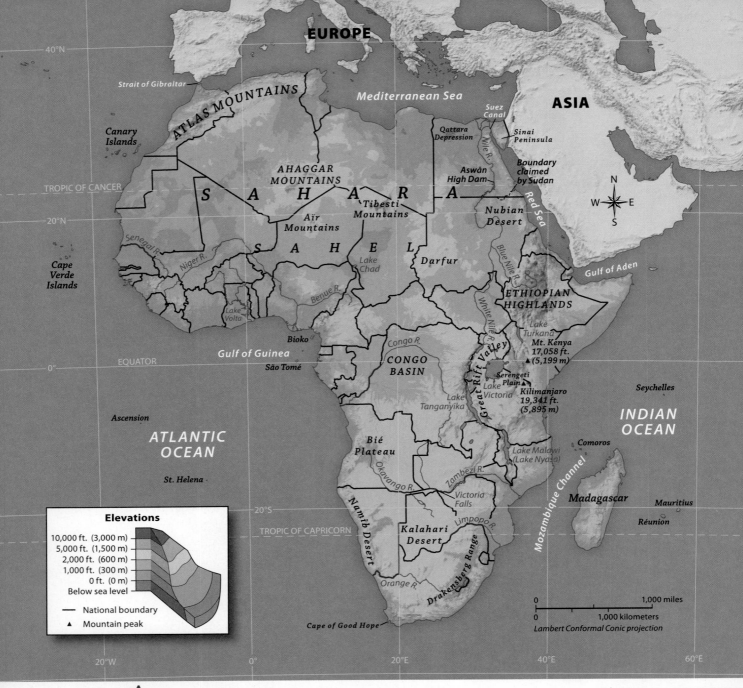

AFRICA

PHYSICAL

MAP SKILLS

1. **PLACES AND REGIONS** What is the major mountain range of North Africa?

2. **THE GEOGRAPHER'S WORLD** Which of Africa's largest lakes lies farthest to the south?

3. **PHYSICAL GEOGRAPHY** What is the highest mountain in Africa?

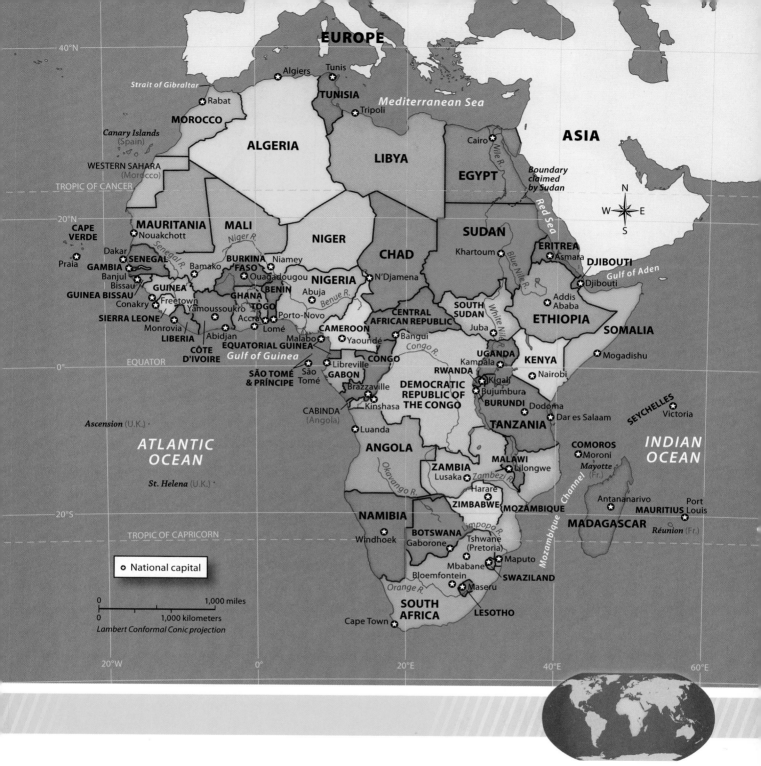

POLITICAL

MAP SKILLS

1. **PLACES AND REGIONS** Which country in the region of North Africa is the largest in area?

2. **PLACES AND REGIONS** Name three island countries of Africa.

3. **THE GEOGRAPHER'S WORLD** Which countries border Namibia in Southern Africa?

Unit 3 257

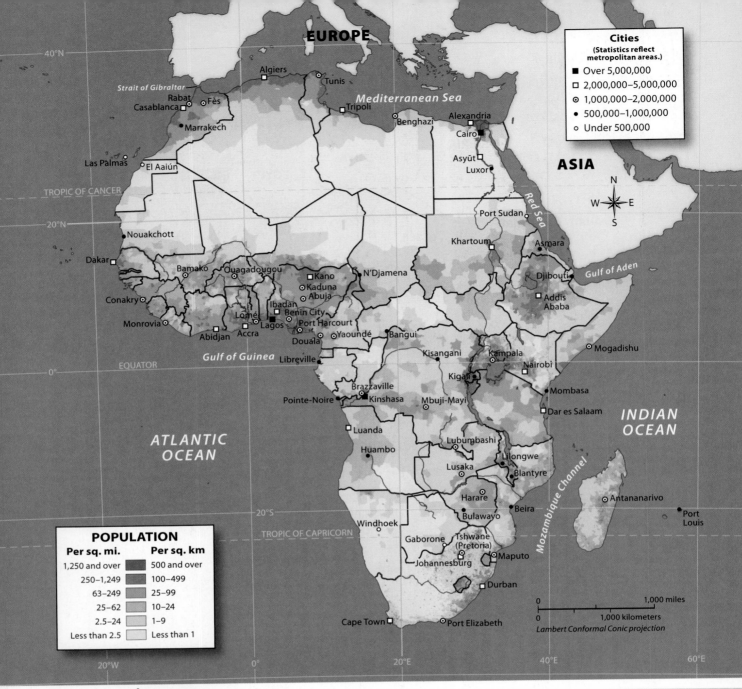

AFRICA

POPULATION DENSITY

MAP SKILLS

1. **PLACES AND REGIONS** Which areas of Africa have the lowest population density?

2. **THE GEOGRAPHER'S WORLD** Generalize about the areas of Africa with the highest population densities. What can you say about these places?

3. **PLACES AND REGIONS** What are the largest cities in Africa?

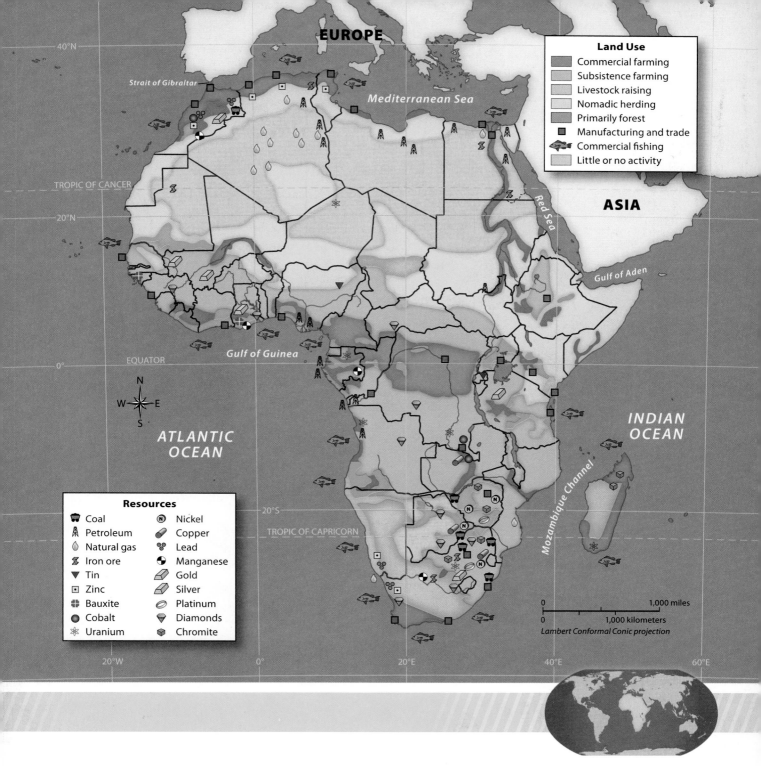

ECONOMIC RESOURCES

MAP SKILLS

1. **ENVIRONMENT AND SOCIETY** Where are precious minerals mined in different areas of Africa?

2. **HUMAN GEOGRAPHY** What is the most common type of farming in Africa?

3. **PLACES AND REGIONS** Why do you think the Nile River valley specializes in commercial farming?

Unit 3 **259**

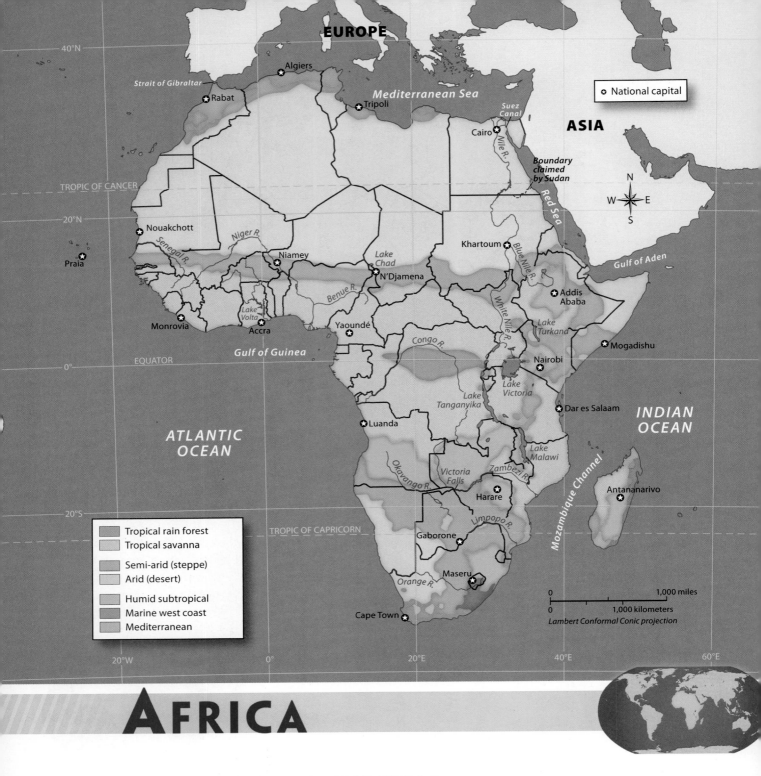

AFRICA

CLIMATE

MAP SKILLS

1. **PLACES AND REGIONS** In the area of Africa south of the Sahara, what is the predominant climate?

2. **PHYSICAL GEOGRAPHY** What climate types are in the area around the Congo River?

3. **PLACES AND REGIONS** What is the climate of Cape Town, South Africa?

NORTH AFRICA

ESSENTIAL QUESTIONS
- How does religion shape society?
- How do people adapt to their environment?
- Why do conflicts develop?

Souks, or markets, such as this one in Esna, Egypt, carry a wide assortment of goods.

networks
There's More Online about North Africa.

CHAPTER 9

Lesson 1
The Physical Geography of North Africa

Lesson 2
The History of North Africa

Lesson 3
Life in North Africa

The Story Matters...

Because of the fertile soil in the Nile River valley, ancient Egyptians created a farming society that developed into an empire thousands of years ago. Their great achievements had a tremendous impact on later civilizations in the region. Other waterways influenced the development of trade and communications among Africa, Asia, and Europe. Waterways and trade routes were also important to the spread of ideas and religion, including Islam.

FOLDABLES
Study Organizer

Go to the Foldables® library in the back of your book to make a Foldable® that will help you take notes while reading this chapter.

Chapter 9
NORTH AFRICA

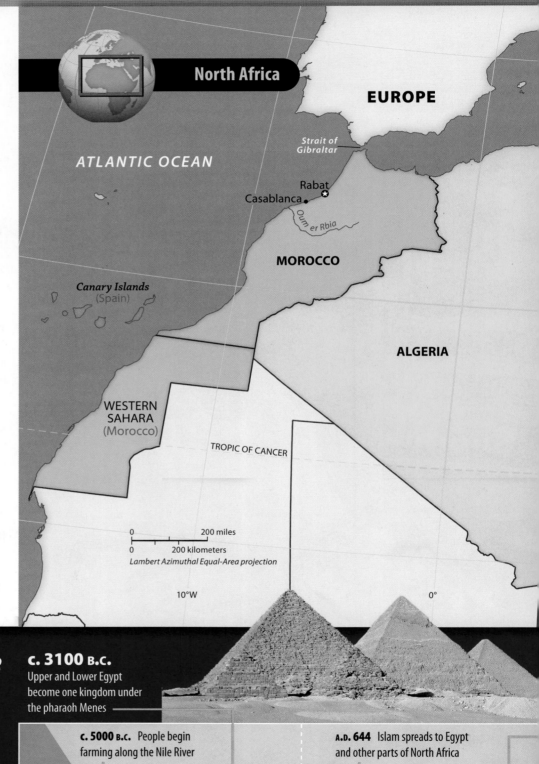

The countries of North Africa border the Mediterranean Sea. They make up one region of the African continent. To their south, the Sahara crosses Africa, and to the south of the Sahara lie the other regions of the second-largest continent.

Step Into the Place

MAP FOCUS Use the map to answer the following questions.

1. **PLACES AND REGIONS** What is the capital of Morocco?

2. **THE GEOGRAPHER'S WORLD** What body of water flows to the east of Egypt?

3. **THE GEOGRAPHER'S WORLD** Which North African country has the largest land area?

4. **CRITICAL THINKING Analyzing** What do the capitals of the North African countries have in common geographically?

Step Into the Time

TIME LINE Choose an event from the time line and write a paragraph explaining the social, economic, or environmental effects of that event on the region and the world.

- **c. 5000 B.C.** People begin farming along the Nile River
- **c. 3200 B.C.** Ancient Egyptians develop hieroglyphic writing
- **c. 3100 B.C.** Upper and Lower Egypt become one kingdom under the pharaoh Menes
- **A.D. 644** Islam spreads to Egypt and other parts of North Africa

networks
There's More Online!

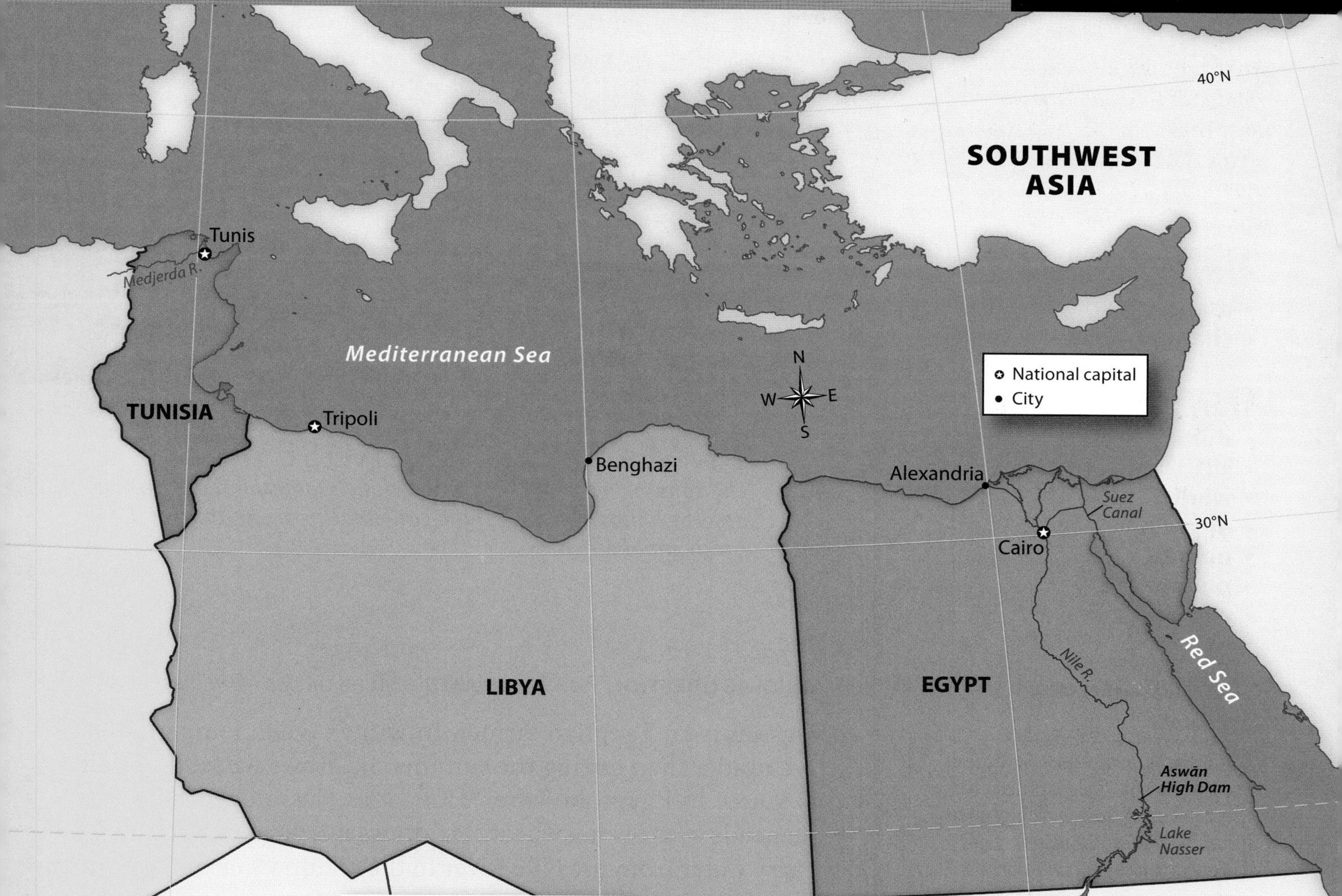

40°N
SOUTHWEST ASIA
Tunis
Medjerda R.
Mediterranean Sea
TUNISIA
Tripoli
Benghazi
N
W
E
S
National capital
City
Alexandria
Suez Canal
30°N
Cairo
LIBYA
EGYPT
Nile R.
Red Sea
Aswān High Dam
Lake Nasser
20°N
10°E

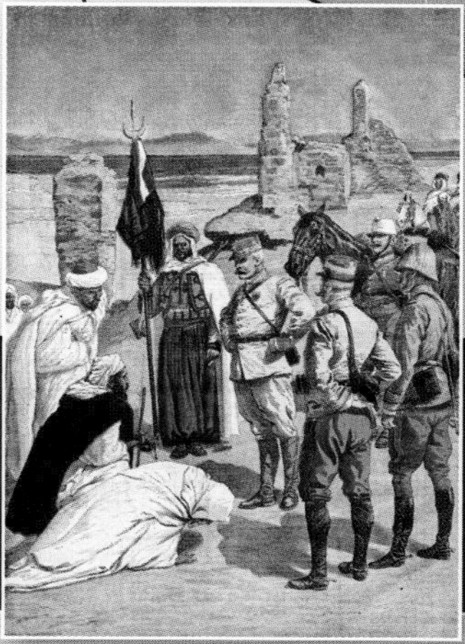

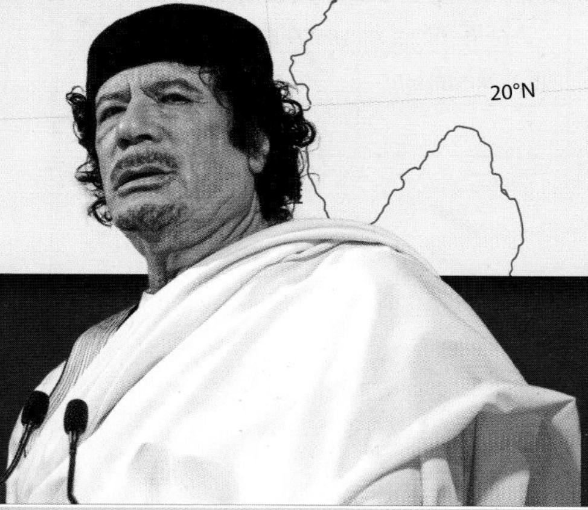

1830 Algeria, Tunisia, and Morocco become part of the French Empire

1859–1869 The Suez Canal is built, linking the Mediterranean and Red seas

1900

1956 Oil is discovered in Libya

1969 Muammar al-Qaddafi seizes power in Libya

1970 The Aswān High Dam is completed

2000

2003 Earthquake in northern Algeria leaves 200,000 people homeless

2010–2011 Pro-democracy revolts known as the Arab Spring take place in Tunisia, Egypt, and Libya

2011 Egyptian president Mubarak leaves office

networks

There's More Online!

- ☑ **IMAGE** Berber Homes
- ☑ **MAP** Mediterranean Climates
- ☑ **VIDEO**

Reading HELPDESK

Academic Vocabulary

- margin
- channel

Content Vocabulary

- delta
- silt
- wadi
- erg
- nomad
- phosphate
- aquifer

TAKING NOTES: Key Ideas and Details

Organize Information As you read about North Africa's physical geography, take notes using the graphic organizer below. Add rows to list more features and their characteristics.

Feature	Characteristic
Atlas Mountains	

Lesson 1
The Physical Geography of North Africa

ESSENTIAL QUESTION • *How do people adapt to their environment?*

IT MATTERS BECAUSE
The Sahara, in North Africa, is the world's largest hot desert. The desert extends over almost the entire northern one-third of the continent of Africa.

Landforms and Waterways

GUIDING QUESTION *How have physical features shaped life in the region?*

Hassan is an Egyptian farmer. In winter, when temperatures are milder than during the summer, he grows wheat. Rainfall is scarce in Egypt, however. How does Hassan get the water he needs to grow his wheat? He draws it from a canal. Canals carry water from the Nile River to the country's farms around the river. Just as they did thousands of years ago, Egypt's farmers still depend on the waters of the Nile.

Countries of the Region

Egypt is the easternmost country in North Africa. The Sinai Peninsula, a triangle of land across the Red Sea from Africa, belongs to Egypt but it is considered a part of Southwest Asia.

North Africa includes five countries. All, like Egypt, sit on the southern shore of the Mediterranean Sea. Libya is to Egypt's west. Tunisia and Algeria are west of that country. Farthest west is Morocco, which has a small Mediterranean coast and a longer coast along the Atlantic Ocean. South of Morocco lies an area called Western Sahara. Morocco claims this area, although the United Nations does not recognize its ownership of this land.

North Africa is a large region. If you placed North Africa over the 48 connected states of the United States, it would reach from Maine to Washington state and cover the northern half of the country.

Coastal Plains and Mountains

In North Africa, low, narrow plains sit on the **margins**, or edges, of the Mediterranean and Atlantic coasts. In the west, the high Atlas Mountains rise just behind this coastal plain. These mountains extend about 1,200 miles (1,931 km) across Morocco and Algeria into Tunisia. They form the longest mountain chain in Africa and greatly influence the region's climate.

The Atlas Mountains are actually two sets of mountains that run alongside each other. A high plateau sits between them. The southern chain is generally higher than the one to the north. It includes Mount Toubkal in Morocco. At 13,665 feet (4,165 m), it is the highest peak in North Africa.

South of these mountains is a low plateau that reaches across most of North Africa. The land rises higher in a few spots formed by isolated mountains. In Egypt, the southern reaches of the Nile River cut through a highland area to form a deep gorge, or valley. Southeastern Egypt has low mountains on the shores of the Red Sea. The southern part of Egypt's Sinai Peninsula is also mountainous. This area includes Egypt's highest point, Gebel Katherina. It reaches 8,652 feet (2,637 m) high. Another set of low mountains lies in northeastern Libya, near the coast.

Lowlands

Northwestern Egypt has a large area of lowland. Called the Qattara Depression, it sinks 440 feet (134 m) below sea level. This area is nearly the size of New Jersey. Marshes and lakes prevent cars and trucks from passing through it.

Academic Vocabulary

margin an edge

A village stands at the foot of a large granite formation in the Atlas Mountains of Morocco.
▶ **CRITICAL THINKING**
Describing How many sets of mountains make up the Atlas Mountains? What landform separates the sets of mountains?

Lush green farmland contrasts sharply with the vast desert areas that stretch for hundreds of miles on either side of the Nile River.

▶ **CRITICAL THINKING**

Analyzing How is the relationship of Egyptians to the Nile River today different from the relationship of ancient Egyptians to the river?

Academic Vocabulary

channel course

Waterways

For centuries, North Africa has been linked by the Mediterranean Sea to other lands. The sea has brought trade, new ideas, and conquering armies.

Next to the Mediterranean, the most important body of water in the region is the Nile River. At 4,160 miles (6,695 km), the mighty Nile is the longest river in the world. It begins far south of Egypt at Lake Victoria in East Africa. That lake sits on the border of Uganda and Tanzania. The river flows northward, joined by several tributaries. The most important of them is the Blue Nile, which begins in the highlands of Ethiopia.

The Nile has a massive delta at its mouth. A **delta** is an area formed by soil deposits that build up as river water slows down. Many deltas form where a river enters a larger body of water. The Nile delta is found where the Nile meets the Mediterranean Sea. Here, at the mouth of the river, the Nile's delta covers more than 9,500 square miles (24,605 sq. km)—larger than the size of New Hampshire. The river once took seven different **channels**, or courses, to reach the sea. Today, only two remain. The others have been filled with soil.

The Nile brings life to dry Egypt. In ancient times, filled by rains to the south, the Nile flooded each year. These floods left **silt**—a fine, rich soil that is excellent for farming—along the banks of the river and in the delta. Farmers used the soil to grow crops. Because they could grow large amounts of food, they were able to support the growth of a great civilization. Ancient Egypt was called "the gift of the Nile."

Today, several dams control the floods. The largest is Aswān High Dam. These dams hold back the high volume of water produced in the rainy season. The water can then be released during the year. An important benefit is that Egypt's farmers today can grow crops year-round. This is the water the farmer Hassan uses to grow his wheat. Another benefit of the dams is that people in Egypt have security from floods. One negative consequence of the dams is that the silt no longer settles on the land and enriches the soil.

Egypt controls another important waterway. This one, the Suez Canal, is human-made. The canal connects the Mediterranean Sea to the Red Sea. As a result, it links Europe and North Africa to the Indian and Pacific oceans. International trade depends on this canal. Using it enables ships traveling between Asia and Europe to avoid going all the way around Africa. The Suez Canal saves many days of travel time and much costly fuel.

✓ **READING PROGRESS CHECK**

Citing Text Evidence Why was ancient Egypt called "the gift of the Nile"?

Climate

GUIDING QUESTION *How do people survive in a dry climate?*

What would it be like if it hardly ever rained? That is the situation that many North Africans face. Large areas of the region receive only a few inches of rainfall each year—if that much.

A container ship passes through Egypt's Suez Canal, one of the world's most heavily used shipping lanes. Opened in 1869, the canal has been enlarged over the years to handle much bigger ships.

▶ **CRITICAL THINKING**

Describing What advantages does the Suez Canal provide for ship travel?

Causes of North Africa's Climates

The Atlas Mountains play a major role in controlling the climate in the western part of North Africa. These mountains create the rain shadow effect. Moist air blows southward from the Atlantic Ocean and the Mediterranean Sea toward the mountains. As the air rises up the northern slopes, it cools and releases rain. By the time it passes over the mountains, the air is dry. This dry air reaches the interior. Inland areas, then, remain arid.

The vast inland area of North Africa is dry for another reason. High-pressure air systems descend over areas to the south of the region for much of the year. They send hot, dry air blowing to the north. This air mass dries out the land. On the rare occasions when it does rain in the desert, the southern winds soon follow. They dry the land and leave behind **wadis**, or dry streambeds.

Desert and Semiarid Areas

Much of North Africa, then, is covered by a desert: the Sahara. Imagine a vast expanse of space, like an ocean, but covered in sand and rock. That is what the Sahara looks like. Spreading across more than 3.5 million square miles (9.1 million sq. km), the Sahara is as large as the entire United States. It covers most of North Africa and spills into three other regions of Africa, as well.

The Sahara's vast stretches of sand are called **ergs**. Strong winds blow the sand about, creating huge dust storms that choke people and animals that are caught outside. The winds also build towering sand dunes. When new winds blow, they can change the shape and size of those dunes.

Landscapes in the Sahara include rugged mountains, stony plains, and large sand dunes. A traveler (left) leads a camel caravan past towering dunes in the Moroccan part of the Sahara. Farther east, Egypt's part of the Sahara (right)—called the Libyan Desert—has rocky surfaces.

▶ **CRITICAL THINKING**
Describing How are sand dunes formed?

Coastal areas of North Africa have a Mediterranean climate that is well suited for growing cereal crops, citrus fruits, grapes, olives, and dates.

Ergs cover only about a quarter of the Sahara. In other areas, rocky plateaus called *hamadas* and rocks eroded by wind are common. Some areas contain oases, areas fed by underground sources of water. Plants can grow in oases and trade caravans that cross the desert stop at them for needed water. **Nomads**, people who move about from place to place in search of food, rely on these oases during their travels. They use the plants to graze herds of sheep or other animals. Some people live on oases and grow crops.

In the North African part of the Sahara, temperatures soar during the day in the summer. They can reach as high as 136°F (58°C). During the winter, though, daytime temperatures can drop as low as 55°F (13°C).

Mediterranean and Other Climates

North of the desert are different climate zones. A band of steppes encircle the desert immediately to the north. Temperatures here are high, and rainfall is slightly greater than in the desert. This band extends to the eastern coast. Coastal cities in Libya receive only 10 inches to 15 inches (25 cm to 38 cm) of rain per year. Alexandria, near Egypt's coast, generally receives only 7 inches (18 cm) of rainfall per year.

A Mediterranean climate dominates the western coast. This climate gives the region warm, dry summers and mild, rainy winters. More rain falls along the coast than in the dry interior. Rain amounts are higher in the west than in the east. In the west, they are higher on the mountain slopes than along the coast. Coastal areas of Morocco receive 32 inches (81 cm) or less of rain per year.

Think Again

Was the Sahara always a desert?

No. Thousands of years ago, the Sahara received more rainfall than it does now. Over time, though, the climate changed. The Sahara became dry—as it remains today.

Chapter 9 **269**

As North Africa's population grows, the demand for water increases. This pump provides water from deep underground to people living in a Sahara environment.

▶ **CRITICAL THINKING**
Describing Why is so much water available underground in parts of the Sahara?

Mountain areas with highland climates also receive more rainfall—as much as 80 inches (203 cm) per year. Highland climates are found within the mountains. Morocco's Atlas Mountains often are covered by snow in the winter. As hard as it might be to believe, just a few hundred miles north of the Sahara, people can snow ski.

✓ **READING PROGRESS CHECK**

Analyzing Where do you think most people in North Africa live? Explain why this might be so.

Resources

GUIDING QUESTION *What resources does North Africa have?*

Oil and natural gas are resources that we use to power our cars and trucks and to generate electricity and heat. Some countries of North Africa have these resources in large quantities. All five countries in the region, though, struggle to get enough of another precious resource—water.

Oil, Gas, and Other Resources

Libya is the most oil-rich country in North Africa. Its oil reserves are ranked ninth in the world and it exports more oil than all but 15 other countries. Libya also has natural gas, but in lesser amounts. The money Libya earns from oil has fueled its economy.

Algeria has large reserves of natural gas—more than all but nine other countries. It also has large supplies of oil. These two resources make up nearly all of its exports.

270 *Chapter 9*

Like Algeria, Egypt has larger reserves of natural gas than oil. Still, it has enough oil to supply most of what it consumes each year. Egypt even sells a small amount to other countries.

Tunisia's main resources are iron ore and phosphates. **Phosphates** are chemical compounds that are often used in fertilizers. These products are important in Morocco, as well. In addition, rich fishing grounds off Morocco's coast are a vital resource. Fish is one of that country's leading exports.

Water

Limited rainfall and high temperatures in this region leave little freshwater on the surface. Rains can be heavy when they come, but the sandy soil soon absorbs the water. Dry winds evaporate the rest. Only the Nile is a reliable source of water for farming throughout the year.

How vital is the Nile? Ninety-five out of every 100 Egyptians live within 12 miles (19 km) of the Nile River or its delta. Yet this narrow river valley and the large delta make up only a small part of Egypt's total area. Without the waters of the Nile, Egypt's people could not survive.

Outside of the Nile valley, most of the region's water needs are met with water that comes from oases and aquifers. **Aquifers** are underground layers of rock in which water collects. People use wells to tap into this water. Libya, for instance, relies on aquifers to meet almost all of its water needs. However, nearly half of Libya's people have no access to water that has been treated to be sure it meets health standards.

A growing population in this region poses problems for the future. Demand for the water in an aquifer shared by Algeria, Libya, and Tunisia has increased ninefold in recent years. In North Africa, aquifers take a long time to refill. If people continue to take water out at a high rate, the aquifers might not be able to refill quickly enough and the region's water problem will become much worse.

Include this lesson's information in your Foldable®.

☑ **READING PROGRESS CHECK**

Analyzing Why would aquifers take a long time to fill up in North Africa?

LESSON 1 REVIEW

Reviewing Vocabulary
1. In which desert feature can people live year-round, a *wadi* or an oasis? Why?

Answering the Guiding Questions
2. *Describing* How has the Mediterranean Sea affected the region?
3. *Analyzing* Does the northern or the southern chain of the Atlas Mountains receive more rainfall? Why?
4. *Determining Central Ideas* Which nations in the region are likely to import energy resources? Why?
5. *Analyzing* How can governments in the region prevent aquifers from being used up?
6. *Informative/Explanatory Writing* Write a paragraph comparing and contrasting the climates of Egypt and Morocco.

networks

There's More Online!

- ☑ **IMAGES** Islam
- ☑ **MAP** The Punic Wars
- ☑ **ANIMATION** How the Pyramids Were Built
- ☑ **SLIDE SHOW** Egyptian Artifacts
- ☑ **VIDEO**

Lesson 2
The History of North Africa

ESSENTIAL QUESTION • *How does religion shape society?*

Reading HELPDESK

Academic Vocabulary
- project
- demonstrate

Content Vocabulary
- **pharaoh**
- **myrrh**
- **hieroglyphics**
- **convert**
- **monotheism**
- **caliph**
- **regime**
- **fundamentalist**
- **civil war**

TAKING NOTES: *Key Ideas and Details*

Summarize As you read about the history of North Africa, note key events and their importance using a graphic organizer like the one below.

Year, Event	Importance

IT MATTERS BECAUSE
One of the world's first civilizations arose in North Africa thousands of years ago.

Ancient Egypt

GUIDING QUESTION *Why was ancient Egypt important?*

Egypt, in North Africa, was one of the earliest known civilizations. Egyptian civilization arose along the Nile River, and Egyptians depended on the Nile for their livelihood. They built cities, organized government, and invented a writing system to keep records and create literature.

The Rise of Egypt

People have been living along the banks of the Nile River for thousands of years. As many as 8,000 years ago, people settled in the area to farm. The rich floodwaters of the Nile allowed farmers to produce enough food to support a growing population. Over time, some members of this early society began to do other things besides farming. Some made pottery. Others crafted jewelry. Some became soldiers. A few became kings.

About 5,000 years ago, two kingdoms along the Nile were united into one. For most of the next 3,000 years, kings called **pharaohs** ruled the land. The great mass of people farmed the land. They paid a share of their crops to the government. The government's leaders also made them work on important **projects**, or planned activities. These projects included building temples and other monuments. Sometimes the people had to fight in the pharaoh's armies.

272

The Expansion of Egypt

For centuries, Egypt traded with nearby lands. Merchants carried Egyptian grain and other products to the south. There they traded for luxury goods like gold, ivory, and incense. They also traded to the east for wood from what is now Lebanon.

Around 1500 B.C., the Egyptians decided to expand their area. They took control of lands to the south that held gold and seized areas along the Red Sea that had **myrrh**. This plant substance gives off a pleasing scent. Priests burned it in religious ceremonies. Egypt also conquered the eastern shores of the Mediterranean. That gave them control of the timber there. Egypt's kings gained wealth by taxing conquered peoples.

Religion and Culture in Ancient Egypt

The pharaoh was the head of Egyptian society. He was seen as more than a man. He was thought to be the son of the sun god. The Egyptians practiced polytheism, which is the belief in many gods. The sun god was one of the most important of their gods. His daily journey through the sky brought the warmth needed to grow crops. The pharaoh, Egyptians believed, connected them to the gods. He made sure that they would flourish as a people.

Academic Vocabulary

project a planned activity

One of the most famous Egyptian pharaohs was the boy-king Tutankhamen. At 10 years of age, Tutankhamen became ruler of Egypt, but he died unexpectedly nine years later.

▶ **CRITICAL THINKING**

Describing Based on the map, describe the area controlled by ancient Egypt.

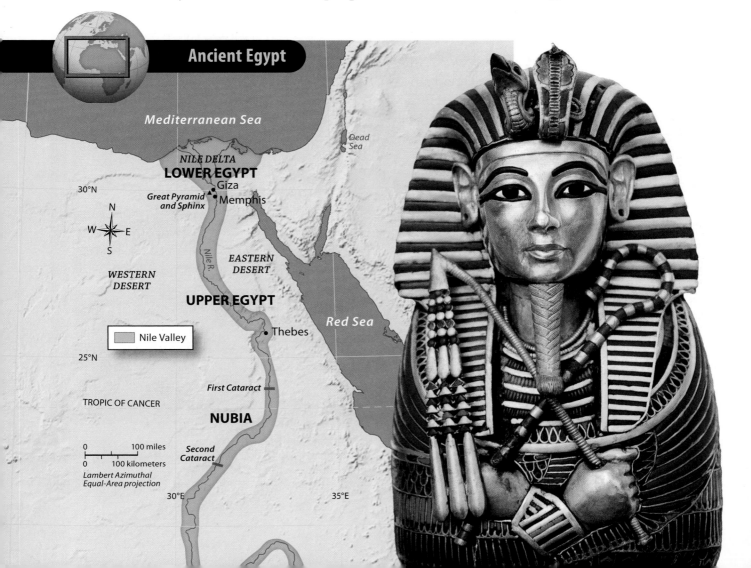

Egyptians believed in life after death. Because of this belief, the pharaohs had vast tombs built for themselves. The tombs were filled with riches, food, and other goods. These goods were meant to support the pharaohs in the afterlife. When the pharaoh died, his body was preserved as a mummy and placed in the tomb.

At first the tombs were low structures built of bricks. Around 2600 B.C., the first pyramid was built as a tomb. These huge tombs, made of rock, were built by thousands of workers. Later, the pharaohs stopped building pyramids. Instead, workers carved their tombs out of rocky cliffs.

Historians know much about ancient Egypt because the Egyptians had a system of writing. The system, called **hieroglyphics**, used pictures to represent sounds or words.

Influence of Ancient Egypt

The Egyptians made many advances in mathematics and science. They used mathematics to measure farm fields and to figure out taxes. Their studies of the stars and planets led to advances in astronomy. They were masters of engineering as **demonstrated** by their great pyramids and temples.

Academic Vocabulary

demonstrate to show

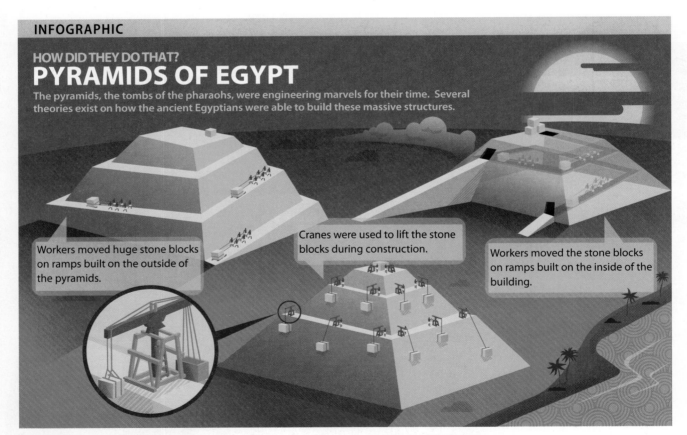

INFOGRAPHIC

HOW DID THEY DO THAT?
PYRAMIDS OF EGYPT

The pyramids, the tombs of the pharaohs, were engineering marvels for their time. Several theories exist on how the ancient Egyptians were able to build these massive structures.

- Workers moved huge stone blocks on ramps built on the outside of the pyramids.
- Cranes were used to lift the stone blocks during construction.
- Workers moved the stone blocks on ramps built on the inside of the building.

BUILDING THE PYRAMIDS
Thousands of people were involved in building a pyramid. Much of the work was done by farmers during the Nile floods, when they could not tend their fields. Surveyors, engineers, carpenters, and stonecutters also lent their skills.

▶ **CRITICAL THINKING**

Analyzing How might the building of the pyramids have led to advances in science and mathematics?

The waters of a Roman bath reflect the ruins of the city of Leptis Magna in Libya. The Romans made Leptis Magna one of the most beautiful cities in North Africa during the A.D. 100s.

Identifying What ancient city fought Rome for control of much of North Africa?

Some of this knowledge was spread to other areas through trade and conquest. Later, Egypt had one of the world's earliest libraries. It was built in the 200s B.C., when Greece conquered and ruled Egypt. The library stored many important works of ancient literature.

☑ **READING PROGRESS CHECK**

Determining Central Ideas Why is it important to know about ancient Egypt?

The Middle Ages

GUIDING QUESTION *How was North Africa connected to other areas?*

Today, people use the Internet to contact each other anywhere in the world. In ancient times, people had to make contact in person. The people of North Africa used the Mediterranean Sea to make this contact with other peoples. Sometimes they were joined by trade. Other times they were joined by conflict.

Carthage and Rome

Western North Africa was first visited by other Mediterranean peoples in the 600s B.C. At that time, traders from what is now Lebanon sailed southwest across the Mediterranean. They built new settlements in many areas. One was a city in what is now Tunisia. They called it Carthage. Within about 200 years, the city had grown powerful. It controlled North Africa from modern Tunisia to Morocco. It also ruled parts of modern Spain and Italy.

In the 200s B.C. and 100s B.C., Carthage fought three wars with the Roman Empire. In the last war, Rome defeated Carthage and destroyed the city. Rome, then, came to control western North Africa. Eventually, Rome conquered Egypt, as well.

Chapter 9 **275**

Religion plays a central role in the lives of most North Africans today. These Muslim women gather for prayer in the main square of El Mansûra, a city in Egypt's Nile delta.

▶ CRITICAL THINKING

Describing How did Islam develop during the century after Muhammad?

During Roman times, many North Africans **converted**, or changed, religions. Because the Roman Empire had adopted Christianity, many North Africans converted to this religion. Others followed their native religions. Except for religion, Roman rule had little effect on native North Africans. Most people continued to live as before. Millions of Berbers who live in western North Africa today are descended from these native people.

Rise of Islam

The Roman Empire fell in the A.D. 400s. Afterward, several local kingdoms formed in North Africa. In the A.D. 600s, though, a new influence emerged in the region. The religion of Islam was founded on the Arabian Peninsula by the prophet Muhammad in A.D. 632. Followers of this religion—called Muslims—began to conquer other lands. By A.D. 642, they had conquered Egypt. By A.D. 705, they ruled all of North Africa. Islam, like Judaism and Christianity, is a monotheistic religion. **Monotheism** means belief in just one god.

Islamic Rule

The Muslim empire was ruled by the **caliph**. This figure had political and religious authority. Caliphs had trouble keeping control over North Africa, however. By the A.D. 800s, separate Berber kingdoms had arisen in parts of the region. These kingdoms often fought one another. Some gained control of most of North Africa. Others only ruled parts of the area.

An Islamic group known as the Fatamids arose in Egypt in the A.D. 1000s. Its rulers expanded Cairo and made it their capital. The city became a center of Muslim learning and trade.

Islamic Culture

At first, Berbers and Egyptians resisted the Islamic religion. By the A.D. 1000s, though, most of them had converted. They also adopted the Arabic language. This language and Islamic learning linked North Africa to the Muslim world. It also helped unite the cultures and people of North Africa and Southwest Asia. Considerable similarities between the regions exist to this day, more than 1,000 years later.

☑ READING PROGRESS CHECK

Identifying Point of View Did the Roman or the Islamic empire have more impact on North Africa? Why do you think so?

The Modern Era

GUIDING QUESTION *What leads people to revolt against a government?*

North Africans formed their own countries in the late 1900s. In recent decades, these countries have changed in far-reaching ways. Often, unrest accompanied the changes.

Foreign Rule

In the 1500s, North Africa began to fall under the rule of foreign armies. The Portuguese and Spanish captured parts of Morocco. The Ottoman Empire, based in modern Turkey, took the rest.

The 1800s saw Ottoman power weaken and Europeans move into North Africa. France began to conquer Algeria in 1830. Although it took several decades, by the late 1800s France controlled that area and Tunisia, too. Some Europeans who settled in these areas grew wealthy. Muslim natives, though, were largely poor. In the early 1900s, France and Spain split control of Morocco. At about the same time, Italy seized Libya.

Egypt kept its independence for much of the 1800s. Its kings tried to build a more modern state. One of the accomplishments was completing construction of the Suez Canal in 1869.

MAP SKILLS

1. **PLACES AND REGIONS** Which North African country was the first to become independent?

2. **HUMAN GEOGRAPHY** How was Libya governed before independence?

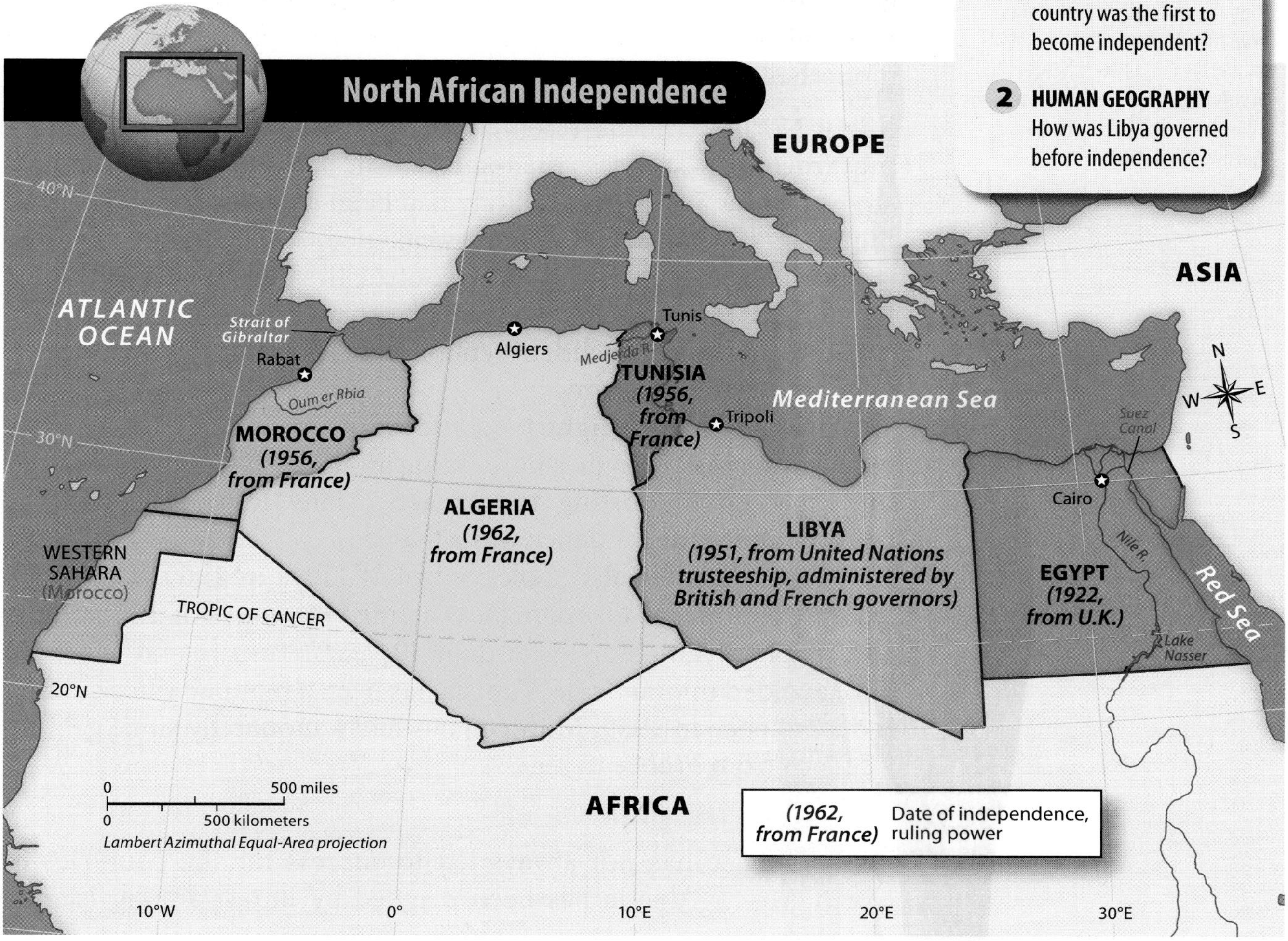

North African Independence

- MOROCCO (1956, from France)
- WESTERN SAHARA (Morocco)
- ALGERIA (1962, from France)
- TUNISIA (1956, from France)
- LIBYA (1951, from United Nations trusteeship, administered by British and French governors)
- EGYPT (1922, from U.K.)

(1962, from France) Date of independence, ruling power

Chapter 9 **277**

In 1987 soldiers marched in a parade in Tripoli, Libya's capital, to celebrate the rule of Muammar al-Qaddafi. Opponents finally overthrew the military dictator in 2011.

▶ CRITICAL THINKING

Analyzing Why was Qaddafi able to rule Libya for so long? Why was he finally overthrown?

The Suez Canal quickly became a vital waterway. Because of the canal's importance, though, other nations wanted to control Egypt. In 1882 Britain sent troops to Egypt. Kings continued to rule, but the British were the real power in the country.

Independence

Many North Africans resented European control. Independence movements arose across the region in the early 1900s. They gained strength after World War II. Italy had been defeated in the war, and France and Britain were severely weakened.

Egypt broke free of foreign control first. In 1952 a group of Egyptian army officers revolted against the king and the British. They created an independent republic, and they put the government in charge of the economy.

Algerians had to fight long and hard for independence. They rebelled against French rule starting in 1954. Not until 1962 did they succeed in ousting the French. Many Europeans fled the country after independence was achieved.

Military leaders also took control of Libya in 1969. They were led by Muammar al-Qaddafi. He remained in control of the nation—and its oil wealth—for more than 40 years. Tunisia and Morocco have avoided military rule. Tunisia has been a republic since gaining independence in 1959. Morocco has had a monarchy since gaining freedom from France in 1956.

Recent Decades

Independence has not always led to success for the countries of North Africa. Algeria has been plagued by unrest among Islamic political groups. Tunisia's government was often accused by the U.S. government of neglecting the rights of the nation's people. Libyan leader Qaddafi had a harsh **regime**, or style of government. Dissent was suppressed, and the government controlled all aspects of life. Qaddafi angered other nations by supporting terrorist groups.

Meanwhile, other problems built up in these nations. High population growth strained their economies. Corrupt governments fueled unrest. In recent years, Muslim **fundamentalists** have led a movement for the people and government to follow the strict laws of Islam. They also reject Western influences on Muslim society.

These problems came to a head in late 2010 in a series of revolts called the Arab Spring. The revolts began in Tunisia, where widespread unrest succeeded in convincing the longtime president to step down from power early in 2011. Tunisians celebrated as a new government took office.

Emboldened by this success, many Egyptians took to the streets. For more than two weeks, thousands of Egyptians turned out every day in Cairo and other cities to protest the government. This revolt also succeeded. In February 2011, Egypt's longtime president Hosni Mubarak gave up power. A group of officers took control and promised to create a new government run by civilians. In 2012 Egyptians voted in the first free presidential election in the country's history.

Unrest also arose in Morocco. There, the king agreed to several reforms that would give more power to the people.

The Arab Spring revolt also reached Libya. The government cracked down on protests. That response angered more Libyans. A **civil war**, or a fight for control of the government, broke out. After months of fighting, the rebels succeeded in taking control of the country. In October of 2011, they killed Qaddafi, and his remaining supporters gave up.

✓ **READING PROGRESS CHECK**

Determining Central Ideas How did the people of North Africa react to European control of the region? Compare that reaction to how North Africans reacted to rule by the Islamic Empire.

Include this lesson's information in your Foldable®.

LESSON 2 REVIEW CCSS

Reviewing Vocabulary
1. How were the *pharaohs* of ancient Egypt and the *caliphs* of the Muslim empire similar? How were they different?

Answering the Guiding Questions
2. *Identifying Point of View* Why did the people of Egypt not revolt against the pharaoh even though they had to pay high taxes and work on major building projects?

3. *Integrating Visual Information* Look at a map of the world. What routes do you think the people of North Africa traveled to trade with the people of Southwest Asia in the Middle Ages?

4. *Determining Central Ideas* What has caused unrest in North Africa in recent years?

5. *Informative/Explanatory Writing* Write a summary of the events and results of the Arab Spring.

networks

There's More Online!

- ☑ **IMAGES** Cuisine of North Africa
- ☑ **VIDEO**

Lesson 3
Life in North Africa

Reading HELPDESK CCSS

Academic Vocabulary
- emphasis
- factor

Content Vocabulary
- souk
- fellaheen
- couscous
- diversified
- constitution

TAKING NOTES: *Key Ideas and Details*

Summarize As you read about daily life, culture, and society in the region, take notes using the graphic organizer below.

Daily Life	Culture	Society
•	•	•
•	•	•

ESSENTIAL QUESTION • *Why do conflicts develop?*

IT MATTERS BECAUSE
North Africa is experiencing political changes.

Culture of North Africa

GUIDING QUESTION *What is daily life like in North Africa?*

The vast majority of people in North Africa practice the Islamic religion. Five times a day, the call to prayer rings out from mosques across North Africa, and devout Muslims stop what they are doing to say prayers. Each week on Friday, millions assemble in the mosques for Friday prayer and to hear a sermon. Once a year during Ramadan, the ninth month of the Islamic calendar, Muslims fast (do not eat) from dawn to dusk.

The People

Three main groups—Egyptians, Berbers, and Arabs—make up the population of North Africa. The region has a varied culture. Egypt's ancient heritage looms over that nation just as the pyramids tower over some of its cities. French influence can be seen from Morocco to Tunisia. Although Arab Muslim culture dominates, some Berber traditions continue.

Although most people are Muslims, some Christians and Jews also live in the region. One in 10 of Egypt's people are Christians. Most of them belong to the Coptic Christian church, which formed in the A.D. 400s.

Of the North African nations, Libya has the highest rate of urbanization. More than three of every four Libyans live in an urban area. Only about half of Egypt's people are city dwellers.

280

Daily Life

Patterns of daily life differ between the cities and the countryside. The region's cities tend to be busy, bustling centers of industry and trade. They also are a blend of traditional cultures and modern life.

Towns and cities of North Africa show no signs of having been planned. Instead, they have grown steadily over the centuries. Streets are narrow and curving. Some built-up areas extend into the surrounding rural farming areas.

Cairo, Egypt, is by far North Africa's largest city, with more than 9.3 million people. The next three largest cities are Algiers, Algeria; Casablanca, Morocco; and Tunis, Tunisia. Combined they have fewer people than Cairo.

Cairo's buildings reflect its more than 1,000-year history. The waterfront along the Nile River boasts gleaming modern skyscrapers and parks. Throughout the city are historic mosques—Islamic places of worship. Tourists flock to the city's famous museums, though they have to endure traffic jams to get there. A jumble of old apartment buildings spreads to the west. Beyond them, a million or so people live in mud huts in a massive poor neighborhood called "the City of the Dead."

An important feature of North African cities is the **souk**, or open-air market. Here, businesspeople set up stalls where they sell food, craft products, and other goods. Singers and acrobats perform here and there in the markets, especially at night.

Life in rural areas follows a different pattern. Farming villages in rural Egypt can be as small as 500 people. Families live in homes built of mud brick with few windows. Each morning, the **fellaheen**—poor farmers of Egypt—walk to work in the fields outside the village.

Spices are among the many products sold at the Khan el-Khalili, the largest souk in Cairo, Egypt. Founded in 1382, the marketplace is a network of streets lined with shops, coffeehouses, and restaurants.

▶ **CRITICAL THINKING**

Describing How did cities in North Africa develop?

A family enjoys a meal at home in the town of Matmata in southern Tunisia. The Berber town is known for its dwellings that are built underground in cave-like structures. To create a home, a resident digs a wide pit in the ground and then hollows out caves around the pit's edge. The caves serve as rooms, which are connected by trench-like corridors.

Many use hand tools and rely on muscle power or animal power. At day's end, they return home.

Farms in Libya are clustered around oases. These communities are small because so little land can be farmed. In Morocco, many farmers live in the well-watered highland areas. They build terraces on steep hillsides to plant their crops.

Some rural dwellers still live like nomads. This is the same kind of life Berbers have followed for centuries. They tend herds of sheep, goats, or camels. They move from place to place in search of food and water for their herds. Some settle in one area for part of the year to grow grains.

Food

Moroccan food has gained fame around the world for its rich and complex flavors. The base of many Moroccan meals is **couscous**, small nuggets of semolina wheat that are steamed. Rich stews of meat and vegetables are poured over it. This style of cooking is also common in Algeria and Tunisia.

Sandwiches in this region are often made with flat pieces of pita bread. They might include grilled pieces of lamb, chicken, or fish. Falafel is made from ground, dried beans and formed into cakes and fried. Pigeon is also popular in Egypt and Morocco.

Arts

The arts in North Africa reflect the influence of Islam. The Islamic religion forbids art that shows the figures of animals or humans. Folk art, like weaving and embroidery, has intricate patterns but no figures. These patterns are also used to decorate buildings.

Many young people in North Africa are attracted to Western

music and movies. This has provoked an angry response among some strict Muslims. In Algeria, some artists have left the country because of harsh criticism. Egypt has long been a center of television and film production. Its shows and movies are seen throughout the Arab world.

Languages and Literature

Arabic is the official language of all five countries in North Africa. French is prominent in Morocco, Algeria, and Tunisia. French and English are most often heard in the region's cities, but Berber languages are more common in rural areas.

As the largest Arabic-speaking country, Egypt has played an important part in the literature of the region. Egyptian writers have explored themes like the impact of influences from Western culture. Novelist Naguib Mahfouz, who wrote more than 30 novels and hundreds of stories, achieved worldwide recognition when he won the Nobel Prize for Literature in 1988.

Academic Vocabulary

emphasis importance

☑ **READING PROGRESS CHECK**

Identifying What is an example of the influence of Islam on daily life in North Africa?

Challenges in North Africa

GUIDING QUESTION *What challenges face North Africa?*

Standards of living vary widely across the region and even within countries. In addition to economic issues, the region faces significant social challenges.

Economic Issues

When oil was discovered in Libya, Muammar al-Qaddafi, the leader of the country, said that a major goal was to provide social benefits to everyone. That did not happen. The income gained from selling oil did not reach most of the country's people. When Qaddafi fell from power in 2011, Libyans hoped that their lives would improve, but progress started slowly.

Algeria has tried to shift its economy away from the **emphasis** on the sale of oil and natural gas. The government keeps tight control of businesses, however. As a result, companies from other countries are not willing to invest there.

A craftsperson in Cairo, Egypt, uses copper thread to embroider Arabic writing onto fabric. Muslims prize the art of beautiful writing, which they use to express the words of the Quran, the Islamic holy book.

▶ **CRITICAL THINKING**
Determining Central Ideas
How has Islam influenced the arts of North Africa?

Morocco's economy is the most **diversified**. A diversified economy includes a mix of many different economic activities. The people of the country engage in mining, some manufacturing, farming, and tourism. Poverty and unemployment are widespread in Morocco, however.

In recent years, thousands have left the region for Europe. They move mostly to Spain and France looking for jobs. Morocco, Algeria, and Tunisia have lost the most people.

Social Issues

High population growth is a major concern in Libya and Egypt. This growth rate contributes to crowding and inadequate health care, as well as poverty. A large share of the population in the region is 14 years old or younger. This is especially true in Egypt and Libya. These countries will have to work hard to develop their economies so that today's young people can find jobs in the future.

In February 2012, U.S. Secretary of State Hillary Clinton addressed young people in Tunisia and across the region. She cited the work they did to bring about the massive changes of the Arab Spring. Clinton warned, though, that it would take a long time and hard work to build the country's economy and increase jobs for young people. The U.S. government has pledged money to several countries to help them accomplish these goals.

Another issue is literacy. Libya has the highest literacy rate in the region: 89 percent of Libyans can read and write. The literacy rate is

Workers load freight wagons at a phosphate mine in southern Tunisia. North Africa has rich deposits of phosphates, which are mineral salts used to make fertilizer.

▶ CRITICAL THINKING
Analyzing What social and economic challenges does North Africa face?

much lower in the other North African countries. Literacy is most serious in Morocco, where little more than one-half of Moroccans can read and write. A very low literacy rate among women is a major **factor**, or cause, for this trend. More than 65 percent of Moroccan men can read and write; less than 40 percent of that nation's women can. Literacy among women is about 20 percent lower than among men in the other four countries of the region, as well. This gap hinders the ability of the countries to build strong economies.

☑ **READING PROGRESS CHECK**

Identifying Point of View What might happen in North Africa if young people grow impatient with the slow rate of economic growth? Why?

Women students meet for class at Cairo, Egypt's Al-Azhar University, the world's chief center for Islamic learning. In Egypt, women may work outside the home, attend universities, vote, and run for office. However, opportunities for women still lag behind those for men in education and the labor market.

▶ **CRITICAL THINKING**
Analyzing How does the issue of women's literacy affect economies in North Africa?

North Africa's Future

GUIDING QUESTION *How will North Africa address the problems it faces?*

Powerful new social movements have swept through the region of North Africa in recent years. They have led to major political changes in three countries and put pressures on the governments of the other two.

Political Issues

Two political forces are strong in the region of North Africa. One is a push for democracy. Many North Africans have grown more and more frustrated with their leaders. They think the leaders focused more on building their own power than on building the economy and improving their countries. Many question the government's harsh treatment of people who criticize their countries' leaders. Some leaders are calling for the different groups to learn to work together to avoid the conflicts that pull societies apart.

Academic Vocabulary

factor a cause

Young people in Benghazi, Libya, yell protest slogans against dictator Muammar al-Qaddafi. Clashes between street demonstrators and armed government forces in February 2011 led to much bloodshed. A civil war broke out, and rebel groups eventually overthrew Qaddafi.

▶ **CRITICAL THINKING**

Describing What happened in Libya after the fall of Qaddafi?

The second force was an increase in Islamic fundamentalism. Some strict Muslims want laws changed to conform to the rules of Islam. They want to see an end to Western influences on their culture. The political party of the Muslim Brotherhood gained a majority in Egypt's parliament in the 2011 elections. It also won a majority in Morocco and a large share of seats in Tunisia. These forces helped bring about the Arab Spring of 2010 and 2011. They have left conditions across the region uncertain.

Egypt began writing a new **constitution** in 2012. A constitution is a set of rules for a nation and its government. Egypt's new government could give more power to the parliament, the lawmaking body. It is not clear how well this new government will work or what groups will control it, though.

For nearly 20 years, Algeria has undergone brutal conflict between Islamist groups and the government and its forces. As many as 100,000 people have died in the fighting. As in Morocco, the government was able to keep power after the Arab Spring, but it had to promise to reform the political system.

By 2012, Libya's victorious rebels were working on making a new government. They also faced the need to rebuild much of the country after the civil war. In 2012, leaders in eastern Libya said they wanted self-rule in their part of the country. Although they said that they did not wish to divide the country or to keep their area's oil wealth for themselves, the move raised the possibility of continued conflict in Libya.

Islam in the Modern World

Many Muslims in the region worry about the impact of Western culture on their lands. They think that Western entertainment conflicts with Islamic values. They also disagree with Western ideas about women's rights.

Women in North Africa generally have more rights than those in other Muslim lands. In Tunisia, for instance, they can own businesses and have their own bank accounts. About half of all university students in Tunisia are women. Women may lose some of these rights if extreme Muslim leaders take control of the governments.

Several million of Egypt's Coptic Christians have grown more worried about their position in recent years as well. Some Muslim extremists have attacked them and bombed churches. Early in 2012, the longtime head of the Coptic church died. He had led the church for nearly 40 years in relative peace until near the end of his life. His death increased the uncertainty for Copts in that area.

Relations with Other Nations

Egypt broke ranks with other Muslim nations in 1979 when it signed a peace treaty with Israel. It has also developed close ties with the United States since then. That friendship has come under increasing criticism from Muslim fundamentalists. Morocco has also had close relations with the United States. Its government has been criticized for this as well.

These situations raise more questions about what will happen if Muslim conservatives gain power. Will the new governments reject close ties with the United States? Will they take steps against Israel?

The situations in Algeria and Libya also are uncertain. Will new governments there be less willing to sell oil to the United States? For what purposes will they use the money they earn from selling oil? The answers to these questions will help to shape the future of North Africa and the world.

Include this lesson's information in your Foldable®.

☑ **READING PROGRESS CHECK**

Analyzing Why were the results of the Arab Spring different in Algeria and Morocco compared with the other countries of the region?

LESSON 3 REVIEW CCSS

Reviewing Vocabulary
1. Is it important for an economy to be *diversified*? Why or why not?

Answering the Guiding Questions
2. ***Determining Central Ideas*** Why do you think many Muslims worry about the impact of Western culture on their lands?

3. ***Describing*** How is the relatively young population connected to the economic issues in these nations?

4. ***Analyzing*** About half of Egypt's people live in rural areas. Most of them are farmers. What impact does that have on Egypt's economy? Why?

5. ***Identifying Point of View*** Why is the political situation in North Africa important to the United States?

6. ***Argument Writing*** Do you think the most serious issues facing North Africa are political, social, or cultural? Write a paragraph explaining why.

Chapter 9 ACTIVITIES

Directions: Write your answers on a separate piece of paper.

1 Use your FOLDABLES to explore the Essential Question.
INFORMATIVE/EXPLANATORY WRITING Briefly describe the population distribution in the region of North Africa and suggest a likely reason for the distribution.

2 **21st Century Skills**
INTEGRATING VISUAL INFORMATION With a partner, create a set of flash cards showing an outline of the five North African countries combined and outlines of the individual countries. Make five separate cards with the names of the capital cities. Devise a game using the flash cards, and exchange games with another pair of classmates. After playing both games, work with the other pair to turn the flash card games into a computer game.

3 **Thinking Like a Geographer**
INTEGRATING VISUAL INFORMATION On an outline map of North Africa, indicate the region's climates. Include the rain shadow areas on the map key.

4 **GEOGRAPHY ACTIVITY**

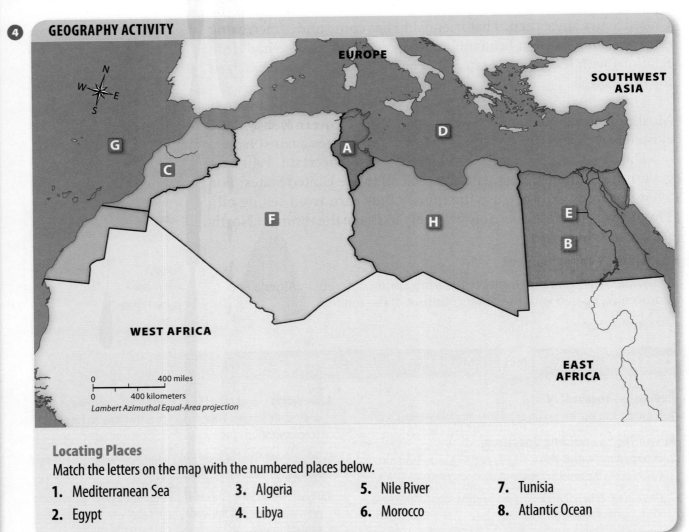

Locating Places
Match the letters on the map with the numbered places below.
1. Mediterranean Sea
2. Egypt
3. Algeria
4. Libya
5. Nile River
6. Morocco
7. Tunisia
8. Atlantic Ocean

288 Chapter 9

Chapter 9 Assessment

REVIEW THE GUIDING QUESTIONS

Directions: Choose the best answer for each question.

1. One nickname for ancient Egypt was
   A. serpent of the sea.
   B. wadi of the floods.
   C. gift of the Nile.
   D. delta dawn.

2. Of the North African countries, Libya has the most
   F. olives.
   G. oil.
   H. water.
   I. cedarwood.

3. Hieroglyphics were
   A. equipment used for building pyramids.
   B. Muslim political and religious leaders.
   C. pictures that represented sounds or words.
   D. spices burned for religious ceremonies.

4. The Berbers and Egyptians became linked to the Muslim world in the A.D. 1000s by
   F. the defeat at Carthage.
   G. the power of the caliphs.
   H. the Arabic language and Islamic learning.
   I. the pharaoh's desire for more territory.

5. The official language of the five North African nations is
   A. Coptic.
   B. French.
   C. Afrikaner.
   D. Arabic.

6. Today's Islamic fundamentalists in North Africa want to
   F. return to their families' farms.
   G. convert the citizens of Israel.
   H. end Western influence on the Islamic culture.
   I. turn around the Tunisian economy.

Chapter 9 ASSESSMENT (continued)

DBQ ANALYZING DOCUMENTS

7 CITING TEXT EVIDENCE Read the following passage about recent changes in Libya's government:

"*In March 2011, a Transitional National Council (TNC) was formed . . . with the stated aim of overthrowing the Qaddafi regime and guiding the country to democracy. . . . Anti-Qaddafi forces in August 2011 captured the capital, Tripoli. In mid-September, the [United Nations] General Assembly voted to recognize the TNC as the legitimate interim governing body of Libya.*"

—from CIA World Factbook, "Libya"

How did the Transitional National Council view its role in Libya?

A. as an ally of Qaddafi
B. as supporters of Libya's former king
C. as a temporary government
D. as social reformers

8 ANALYZING What happened to the Qaddafi government in 2011?

F. It remained in control of the country.
G. It relocated to a new capital.
H. It forged an alliance with the TNC.
I. It fell out of power.

SHORT RESPONSE

"*Morocco, on account of the invasions of Arabs and the exterior adventures of Moorish kings, was strongly influenced by Middle Eastern culture and the culture of the Andaluz [Muslim-ruled Spain]. The Arabs learned [cooking] secrets from the Persians and brought them to Morocco; from Senegal and other lands south of the Sahara came caravans of spices. Even the Turks made a contribution.*"

—from Paula Wolfert, *Couscous and Other Good Food From Morocco* (1973)

9 DETERMINING CENTRAL IDEAS What is the main idea of this passage?

10 DESCRIBING What kinds of contact by different groups led to these influences on Moroccan culture?

EXTENDED RESPONSE

11 INFORMATIVE/EXPLANATORY WRITING Research the Arab Spring of 2011. Write an essay contrasting the current situation in the countries that were involved in the Arab Spring with their situation before the upheaval. Did the "spring" last? Did the citizens of these countries gain or lose what they were trying to achieve? Are their lives better or worse today than they were before 2011? Do any of them now have a stable, democratic government?

Need Extra Help?

If You've Missed Question	1	2	3	4	5	6	7	8	9	10	11
Review Lesson	1	1	2	2	3	3	2	2	2	2	2

EAST AFRICA

ESSENTIAL QUESTIONS • How does geography influence the way people live?
• Why do people trade? • Why does conflict develop?

networks
There's More Online about East Africa.

CHAPTER 10

Lesson 1
Physical Geography of East Africa

Lesson 2
History of East Africa

Lesson 3
Life in East Africa

The Story Matters...

Some of Africa's earliest kingdoms developed in East Africa, where trade in gold and ivory brought great wealth. Since ancient times, thriving trade has fostered interaction among different cultures, influencing language and religion and creating much ethnic diversity across the region. The landscape of East Africa also has great diversity—from the Serengeti Plain and the Great Rift Valley to the highlands in Ethiopia and Kilimanjaro in Kenya.

Go to the Foldables® library in the back of your book to make a Foldable® that will help you take notes while reading this chapter.

Teenage girl from the East African country of Somalia

Chapter 10
EAST AFRICA

Some of Africa's important early civilizations flourished in East Africa. Many of the countries have been scarred by conflict in recent years.

Step Into the Place

MAP FOCUS Use the map to answer the following questions.

1. **THE GEOGRAPHER'S WORLD** Which three East African countries share Lake Victoria?

2. **PLACES AND REGIONS** What is the capital city of Kenya?

3. **THE GEOGRAPHER'S WORLD** The Tekeze is a major river in what country?

4. **CRITICAL THINKING** Integrating Visual Information What country is cut off from the sea by Eritrea, Djibouti, and Somalia?

INACTIVE VOLCANO Snowcapped Kilimanjaro looms over savanna plains near the border of Tanzania and Kenya. The mountain is made up of three volcanic cones, all inactive.

WAR-TORN CITY Ruined buildings line an Indian Ocean beach in Mogadishu, the capital of Somalia. Since the early 1990s, various armed groups have fought over Somalia.

Step Into the Time

TIME LINE Using at least two events on the time line, write a paragraph describing how trade influenced the development of East Africa.

30,000–20,000 B.C. Ancient people live in what is now Sudan

800 B.C. Kingdom of Kush develops along the Nile River

A.D. 400 Kingdom of Aksum prospers from trade

1100s Muslim settlements multiply in East Africa

networks

There's More Online!

- ☑ **IMAGES** Glaciers in East Africa
- ☑ **MAP** Desertification of the Sahel
- ☑ **SLIDE SHOW** The Nile River's Source
- ☑ **VIDEO**

Reading HELPDESK

Academic Vocabulary
- consist

Content Vocabulary
- rift
- desertification
- hydroelectric power
- geothermal energy

TAKING NOTES: *Key Ideas and Details*

Identifying As you study the lesson, use a web diagram like this one to list information about the land and water features of the region.

Lesson 1
Physical Geography of East Africa

ESSENTIAL QUESTION • *How does geography influence the way people live?*

IT MATTERS BECAUSE
East Africa offers a rugged, beautiful landscape and different climates. The region provides variety, potential, and considerable challenges for economic development.

Land and Water Features

GUIDING QUESTION *What makes the ecosystem of East Africa diverse?*

The region of East Africa **consists** of 11 countries. Sudan and South Sudan dominate the northern part of the region. Eritrea, Djibouti, Somalia, and Ethiopia are located in the northeast. This area is called the Horn of Africa because it is a horn-shaped peninsula that juts out into the Arabian Sea. Three countries occupy the central and southern parts of the region: Kenya, Tanzania, and Uganda. Finally, in the western sector lie the landlocked countries of Rwanda and Burundi. East Africa offers a rugged, beautiful landscape that has great variety.

Landforms
The Great Rift Valley is the most unusual feature of East Africa's physical geography. Sometimes it is called the Great Rift system because it is not one single valley. Rather, it is a series of large valleys and depressions in Earth's surface. These are formed by long chains of geological faults. The Great Rift started forming about 20 million years ago when tectonic plates began to tear apart from one another. Africa was once connected to the Arabian Peninsula. But as the two **rifted** apart, or separated from one another, the land in between

sank and was filled by the Red Sea. Eventually, all of East Africa will separate from the rest of Africa, and the Red Sea will fill the rift.

The Great Rift system's northern end is in Jordan in Southwest Asia. From Jordan, it stretches about 4,000 miles (6,437 km) to its southern end in Mozambique in southeastern Africa. The rift has an average width of 30 miles to 40 miles (48 km to 64 km).

The rift system has an eastern and western branch in East Africa. The eastern Rift Valley—the main branch—runs from Southwest Asia along the Jordan River, Dead Sea, and Red Sea. It continues through the Danakil plain in Ethiopia. It is one of the hottest and driest places on Earth, and earthquakes and volcanic activity occur here regularly. Long, deep cracks develop in Earth's surface as the tectonic plates rift apart.

As the eastern Rift Valley continues south from the Danakil plain, the conditions are not as severe. It takes the form of deep valleys as it extends into Kenya and Tanzania, and down to Mozambique. The shorter western Rift Valley stretches from Lake Malawi in the south through Uganda in the north through a series of valleys. A chain of deep lakes that includes Lake Tanganyika, Lake Edward, and Lake Albert marks the western rift's northward path.

Along the branches of the Great Rift Valley, much volcanic and seismic activity occurred. The largest volcanoes are located on the eastern Rift. These include Mount Kenya and Kilimanjaro. Kilimanjaro is on the border between Kenya and Tanzania. With a summit of 19,341 feet (5,895 m), Kilimanjaro is the tallest mountain in Africa. Its summit is covered with snow year-round, even though the mountain is near the Equator.

Sudan is home to vast plains and plateaus. The northern part of the country is desert covered in sand or gravel. Somalia lies in the eastern part of the region, along the Indian Ocean.

Academic Vocabulary

consist to be made up of

This aerial view shows a section of the floor of the eastern Rift Valley in Kenya. Many fault lines appear in the valley. Hardened lava from volcanoes and openings in the ground also mark the landscape.

▶ **CRITICAL THINKING**

Describing How will East Africa eventually be affected by the Rift's tectonic plate activity?

The palace of Sudan's president in Khartoum stands near where the Blue Nile joins the White Nile. The White Nile is named for the light-colored clay sediment found in its waters. The Blue Nile's name comes from the river's appearance during flood season when the water level is high.

▶ **CRITICAL THINKING**
Describing Where do each of the two Nile tributaries begin?

Somalia is also an extremely dry area. The country is made up largely of savanna and semidesert. To the north of Somalia lies the small country of Djibouti. Located on the coast between the Red Sea and the Gulf of Aden, Djibouti displays a highly diverse landscape. It has rugged mountains and desert plains.

South of Sudan, at the western edge of Uganda, the Ruwenzori Mountains divide that country from the Democratic Republic of the Congo. These peaks are sometimes called the "Mountains of the Moon." Mountains give way to hills in small, landlocked Rwanda. It is known as the "land of a thousand hills" for its beautiful landscape.

Bodies of Water

The longest river in the world is the Nile (4,132 miles or 6,650 km). The Nile Basin includes parts of many countries in the East African region: Tanzania, Burundi, Rwanda, Kenya, Uganda, Ethiopia, South Sudan, and the Sudan. Beginning in the 1800s, European explorers made numerous expeditions in attempts to find the source of the Nile River. The great river was discovered to have two sets of headwaters. One of them, the Blue Nile, rises in the northern highlands of Ethiopia. The other source, the White Nile, begins in Lake Victoria and runs through Lake Albert. The White Nile then passes through the swampy wetlands of central South Sudan, a huge area called the Sudd.

In northern Sudan, the Blue Nile and the White Nile meet at the city of Khartoum. The great river then runs northward through Egypt and empties into the Mediterranean Sea. Other than the Nile,

East Africa has few important rivers. This is due to the intermittent rainfall and the high temperatures in many areas of the region.

In the late 1970s, the swampy Sudd was the focus of a huge construction project called the Jonglei Canal. This channel was designed to avoid the Sudd. The goal was to allow the headstreams of the White Nile to flow more freely. Instead of the water spreading across the Sudd and slowly moving through it, the canal would allow more water to flow downstream and reach Sudan and Egypt. That would support more agriculture and better city services in those countries. But it would also damage the wetland environment of the Sudd. Fisheries could collapse and go extinct. Construction was suspended in 1983. The project could not continue because civil war in Sudan made it too dangerous.

Many of the lakes in East Africa are located near the Great Rift Valley. The largest lake on the continent of Africa is Lake Victoria. This lake lies between the western and the eastern branches of the Great Rift. The lake stretches into three countries: Uganda and Kenya in the north and Tanzania in the south. With an area of 26,828 square miles (69,484 sq. km), Lake Victoria is the second-largest freshwater lake in the world, after Lake Superior in the United States. For such a large body of water, Lake Victoria is relatively shallow. Its greatest known depth is about 270 feet (82 m). The lake is home to more than 200 species of fish. Of these, tilapia has the most economic value.

Another important lake in the region is Lake Tanganyika. This long, narrow body of water is located south of Lake Victoria, between Tanzania and the Democratic Republic of the Congo. The lake is only 10 to 45 miles (16 km to 72 km) wide, but very long. Measuring 410 miles (660 km) north to south, it is the world's longest freshwater lake. With a maximum depth of 4,710 feet (1,436 m), it is also the second deepest. Only Lake Baikal in Russia is deeper than Lake Tanganyika.

Farther south is Lake Malawi. It is the third-largest lake in the East African Rift Valley. The lake lies mainly in Malawi and forms part of that country's border with Tanzania and Mozambique.

Fishers leave the eastern shore of Lake Victoria by boat early in the morning to fish for tilapia and Nile perch. With its many fish species, Lake Victoria supports Africa's largest inland fishery.

✓ **READING PROGRESS CHECK**

Identifying What caused the striking physical features of the Great Rift Valley in East Africa?

Climates of East Africa

GUIDING QUESTION *How does climate vary in East Africa?*

Climate varies widely in the East African region. Temperature and rainfall can be quite different from one local area to another. The major factors explaining these variations include latitude, altitude, distance from the sea, and the type of terrain, such as mountains, highlands, desert, or coastal plains.

Temperatures

The diverse physical features of East African geography are matched by an extremely varied climate. In general, temperatures tend to be warmer toward the coast and cooler in the highlands. Sudan, Djibouti, and Somalia have high temperatures for much of the year. High mountains such as Kilimanjaro and the peaks of the Ruwenzori Range have had glaciers for thousands of years. Due to climate change, however, these glaciers are melting. Some experts predict that the glaciers of Kilimanjaro will completely disappear over the next 20 years.

The climate is always spring-like in the highlands of Kenya and Uganda. As a whole, however, Kenya and Uganda display considerable variations in climate. These variations depend on factors such as latitude, elevation, wind patterns, and ocean currents.

Rainfall

In many parts of East Africa, rainfall is seasonal. This is especially true close to the Equator. Wet seasons alternate with dry ones. For example, on the tropical grasslands, or savannas, of Kenya and

The Savoia glacier is located along the border of Uganda and the Democratic Republic of the Congo. Many scientists are concerned about the effects of climate change on the Ruwenzori glaciers. Around 1900, some 43 glaciers were distributed over 6 mountains in the range. Today, fewer than half of these glaciers still exist, on only 3 of the mountains. The rest have melted.

▶ **CRITICAL THINKING**
Analyzing Why do temperatures tend to be cool in inland East Africa despite the region's closeness to the Equator?

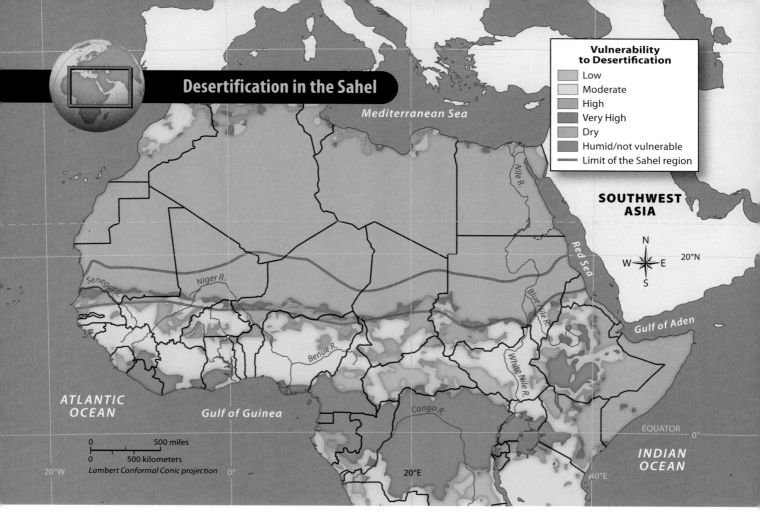

Desertification in the Sahel

Tanzania, two rainy seasons occur in most years. These are the "long rains" of April and May and the "short rains" of October and November. The months in between these periods are dry, with little or no rainfall.

Rainfall in the region, however, can be unpredictable. Sparse rainfall can result in severe drought. In 2011, for example, Somalia suffered one of the worst droughts in its history. Political instability in that country made the effects of the drought especially severe. Observers estimated that 13 million people struggled to survive in the countries of Somalia, Ethiopia, Djibouti, and Kenya.

Another urgent issue in the region is **desertification**, or the process by which agricultural land is turned into desert. This process occurs when long periods of drought and unwise land use destroy vegetation. The land is left dry and barren. During the past half century, desertification has affected much of the Sahel. The Sahel is the "edge," or border area, between the Sahara and the countries farther to the south. Two such border nations in East Africa are Sudan and South Sudan.

✓ **READING PROGRESS CHECK**

Determining Central Ideas What generalization can you make about the variations in temperature in East Africa?

MAP SKILLS

1. **PLACES AND REGIONS** Based on the legend, most of the Sahel is at what level of desertification?

2. **THE GEOGRAPHER'S WORLD** What causes desertification?

Chapter 10 299

Workers collect salt at Lake Assal in Djibouti. Salt covers everything, so very little vegetation is able to grow along the lake's shoreline. Located in the hot desert, the lake's area has summer temperatures as high as 126°F (52°C).

Identifying What other mineral resources are found in East Africa?

Resources of East Africa

GUIDING QUESTION *Which natural resources are important in East Africa?*

The natural resources of a region are closely linked to its economy and people's way of life. Settlement patterns in a geographical area have often been shaped by that area's natural resources. Important resources in East Africa are minerals, energy sources, landscapes, and wildlife. The ability of some countries to exploit these resources, however, has been hampered by political issues.

Mineral Resources

Mineral resources in East Africa include small gold deposits along the rifts in Kenya, Uganda, and Tanzania; gemstones like sapphires and diamonds in Tanzania; and tin in Rwanda. Ethiopia and Uganda produce lumber. Lake Assal in Djibouti, located about 500 feet (152 m) below sea level, is the world's largest salt reserve, with more than 1 billion tons of salt. This lake is located at the lowest point in Africa.

Energy Resources

Energy resources in East Africa include coal in Tanzania, as well as petroleum in Uganda, South Sudan, and northwestern Kenya. East Africa's energy potential has yet to be realized, though. For example, Sudan has the opportunity to develop **hydroelectric power**, or the production of electricity through the use of falling water. Hydroelectric power is already used in Kenya and Tanzania.

Likewise, Kenya and Djibouti are favorable locations for the development of **geothermal energy**. This type of energy comes from underground heat sources, such as hot springs and steam. In Kenya, an international group of companies is working with the government to develop geothermal energy sources. If they are successful, 30 percent of the country's energy needs could be met by geothermal energy by the year 2030. In Djibouti, geothermal energy production is expected to begin by the year 2014.

In East Africa, management of energy resources and energy use often has been inconsistent and uneven. Major cities gobble up much of the energy that is produced. Energy is often unavailable in rural areas.

Land and Wildlife

Besides mineral and energy resources, East Africa's land and wildlife are important assets. The soils in the region are not especially rich for agriculture, and farming is challenging. The breathtaking scenery of the Great Rift Valley, however, is an important tourist resource.

East Africa is also home to the greatest assemblage of wildlife in the world. Many national parks and wildlife sanctuaries are found in the region. Perhaps the most well-known wildlife reserves are located in Kenya and Tanzania. An outstanding example is the Serengeti Plain; this vast area, larger than the state of Connecticut, consists of tropical savanna grasslands. Two internationally famous national parks are located in East Africa—Serengeti National Park in Tanzania and the Masai Mara National Reserve in Kenya. These parks harbor lions, leopards, cheetahs, giraffes, zebras, elephants, and dozens of species of antelope.

Every year, thousands of tourists pour in from all over the world to see the marvel of the Great Migration. In this mass movement, more than 1 million animals travel hundreds of miles in search of fresh grazing land. The spectacular wildlife of East Africa makes an important contribution to the economy of the region.

Think Again

Animals involved in the Great Migration on the Serengeti Plain travel together.

Not True. Nature employs a more sophisticated system. The three major migrating species are zebras, wildebeests, and Thomson's gazelles. These species migrate in a succession. First come the zebras. They consume crude, coarse, high grasses. Then the wildebeests follow, grazing on the lower shoots exposed by their predecessors. Last are the smaller Thomson's gazelles, antelopes that eat tender, fine shoots close to the ground.

☑ **READING PROGRESS CHECK**

Identifying What two promising alternatives might help improve energy supplies in the East African region?

Include this lesson's information in your Foldable®.

LESSON 1 REVIEW

Reviewing Vocabulary
1. What causes the process of *desertification*?

Answering the Guiding Questions
2. ***Describing*** What are the differing characteristics that make Lake Victoria and Lake Tanganyika noteworthy bodies of water, both in East Africa and in the world as a whole?

3. ***Analyzing*** How might desertification affect the economy in a region?

4. ***Identifying*** How are energy supplies distributed in East Africa?

5. ***Informative/Explanatory Writing*** Write a letter to a friend or a relative explaining why you want to visit East Africa to see the region's wildlife.

networks

There's More Online!

- ☑ **IMAGE** British at Omdurman
- ☑ **MAP** African Trade Routes and Goods
- ☑ **SLIDE SHOW** Ancient Africa
- ☑ **VIDEO**

Reading HELPDESK

Academic Vocabulary
- impact

Content Vocabulary
- tribute
- imperialism
- genocide
- refugee

TAKING NOTES: *Key Ideas and Details*

Organizing As you study the lesson, use a chart like this one to list important facts about the places.

Place	Facts
Nubia/Kush	
Aksum	
Coastal City-States	

Lesson 2
History of East Africa

ESSENTIAL QUESTION • *Why do people trade?*

IT MATTERS BECAUSE
East Africa has been a center of trade since ancient times. Throughout much of its history, East Africa has attracted people from many other continents.

Kingdoms and Trading States

GUIDING QUESTION *How has the history of trade impacted the region?*

Trade was important in the ancient kingdoms in East Africa. Contact between East Africa and other areas brought together people from different civilizations. Trade also resulted in the spread of Christianity and Islam into the region.

Ancient Nubia

The ancient region of Nubia was located in northeastern Africa, below ancient Egypt. The region stretched southward along the Nile River valley almost to what is now the Sudanese city of Khartoum. The region was bounded by the Libyan Desert in the west and by the Red Sea in the east. The Nile River was the pathway by which Nubia and the powerful empire of Egypt interacted.

In about 1050 B.C., a powerful civilization arose in Nubia. This was known as Kush. The Egyptians traded extensively with the Kushites, purchasing copper, gold, ivory, ebony, slaves, and cattle. The Kushites, in turn, adopted many Egyptian customs and practices. For example, they built pyramids to mark the tombs of their rulers and nobles.

During the final centuries of their civilization, the Kushites were isolated from Egypt. As a result, they turned increasingly to other African people south of the Sahara for trade and cultural contact. Around A.D. 350, Kush was

302 Chapter 10

conquered by Aksum, a powerful state in what is now northern Ethiopia.

Aksum

The date of Aksum's establishment is uncertain but it might have been around 1000 B.C. The people of Aksum derived their wealth and power primarily from trade. Aksum was strategically located, and it controlled the port city of Adulis on the Red Sea. At its height of power, Aksum was the most important trading center in the region. Its trading connections extended all the way to Alexandria on the Mediterranean Sea. Aksum traders specialized in sea routes that connected the Red Sea to India.

Through the port of Adulis flowed gold and ivory, as well as raw materials. It is possible that Aksum sold captives for the slave trade. Aksum traded glue, candy, and gum arabic, a substance from acacia trees that today is used in the food industry. Christianity spread from its origin in Jerusalem along the trade routes. The Aksum kings adopted Christianity as their religion.

Trade Cities

Beginning around the A.D. 900s, after the decline of Aksum, Arabs settled on the East African coast of the Indian Ocean. The religion of Islam grew steadily more important in the region. At the same time, the Arabic and Bantu languages mingled to create a new language. This language is known as Swahili. The name comes from an Arabic word meaning "coast dwellers." Swahili is widely spoken today in Tanzania and Kenya, as well as in some other countries.

Gradually, the coastal settlements formed independent trading states. From coastal Somalia southward, along the shores of Kenya and Tanzania, these city-states prospered. They included Mogadishu, Lamu, Malindi, and Mombasa.

Many of the pyramids of ancient Kush still stand in present-day Sudan. Near the pyramids, the Kushites built a capital city called Meroë. Archaeologists have uncovered some of the remains of Meroë, including a royal palace, temples, and mud-brick homes.

▶ **CRITICAL THINKING**
Determining Central Ideas What do the pyramids of Meroë reveal about Kushite culture?

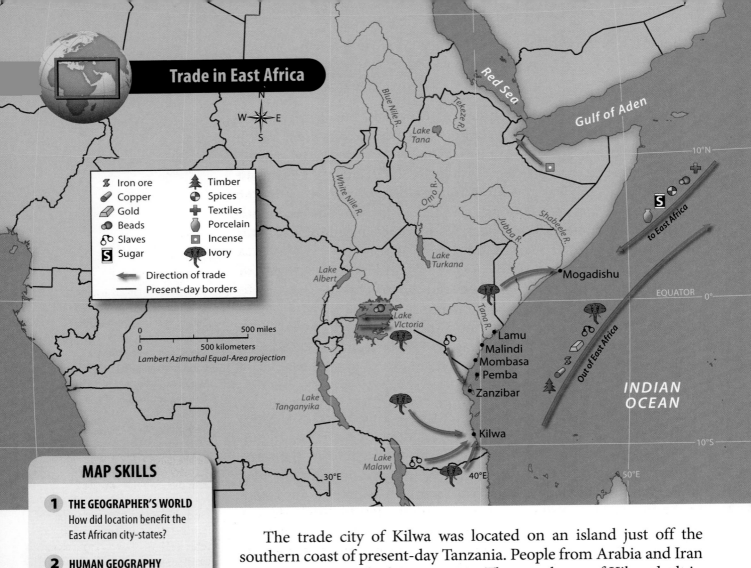

Trade in East Africa

MAP SKILLS

1. **THE GEOGRAPHER'S WORLD** How did location benefit the East African city-states?
2. **HUMAN GEOGRAPHY** What part did inland Africa play in the region's trade?

The trade city of Kilwa was located on an island just off the southern coast of present-day Tanzania. People from Arabia and Iran founded Kilwa in the late A.D. 900s. The merchants of Kilwa dealt in copper, iron, ivory, and gold. They exchanged these goods for products from many lands, including Chinese porcelain and Indian cotton.

Kilwa was a walled city. Its ruler lived in an impressive palace. For two centuries, the city was probably the wealthiest trading center in East Africa. The fourteenth-century traveler Ibn Battuta praised Kilwa as a beautiful city. At the time of Ibn Battuta's visit, Kilwa was ruled by Abu al-Mawahib. The sultan was so generous that people called him "the father of gifts."

✓ READING PROGRESS CHECK

Identifying Compare the economies of the coastal city-states in East Africa to those of the kingdom of Aksum.

The Colonial Era

GUIDING QUESTION *What was the effect of colonization on East Africa?*

Until the late 1800s, most Europeans knew little or nothing about Africa. Two of the continent's most famous explorers were Henry Morton Stanley and David Livingstone. In 1878 Stanley published a popular travel book about his adventures in Africa. The book's title was *Through the Dark Continent*. The goal of Stanley's journey was to locate Livingstone, a medical missionary. Livingstone had traveled to Africa in the hope of locating the source of the Nile River.

European Traders

Just before 1500, the European age of discovery began to **impact** East Africa. Among the European countries, Portugal took the lead in overseas exploration. Along with other Europeans, the Portuguese established a sea route to India. From Europe, they sailed south along the west coast of Africa and then along the east coast of Africa. Then, they sailed along the coast of Arabia and on to India. This was a much easier and less expensive way to trade with India than any of the overland trade routes. In this way, the Portuguese were able to bring back many valuable spices from India.

As trade increased, the Portuguese began to demand **tribute**, or a regular tax payment, from the East African trading cities. The Portuguese had religious as well as economic motives; they believed that Christianity should replace Islam as the region's religion. Portuguese influence in the region did not last long, however. The Portuguese could not withstand attacks by African groups in the region. Other European countries became interested in colonizing Africa.

European Colonial Rule

In the late 1800s, European leaders set out a plan to dominate and control the continent of Africa. The action by which one nation is able to control another smaller or weaker nation is known as **imperialism**.

Academic Vocabulary

impact an effect or an influence

The Battle of Omdurman was fought in Sudan in 1898. In this battle, British and Egyptian forces—equipped with modern guns—defeated a much larger Mahdist army that used older weapons.

▶ **CRITICAL THINKING**
Integrating Visual Information
How does Hale's painting present the battle scene? What view of imperialism does it seem to support?

A painting in traditional Ethiopian style shows King Menelik II receiving ammunition for his army. Menelik worked to bring modern ways to Ethiopia. He especially wanted to prepare his army to successfully resist European invaders.

Africa was carved up into colonies. The reasons for colonization included economic profit, access to raw materials, and the opening of new markets. These reasons also included national pride, the protection of sea routes, the maintenance of the balance of power, and a quest to convert Africans to Christianity.

Occasional rebellions challenged European colonial rule. An especially bloody rebellion occurred against British and Egyptian domination in Sudan. Muhammad Ahmad, a religious and military leader, declared that he was the Mahdi, or redeemer of Islam. Mahdist forces succeeded in capturing Khartoum, the Sudanese capital. They established a new state there. In 1898 the British succeeded in reasserting their control of the region.

Independent Ethiopia

The revolt against foreign influence in Sudan eventually resulted in failure. In Ethiopia, however, the desire for independence prevailed. Italy had colonized the neighboring territory of Eritrea along the Red Sea coast. In 1889 the Italians signed a treaty with the Ethiopian emperor, Menelik II. Over the next few years, Italy claimed that, according to one provision of this treaty, it had the right to establish a "protectorate" in Ethiopia.

Menelik firmly denied these claims. He rejected the treaty in 1893. The Italian governor of Eritrea finally launched a major military attack in response in 1896. At the Battle of Adwa on March 1 of that year, Menelik defeated the Italian army. This conflict was one of the most important battles in African history. After the Battle of Adwa, the European powers had no choice but to recognize Ethiopia as an independent state. Physical geography played an important role in Ethiopia's ability to remain independent. Rugged mountains with difficult terrain provided a barrier that was difficult for attacking forces to overcome.

✅ **READING PROGRESS CHECK**

Explaining What was the significance of Menelik II's victory at the Battle of Adwa in 1896?

Independence

GUIDING QUESTION *How did the countries of East Africa gain their independence?*

After the end of World War II in 1945, a movement ensued to end colonialism in Africa, Asia, and Latin America. In East Africa, particularly, Europeans were seen as disrupting traditional life. In addition, European countries were weakened by the fighting in World War II. Because of these pressures, Europeans granted East African colonies their independence in the 1960s. However, many of the former colonies faced difficulties in establishing their own countries.

New Nations Form

The early 1960s was a turning point for East Africa. During the period from 1960 to 1963 alone, six East African countries obtained independence: Somalia, Kenya, Uganda, Tanzania, Rwanda, and Burundi.

The achievement of independence in Kenya and Tanzania was especially important. Kenya had been a British colony for about 75 years. British plantation owners dominated the economy. They disrupted the traditional East African agricultural system. Local village agriculture was replaced by the production of cash crops, such as coffee and tea, on a large scale. Native people, such as the Kikuyu, were driven off the land. The British also controlled the government.

A nationalist named Jomo Kenyatta led the political protest movement in Kenya and negotiated the terms of independence for his country. In late 1963, Kenya became independent. Jomo Kenyatta served as the country's first prime minister and later as its president.

Tanzania also sought independence. Before independence, the country was called Tanganyika.

As independent Kenya's first leader, Jomo Kenyatta brought stability and economic growth to the country. When appearing in public, Kenyatta often carried a fly whisk, a symbol of authority in some traditional African societies.

▶ **CRITICAL THINKING**
Describing How did Kenya win its independence from British rule?

Villagers in South Sudan try to put out fires after warplanes from neighboring Sudan raided the area in early 2012. A year earlier, South Sudan had gained independence from Sudan following years of civil war. However, tensions remained high and conflict continued.

▶ **CRITICAL THINKING**
Describing Why have some African countries after independence faced civil wars and conflicts with neighboring countries?

When Germany was defeated in World War I, Tanganyika came under British control. Independence was the ultimate goal for Tanganyika—a goal it reached in late 1961. Three years later, the country merged with Zanzibar, and its name was changed to Tanzania.

Highland Countries

The Highland areas had a difficult road to independence. Many ethnic groups in the former colonies were often in conflict with one another. Ethnic tensions have long simmered in Rwanda and Burundi. These countries are home to two rival ethnic groups. The Hutu are in the majority there, and the Tutsi are a minority. In the 1990s, the Hutu-dominated government of Rwanda launched an attack on the Tutsi that amounted to **genocide**—the slaughter of an entire people on ethnic grounds. Hundreds of thousands of people were killed.

Bloodshed also stained the history of Uganda after independence. From 1971 to 1979, the country was ruled by the military dictator Idi Amin. Cruelty, violence, corruption, and ethnic persecution marked Amin's regime. Human rights groups estimate that hundreds of thousands of people lost their lives under his rule. Amin was finally forced to flee into exile. He died in 2003.

The Horn of Africa

The history of Somalia since independence in 1960 offers another example of the problems East African countries have faced. Since the 1970s, Somalia has been scarred by civil war. Border disputes with Ethiopia have also increased instability. Rival clan factions have engaged in bitter feuds. Drought has brought famine to much of the country. In late 1992, the United States led a multinational intervention force in an effort to restore peace to the country. The civil war in Somalia, however, remained unresolved.

The instability, misery, and violence in Somalia also have affected neighboring countries. Thousands of **refugees**, for example, have made their way into Kenya. A refugee is a person who flees to another country for safety.

Elsewhere in the Horn of Africa, more than 30 years of fighting have marked the recent history of Eritrea. This country achieved independence in 1993 after a long struggle with Ethiopia. Access to the sea was an important territorial issue in this conflict. In the years since independence, Eritrea has undertaken military conflicts with Yemen and resumed attacks on Ethiopia. The country is unable to provide enough food for its people. Furthermore, economic progress has been limited because many Eritreans serve in the army rather than in the workforce.

A New Nation

Africa's newest country emerged as a result of civil war. Sudan won independence from Egyptian and British control in 1956. Leaders in southern Sudan were angered because the newly independent Sudanese government had failed to carry out its promise to create a federal system. Southern leaders also feared that the new central government would try to establish an Islamic and Arabic state.

Religion was also an issue that generated conflict. Most people in Sudan are Muslim, but in the southernmost 10 provinces, most people follow traditional African religious practices or the Christian religion. Economic issues are also a problem. The southern provinces hold a large share of the area's petroleum deposits. As a result of the civil war, the country of South Sudan became independent from Sudan in 2011.

Include this lesson's information in your Foldable®.

✓ **READING PROGRESS CHECK**

Determining Central Ideas How has civil war played an important part in the recent history of East Africa?

LESSON 2 REVIEW

Reviewing Vocabulary
1. What were some of the factors that led European nations to practice *imperialism* in Africa?

Answering the Guiding Questions
2. **Identifying** Discuss two important events that occurred in the history of the Ethiopian kingdom of Aksum.

3. **Identifying** Which two countries took the lead in the European colonization of East Africa in the late 1800s?

4. **Describing** What have been some of the major problems that East African countries have faced in building their nations after achieving independence?

5. **Narrative Writing** You are a modern-day Ibn Battuta, traveling through East Africa. Write a series of journal or diary notes telling about the people you meet and the sights you see there.

networks

There's More Online!
- ☑ **IMAGES** Animal Poaching
- ☑ **MAP** Museums: Preserving Kenya's Heritage and Culture
- ☑ **VIDEO**

Reading HELPDESK

Academic Vocabulary
- diverse

Content Vocabulary
- population density
- clan
- subsistence agriculture
- oral tradition
- poaching

TAKING NOTES: *Key Ideas and Details*

Summarizing As you read about East African populations, daily life, culture, and challenges today, use a web diagram like the one here to list facts and details about each important idea.

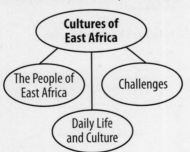

Lesson 3
Life in East Africa

ESSENTIAL QUESTION • *Why does conflict develop?*

IT MATTERS BECAUSE
East Africa is a region of great diversity in ethnicity, religion, and language—not only across the region, but also within individual countries.

The People of East Africa

GUIDING QUESTION *What ethnic groups contribute to the diversity of the population?*

East African countries typically are home to many ethnic groups. Another striking feature in this region is the split between urban and rural populations. Languages and religions make up a mosaic of many different elements.

Where People Live

The population of East Africa is split between large cities and rural areas. Many large cities are on or near the coast of the Indian Ocean (for example, Mogadishu in Somalia, Mombasa in Kenya, and Dar es Salaam in Tanzania). Some large cities, however, developed from important trading centers. Such cities include Nairobi, the capital of Kenya, and Addis Ababa, the capital of Ethiopia.

Of the 11 countries in the region, Ethiopia has the largest population (about 80 million), and Djibouti has the smallest (about 1 million). People are distributed unevenly in East Africa. **Population density** measures how many people live in a given geographical area. A thickly settled area has a high population density. In thinly settled areas, the density is low. In Tanzania, population density varies greatly from one area to another. Overall, Rwanda has the highest population density in the region. Somalia has the lowest.

310 Chapter 10

In Ethiopia, the majority of people live in the central highlands. The warmer and drier areas of lower elevations are thinly inhabited. In Sudan, most people live along the Nile River. Arid parts of the country are thinly populated. In Somalia, most people are nomadic or seminomadic.

Ethnic Groups

The populations of Kenya, Tanzania, and Ethiopia are **diverse** in terms of ethnicity. Sometimes competition among different ethnic groups has led to political and economic conflict. Ethnic identity is closely linked to language and also to geography.

In Kenya, for example, the Kikuyu, Kamba, Meru, and Nyika people inhabit the fertile highlands of the Central Rift. The Luhya live in the Lake Victoria basin. The rural Luo people are located in the lower parts of the western plateau. The Masai people tend their herds of cattle in the south, along the Kenya-Tanzania border. Like the Masai, the Samburu and the Turkana are pastoralists. They live in the arid northwestern region of Kenya.

Another type of ethnic identity is the **clan**. A clan is a large group of people sharing a common ancestor in the far past. A group of related clans is called a clan family. Smaller groups of related people within a clan are called subclans. In Somalia, the basic ethnic unit is the clan.

Academic Vocabulary

diverse having or exhibiting variety

Nairobi, the capital of Kenya, was founded in 1899 as a railway stop between plantations in Uganda and ports on the Kenyan coast. Today, Nairobi is one of East Africa's largest cities, with a population of about 3 million.

▶ **CRITICAL THINKING**
Describing Why are most East African cities located either along the Indian Ocean coast or in inland, highland areas?

In the A.D. 1100s, an Ethiopian king had the Church of St. George carved from solid red volcanic rock. Today, St. George and 10 similar churches in the town of Lalibela attract Ethiopian Christian worshippers as well as tourists from around the world.

Identifying What are the major religions in East Africa today?

In countries that have many diverse ethnic groups, building a sense of national identity is difficult. People often feel a stronger attachment and allegiance to their ethnic group than to their country. A Somali, for example, might feel a greater attachment to his or her clan than to the country of Somalia.

Languages

East Africa is a region where many African languages are spoken. For example, Ethiopians speak about 100 distinct languages. Kenya also has a wide variety of spoken languages. Swahili and English are used by large numbers of people to communicate. Those two languages are the official languages of the Kenyan legislature and of the courts.

Swahili is almost universal in Tanzania. The geographical location and colonial history of East African countries have often made an impact on the languages spoken there. For example, in Somalia the official language is Somali. However, Arabic is widely spoken in the northern area of the country, and Swahili is widespread in the south. In Somalia's colleges and universities, it is not uncommon to hear people speaking English or Italian. In Djibouti, Arabic and French are important languages.

Religion

Most people of East Africa follow either the Christian or Muslim faith. However, a number of traditional African religions also thrive in the region. Traders and missionaries from the Mediterranean region brought Christianity to Ethiopia in the A.D. 300s. The Ethiopian Orthodox Church is one of the world's oldest Christian churches. Today, about 60 percent of Ethiopians are Christians.

In Kenya, the constitution guarantees freedom of religion. Christianity first arrived in Kenya with the Portuguese in the 1400s. But the religion was not practiced for several hundred years, until colonial missionaries arrived in Kenya in the late 1800s. Muslims are an important religious minority in Kenya. Today, Christianity is practiced by more than two-thirds of Kenya's population.

Tanzania is evenly split among Christianity, Islam, and traditional African religions. About one-third of the population follows each one of these three religious traditions.

✓ **READING PROGRESS CHECK**

Analyzing In a region with such diverse languages, how do you think East Africans can communicate with people outside their own language group?

A Masai mother and son (top) stand outside their home built of mud, sticks, and grass. The Masai people herd cattle on the inland plains of Kenya and Tanzania. A mosque and Islamic-style buildings (bottom) crowd the harbor of Mombasa, a city on Kenya's Indian Ocean coast.

▶ **CRITICAL THINKING**
Describing How do ways of life differ in East Africa depending on location and culture?

Life and Culture

GUIDING QUESTION *What is daily life like for people in East Africa?*

In East Africa, traditional customs, as well as the impact of modernization, can be seen in daily life and culture. Culture in East Africa often displays a blend of African and European ways of life.

Daily Life

The rhythms of daily life are varied in East Africa. One factor is where people live: in cities or in rural areas. Most East Africans live in the countryside. But cities are growing rapidly, due to the economic opportunities they provide.

Nairobi is Kenya's capital and most important industrial city. The city is home to more than 3 million people. This makes Nairobi the most populous city in East Africa. It is a city of contrasts. High-rise business and apartment buildings sit near slums built of scrap material.

Daily life in rural areas is quite different from life in the cities. A rural family's housing, for example, might consist of a thatched-roof dwelling with very little in the way of modern or sanitary conveniences. Often, no electricity is available. Some rural people practice **subsistence agriculture**, growing crops to feed themselves and their families. Other rural people grow cash crops to sell.

Chapter 10   313

In Tanzania, groups such as the Sukuma farm the land south of Lake Victoria. The Chaggas grow coffee in the plains around Kilimanjaro.

The Masai are a nomadic people who live in Tanzania and Kenya. They wander from place to place throughout the year as they tend herds of cattle. Their cattle provide the Masai with most of their diet.

The Masai have developed a unique way of living. Groups of four to eight families build a kraal, or a circular thornbush enclosure. The kraal shelters their herds of livestock. The families live in mud-dung houses inside the kraal.

A tarab orchestra performs in Zanzibar, an Indian Ocean island that is part of Tanzania. Tarab is a form of music that began in Zanzibar and spread to other areas. The musician (left) plays a *qanun*, a stringed instrument believed to have been first used in Islamic Persia during the A.D. 900s.

▶ CRITICAL THINKING
Determining Central Ideas
What does a form of music like tarab reveal about East African culture?

The governments of Kenya and Tanzania have set up programs to persuade the Masai to abandon their nomadic lifestyle. The governments want to conserve land and protect wildlife, but the Masai have resisted. They want to preserve their way of life.

Arts and Culture

East African culture is deeply influenced by **oral tradition**. This means that stories, fables, poems, proverbs, and family histories are passed by word of mouth from one generation to the next. Folktales and fables offer good examples of oral tradition. In Kenya, the oral tradition functioned in a political way. Hymns of praise were passed on to support independence.

The small country of Djibouti is well known for its colorful dyed clothing. This includes a traditional piece of cloth that men wear around their waist like a skirt. It is common clothing for herders.

A leading novelist in East Africa is Kenya's Ngugi wa Thiong'o. His novel *Weep Not, Child* (1964) is considered the first important English-language novel written by an East African. This book is a story about the effects of conflict on families in Kenya. He also has authored works in the Bantu language of Kenya's Kikuyu people.

In Tanzania, an appealing and popular form of music is *tarab*. This type of music combines African, Arab, and Indian elements and instruments. Tarab has developed an international following. In Kenya, a popular musical style is *benga*. This pop style emerged in the 1960s in the area near Lake Victoria, which is inhabited by the Luo ethnic group.

East Africa is also linked to important findings in the fields of anthropology and ecology. Evidence indicates that East Africa is where human beings originated. The earliest known human bones come from Kenya and Ethiopia. The fossil beds of Olduvai Gorge in northern Tanzania have furnished us with an important record of 2 million years of human evolution.

In the domain of ecology, the national park systems of East Africa have no equal in the world. Protected areas like the Masai Mara National Reserve and Samburu National Reserve in Kenya, the Serengeti National Park in Tanzania, Queen Elizabeth National Park in Uganda, and Volcanoes National Park in Rwanda are preserving a precious inheritance.

✓ **READING PROGRESS CHECK**

Describing Compare and contrast urban and rural daily life in East Africa.

Challenges

GUIDING QUESTION *How do economic, environmental, and health issues affect the region today?*

Today, the people of East Africa face many complex, challenging issues. Some of the most important challenges involve economic development, the environment, and health.

Economic Development

Agriculture is the main economic activity in East Africa. Farmers in the region, however, face difficult challenges. First, the soils in East Africa are not especially fertile. Second, climate conditions are often unpredictable. Rainfall can be intermittent. Drought can severely damage crops.

Government policies in some countries of East Africa also favor the production of cash crops such as coffee for export. Such policies harm subsistence farmers who attempt to produce enough food to meet local needs. Much of this pattern of growing cash crops results from colonialism. Even after the countries of East Africa gained independence, the practice of growing cash crops for sale continued.

Self-sufficiency is a challenge in East Africa. The region is one of the poorest in the world. In addition, the population of many countries there is growing at a faster rate than the world's average. Industrialization has come slowly for East Africa.

Students view an exhibit of African wildlife at the Kenya National Museum in Nairobi. The museum's purpose is to collect, preserve, study, and present Kenya's cultural and natural heritage.

▶ **CRITICAL THINKING**
Describing How do East Africans protect living wildlife?

Scores of elephant tusks, seized from illegal poachers, are burned in Kenya. The purpose of the burning was symbolic: to point out the need to keep ivory from reaching international markets and to stop the illegal killing of elephants for their tusks.

Necessary resources such as trained workers, new facilities, and equipment have been lacking. In Ethiopia, for example, manufacturing amounts to only about 10 percent of the economy. Most of Ethiopia's exports are agricultural products. Its most important export is coffee.

The emphasis on primary industries that harvest or extract raw material, such as farming, mining, and logging, is also derived from colonialism. Colonial powers developed their colonies to provide products for the powers. Even after independence, the former colonies continue to produce the same products.

In Tanzania, the economy is mostly agricultural. Many farmers practice subsistence agriculture. Corn (maize), rice, millet, bananas, barley, wheat, potatoes, and cassava are among the important crops. Coffee and cotton are the most important cash crops. Gold is Tanzania's most valuable export.

In parts of East Africa, the economy has suffered because of civil war and political instability. The economy also is linked to the availability of transportation, communication, and education. One key indicator of progress in education is a country's literacy rate. Literacy rates across the region range from a low of 38 percent in Somalia to a high of 87 percent in Kenya.

Environmental Issues

East Africa faces challenging issues related to the environment. The region's lack of electric power has quickened the pace of deforestation. People are cutting down trees to meet their energy needs at an alarming rate. They use the wood to cook food and heat their homes. Along with deforestation, desertification poses serious problems in countries like Sudan.

By setting up national parks and wildlife sanctuaries, the countries of East Africa are hoping that this will boost their economies and preserve their heritage. Ecotourism is tourism for the sake of enjoying natural beauty and observing wildlife. Revenue from ecotourism is important to the East African economy.

Wild animals such as elephants and lions also face the threat of **poaching**. Poaching is the trapping or killing of protected wild animals for the sake of profit in the illegal wildlife trade. African elephants are especially vulnerable to poaching; they are killed for their ivory tusks.

Health Issues

In East Africa, poor nutrition continues to be a difficult problem to overcome. One of the main causes of hunger and malnutrition in the region has been war. Since 1990, conflict in several East African countries has halted economic development and caused widespread starvation. Large numbers of refugees have poured across international borders.

HIV/AIDS is a serious and often fatal disease affecting people in the region. AIDS is an abbreviation that stands for "acquired immune deficiency syndrome." AIDS is caused by a virus that spreads from person to person. This disease continues to be a major health issue in Kenya, Tanzania, and Ethiopia. The resources required for medical education and treatment have put a further strain on East African economies.

Deaths from AIDS have cut the average life expectancy in East Africa. Drought and famine also have an impact on life expectancy. In East Africa, life expectancy at birth is 58 years in Rwanda and 62 years in Sudan. In Kenya, it is 63 years. By contrast, life expectancy in the United States is now about 78.5 years.

Include this lesson's information in your Foldable®.

✓ **READING PROGRESS CHECK**

Citing Text Evidence What is one major cause of deforestation in the region of East Africa?

LESSON 3 REVIEW CCSS

Reviewing Vocabulary
1. How might *poaching* affect the economies of some East African countries?

Answering the Guiding Questions
2. ***Determining Central Ideas*** What general statements can you make about the ethnic groups and where people live in East Africa?

3. ***Identifying*** Identify two ways in which trade has played a central role in the history of East Africa.

4. ***Describing*** What are two of the most important challenges confronting East Africa today?

5. ***Informative/Explanatory Writing*** Write a paragraph or two in which you explain some of the environmental issues that confront East Africa today.

GLOBAL CONNECTIONS

Sudan: Refugees and Displacement

Sudan has been involved in civil war for many years. Most people in the northern part of Sudan are Arab Muslim and live in cities. People in the southern part are African, rural farmers, and follow either African traditional religions or Christianity.

Geography Sudan is the sixteenth-largest country in the world in area and the third-largest country in Africa. It was the largest before South Sudan gained independence. Sudan's population is 33.4 million. South Sudan has approximately half that number. In land area, South Sudan ranks forty-fourth in the world.

Northern Control As an independent country, northern Sudan and its leaders controlled the government. They wanted to unify Sudan under Arabic and Islamic rule. In opposition were non-Muslims and the people of southern Sudan.

> By the end of 2010, about 43.7 million people of the world did not have a home.

Violence Continues When South Sudan became an independent country on July 9, 2011, many people hoped to start a new, peaceful life. However, several violent conflicts broke out, including continued conflict in Darfur.

Conflict in Darfur Darfur is a region in western Sudan. In 2003 Darfur rebel groups rose up against the Sudanese government. The rebels demanded that the government stop its unjust social and economic policies. The government reacted by raiding and burning villages. In the long conflict that followed, thousands were killed, and many were forced from their homes.

Refugees and IDPs Refugees are people who have left their country because they are in danger or have been victims of persecution. A major problem also exists with internally displaced persons (IDPs). An IDP is someone who is forced to flee his or her home because of danger, but who remains in his or her country.

World Refugee Day The United Nations (UN) World Refugee Day is observed every year on June 20. The events call attention to the problems refugees face.

People of South Sudan move to a new refugee camp to escape conflict and hunger. ▶

THERE'S MORE ONLINE

SEE a timeline of the crisis in Darfur • WATCH the Red Cross help refugees

Chapter 10 319

GLOBAL CONNECTIONS

These numbers and statistics can help you learn about the problems the Sudanese people face.

$1.25 a day

Although a peace agreement in 2005 brought some stability to the people of South Sudan, many terrible problems exist. More than 80 percent of the residents live on less than $1.25 a day. The country has the world's highest maternal mortality rate. About 50 percent of elementary school-age children do not attend school.

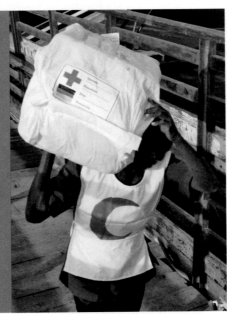

300,000

In 2003 rebellion broke out in Darfur, a region in western Sudan. Government militia attacked Darfur and the rebels. The United Nations estimates that as many as 300,000 people died in five years of conflict in Darfur. Violence still erupts at times, breaking the fragile peace.

More than 500,000

Refugees are people who have left their country because they are in danger or have been victims of persecution. In January 2011, 178,000 refugees were in Sudan, and 387,000 Sudanese who were living in other countries.

one point six million

In January 2011, more than 1.6 million Sudanese people were internally displaced.

two million

South Sudan seceded from Sudan in 2011 as a result of a peace treaty that ended decades of war that had killed 2 million people. The two countries have come close to war again. Disputes over control of territory led to armed conflict.

75%

Oil is a source of conflict between Sudan and South Sudan. About 75 percent of the oil is in South Sudan, but all the pipelines run north to Sudan. When disputes over oil erupted in 2012, Sudan bombed oil fields in South Sudan.

43.7 MILLION

By the end of 2010, about 43.7 million people of the world did not have a home. The number of refugees is the highest in 15 years.

320 Chapter 10

GLOBAL IMPACT

MIGRATION OF REFUGEES Refugees are people who flee to another country because of wars, political unrest, food shortages, or other problems. The graph on the left lists the 10 major source countries of refugees and the number of refugees who emigrated from those countries in 2011.

The graph on the right lists the 10 major host countries. A host country is the country a refugee moves to. For example, more than 1.7 million refugees immigrated to Pakistan in 2011.

Sudan and South Sudan

The map shows the two countries, their national capitals, and disputed areas.

Thinking Like a Geographer

1. **The Geographer's World** What are the major differences between Sudan and South Sudan?

2. **Human Geography** Why is oil a major factor in the conflict between Sudan and South Sudan?

3. **Human Geography** Imagine you are from Sudan and you come to live in the United States. Write a story about how you and your family learn to live in an American community.

Chapter 10 ACTIVITIES CCSS

Directions: Write your answers on a separate piece of paper.

1 Use your FOLDABLES to explore the Essential Question.
INFORMATIVE/EXPLANATORY WRITING Review the population map of East Africa at the beginning of the chapter. In two or more paragraphs, explain why people have settled in the locations indicated on the map.

2 21st Century Skills
INTEGRATING VISUAL INFORMATION Conduct research and write a paragraph about one of the national park systems in East Africa. Review a partner's paragraph using these questions to guide you: Did the paragraph include relevant details? Was there anything missing that you expected to find in the paragraph? Discuss the review of your paragraph with your partner. Revise your paragraph as needed.

3 Thinking Like a Geographer
IDENTIFYING Choose 1 of the 11 countries of East Africa. In a graphic organizer like the one shown, identify the capital city of the country and write two geographical facts about the country.

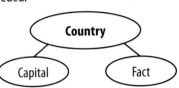

4 GEOGRAPHY ACTIVITY

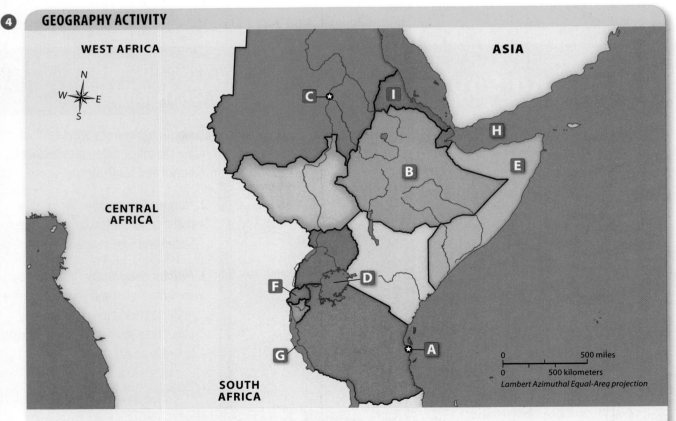

Locating Places
Match the letters on the map with the numbered places listed below.

1. Somalia
2. Eritrea
3. Lake Victoria
4. Khartoum
5. Rwanda
6. Gulf of Aden
7. Ethiopia
8. Lake Tanganyika
9. Dar es Salaam

Chapter 10 ASSESSMENT CCSS

REVIEW THE GUIDING QUESTIONS

Directions: Choose the best answer for each question.

1. The physical geography and landscape of East Africa are dominated by a series of geological faults collectively known as
   A. the Ruwenzori Mountains.
   B. the Great Rift Valley.
   C. Kilimanjaro.
   D. Jonglei.

2. Because of long periods of drought and overgrazing, agricultural land has turned into desert in a process called
   F. irrigation.
   G. desertification.
   H. urbanization.
   I. defoliation.

3. Throughout history, the countries of East Africa have been centers of
   A. trade.
   B. revolution.
   C. oil exploration.
   D. the slave trade.

4. Most countries in East Africa earned their independence from European colonial powers during which decade of the twentieth century?
   F. the 1950s
   G. the 1980s
   H. the 1940s
   I. the 1960s

5. What is the name of the nomadic people who herd cattle and build their mud-dung houses inside a kraal?
   A. Samburu
   B. Masai
   C. Kamba
   D. Meru

6. What is the main economic activity in East Africa?
   F. manufacturing
   G. tourism
   H. agriculture
   I. oil and gas production

Chapter 20 ASSESSMENT (continued)

DBQ ANALYZING DOCUMENTS

7 ANALYZING Read the following passage about economies in Africa:

> "Kenyan farmers, mostly small, are responsible for $1 billion in annual exports of fruits, vegetables, and flowers, a figure that dwarfs the country's traditional coffee and tea exports.... Rwanda,... long an importer of food, now grows enough to satisfy the needs of its people, and even exports cash crops such as coffee for the first time."
>
> —from G. Paschal Zachary, "Africa's Amazing Rise and What It Can Teach the World" (2012)

What statement best explains the success of Kenya's farmers?
A. They produced a variety of crops that were in demand.
B. They produced more coffee and tea.
C. They exported cash crops for the first time.
D. They imported food from Rwanda.

8 CITING TEXT EVIDENCE Which sector of Rwanda's economy has seen success?
F. agriculture
G. industry
H. mining
I. service industries

SHORT RESPONSE

> "The world's biggest refugee camp, Dadaab, in northeastern Kenya marks its 20th anniversary this year. The camp, which was set up to host 90,000 people, now shelters nearly one-half million refugees.... The [United Nations] set up the first camps in Dadaab between October 1991 and June 1992, following a civil war[in Somalia] that continues to this day."
>
> —from Lisa Schlein, "World's Biggest Refugee Camp in Kenya Marks 20th Anniversary" (2012)

9 DETERMINING CENTRAL IDEAS Why was a refugee camp needed?

10 ANALYZING What kinds of facilities would officials need to create to take care of tens of thousands of people?

EXTENDED RESPONSE

11 INFORMATIVE/EXPLANATORY WRITING In an essay, compare and contrast the countries of Somalia and Kenya. Use your text and Internet research to examine each country's physical geography, culture, average income, education levels, type of government, employment, and other factors that affect the way people live. Of the two, which country would you rather live in?

Need Extra Help?

If You've Missed Question	1	2	3	4	5	6	7	8	9	10	11
Review Lesson	1	1	2	2	3	3	3	3	2	2	3

CENTRAL AFRICA

ESSENTIAL QUESTIONS • *How do people adapt to their environment?*
• *How does technology change the way people live?* • *What makes a culture unique?*

A Congolese woman wears the traditional hairstyle that originated in the area long ago.

networks

There's More Online about Central Africa.

CHAPTER 11

Lesson 1
Physical Geography of Central Africa

Lesson 2
History of Central Africa

Lesson 3
Life in Central Africa

The Story Matters...

The region of Central Africa straddles the Equator, which creates tropical climates that support the growth of rain forests and savannas. The Congo River—Africa's second-longest river—has so many tributaries that it forms Africa's largest system of waterways. Abundant natural resources in Central Africa greatly influenced its history. Today, the resources are vital to helping nations in the region achieve and maintain stability.

FOLDABLES
Study Organizer

Go to the Foldables® library in the back of your book to make a Foldable® that will help you take notes while reading this chapter.

325

Chapter 11
CENTRAL AFRICA

Central Africa's many rivers are a source of life for people of the region. The geography of the region is dominated by the rain forest basin of the Congo River.

Step Into the Place

MAP FOCUS Use the map to answer the following questions.

1. **THE GEOGRAPHER'S WORLD** How many countries make up the region of Central Africa?

2. **THE GEOGRAPHER'S WORLD** What island country is part of the region?

3. **PLACES AND REGIONS** What is unusual about the capitals of Congo and the Democratic Republic of the Congo?

4. **CRITICAL THINKING**
   Analyzing Which cities in Gabon and Cameroon are located along the coasts? Think about their locations. What economic activities do you think are important in those cities?

ISLAND PARADISE A highway circles the scenic coast of São Tomé, a volcanic island, off the western coast of Central Africa. São Tomé forms part of São Tomé and Príncipe, a small island nation.

SULTAN'S COURT MUSICIANS Musicians perform traditional music at a palace of a sultan in Foumban, Cameroon. Music and dance are an important part of ceremonies and social gatherings in Central Africa.

Step Into the Time

TIME LINE Choose at least two events from the time line to describe the cause-and-effect relationship between natural resources and the slave trade in Central Africa.

1000 B.C. Iron Age spreads to Central Africa

1000 B.C.–A.D. 1100 Bantu people migrate to Congo rain forest

1470s São Tomé becomes the first port of the Atlantic slave trade

1700s Slave trade spreads to interior of continent

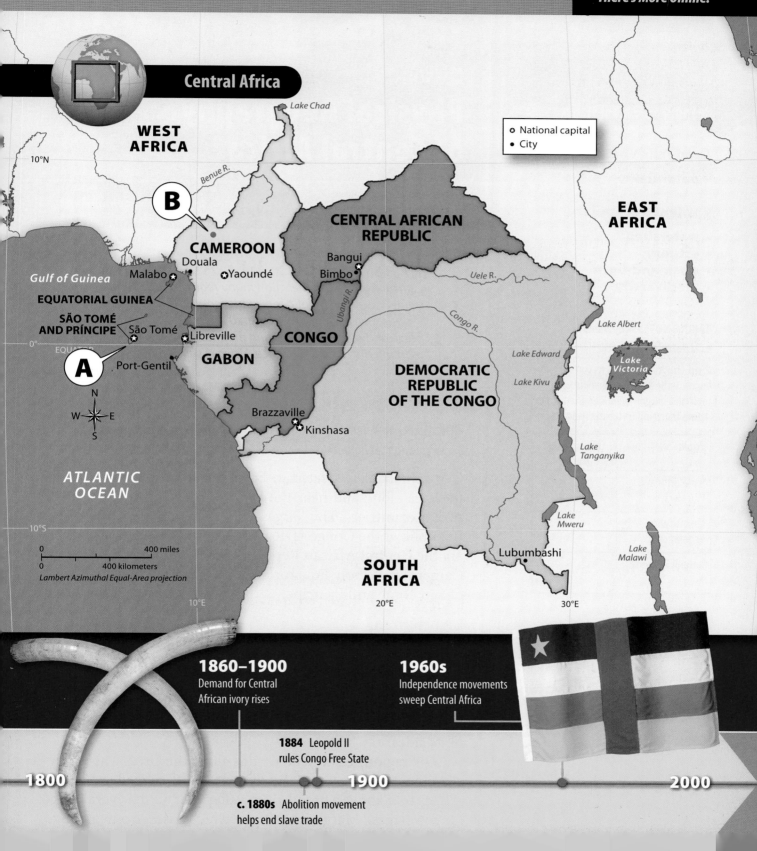

Central Africa

- National capital
- City

1860–1900 Demand for Central African ivory rises

c. 1880s Abolition movement helps end slave trade

1884 Leopold II rules Congo Free State

1960s Independence movements sweep Central Africa

networks

There's More Online!

☑ **IMAGES** Slash and Burn Agriculture

☑ **VIDEO**

Reading HELPDESK

Academic Vocabulary
- potential

Content Vocabulary
- watershed
- estuary
- slash-and-burn
- biodiversity

TAKING NOTES: *Key Ideas and Details*

Find the Main Idea As you study the lesson, write a main idea about each topic on a graphic organizer. Then write details that support the main idea.

Topic	Main Idea/ Detail
Landforms and Waterways	
Climate and Vegetation	
Natural Resources	

Lesson 1
Physical Geography of Central Africa

ESSENTIAL QUESTION • *How do people adapt to their environment?*

IT MATTERS BECAUSE
Central Africa is smaller than many regions, but it holds a tremendous variety of geographic features. These features include a vast rain forest–covered basin, one of the world's greatest river systems, soaring mountains, and a deep rift valley marking the line along which Africa is splitting apart.

Landforms and Waterways

GUIDING QUESTION *What makes some landforms and waterways so important to the region?*

Central Africa is located in Earth's equatorial zone—that is, the area along and near the Equator. The region consists of seven countries. The largest of these is the Democratic Republic of the Congo (DRC). It dwarfs its neighbors, which are the Central African Republic, the Republic of the Congo, Cameroon, Gabon, Equatorial Guinea, and the island country of São Tomé and Príncipe.

Landforms

The dominant landform of Central Africa is the **watershed** of the Congo River. A watershed is the land drained by a river and its system of tributaries. At the center of the watershed is a depression called the Congo Basin. A rolling plain spreads across the center of the basin, and high plateaus rise on most of its sides.

The region's eastern edge runs along the Great Rift Valley, also known as the Great Rift System. Here, rugged mountain ranges soar above a broad, deep valley that holds several long, narrow lakes. Margherita Peak, which rises from a range

328

called the Ruwenzori (ROO-un-ZO-ree), reaches the lofty height of 16,763 feet (5,109 m). Margherita Peak is the highest summit in the region and the third highest on the entire continent, ranking after Kilimanjaro and Mount Kenya.

Along the Atlantic coast of Central Africa stretches a narrow lowland. Off the coast lie several important islands. Two of these islands form the country of São Tomé and Príncipe. Two other islands, called Bioko and Pagalu, belong to Equatorial Guinea. This country also includes several smaller islands, as well as a territory on the mainland known as Mbini.

Waterways

Six of the seven countries in Central Africa have coasts on the Atlantic Ocean. The Central African Republic is the region's only landlocked country.

The Congo River and its tributaries account for most of Central Africa's inland waterways. The source of the Congo lies in East Africa between Lake Tanganyika and Lake Malawi (also called Lake Nyasa). From there, the river flows about 2,900 miles (4,667 km) to its mouth on the Atlantic Ocean. Among African rivers, only the Nile is longer than the Congo. When measured by water flow, the Congo tops every river in the world except South America's Amazon.

Difficult Navigation

No other river system in Africa offers as many miles of navigable waterways as the Congo and its tributaries. A navigable river is one on which ships and boats can travel. The Congo River is not navigable for its entire course, however. Several series of cataracts and rapids interrupt the passage of ships on the river.

Perhaps the most significant of these interruptions occurs rather close to the mouth of the Congo River. Only about 100 miles (161 km) from the Atlantic Ocean lie cataracts that block seagoing ships from traveling farther inland. The seaport city of Matadi is found here. Downstream from Matadi, in the final part of its journey, the river widens into an **estuary**, a passage in which freshwater meets salt water.

Fishers in the Democratic Republic of the Congo use bamboo supports to lower their nets into the waters along the rapids of the Congo River.
▶ CRITICAL THINKING
Describing What makes navigation difficult on parts of the Congo River?

A scientist and animal trackers analyze camera trap video of gorillas in a Central African rain forest. A camera trap is a camera equipped with a special sensor that captures images of wildlife on film, when humans are not present.

▶ CRITICAL THINKING
Analyzing How might camera traps aid in protecting Central African wildlife?

The Congo River is important to Central Africa for several reasons. First, it provides a livelihood for people who live along its banks. They use the river's water for agriculture and depend on its fish for food. Second, the river is a vital transportation artery. Although the cataracts and rapids prevent ships from navigating the entire river, ship traffic connects people and places along various sections of the river. Finally, dams on the river generate hydroelectric power.

☑ **READING PROGRESS CHECK**

Identifying Give two reasons for the Congo River's importance to Central Africa.

Climate and Vegetation

GUIDING QUESTION *What are the prevailing climates in Central Africa?*

Because Central Africa is centered on the Equator, the climate in much of the region is tropical. Temperatures are warm to hot, and rainfall is plentiful. The amount of rainfall generally decreases as distance from the Equator increases. In the northern and southern parts of the region, dry seasons alternate with wet seasons.

Climate Zones

The belt of Central Africa that lies along the Equator has a warm, wet climate. Because of its location, this zone experiences little seasonal variation in weather and length of daylight. The midday sun is directly or almost directly overhead every day, and daytime temperatures are always high. Rainfall is abundant throughout the year, with totals greater than 80 inches (203 cm) in some areas.

To the north and the south of the region's equatorial zone lie tropical wet-and-dry climate zones. As the name suggests, these zones have both rainy and dry seasons. There is great climate variation within the zones. In the areas nearest the Equator, the dry season typically lasts about four months. In the areas farthest from the Equator, it might last as long as seven months.

Rain Forest

A tropical rain forest, the second largest in the world, covers more than half of Central Africa. In this rain-soaked realm, closely packed trees soar as high as 15-story buildings. The forest also holds a tremendous variety of other plants, some of which are used in traditional medicines. Scientists think that only a fraction of the plant species in the forest have been identified. Most remain to be discovered.

The trees and other plants in the rain forest compete for survival. The interwoven crowns of the trees create a dense canopy, or roof, that blocks nearly all of the sunlight, leaving the lower levels in gloom. A tree seedling or sapling can only grow tall if a mature tree dies or if wind blows down some of its branches, creating an opening in the canopy. Because of the lack of sunlight, the forest floor has few of the small plants found in other types of forests. Instead, the forest floor is covered by dead leaves that decompose rapidly in the warm, wet conditions.

Savannas

In the northern and southern areas of Central Africa, rain forests give way to savannas, or areas with a mixture of trees, shrubs, and grasslands. The exact mix of vegetation depends on the length of the dry seasons. Human activity over thousands of years might be responsible for an increase in the size of the savanna areas. Experts believe that the change of forestland into savanna could be the result of clearing land through **slash-and-burn** agriculture.

This form of farming involves turning forestland into cropland. Trees and shrubs are cut down and burned in order to make the soil more fertile, at least temporarily.

Garamba National Park in the Democratic Republic of the Congo is home to the few surviving Northern white rhinoceroses.

Miners search for gold in an open-pit mine in the northeastern part of the Democratic Republic of the Congo. For more than a decade, rival groups in the DRC have fought each other for control of the country's rich natural resources, such as gold, diamonds, and timber.

Within a few years, the soil becomes depleted and farmers move on. The plots of land are left so that trees and shrubs can return. Then the farmers return and repeat the process of cutting, burning, and farming. Sometimes plots remain fallow—that is, not planted—for as long as 20 years.

Slash-and-burn agriculture provides a livelihood for many people. Others, however, point out that it destroys plant and animal habitat, creates air pollution, and is causing the world's tropical forests to shrink at an alarming rate.

Those who support protecting the environment worry that activities that harm rain forests threaten biodiversity. **Biodiversity** refers to the wide variety of life on Earth. These conservationists argue that if ecosystems are wiped out, numerous valuable species will be lost forever.

☑ **READING PROGRESS CHECK**

Describing Imagine that you live in northern Cameroon. You decide to move to a location in the Republic of the Congo very near the Equator. What changes in climate can you expect?

Natural Resources

GUIDING QUESTION *Which natural resources are important in Central Africa?*

Central Africa is rich in mineral, energy, and other resources. However, many of these resources have not yet been developed.

Mineral Resources

The greatest abundance of mineral resources in Central Africa is found in the Democratic Republic of the Congo. Within that country, the area richest in mineral resources is a province called Katanga. Katanga holds deposits of more than a dozen minerals, including cobalt, copper, gold, and uranium. Minerals mined in other areas of the country include diamonds, iron ore, and limestone.

Gabon produces more than a tenth of the world's supply of manganese. This hard, silvery metal is used in the manufacture of iron and steel. The country also produces uranium, diamonds, and gold, and it holds reserves of high-quality iron ore.

Bauxite and cobalt are among Cameroon's most significant mineral resources. Equatorial Guinea has deposits of uranium, gold, iron ore, and manganese. Most of these deposits have yet to be exploited. Diamond mining is an important industry in the Central African Republic. Rich deposits of uranium, gold, and other minerals could bring the country wealth in the future.

Why have the mineral resources in some parts of Central Africa remained underdeveloped? Political instability, civil conflict, and the high costs of investment have played a role. Perhaps the most important reason, however, is that the region lacks good transportation networks. In the Democratic Republic of the Congo, for example, the Congo River still serves as the major transportation artery. The landlocked Central African Republic also must use rivers for transport. It has few paved roads and no railways. Similarly, most roads in Equatorial Guinea are unpaved and there is no railway system.

Other Resources

The region is rich in resources other than minerals. In the Democratic Republic of the Congo, for example, rapids and waterfalls on the Congo River and its tributaries offer vast **potential** for hydroelectric power. The forest reserves of the DRC are rivaled by few countries in the world. Fish from the ocean and its fresh water systems provide another important resource.

Developing these resources, however, will affect the environment. Damming rivers for hydroelectric power results in large changes to river ecosystems. Deforestation and habitat change occur when forests are cut.

In the 1990s, large reserves of petroleum and natural gas were discovered under the seafloor off Equatorial Guinea's Atlantic Coast. The export of oil and gas products have boosted the country's economy.

Likewise, petroleum has been Cameroon's leading export since 1980. The country also has natural gas deposits, but the high cost of development has kept them untapped. Nearly all of Cameroon's energy comes from dams that generate hydroelectricity.

Academic Vocabulary

potential possible; capable of becoming

Include this lesson's information in your Foldable®.

☑ **READING PROGRESS CHECK**

Identifying Central Issues What are two of the factors that have slowed development of Central Africa's rich natural resources?

LESSON 1 REVIEW

Reviewing Vocabulary
1. Why might *slash-and-burn agriculture* be harmful to a country's land in the long term?

Answering the Guiding Questions
2. ***Analyzing*** What comparisons can you make between the Congo River in Central Africa and two of the other great rivers of the world: the Nile and the Amazon?

3. ***Describing*** What conclusion can you make about the prevailing climate conditions in areas near the Equator?

4. ***Analyzing*** How would you evaluate Central Africa's hydroelectric potential?

5. ***Argument Writing*** Write a letter to a friend to persuade him or her to invest in mineral production in Central Africa. In your letter, use some of the information you have learned in this lesson.

networks

There's More Online!
- VIDEO

Reading HELPDESK

Academic Vocabulary
- foundation

Content Vocabulary
- millet
- palm oil
- cassava
- colonialism
- missionary
- coup

TAKING NOTES: *Key Ideas and Details*

Organize Use a diagram like this one to note important information about the history of Central Africa, adding one or more facts to each of the boxes.

Topic	Information
Products	Millet Sorghum
European contact	Slave trade

Lesson 2
History of Central Africa

ESSENTIAL QUESTION • *How does technology change the way people live?*

IT MATTERS BECAUSE
Central Africa's past is a fascinating and often tragic story involving migrations, slavery, exploitation by foreign powers, and the struggle for independence, stability, and prosperity.

Early Settlement

GUIDING QUESTION *How did agriculture and trade develop in Central Africa?*

Through most of the prehistoric period, the people of Central Africa were hunters and gatherers who survived on wild game and wild plants. Eventually, climate changes forced the people to develop new ways of life.

Development of Agriculture

Around 10,000 years ago, Earth's climate entered a dry phase. Vegetation patterns changed in response and led to the movement of people. The changes also intensified the struggle for survival. The region's inhabitants were forced to find ways to get more food from a smaller area of their environment.

Gradually, a transformation that historians call the agricultural revolution swept through the region, beginning in the north. People began to collect plants—especially roots and tubers—on a more regular basis. They developed and refined tools such as stone hoes that were specially designed for digging. They discovered that if they planted a piece of a root or tuber in fertile soil, a new plant would grow from it. Over time, the hunters and gatherers turned into farmers.

Cereal farming was the next agricultural development in Central Africa. In the savannas of the north, people began cultivating **millet** and sorghum, two wild grasses that produce edible seeds. Millet proved to be especially well-suited to the

334

area's climate. The crop thrives in high temperatures and is resistant to drought.

In addition to growing crops, Central Africa's early farmers cultivated trees and gathered their fruit. From the fruit of oil palms they made a cooking oil that was rich in proteins and vitamins. The nutrition boost provided by **palm oil** helped people to become healthier, and improved health brought about population growth.

The increase in the food supply from the practice of agriculture meant that people could live in larger, more settled communities. The agricultural revolution laid the **foundation** for village life and also for the development of items used in daily life.

Academic Vocabulary

foundation the basis of something

Using Mineral Resources

Since early times, people in Central Africa have made tools from stone. Around 3,000 years ago, they began using a new material that was far better in many ways: iron. Iron tools were expensive and could only be made by skilled artisans, but they were far more efficient and less brittle than stone tools.

In addition to iron, people in ancient Central Africa made use of other minerals, especially copper and salt. These resources played an important role in trade.

☑ **READING PROGRESS CHECK**

Determining Central Ideas Why was the agricultural revolution so important for the development of Central Africa as a region?

European Contact and Afterward

GUIDING QUESTION *How did colonization by foreign countries affect Central Africa?*

Regular contact with Europeans, which began in the 1400s, marked the start of a new era in Central Africa's history.

The Slave Trade

As European ships began reaching the Atlantic coast of Central Africa in the 1400s, the region began developing into one of the busiest hubs of the slave trade. The trade was driven because European colonizers demanded a large workforce for their huge plantations in the Americas. The slave trade would continue and grow for more than three centuries.

The first European country to become actively involved in the slave trade was Portugal. In the late 1400s, the Portuguese established a colony on the island of São Tomé in order to grow sugar.

A woman in Cameroon carries fruit from oil palm trees to market. The pulp from this fruit is used to make cooking oil. Some environmentalists fear that the creation of more palm oil plantations in the country will endanger the livelihood of small farmers and lead to the destruction of existing rain forests.

Chapter 11 **335**

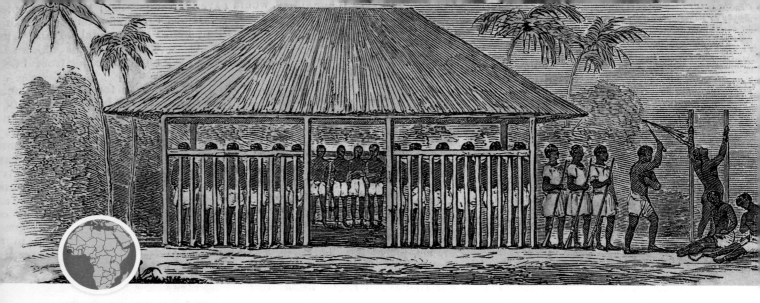

Slave traders held people who were captured for slavery in *barracoons*, or sheds. There, the captives stayed for several months before they were sold and shipped to the Americas. During imprisonment, the captives were chained by the neck and legs and often beaten.
▶ **CRITICAL THINKING**
Determining Word Meaning What was the Middle Passage?

Later, this island became a staging area for the transportation of African slaves to Portugal's main conquest in the Western Hemisphere: Brazil in South America.

Gabon served as one of the most important centers of the slave trade. Slaves were gathered in the country's interior and taken on boats to a coastal inlet called the Gabon Estuary. Some of these slaves were people who had been cast out of their own societies. Others had been captured in warfare. At settlements on the estuary, the slaves were held in enclosures known as barracoons until European ships arrived.

The slave trade was part of what is sometimes called the "triangular trade," named for the triangular pattern formed by the three stages of the trade. In the first stage, ships would sail to Africa with cargoes of manufactured goods such as cloth, beads, metal goods, guns, and liquor. These goods would be traded for slaves. In the second stage, known as the Middle Passage, the ships would carry their human cargo to the Americas. There, the slaves would be exchanged for goods such as rum, tobacco, molasses, and cotton that were produced on slave-labor plantations. In the third stage, the ships would return to Europe. Ships of the time were powered by wind, and each stage followed the direction of prevailing winds.

Adoption of New Crops

After European countries established colonies in the Americas, they brought some of the native plants back to Europe and to Africa. Two plants in particular had an important effect on farming and diet in Central Africa. These were cassava and maize.

The **cassava** plant has thick, edible roots known as tubers. Rich in nutritious starch, the tubers are used to make flour, breads, and tapioca. The plant thrives in hot, sunny climates and is able to survive droughts and locust attacks. Historians believe that Portuguese ships brought cassava to Africa from Brazil. Today, cassava is a staple food for many of Africa's people.

Maize, often called corn, is one of the most important staple grains of the Western Hemisphere. Today it is the most widely grown grain crop in the Americas. Maize was domesticated in prehistoric times, probably in Central America. It was carried around the world by Europeans after their discovery of the Americas.

Colonialism

Colonialism is the political and economic rule of one region or country by another country, usually for profit. European countries began to practice colonialism in the 1500s and 1600s. They first founded colonies in the Americas for economic gain.

Colonialism came much later to Central Africa. Exploration and settlement by Europeans was impeded by the difficulty of transportation, the presence of tropical diseases like malaria, and other challenges. In the second half of the 19th century, however, European presence in the region began to grow sharply.

In 1884–1885, Germany hosted a landmark conference of European countries in the city of Berlin. The countries attending the conference agreed on a plan for dividing Africa into colonies that could be exploited for European profit.

King Leopold II of Belgium was among the strongest supporters of the conference. He believed that a fortune could be made from rubber plants. Rubber was one of the most plentiful and valuable natural resources of Central Africa. After 1885, King Leopold took over a vast area that came to be known as the Congo Free State. The king held the area as a personal possession, and he did indeed make a fortune.

Belgium's King Leopold II (oval) turned his privately owned Congo territory into a large labor camp for the harvesting of rubber. African workers (below) were overseen by European officials. Nearly 10 million Africans died of overwork or cruelty under Leopold's direct rule.

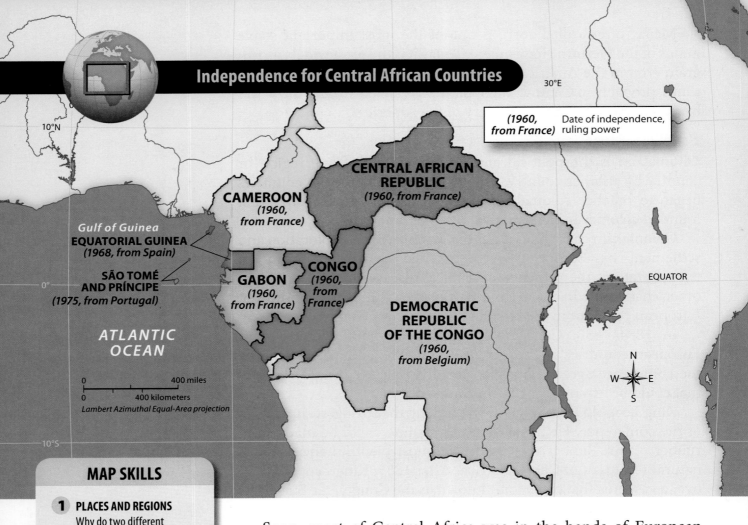

Independence for Central African Countries

(1960, from France) Date of independence, ruling power

MAP SKILLS

1. **PLACES AND REGIONS** Why do two different countries have Congo as part of their name?

2. **THE GEOGRAPHER'S WORLD** What country of Central Africa is landlocked?

Soon, most of Central Africa was in the hands of European colonizers. France gained control of what is now the Republic of the Congo, Gabon, and the Central African Republic. Spain colonized what is now Equatorial Guinea, while Germany ruled Cameroon. Portugal retained possession of São Tomé and Príncipe.

Europeans often justified their economic exploitation of Africa by claiming that their goal was to promote civilization and to spread Christianity. Europeans sent many **missionaries** to Africa in order to convert the native people.

At the same time, Europeans often treated the African workers under their control harshly. By the early 1900s, small revolts against French rule and the plantation-based economy were common. In the Congo Free State, the people suffered severe hardships and cruel treatment under King Leopold. Pressured by growing outrage from around the world, the Belgian parliament took over the vast area from King Leopold. It became an official colony of Belgium and was known as the Belgian Congo.

✓ **READING PROGRESS CHECK**

Determining Central Ideas How was Central Africa affected by a conference held in Germany in the mid-1880s?

Independent Countries

GUIDING QUESTION *What effects did gaining independence have on the countries of Central Africa?*

Near the middle of the 20th century, European countries became willing to grant independence to their African colonies. All seven of Central Africa's countries gained their independence in the period from 1960 to 1975.

A Wave of Independence

In 1960 France was the most important European colonial power in Central Africa. That year witnessed the independence of four French colonies: Gabon, the Republic of the Congo, the Central African Republic, and Cameroon, which France had gained from Germany during World War I. In the same year, the Democratic Republic of the Congo won independence from Belgium.

Many of these new countries experienced hard times after independence. Their people suffered through periods of ethnic conflict, harsh rule, and human rights abuses. In the Central African Republic, military officer Jean-Bédel Bokassa staged a **coup**. He ruled as a dictator and proclaimed himself the country's emperor. He brutally punished anyone who protested against his rule.

There have also been success stories, such as Gabon. Thanks in large part to its plentiful natural resources, Gabon has become one of the wealthiest and most stable countries in Africa.

Smaller Countries

Equatorial Guinea won independence from Spain in 1968. The country's first president, Francisco Macías Nguema, soon took over the government and became a ruthless dictator. In 1979, he was ousted by his nephew, who also ruled with an iron hand.

Portugal granted independence to São Tomé and Príncipe in 1975. With independence came great hope for the future. Like many of its neighbors, however, the country has been plagued by political instability and corruption.

Include this lesson's information in your Foldable®.

✓ **READING PROGRESS CHECK**

Identifying Since the countries of the region gained independence, list at least two factors that have limited their political and economic progress.

LESSON 2 REVIEW (CCSS)

Reviewing Vocabulary
1. Why was *millet* a suitable grain for planting in Central Africa?

Answering the Guiding Questions
2. ***Determining Central Ideas*** Why might tools made from iron be superior to tools made from stone?

3. ***Identifying*** What were two major motivations for the European colonization of Central Africa?

4. ***Analyzing*** In what way was the rule of Bokassa in the Central African Republic similar to the rule of Nguema in Equatorial Guinea?

5. ***Informative/Explanatory Writing*** Write a few paragraphs explaining the development of the slave trade in Central Africa.

networks

There's More Online!

- ☑ **SLIDE SHOW** Types of Housing in Central Africa
- ☑ **VIDEO**

Reading HELPDESK

Academic Vocabulary
- characteristic
- depict

Content Vocabulary
- refugee
- trade language

TAKING NOTES: *Key Ideas and Details*

Find the Main Idea As you study the lesson, create a chart like this one, and fill in at least two key facts about each topic.

Topic	Fact #1	Fact #2
People		
Daily Life and Culture		

Lesson 3
Life in Central Africa

ESSENTIAL QUESTION • *What makes a culture unique?*

IT MATTERS BECAUSE
Central Africa is characterized by tremendous diversity in terms of its people, population patterns, languages, arts, and daily life.

The People of Central Africa

GUIDING QUESTION *What are some of the differences found among the people of Central Africa?*

Many different ethnic groups live in Central Africa. Each group is united by a shared language and culture.

Makeup of the Population

Central Africa is home to around 105 million people, which is roughly one-third as many as the United States has. The Democratic Republic of the Congo is by far the most populous country: It holds more than two-thirds of the region's people. Compared to other parts of the world, Central Africa does not have a large population or a high population density. However, its countries have high population growth rates. In all seven countries, the median age is below 20, which means that children and teenagers make up about half of the population. By comparison, the median age in the United States is about 37.

Life expectancy at birth for the people of the region varies considerably. In the Central African Republic, for example, life expectancy is about 50 years, while in Equatorial Guinea and São Tomé and Príncipe, it is around 63.

Hundreds of ethnic groups live in the region. Cameroon and the DRC each are home to more than 200 different groups. Some groups spread into two or more countries. One

340

such group is called the Fang. Historians believe the Fang once dwelled on the savanna. In the late 18th century, they began a migration into the rain forests. Today they live in mainland Equatorial Guinea, northern Gabon, and southern Cameroon.

Another people found in the region is the Bambuti, sometimes called the Mbuti. The Bambuti live in densely forested areas of the Democratic Republic of the Congo. They are extremely short in stature: Adults average less than 4 feet 6 inches (137 cm) in height. They were probably the earliest inhabitants of an area known as the Ituri Forest. Historical records show that the Bambuti have lived in this area for at least 4,500 years as nomadic hunters and gatherers.

The population of Central Africa also includes many refugees. **Refugees** are displaced people who have been forced to leave their homes because of war or injustice. Between 1997 and 2003, for example, a brutal civil war devastated the Democratic Republic of the Congo. A huge number of refugees fled the conflict.

City and Country

Most of Central Africa's people live in the rural areas and make their living through subsistence farming. In the DRC, for example, almost two-thirds of the people live in the country.

Large urban areas dot the region, however. More importantly, the ratio of city-dwellers to rural residents is changing rapidly. For example, every year, the DRC's urban areas gain about 4.5 percent of the population. Many of the chief cities of the region are capitals of countries. Kinshasa, the capital of the DRC, has grown into a sprawling metropolis with around 9 million inhabitants. Just across the Congo River from Kinshasa sits Brazzaville, the capital of the Republic of the Congo. Brazzaville is home to about 1.6 million people.

Government buildings surround a busy central square in Brazzaville, capital of the Republic of the Congo. Brazzaville is located on the north bank of the Congo River, just across from Kinshasa, the capital of the neighboring Democratic Republic of the Congo. This is the only place in the world where two national capital cities are located on opposite sides of a river, within sight of each other.

Village women of the Hutu ethnic group prepare a meal in the eastern part of the Democratic Republic of the Congo. The Hutu are one of hundreds of ethnic groups in Central Africa.

▶ **CRITICAL THINKING**
Describing Why did trade languages emerge in Central Africa?

In a few countries in this region, more than half the people live in cities. In Gabon, for example, city-dwellers account for 86 percent of the total population. In the Central African Republic, the figure is 62 percent, and in Cameroon it is 58 percent. Cameroon's two major cities, Douala and Yaoundé, each have populations of around 2 million, roughly the population of Houston or Philadelphia.

Language and Religion

Because Central Africa's population is made up of hundreds of different ethnic groups, it is not surprising that hundreds of different languages are spoken across the region. Cameroon, which has been described as an ethnic crossroads, serves as an example of the region's linguistic diversity. Three main language families are used in Cameroon. In the north, where Islam has been a significant influence, the languages spoken are in the Sudanic family. The Sudanic-speaking people in this part of Cameroon include the Fulani, the Sao, and the Kanuri. The Fulani are Muslims. They began migrating into Cameroon from what is now Niger more than 1,000 years ago. In the southern part of Cameroon, people speak Bantu languages. In the west are found semi-Bantu speakers, such as the Bamileke and the Tikar.

During the colonial era, people searched for common linguistic ground, especially when they traded with one another. Certain **trade languages** emerged. Because France and Belgium had the most widespread interest in the region, French became the most common trade language of Central Africa.

Central Africa is also diverse in terms of religion. In the Democratic Republic of the Congo, for example, roughly 50 percent of the people are Roman Catholic, 20 percent are Protestant, 10 percent are Muslim, and 10 percent belong to a local sect called the Kimbanguist Church. The remaining 10 percent follow traditional African religions, which are based on a core set of beliefs, including the existence of a supreme being, the presence of spirits in the natural world, and the power of ancestors and magic.

☑ **READING PROGRESS CHECK**

Describing Where do most of Central Africa's people live? How do they make their living?

How People Live

GUIDING QUESTION *What are some of the key aspects of daily life and culture in Central Africa?*

In Central Africa, daily life and culture reflect the variety of influences to which the region has been exposed.

Daily Life

Daily life in Central Africa is a blend of traditional and modern **characteristics**. This combination results mainly from the impact of colonialism. It also reflects the urban-rural division of the population.

In the countryside, most people practice subsistence farming. On small plots of land, they grow crops such as cassava, maize, and millet, and they raise livestock. They strive to grow enough crops and raise enough animals to feed themselves and their families. If any surplus remains, they may sell it for small amounts of cash. Most people in the countryside live in small houses that they have built themselves. Building materials include mud, sundried mud bricks, wood, bark, and cement. For roofs, people typically use palm fronds woven together or sheets of corrugated iron.

One of the most common methods of constructing houses is known as wattle-and-daub. Poles driven into the ground are woven with slender, flexible branches or reeds to create the wattle. The finished wattle is plastered with mud or clay, known as daub. Then, a palm-frond roof is placed on top of the house.

In much of rural Central Africa, women carry out the gathering, production, and preparation of food for the household. The men hunt, trap, and fish. In addition to growing food to eat, men and women may also grow crops for sale, such as coffee, cotton, and cocoa.

Okra, corn, and yams are staple vegetables in the diets of many Central Africans. Other important foods are cassava, sweet potatoes, rice, beans, and plantains. Game is popular, as are fish-based dishes. Peanuts, milk, and poultry add protein to many dishes.

Academic Vocabulary

characteristic a quality or an aspect

Many food-related customs of Central Africa may seem surprising. In some areas, for example, males eat in one room while females eat in another. Generosity and hospitality are so deeply ingrained in some cultures that hosts might offer guests an abundance of food while remaining hungry themselves.

In many Central African cities, traditional ways of life are yielding to contemporary lifestyles. Modernization, however, has not always come easily. Central Africa has little infrastructure, which includes the fundamental facilities and systems that serve a city, an area, or a country. In the Central African Republic, for example, only a few cities and towns have modern health care facilities. Other obstacles to modernization include poor transportation and educational systems.

Culture and Arts

The visual, verbal, and performing arts are important to the cultures of many ethnic groups in Central Africa. In the southwestern part of the DRC, for example, the people known as the Kongo produce wooden statues in which nails and other pieces of metal are embedded. The Yaka create highly decorative masks and figurines. The Luba people of the southeastern part of the country are known for their skillful carvings that **depict** women and motherhood. The Mangbetu people of the northeast are known internationally for their pottery and sculpture.

In the field of literature, several modern Congolese authors are internationally recognized as poets, playwrights, and novelists. They include Clémentine Madiya Faik-Nzuji, Kama Kamanda, Ntumb Diur, and Timothée Malembe.

The DRC's capital city of Kinshasa is known around the world for its thriving music scene. The most popular style is African jazz, known as OK jazz. This style originated in the nightclubs of Kinshasa in the 1950s.

In northern Cameroon, the Fulani people decorate leather items and gourds with elaborate geometric designs. In music, the country's southern forest region is known for its drumming. In the north, the focus is on flute music.

Equatorial Guinea was formerly a colony of Spain, and Spanish influence can be tasted in the country's cuisine. In Malabo, the capital city, Spanish styles are found in the architecture. The country has produced several writers whose Spanish-language works have become known around the world.

Musicians of the Congolese Symphony Orchestra perform in Kinshasa, capital of the DRC.
▶ CRITICAL THINKING
Describing Why is Kinshasa widely regarded as one of the music centers of the world?

☑ READING PROGRESS CHECK
Determining Word Meanings What is the main goal of subsistence farmers?

Regional Issues

GUIDING QUESTION *What are the greatest challenges confronting Central Africa?*

Central Africa's people and governments face many complex issues. Among the issues are the economy, the environment, political stability, and population growth.

Growth and the Environment

With great population growth comes the need for economic development. The economies of many countries in the region depend heavily on agriculture, logging, and mining. Central African countries such as Gabon and Equatorial Guinea export significant amounts of valuable hardwoods, including mahogany, ebony, and okoumé.

In general, economic development depends on economic growth. Economic growth usually means that more resources are used and more pollution and waste are produced. In Central Africa, economic activities such as mining and logging are taking place in areas of high biodiversity. Some conservationists fear that this biodiversity is being lost in the rush to exploit resources.

The result is tension between the forces of economic growth and the forces of environmental conservation. In some areas, these tensions have at times led to conflict between local people and outside groups.

Development in Central Africa also raises other questions. Should the profits from economic activities go to foreign corporations and investors, or should they remain with the national governments? How should the profits and economic benefits be shared by the people?

✅ **READING PROGRESS CHECK**

Describing Why are some people critical of mining and logging activities in the region?

A nurse checks a patient's treatment chart at a clinic in Brazzaville, Republic of the Congo.

Include this lesson's information in your Foldable®.

LESSON 3 REVIEW

Reviewing Vocabulary
1. Why was the use of a trade language helpful in Central Africa?

Answering the Guiding Questions
2. ***Describing*** What are two distinctive features of the Bambuti people, who live in the rain forests of the Democratic Republic of the Congo?

3. ***Determining Central Ideas*** What are some of the customs in Central Africa related to food?

4. ***Analyzing*** Why has the lack of political stability posed major problems for development in some Central African countries?

5. ***Informative/Explanatory Writing*** Write a paragraph or two in which you explain some of the economic issues confronting Central Africa.

Chapter 11 **345**

What Do You Think?

Has the United Nations Been Effective at Reducing Conflict in Africa?

The 53 independent states of Africa share one continent, but the vast land holds different people and varied resources and economies. During the past six decades, armed political conflict, either within a nation or between nations, has been the norm rather than the exception for many Africans. Since 1945, the United Nations (UN) Security Council, in its resolve to maintain international peace and security, has mounted 29 peacekeeping missions in Africa. Have the efforts of the United Nations been effective?

No!
PRIMARY SOURCE

" Africa has thus been a giant laboratory for UN peacekeeping and has repeatedly tested the capacity and political resolve of an often self-absorbed Security Council. . . . Under the loose heading of peacekeeping, the UN launched an unprecedented number of missions in the post–Cold War era. But . . . hard times appeared after disasters in Angola in 1992, when warlord Jonas Savimbi brushed aside a weak UN peacekeeping mission to return to war after losing an election; in Somalia in 1993, when the UN withdrew its peacekeeping mission after the death of eighteen U.S. soldiers; and in Rwanda in 1994, when the UN shamefully failed to halt genocide against about eight hundred thousand people and instead withdrew its peacekeeping force from the country. These events scarred the organization and made its most powerful members wary of intervening in Africa: an area generally of low strategic interest to them. "

—Adekeye Adebajo, Director, Centre for Conflict Resolution, University of Cape Town, South Africa, in *UN Peacekeeping in Africa: From the Suez Crisis to the Sudan Conflicts*

A militiaman stands guard with a machine gun in Somalia. The East African country has had political upheaval since 1991. Islamic militia forces and UN-backed government forces compete for control of Somalia.

Troops from Thailand have contributed to UN–African Union peacekeeping efforts in the war-torn Darfur region of Sudan. In Darfur, the Thai troops conducted patrols and provided protection to civilians and humanitarian aid workers.

Yes!

PRIMARY SOURCE

"Bearing in mind the relatively young period of existence of the PBC [Peacebuilding Commission] (and the fact that inevitably it was experimenting in its first two years), Sierra Leone and Burundi—the first two countries placed on the PBC agenda—seem to have achieved some real success in consolidating peace. Burundi has made headway with its peace process, including the gains relating to its inclusive political dialogue. Sierra Leone has also chalked up significant milestones and emerged as a post conflict state on the tracks of peace consolidation, with clear reforms in socioeconomic and security sector aspects. Both countries have also had to deal with some potentially serious setbacks and in both cases it seems that the proactive involvement of the PBC has added value in overcoming those problems."

—Security Council Report, Special Research Report No. 2, The Peacebuilding Commission, 17 November 2009

What Do You Think? DBQ

1 Identifying Point of View What evidence does the Yes answer contain to support its viewpoint? What evidence does the No answer contain to support its viewpoint?

2 Distinguishing Fact From Opinion The two sources state facts and opinions. Identify one fact and one opinion from each source.

Critical Thinking

3 Analyzing Weigh the evidence. Africa has had 29 peacekeeping missions in roughly 60 years. If the United Nations has to keep going back, have the missions been effective? Why or why not?

Chapter 11 ACTIVITIES

Directions: Write your answers on a separate piece of paper.

1 Use your FOLDABLES to explore the Essential Question.
INFORMATIVE/EXPLANATORY WRITING Some geographers believe the area of Central Africa that is now savanna was at one time forestland. What might have changed the physical composition of this area? Write a short essay discussing this hypothesis. How does this relate to the concept of people adapting to the environment?

2 21st Century Skills
IDENTIFYING Use the text and the Internet to find information about one of the countries of Central Africa. List important resources for that country.

3 Thinking Like a Geographer
ANALYZING Much of Central Africa depends on agriculture as a main economic activity. Why is a good transportation system important to an agricultural society?

4 GEOGRAPHY ACTIVITY

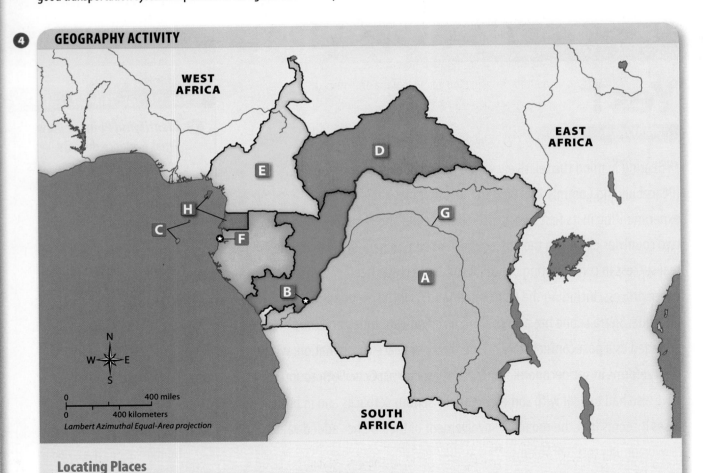

Locating Places
Match the letters on the map with the numbered places below.
1. Libreville
2. Congo River
3. Equatorial Guinea
4. Democratic Republic of the Congo
5. Brazzaville
6. Central African Republic
7. Cameroon
8. São Tomé and Príncipe

Chapter 11 Assessment

REVIEW THE GUIDING QUESTIONS

Directions: Choose the best answer for each question.

1. Choose Central Africa's primary landform.
    A. islands
    B. caves
    C. Congo River watershed
    D. steppes

2. Because Central Africa is centered on the Equator, its climate is
    F. Mediterranean.
    G. tropical.
    H. arid.
    I. mild.

3. Why were Europeans slow to establish colonies in Central Africa?
    A. The area had no worthwhile natural resources to exploit.
    B. They couldn't decide how to share the African countries.
    C. Transportation was difficult, and they feared malaria and other diseases.
    D. They didn't want to convert people.

4. The Republic of the Congo was a colony of which country?
    F. Spain
    G. France
    H. Portugal
    I. England

5. Most of the population of Central Africa makes a living by means of
    A. subsistence farming.
    B. factory labor.
    C. crafting.
    D. foreign trade.

6. Identify an obstacle to developing a modern economy in Central Africa.
    F. diverse populations
    G. commitment to traditional religions
    H. primitive housing construction
    I. insufficient infrastructure, poor transportation, and poor education

Chapter 11 Assessment (continued)

DBQ ANALYZING DOCUMENTS

7 ANALYZING INFORMATION Read the following passage about European colonial rule in Africa:

"The manner in which colonial administrations governed virtually ensured the failure of Africa's transition into independence. Their practice of 'divide and rule'—favoring some tribes to the exclusion of others—served to [heighten] the ethnic divisiveness that had been pulling Africa in different directions for centuries."

—from David Lamb, *The Africans* (1984)

What group does Lamb blame for the problems African nations have had creating stable governments since independence?
- A. the groups that led independence movements in Africa
- B. the wealthy, more-developed nations of the world today
- C. European colonial governments
- D. African leaders since independence

8 IDENTIFYING POINT OF VIEW Why would favoring some tribes over others cause problems?
- F. by draining the country of valuable resources
- G. by making disfavored tribes dependent on the colonial government
- H. by creating inequality that becomes entrenched
- I. by forcing tribes to find allies outside their country

SHORT RESPONSE

"The Congo's economy is based primarily on its petroleum sector, which is by far the country's greatest revenue earner. . . . A new potash mine . . . is expected to produce 1.2 million tons of potash . . . [used in fertilizer] per year by 2013. This will make Congo the largest producer of potash in Africa. . . . [One new iron ore mine] is thought to have ore reserves enough to be the world's third-largest iron ore mine."

—from "Republic of the Congo," State Department Background Notes

9 DETERMINING CENTRAL IDEAS What is the main idea of this passage?

10 IDENTIFYING POINT OF VIEW If oil is so valuable, why would the government of the Republic of the Congo bother developing potash and iron ore mines? Explain your answer.

EXTENDED RESPONSE

11 ANALYZING How was slavery in Central Africa and in the Americas similar?

Need Extra Help?

If You've Missed Question	1	2	3	4	5	6	7	8	9	10	11
Review Lesson	1	1	2	2	3	3	2	2	3	3	2

WEST AFRICA

ESSENTIAL QUESTIONS • How does physical geography influence the way people live? • How do new ideas change the way people live? • What makes a culture unique?

networks
There's More Online about West Africa.

CHAPTER 12

Lesson 1
Physical Geography of West Africa

Lesson 2
The History of West Africa

Lesson 3
Life in West Africa

Young woman wears dress of traditional Kente cloth at a cultural festival in Ghana

The Story Matters...

West Africa's diverse landscape includes desert, steppe, savanna, and tropical rain forests. Ancient rock paintings in Chad reveal details of the early herding societies who lived in this region. With rich deposits of gold and salt, trading kingdoms developed in Ghana and Mali. Trade also spread Islamic culture through much of the region. Gold attracted Europeans, whose colonial rule and slave trade impacted West Africa. Many of the independent nations of West Africa still struggle with civil war.

Study Organizer

Go to the Foldables® library in the back of your book to make a Foldable® that will help you take notes while reading this chapter.

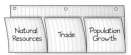

Chapter 12
WEST AFRICA CCSS

West Africa presents a rich variety of ethnic groups who speak many languages. Over the past 50 years, a number of the countries in the region have gained independence.

Step Into the Place

MAP FOCUS Use the map to answer the following questions.

1. **THE GEOGRAPHER'S WORLD** What are the two major lakes shown on the map?

2. **THE GEOGRAPHER'S WORLD** Which of the following countries is landlocked: Senegal, Burkina Faso, or Benin?

3. **PLACES AND REGIONS** What is the capital city of Liberia?

4. **CRITICAL THINKING Integrating Visual Information** If you are traveling from the capital city of Togo to the city of Bamako, in which direction are you traveling?

Step Into the Time

TIME LINE View the time line to answer these questions: In what year did Liberia declare independence? What event occurred in 1787?

1000 B.C. Bantu people of West Africa begin migrations south and east

250 B.C. Mali becomes center for trade

1300s Timbuktu is a center of Islamic culture

1324 Mansa Musa makes pilgrimage to Mecca

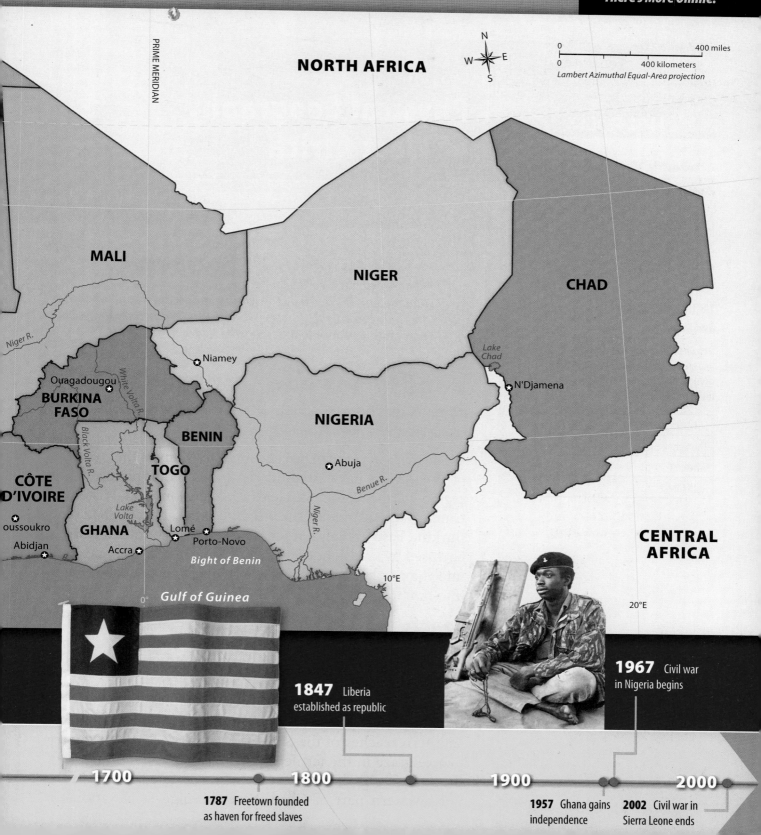

networks

There's More Online!

- ☑ **IMAGE** Lake Chad
- ☑ **SLIDE SHOW** West Africa's Variety of Land
- ☑ **VIDEO**

Reading HELPDESK

Academic Vocabulary
- volume

Content Vocabulary
- basin
- harmattan

TAKING NOTES: *Key Ideas and Details*

Organize As you read the lesson, fill in a chart like this one, discussing the dryness and vegetation in these different regions.

Sahara	
The Sahel	
Savanna	
Rain forest	

Lesson 1
Physical Geography of West Africa

ESSENTIAL QUESTION • *How does physical geography influence the way people live?*

IT MATTERS BECAUSE
West Africa has a varied landscape. Its vast grasslands are bounded on the north by the world's largest desert and on the south by coastal rain forests. West Africa provides an example of how cultures adapt to different climates and landforms.

Landforms and Bodies of Water

GUIDING QUESTION *In what ways do major rivers contribute to the economies of West African countries?*

West Africa is a land of broad contrasts. Many of the countries of the region have Atlantic coastlines. Four places in the region—Cape Verde, St. Helena, Ascension, and Tristan da Cunha—are islands in the Atlantic. Mali, Niger, Chad, and Burkina Faso are landlocked countries, or countries entirely enclosed by land. The southern reaches of the Sahara extend into the northern regions of the steppe, and savannas merge into rain forests in the south.

Landforms

Erosion has worn most of the land of West Africa into a rolling plateau that slopes to sea level on the coasts. West Africa has no major mountain ranges, but it has highland regions and isolated mountains. The Air, also known as the Air Massif, is a group of mountains in central Niger, along the southern reaches of the Sahara. The livestock of the Tuareg people graze in the fertile valleys between the mountains of the Air Massif. The Tibesti Mountains, located mostly in the northwestern part of Chad, have the highest elevations in

West Africa. The highest peak is Emi Koussi, an extinct volcano standing 11,204 feet (3,415 m) above sea level.

Southeast of the Tibesti Mountains, near Chad's eastern border, is an arid desert plateau region called the Ennedi. There are abundant wild game animals in the Ennedi. These animals attract a small population of seminomadic people who live here with their livestock during the rainy season. The Jos Plateau in central Nigeria is mostly open grassland and farmland.

In central Guinea is a highland region of savanna and deciduous forest known as the Fouta Djallon. It extends southeast to become the Guinea highlands, a humid, densely forested region.

Bodies of Water

The Niger River, West Africa's longest and most important river, originates at an elevation of 2,800 feet (853 m) in the Guinea highlands. It flows northeast toward the Sahara. Just beyond the town of Mopti in Mali, the Niger River enters the "inland delta." This is an area where the river spreads out across the relatively flat land into many creeks, marshes, and lakes that are connected to the river by channels. During the rainy season, the area floods completely. In the dry season, however, the waters recede, leaving fertile farmland.

As the Niger flows past Timbuktu, along the southern edges of the Sahara, the inland delta ends and the river returns to a single channel. As it flows through Niger and Nigeria, it joins its most important tributary, the Benue River, which doubles its **volume** of water. Where the Niger River reaches the Gulf of Guinea, it forms the Niger delta. This large area is a great **basin**, a lower area of land drained by a river and its tributaries. Along its 2,600-mile (4,184-km) course, the Niger provides water for irrigation and hydroelectric power. It is the main source of Mali's fishing industry, and it serves as an important route for transporting crops and goods.

Academic Vocabulary

volume an amount

Tuareg people lead their camels and goats to an oasis. The nomadic group tend to their livestock in the valleys of mountains. At one time, the Tuareg controlled the caravan trade routes across the Sahara.

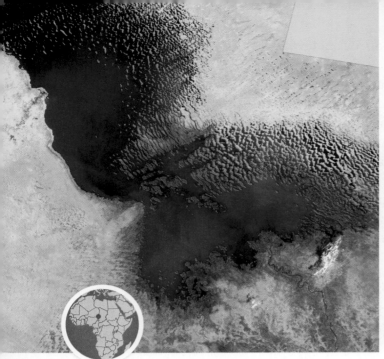
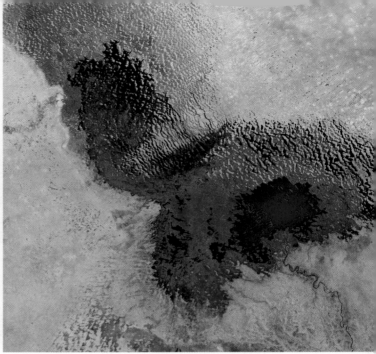

Satellite images show the Lake Chad area. Lake water is shown in blue, vegetation in red, and the surrounding desert in brown. The lake once was one of the largest bodies of water in Africa. The image on the left shows Lake Chad as it appeared in 1973. By 2007 (right), the lake had dramatically decreased in size.

▶ CRITICAL THINKING

Analyzing Taking water from the lake to use for irrigation is one reason that Lake Chad is smaller. What is another major cause of the lake shrinking in size?

The Senegal River rises in the Fouta Djallon in Guinea. Then it flows northwest to the Atlantic Ocean. For about 515 miles (829 km), the course of the Senegal River marks the border between the countries of Mauritania and Senegal.

The Black Volta River and the White Volta River originate in Burkina Faso and flow into Ghana. At a point near where the two rivers once met is the Akosombo Dam, which forms one of the world's largest artificial lakes—Lake Volta. The dam provides most of Ghana's electrical needs, and the lake provides water for irrigation.

Lake Chad covers portions of Niger, Nigeria, Chad, and Cameroon along the south part of the Sahara. The size of the lake varies from season to season, depending on the rainfall that feeds its tributaries. A series of droughts helped cause Lake Chad to shrink. Another cause was the amount of water taken from the lake and from the rivers that feed it to use for irrigating crops. People may be taking more water than water systems will be able to replace.

☑ READING PROGRESS CHECK

Describing Describe the landforms of West Africa.

Climate

GUIDING QUESTION *How do amounts of rainfall differ throughout West Africa?*

The climate of West Africa is diverse, from the harsh, arid Sahara in the north to the lush, coastal rain forests in the south. In between are vast stretches of grassland—the dry, semiarid Sahel and the lush, rainier savanna. The key characteristic of the climates throughout the region is its two distinct seasons: the wet season and the dry season.

Dry Zones

The Sahel is a semiarid region that runs between the arid Sahara and the savanna. Beginning in northern Senegal and stretching west into East Africa, the Sahel has a short rainy season. Annual rainfall ranges from only 8 inches to 20 inches (20 cm to 51 cm). This climate supports low grasses, thorny shrubs, and a few trees. The grasses are plentiful enough to support grazing livestock, such as cattle, sheep, camels, and pack oxen. It is important, however, that the herds do not grow too large. Too many animals leads to overgrazing and permanent damage to the grasslands. Overgrazing and too much farming result in desertification, the process in which semiarid lands become drier and more desert-like.

North of the Sahel is the Sahara. Daytime high temperatures are often more than 100°F (38°C) in the summer, though temperatures at night can drop by as much as 50°F. The few shrubs and other small plants that live in the Sahara must go for long periods without water. Some plants send long roots to water sources deep underground. After a rainfall, plants that have waited many months will suddenly flower, carpeting the desert with color.

From late November until mid-March, the dry, hot wind known as the **harmattan** blows through the Sahara with intensity. It carries thick clouds of dust that can extend hundreds of miles over the Atlantic Ocean, where it settles on the decks of ships. Harmattan winds contribute to the process of desertification.

Wet Zones

South of the Sahel, rains are more plentiful and feed the fuller, lusher plant life of the savanna. The savanna has two seasons—rainy and dry. The rainy season usually extends from April to September, and the rest of the year is dry. Annual rainfall reaches between 31 inches and 59 inches (79 cm and 150 cm). In some locations, though, annual rainfall can be as little as 20 inches (51 cm). This wide variability in rainfall makes human activity difficult.

The hot, dry wind that streams in from the northeast or east in the western Sahara is called the harmattan.

▶ **CRITICAL THINKING**
Describing During what months is the harmattan the strongest?

INFOGRAPHIC

WEST AFRICAN ENERGY

West Africa is relatively rich in energy resources. These resources are not, however, equally developed. Also, each form of energy has benefits and drawbacks.

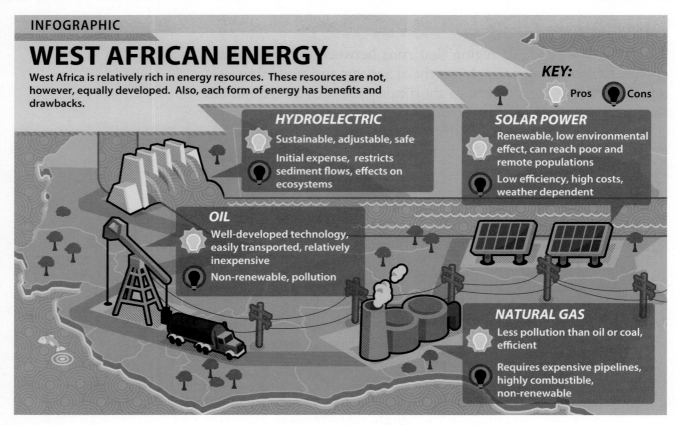

KEY: Pros / Cons

HYDROELECTRIC
- Sustainable, adjustable, safe
- Initial expense, restricts sediment flows, effects on ecosystems

SOLAR POWER
- Renewable, low environmental effect, can reach poor and remote populations
- Low efficiency, high costs, weather dependent

OIL
- Well-developed technology, easily transported, relatively inexpensive
- Non-renewable, pollution

NATURAL GAS
- Less pollution than oil or coal, efficient
- Requires expensive pipelines, highly combustible, non-renewable

West Africa is rich in energy resources. Because of the increasing demand for energy, West Africa's importance as an energy supplier to world markets continues to grow.

▶ **CRITICAL THINKING**
Analyzing What form of energy has little or no effect on the environment?

Farther to the south, tropical rain forests cover the land. The rain forests, known for their broad-leaved evergreen trees and rich biodiversity, receive plenty of rain. Rain forests in Sierra Leone and Liberia can receive as much as 200 inches (508 cm) of rain per year.

☑ **READING PROGRESS CHECK**

Identifying What are the two major grassland areas of West Africa?

Resources

GUIDING QUESTION *How do West African countries provide for their energy needs?*

West Africa contains many resources, but some countries lack the money to develop their resources into industries.

Energy Resources

Nigeria is the region's biggest producer of petroleum. In 2010 Nigeria ranked among the top 10 in the world, producing oil at a rate of almost 2.5 million barrels per day. The petroleum industry has not been as quick to tap the country's vast natural gas reserves.

Chad has oil fields in the area north of Lake Chad. Benin has offshore oil fields, as does Ghana. The oil in Ghana was considered too expensive to extract until increases in oil prices made it profitable to drill.

Ghana's main sources of electricity are two dams on the Volta River: the Akosombo Dam and the dam at Kpong. The Organization for the Development of the Senegal River, which is made up of several countries of the region, manages the river's resources, including hydroelectric stations. The Senegal River provides about half of the energy used in Mauritania today. Hydroelectric power is vital to meeting the energy needs in Togo and Nigeria.

Other Resources

The gold trade attracted Portuguese explorers, who first visited the region in 1471. For centuries afterward, the country was simply called the Gold Coast and Europeans lived and worked there, building trading posts and forts. Gold mining remains important to the economy of Ghana, along with the mining of diamonds, manganese, and bauxite. Ghana also has unmined deposits of limestone and iron ore.

Gold is mined in Mali, Burkina Faso, and Nigeria, too. Togo mines phosphate and limestone, which are used as fertilizers and to make paper, glass, paint, and other everyday products. Togo also has promising gold deposits, but so far no gold-mining industry. Niger has a salt-mining industry. In addition to its important gold industry, Mali also mines salt and limestone, but many of Mali's mineral resources are untapped. These include iron ore and manganese, which are important in making steel.

Mauritania mines copper and iron ore, but many of the iron ore deposits have been depleted. Nigeria mines iron ore, tin, limestone, and small quantities of other minerals. Burkina Faso is one of the world's leading sources of manganese. Benin has deposits of iron ore, limestone, chromium ore, gold, and marble. Benin is a leader in the production of hardwoods, but most of the rain forests where this wood comes from have been cleared.

Include this lesson's information in your Foldable®.

✓ **READING PROGRESS CHECK**

Analyzing Why is Benin's hardwood industry at risk? What is another industry in this region that has faced a similar problem?

LESSON 1 REVIEW

Reviewing Vocabulary
1. How can the *harmattan* winds influence climate and vegetation?

Answering the Guiding Questions
2. *Describing* Why has Lake Chad changed in size over time?
3. *Determining Central Ideas* Why is desertification an issue in West Africa?
4. *Identifying* Why did the government in Ghana change its mind about its offshore oil deposits?
5. *Informative/Explanatory Writing* Explain how desertification occurs and why it is such an important environmental issue in West Africa.

networks

There's More Online!

- ☑ **IMAGES** Freetown, Sierra Leone
- ☑ **MAP** The First Trading Kingdoms
- ☑ **VIDEO**

Reading HELPDESK CCSS

Academic Vocabulary
- displace
- element
- revenue

Content Vocabulary
- imperialism
- secede

TAKING NOTES: *Key Ideas and Details*

Describe On a chart like this one, write at least two different facts about the three ancient empires of Ghana, Mali, and Songhai.

Ghana	Mali	Songhai

360

Lesson 2
The History of West Africa

ESSENTIAL QUESTION • *How do new ideas change the way people live?*

IT MATTERS BECAUSE
West Africa was under the control of a series of wealthy trading kingdoms until the late 1800s, when Europeans seized control of their lands. Regaining their independence from Europe and establishing themselves in the modern world has been a challenge.

Ancient Times

GUIDING QUESTION What opportunities did Muslims from North Africa see in West Africa?

Early civilizations in West Africa learned to thrive in a variety of climates and landscapes, from the Sahara to the tropical rain forests. Throughout its history, the region was open to many migrations and invasions because of its resources.

Ancient Herders

We think of the Sahara as a hot, intensely dry place where few living things can survive. Ten thousand years ago, the Sahara was a much different place. Rock drawings from 8000 B.C. depict a world that looks more like a savanna than a desert. Drawings include lakes, forests, and large animals not seen in the Sahara in modern times: ostriches, giraffes, elephants, antelope, and rhinoceroses. Seminomadic people herded cattle and hunted wild animals. As the climate grew drier, fewer species of plants and animals survived the harsh environment. Many people moved south, following the retreat of the grasslands and the rain. During this period of desertification, people discovered that camels can survive without water for longer periods than cattle, sheep, or goats. Camels also can carry heavy loads for long distances. They were perfect domesticated animals for desert dwellers.

Movement of People

The Bantu people inhabited West Africa in ancient times. They had developed farming as early as 2000 B.C. A vast migration of Bantu people out of West Africa began around 1000 B.C. and continued for hundreds of years. Their farming practices allowed them to expand rather easily into areas occupied by hunter-gatherer groups. The Bantu became the dominant population in most of East and South Africa. The Bantu might have reached the East African coast as early as the A.D. 200s. They **displaced** many of the people who had lived on these lands before them.

Wherever the Bantu moved, they spread their culture, including three vital **elements**. First, the Bantu cultivated bananas, taro, and yams, which were originally from Malaysia. These crops thrived in the humidity of the tropical rain forests. Second, the Bantu spread their languages everywhere they settled. Today, more than 500 distinct Bantu languages are spoken by 85 million Africans. Third, the Bantu brought iron-smelting technology. They could create tools and weapons unlike any the native people had ever seen.

Academic Vocabulary

displace to take over a place or position of others

element an important part or characteristic

Trade Across the Sahara

For thousands of years, the Sahara was a barrier to contact between West Africa and North Africa. By the A.D. 700s, Arab Muslims had crossed the Sahara and conquered most of North Africa. They dominated the southern Mediterranean and controlled trade in that region, as well as Saharan trade routes into West Africa. Arab geographers slowly learned about Africa south of the Sahara. They realized that the region offered opportunities, not only in trade, but also in adding converts to Islam. One of the West African kingdoms they learned about was Ghana.

Ghana controlled the gold trade in the region. It traded gold for salt that Arab traders brought in from the Sahara. Salt was a very important trade good. It was the best preserver of food, and it was rare and difficult to acquire.

The Berber people had lived in North Africa long before the Arabs arrived. The Berbers resisted Islam for a while, but in time they converted. Several groups of Islamic Berbers joined together to become the Almoravids. They were fierce fighters, and they wanted to spread their new faith.

Merchants sell slabs of salt in a market town in central Mali. Salt is plentiful in the Sahara. These conditions gave rise to the African salt trade of ancient times.

▶ **CRITICAL THINKING**
Describing Why was salt a valuable item for trade?

Trading Kingdoms of West Africa

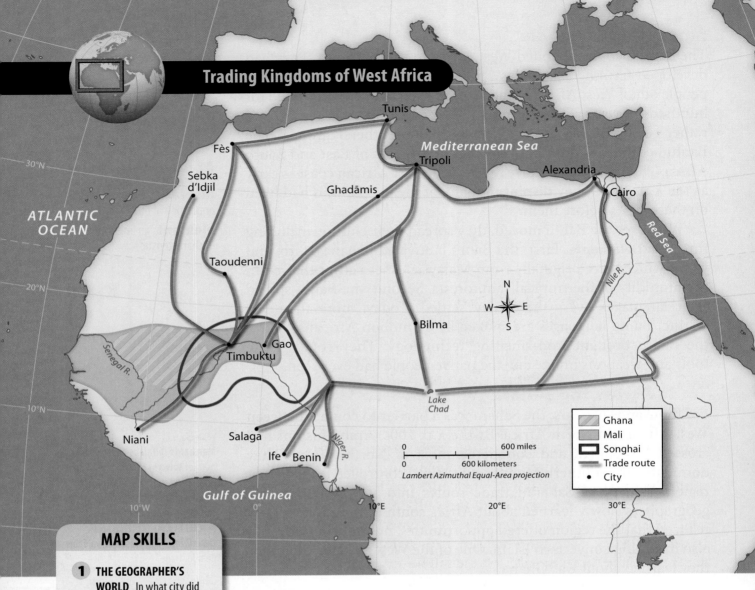

MAP SKILLS

1. **THE GEOGRAPHER'S WORLD** In what city did many of the trade routes merge?

2. **PLACES AND REGIONS** The early African kingdoms developed near gold-rich areas. What did a plentiful supply of gold allow them to do?

The First Trading Kingdom

The Ghana Empire was a powerful kingdom in West Africa during the Middle Ages. It had grown wealthy from its control of the gold trade. As the empire grew wealthy, it conquered many of its neighbors, including gold-rich lands to the south. Ghana built a strong trade with Muslim countries in North Africa.

In the A.D. 1000s, the Almoravids conquered Ghana. Almoravid rule over Ghana lasted only a few years, but that was long enough to damage the trade that kept the empire alive. With the arrival of so many Almoravids, there was now a larger population to feed. The dry climate could not support the sudden increase in agriculture, and this resulted in desertification: Land that at one time was fertile became desert. Ghana became weak, and its conquered neighbors began to break away.

☑ **READING PROGRESS CHECK**

Determining Central Ideas Why did the Ghana Empire collapse?

West African Kingdoms

GUIDING QUESTION *Why was the city of Timbuktu important to different trading kingdoms in West Africa?*

Several West African kingdoms sought to take control of the trade the fallen Ghana Empire had established in the region. Islam was taking hold in northern West Africa, and the most powerful of the new trading kingdoms were Muslim: Mali and Songhai.

Mali

The trading kingdom of Mali came after Ghana. Mali grew rich from the gold-for-salt trade. Its rulers conquered neighboring lands and built an empire.

The empire of Mali reached its height under the emperor Mansa Musa. The city of Timbuktu, which was already an important trading post, became a center for Islamic culture. Mansa Musa went on a historic pilgrimage to Mecca. By the time Mansa Musa returned home, the Islamic world knew there was a new and powerful Islamic kingdom south of the Sahara.

Songhai

After Mansa Musa's death, Songhai replaced Mali as the most powerful West African empire. A well-trained army and a navy that patrolled the Niger River made Songhai the largest of the three trading empires. Songhai now controlled the region's trade routes, had salt mines in the Sahara, and sought to turn their empire into the center for Islamic learning. The Songhai Empire eventually fell to the Moroccans, who seized the salt mines and destroyed the empire by the end of the 1500s.

Coastal Kingdoms

Slavery had been practiced in Africa for centuries. Muslims from North Africa and Asia had been buying enslaved people from south of the Sahara. As European colonists established colonies in the Western Hemisphere, they purchased enslaved people to do their labor. Small African kingdoms along the Atlantic coast became trading partners with Portugal, Spain, and Great Britain. Trade in enslaved persons became highly profitable in these kingdoms. When Europeans outlawed the slave trade, the kingdoms' economies began to fail.

✓ **READING PROGRESS CHECK**

Identifying How did the slave trade in West Africa change after the arrival of the Europeans?

Mansa Musa is shown sitting on his throne in this map of Africa from an atlas created in 1375.

European Domination

GUIDING QUESTION *What were some of the factors that aroused European interest in exploring and colonizing Africa?*

In the late 1800s, several European countries were eager to expand into Africa with its many natural resources. After hundreds of years of self-rule, West Africans lost control of their lands to the Europeans.

Changing Trade

The British government outlawed the slave trade in 1807. Great Britain tried but could not keep other European countries from continuing to export slaves. The British founded the colony of Freetown in Sierra Leone in 1787 as a safe haven for runaway or freed enslaved persons. The Americans followed suit in 1822, founding Liberia as a home for freed American slaves.

The British encouraged development of the palm oil trade to make up for the loss of **revenue** from the slave trade. British traders and missionaries familiarized themselves with the trading culture of the Niger River. The British started to profit from their growing role in West African trade. Before long, France took notice.

Creating Colonies

Before the mid-1800s, the British and French posted military in North Africa, and the British and Dutch had settled in South Africa. In 1869 two events increased Europe's interest in Africa. First, the Suez Canal—an artificial waterway connecting the Mediterranean Sea to the Red Sea—opened. Second, diamonds were discovered in South Africa.

Academic Vocabulary

revenue income

Freetown, the largest city in Sierra Leone, is located in the western part of the country along the Atlantic coast.
▶ **CRITICAL THINKING**
Describing What is unique about how Freetown was founded?

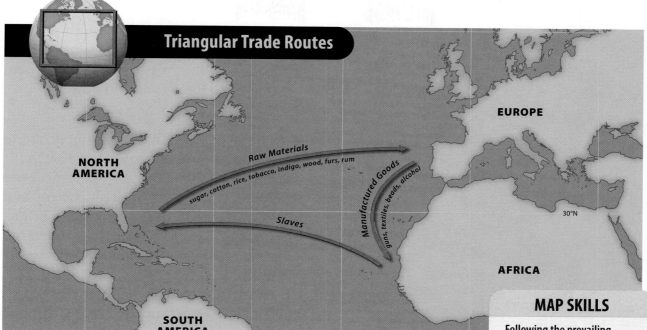

Triangular Trade Routes

MAP SKILLS

Following the prevailing winds, European ships sailed along the triangular trade route.

1. **THE GEOGRAPHER'S WORLD** Why were raw materials from the Americas shipped to Europe rather than to other parts of the world?

2. **PLACES AND REGIONS** Most European trade with Africa was conducted in West Africa. Why do you think this was so?

Before that time, Europeans had been in contact with the coastal kingdoms, but few had ventured into the interior of West Africa. This began to change as Great Britain, France, Portugal, Belgium, and the new, unified country of Germany began planning to carve out empires in Africa. Europeans followed a policy of **imperialism**, or seizing political control of other places to create an empire. At the Berlin Conference of 1884–1885, the European powers established rules for partitioning Africa.

The French and British claimed the most territory in West Africa. The modern countries of Benin, Burkina Faso, Chad, Côte d'Ivoire, Guinea, Mali, Mauritania, Niger, and Senegal were French colonies. Gambia, Ghana, Nigeria, Sierra Leone, and St. Helena were British colonies. Togo started out as a German possession, but Germany lost it in World War I.

Some colonies were ruled more harshly than others, but one thing was true about all of them: Europeans made the important decisions about how Africans lived. Settlers from Europe could set up farms on the most fertile land, even if Africans were forced off that land. Africans resisted, sometimes violently.

☑ READING PROGRESS CHECK

Describing What was the purpose of the Berlin Conference?

New Countries

GUIDING QUESTION *How does a ruler gain and hold political power?*

West African colonies began gaining independence in the late 1950s. The struggle for economic and political success was difficult because of the complex ethnic makeup of these countries.

Another factor working against political stability was the actions taken by the leaders. Often, they seized power and then used force to stay in power.

Ghana Leads the Way

The Gold Coast had been important to European trade since Portugal set up its first fort there in 1487. By the early 1800s, it was under British control, and in 1874 the Gold Coast became a British colony. The most important industries were gold, forest resources, and a newly introduced crop—the cocoa bean from which chocolate is made. By the 1920s, the Gold Coast supplied more than half the world's cocoa.

The leader of Ghana, Kwame Nkrumah (center), meets with a citizen in 1959. Bediako Poku (standing) headed the Convention People's Party (CPP), a political party Nkrumah started. A referendum election in 1964 made the CPP the nation's only legal party, and Nkrumah was installed as president of Ghana for life. The military and police ousted Nkrumah from power in 1966.

After the end of World War II, Dr. Kwame Nkrumah led protests calling for independence. Protests flared throughout the colony. In 1957 the Gold Coast gained its independence and became the Republic of Ghana. For many years, Ghana was troubled by conflict. In 1992 Ghana approved a new constitution that established a multiparty democracy. Since then, presidential elections have been peaceful, and Ghana has become a model of political reform in West Africa.

Former French Colonies

Throughout the 1950s, France made concessions to its West African colonies, many of which had instituted some form of self-rule. The independence of Ghana excited and inspired Africans, however. Cries for independence began erupting all over Africa. French Sudan and Senegal united to become the Republic of Mali, but Senegal broke away the following year, and Mali declared its own independence. In 1960 Mauritania, Niger, Côte d'Ivoire, Gambia, and Burkina Faso declared their independence. By 1961, all of France's colonies in West Africa were independent.

Nigeria

Nigeria had never been a country before the British combined two of their colonies to form the Nigerian Protectorate in 1914. Several different ethnic groups lived in this new territory. The Yoruba and Edo people had distinct territories in the south. The Hausa people were Muslims who lived in the north. The Igbo people were farmers who began to migrate east after British colonization.

When Nigeria gained its independence in 1960, tensions grew between ethnic groups. Because Britain had concentrated most of its development, including building schools, in southern Nigeria, most well-educated Nigerians lived in the south. They became leaders of the new government. Many of them were Igbo, who were mostly Christian. When they ventured north to govern the Hausa, they were met with hostility. Conflicts among the three major ethnic divisions in the country—the Hausa, the Yoruba, and the Igbo— grew violent. Thousands of Igbo living in northern Nigeria were massacred. As many as a million more fled to a region dominated by the Igbo. In 1967 the eastern region **seceded**, or withdrew formally, from Nigeria and announced that it was now the independent republic of Biafra. Nigerian forces invaded Biafra, and after more than two years of war, Biafra was in ruins. Starvation and disease may have killed more than 1 million people.

For long periods since then, the military has controlled the country's government. A new constitution was written in 1978. A year later, a democratically elected civilian government took office. That ended in a military coup in 1983. It was not until 1999 that another democratically elected president was able to rule Nigeria.

Civil War

Sometimes a military coup erupts into civil war. In late 1989, Charles Taylor led an invasion of Liberia. His aim was to depose the president, Samuel Doe. Ethnic conflict was at the heart of this struggle. Taylor and his rebels were of the Mano and Gio peoples. President Doe belonged to the Krahn people. After Doe's arrest and execution, Liberia endured seven years of civil war.

Conflict spilled over Liberia's borders into neighboring Sierra Leone. Thousands of civilians died, and many were forced to leave their homes. The civil war in Sierra Leone did not end until 2002. Estimates are that 50,000 died and another 2 million people lost their homes in the civil war. In 2012 Charles Taylor was brought to trial by a special court. He was found guilty of war crimes and crimes against humanity for his part in Sierra Leone's civil war.

✓ **READING PROGRESS CHECK**

Determining Central Ideas What reason might the French have had for letting their African colonies declare independence?

Thinking Like a Geographer

Religious Rivalry
In the A.D. 700s, Muslim people converted many West Africans to Islam. Over time, Islam became the dominant religion in northern parts of the region. Many West Africans along the coastal areas adopted Christianity after the arrival of the Europeans in the 1500s. Others continue to practice traditional African religions. Relations between these different religious groups have not always been good.

Include this lesson's information in your Foldable®.

LESSON 2 REVIEW

Reviewing Vocabulary
1. How did the Berlin Conference of 1884–1885 help achieve the goals of European *imperialism*?

Answering the Guiding Questions
2. **Determining Central Ideas** How did the Bantu culture influence the parts of Africa to which the Bantu migrated?
3. **Identifying** How did the Islamic world become aware of the wealth of the Mali Empire?
4. **Describing** Why did the British encourage development of the palm oil trade in African kingdoms on the Atlantic coast?
5. **Analyzing** Why did Nigeria face so many challenges when becoming an independent country?
6. **Argument Writing** Write a speech to the French government in the late 1950s about how important the independence of Ghana is and why France should release its colonies.

networks

There's More Online!

- ☑ **SLIDE SHOW** Daily Life in West Africa
- ☑ **VIDEO**

Reading HELPDESK

Academic Vocabulary
- diverse

Content Vocabulary
- pidgin
- creole
- animist
- extended family
- nuclear family
- kente
- infrastructure

TAKING NOTES: *Key Ideas and Details*

Compare and Contrast Use a graphic organizer like the one below to show some of the differences between life in a West African city and life in a rural area of West Africa.

City	Rural
•	•
•	•
•	•
•	•

Lesson 3
Life in West Africa

ESSENTIAL QUESTION • *What makes a culture unique?*

IT MATTERS BECAUSE

West African countries have faced many challenges in the decades since they achieved independence. Even as they struggle to modernize and improve their economies, West Africans have been able to spread their artistic gifts to people around the world.

The People of the Region

GUIDING QUESTION Why are two or more languages spoken in some West African countries?

West Africa is the most populous region of Africa. Many different ethnic groups live here, and each group brings something unique to their country's culture.

Ethnic Groups

European countries that carved up Africa to add to their overseas empires had their own reasons for setting colonial borders where they did. They did not consider the borders of the different ethnic groups of the Africans who already lived on the land they colonized. As a result, the European colonies in Africa contained populations that were ethnically **diverse**. When African countries declared independence, the new national borders closely followed the old colonial borders, preserving that diversity.

Establishing a sense of national unity is difficult, however. Many people have a stronger identification with their ethnic group than with the country they live in. For example, the Hausa are citizens of Nigeria, but many feel closer ties to the people and culture of the Hausa than to the country of Nigeria.

Another factor working against national unity is that a West African ethnic group does not typically live in only one country. Some Yoruba people, for example, live in Nigeria. Other Yoruba live in Benin. No matter where they live, the Yoruba feel a closer connection to other Yoruba people than to Nigeria or Benin.

Academic Vocabulary

diverse comprised of many distinct and different parts

Languages

When West African countries achieved independence, the majority held on to the European language that had been used most in business and government during the colonial period. The European languages—English, Portuguese, and French—became the official languages of West African countries. In some cases, Arabic is also an official language.

Most ethnic groups retain their traditional languages, and many use that language more than they do the official European language. In some places, European and African languages have intermixed to form a **pidgin** language. A pidgin is a simplified language used by people who cannot speak each other's languages but need a way to communicate. Sometimes two or more languages blend so well that the mixture becomes the language of the region. This is called a **creole** language. Crioulo is the language spoken most often in Cape Verde. It is a creole language, part Portuguese and part African dialect.

Religion

Islam was introduced to North Africa in the A.D. 600s, and it spread southward with trade. It was well established in many parts of West Africa long before the arrival of Christian missionaries from Europe. Today, a sizable portion of West African populations are Muslim.

Women talk near the Great Mosque in Djenne, Mali. The Great Mosque is the largest mud-brick building in the world. The first mosque on this site was built in the 1200s. The current mosque was built in 1907.

▶ **CRITICAL THINKING**

Describing In what region of Africa was Islam first introduced?

Lagos is Nigeria's largest city and ranks among the fastest-growing megacities in the world. During the 2000s, about 600,000 people have moved to Lagos every year.

▶ **CRITICAL THINKING**
Describing Are megacities the most common form of settlement in West Africa? Explain.

In some countries, however, Christianity is just as dominant as Islam, or more so. Even in countries where Christianity or Islam is dominant, many people still practice traditional African religions. These religions have their own rituals and celebrations that tie communities together. The beliefs are not written on paper but passed on from one generation to another. Often, the people believe in a supreme creator god and are **animists**, which means they believe in spirits—spirits of their ancestors, the air, the earth, and rivers.

Settlement Patterns

West Africa had few large towns until the colonial period. Even today, the most common settlement patterns in West Africa consist of scattered villages. Villages represent the homesteads of **extended families**, or families made up of parents, children, and other close relatives, often of more than two generations. The size of the villages and the population density depend on how much human activity the land can sustain. Water is an issue in the northern parts of West Africa, so populations are small and spread out.

Most of the largest cities in West Africa are capital cities, like Bamako, Mali, which more than tripled in population between 1960 and 1970, when droughts caused people to migrate from the countryside. The largest city in West Africa is Lagos in Nigeria, with an estimated 10.5 million people. Lagos was Nigeria's capital until 1991; Abuja is now Nigeria's capital city.

☑ **READING PROGRESS CHECK**

Determining Central Ideas What are two ways traditions have survived in West Africa?

Life and Culture in the Region

GUIDING QUESTION *Why are traditions more important in rural areas of West Africa than they are in West African cities?*

A variety of cultures thrive in West Africa, some of them traditional and some contemporary. Countries in the north—such as Mali, Mauritania, and Niger—are more influenced by the culture of North Africa than are their neighboring countries.

Daily Life

West Africa, with its complicated history and rich mixture of ethnicities, has a diverse culture. English or French may be the language of the cities, or of business and politics, but hundreds of other languages are still spoken. People in cities are more likely to wear Western-style clothing and to live and work in Western-style buildings. Far from the city, many people retain the traditions of their ancestors. These people are more likely to wear traditional clothing. Life in rural villages is built around the extended family. In cities, the **nuclear family**—parents with their children—is the more common family structure.

City dwellers are far more likely to deal with a wide variety of people over the course of a day than are rural dwellers. Capital cities teem with people from different ethnic groups, races, and countries. They are more likely to speak the official language because that is the language they all have in common. In rural areas, ethnicity and tradition are still important. Devotion to traditional values keeps those languages and cultures alive even as the country is changing. The downside is that ethnic pride sometimes results in conflicts between neighboring ethnic groups.

The Arts

West Africans have created numerous unique and important works of art in many artistic fields. Traditional artwork, such as carved masks from Nigeria and Sierra Leone, are world famous. Another well-known traditional art form is **kente**, a colorful, handwoven cloth from Ghana. One of West Africa's most important artist-figures is the griot, a musical storyteller who is part historian and part spiritual advisor.

Dance is the most popular form of recreation in West Africa. Workers use dance to celebrate their skills and accomplishments; professional guilds have their own dances. Dance is used for its healing qualities, but people also dance to popular music in clubs.

A worker weaves kente cloth in Ghana (above). The painted wood mask (below) comes from Burkina Faso.

Workers are on the job at a steel plant in Côte d'Ivoire. Like many other countries in the region, Côte d'Ivoire is developing its industries. Major manufacturing industries include bus and truck assembly and shipbuilding.

West African music blends a variety of many sounds, combining traditional and modern instruments. The Arabic influences of North African music mix with the music of sub-Saharan Africa, as well as with American and European rock and pop music.

West Africa has produced some of the world's finest writers. The most widely read of all African novels is *Things Fall Apart* by Chinua Achebe of Nigeria. Senegal's first president, Léopold Senghor, was a famous poet and lecturer on African history and culture.

✓ **READING PROGRESS CHECK**

Identifying Point of View Select an artistic field that you think best shows the culture of West Africa, and explain why you believe it does.

Challenges Facing the Region

GUIDING QUESTION *How did West African countries build up so much debt?*

Many West African countries face serious problems. To improve their situations, they must deal with these challenges and more: bad economies, corrupt governments, out-of-control population growth, disease, and poorly funded schools.

Government and Economics

The European powers that colonized Africa built the colonies' economies on a few key resources, such as petroleum, gold, peanuts, or copper. The **infrastructure**, or underlying framework of the colonies, was built around those key resources. Once the colonies achieved independence, it became important for them to develop a variety of different industries. Otherwise, any drop in the world price of a key resource would greatly affect a country's economy. Many foreign governments invested money in developing the industrial base of the new African countries. In many cases, money from the loans was not used wisely. As a result, the invested money did not yield high returns. West African countries were not only dealing with struggling economies but also faced enormous debt.

The International Monetary Fund and the World Bank have declared that no poor or developing country should have to pay a debt it cannot possibly manage. Debt relief has been important for countries such as Ghana, which suffered from bad economic choices and political instability. Ghana is now a model for economic and political reform in West Africa. Nigeria benefited from debt relief as well, and was able to pay off the remainder of what it owed in 2006.

A major challenge for the region is that, as the population grows, the demand for food and jobs grows. An economy does not have a chance to grow if it cannot meet the needs of a growing population.

Health and Education

Thirty-four million people in the world are living with the HIV virus, and 22.9 million of them live in sub-Saharan Africa. Dealing with such a large number of HIV-infected people presents several challenges. The first challenge is to supply health care to the growing number of people who carry the virus. The second challenge is to reduce the number of new HIV infections, usually through education. The third challenge is to deal with the families and communities hurt by AIDS-related deaths.

West Africa has a high birthrate, and the population is growing quickly, but life expectancy is still short compared to other parts of the world. Health care is an issue in large cities, but it is an even greater issue in rural areas.

The population of West Africa is young. For West Africa to have a sound future, educational systems must effectively prepare young people for economic and social development. A lack of education funding means that some countries cannot afford to make updates or improvements to their schools.

Include this lesson's information in your Foldable®.

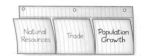

✓ **READING PROGRESS CHECK**

Analyzing Why is education such an important issue for the future of West Africa?

LESSON 3 REVIEW

Reviewing Vocabulary
1. How did the *infrastructure* of West African colonies lead to poor economies when they became independent countries?

Answering the Guiding Questions
2. ***Analyzing*** What are the advantages of a pidgin language?
3. ***Describing*** What is one potential downside to preserving traditional values in rural West Africa?
4. ***Identifying*** What are three major challenges to fighting AIDS in West Africa?
5. ***Narrative Writing*** Select a West African country and write a letter to the government of the country about the importance of investing in education. Explain how education will help solve the problems discussed in this lesson and how it will help preserve West African culture.

Chapter 12 ACTIVITIES

Directions: Write your answers on a separate piece of paper.

1. Use your **FOLDABLES** to explore the Essential Question.
   INFORMATIVE/EXPLANATORY WRITING Choose any three countries in this region and take a closer look at how the people who live there earn their living. Use the CIA Factbook or another Internet resource to list the top five occupations in each of the three countries and the average annual wage people earn for each. Arrange your data in a chart. Are any of those occupations directly related to the physical geography of the area? Explain in one or two paragraphs.

2. **21st Century Skills**
   INTEGRATING VISUAL INFORMATION In small groups, research the problems facing one of the countries of West Africa. Discuss possible solutions to one of the problems. Create a bulletin board display with pictures and captions to illustrate your solution to the problem.

3. **Thinking Like a Geographer**
   DETERMINING CENTRAL IDEAS After reviewing the chapter, choose five of the most important events in the history of West Africa. Place those events and their dates on a time line.

4. **GEOGRAPHY ACTIVITY**

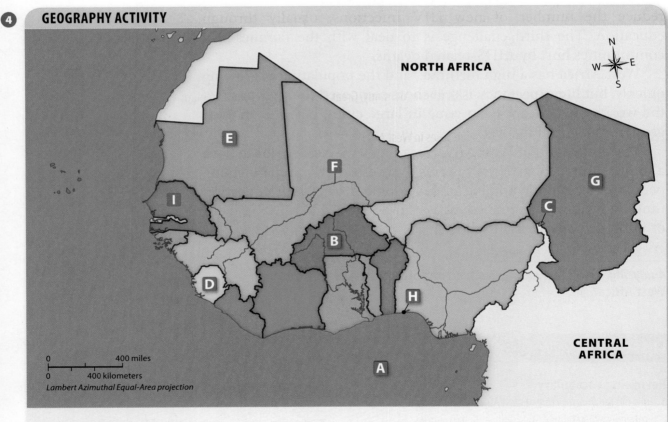

Locating Places
Match the letters on the map with the numbered places below.

1. Chad
2. Niger River
3. Lagos
4. Senegal
5. Mauritania
6. Gulf of Guinea
7. Burkina Faso
8. Lake Chad
9. Sierra Leone

Chapter 12 Assessment

REVIEW THE GUIDING QUESTIONS

Directions: Choose the best answer for each question.

1. The climate of northern West Africa can best be described as
   A. desert.
   B. savannah.
   C. tropical rain forest.
   D. marine west coast.

2. What attracted Portuguese explorers to Ghana in 1471?
   F. salt
   G. petroleum
   H. lithium
   I. gold

3. What did wealthy North African Muslim traders want from the people of West Africa?
   A. gold and converts to Islam
   B. salt and gold
   C. diamonds and emeralds
   D. ivory and camels

4. At the Berlin Conference of 1884–1885, France, Germany, Great Britain, Portugal, and Belgium decided to
   F. fund secondary schools and universities in West Africa.
   G. restrict the trade of enslaved people.
   H. heavily tax coffee and cocoa grown in West Africa.
   I. build empires by carving up and taking political control of African lands for themselves.

5. Traditional African religions are
   A. the dominant religions in West Africa.
   B. no longer practiced.
   C. only practiced in Mali.
   D. practiced, even in countries where Christianity or Islam is dominant.

6. When two or more languages blend and become the language of a region, the result is called a
   F. pidgin language.
   G. blended language.
   H. creole language.
   I. diverse language.

Chapter 12 ASSESSMENT (continued)

DBQ ANALYZING DOCUMENTS

7 DETERMINING WORD MEANINGS Read the following passage about the problem of desertification:

> "Nomads are trying to escape the desert, but because of their land-use practices, they are bringing the desert with them. It is a misconception that droughts cause desertification. Droughts are common in arid and semiarid lands. Well-managed lands can recover from drought when the rains return. Continued land abuse during droughts, however, increases land degradation."
>
> —from United States Geological Survey, "Desertification"

What does the passage mean in saying that nomads "are bringing the desert with them"?

A. Nomads bring their desert customs wherever they move.
B. Nomads create desert in new areas because of their land-use practices.
C. Nomads have the skills they need to survive in the desert.
D. Nomads can teach their way of life to other people.

8 IDENTIFYING What evidence does the passage give that droughts alone do not cause desertification?

F. the movement of nomads to new areas
G. the rate at which desertification takes place
H. droughts prevent desertification
I. the ability of lands to recover from drought

SHORT RESPONSE

> "The people of Ghana have . . . put democracy on a firmer footing, with repeated peaceful transfers of power. . . . This progress . . . will ultimately be more significant [than the struggle for independence]. For just as it is important to emerge from the control of other nations, it is even more important to build one's own nation."
>
> —from President Barack Obama, "Remarks to the Ghanian Parliament" (2009)

9 DETERMINING CENTRAL IDEAS How do "repeated peaceful transfers of power" show that democracy in Ghana is on a "firmer footing"?

10 IDENTIFYING POINT OF VIEW Why does President Obama think this achievement is more important than winning independence?

EXTENDED RESPONSE

11 ARGUMENT WRITING Ecotourism is often mentioned as a way that developing nations can use their natural resources to produce income while preserving those resources for future generations. Research the pros and cons of the topic, and decide whether or not ecotourism would be good for the countries of West Africa. Present your response in an essay.

Need Extra Help?

If You've Missed Question	1	2	3	4	5	6	7	8	9	10	11
Review Lesson	1	1	2	2	3	3	1	1	2	2	3

SOUTHERN AFRICA

ESSENTIAL QUESTIONS • How does geography influence the way people live?
• How do new ideas change the way people live?

Miner from Johannesburg, South Africa

networks

There's *More Online* about Southern Africa.

CHAPTER 13

Lesson 1
Physical Geography of Southern Africa

Lesson 2
History of Southern Africa

Lesson 3
Life in Southern Africa

The Story Matters...

From the steep slopes of the Great Escarpment to the plunging Victoria Falls, Southern Africa is filled with magnificent scenery and wildlife, which draw tourists from around the world. Southern Africa is also the continent's richest region in natural resources, including gold and diamonds. Many of Southern Africa's natural resources have become important to the global economy. Control over these vital resources has brought many great economic, political, and social changes to the region.

Go to the Foldables® library in the back of your book to make a Foldable® that will help you take notes while reading this chapter.

377

Chapter 13
SOUTHERN AFRICA

Most of inland Southern Africa is rich in resources and home to a wide variety of ethnic groups. The region's coastal and island countries are struggling to develop their economies.

Step Into the Place

MAP FOCUS Use the map to answer the following questions.

1. **PLACES AND REGIONS**
   Luanda is the capital city of what country?

2. **THE GEOGRAPHER'S WORLD**
   What country is located on the southern tip of the African continent?

3. **THE GEOGRAPHER'S WORLD**
   What countries share a border with Zimbabwe?

4. **CRITICAL THINKING**
   Integrating Visual Information Which of these places is the smallest in area: Lesotho, Gabarone, or Malawi?

PROVINCIAL CAPITAL The flat-topped Table Mountain overlooks the city of Cape Town, South Africa. Cape Town serves as the capital of the Western Cape province.

DESERT MAMMAL The meerkat is a member of the mongoose family. Only about 1 foot tall (30 cm), the meerkat stands by using its long tail for balance.

Step Into the Time

TIME LINE Based on events on the time line, predict the effects of Southern Africa's colonial past on the economy, government, and culture of the region.

A.D. 900 Kingdom of Great Zimbabwe established

1400s Mutapa Empire flourishes

1652 Dutch establish Cape Colony

1806 Britain gains control of Cape Colony

1800

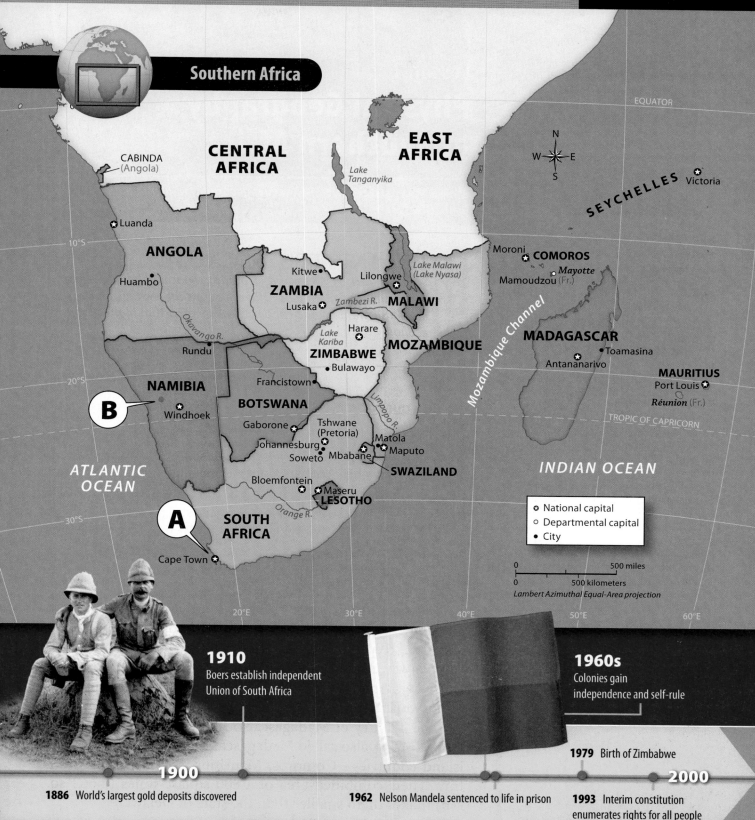

networks

There's More Online!

- ☑ **IMAGES** Etosha Pan
- ☑ **SLIDE SHOW** Diamonds
- ☑ **VIDEO**

Reading HELPDESK

Academic Vocabulary

- network

Content Vocabulary

- escarpment
- landlocked
- reservoir
- blood diamonds
- poaching

TAKING NOTES: *Key Ideas and Details*

Summarize As you read, list important details about two countries of Southern Africa in a graphic organizer like the one below.

Country:		
Physical Geography		
Climate		
Natural Resources		

380

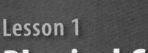

Lesson 1
Physical Geography of Southern Africa

ESSENTIAL QUESTION • *How does geography influence the way people live?*

IT MATTERS BECAUSE

Southern Africa is the world's leading producer of gold, platinum, chromium, and diamonds. Other minerals the region supplies, including uranium and copper, are also important in the global economy.

Landforms and Bodies of Water

GUIDING QUESTION *What are the dominant physical features of Southern Africa?*

The region of Southern Africa consists of the 10 southernmost countries on the African continent. It also includes four independent island countries and two French island territories in the Indian Ocean off Africa's east coast.

Southern Africa is bordered by the Indian Ocean on the east and the Atlantic Ocean on the west. The Cape of Good Hope at the southern tip of the continent is considered the place where the two oceans meet.

Several of the region's countries are fairly large. Angola and South Africa are each nearly the size of Western Europe and are the continent's seventh- and ninth-largest countries, respectively. Along with Namibia, Mozambique, Zambia, and Botswana, Angola and South Africa rank in the top 25 percent of the world's countries in land area.

The country of Madagascar occupies the world's fourth-largest island, also called Madagascar. The region's three other island countries—Comoros, Mauritius, and Seychelles—are tiny. Their combined area of 1,800 square miles (4,662 sq. km) makes them smaller than the state of Delaware.

Landforms

If Southern Africa's physical geography had to be described with one word, that word would be *high*. A series of plateaus that range in elevation from 3,000 feet to 6,000 feet (914 m to 1,829 m) cover most of the region. The northern plateaus extend from Malawi across Zambia and Angola. These plateaus are largely forested. Farther south, the plateaus are covered mainly by grasslands.

The plateau's outer edges form a steep slope called the Great Escarpment. In Angola, the **escarpment**, a steep cliff between a higher and a lower surface, runs parallel to the Atlantic Coast and continues through Namibia. Between the escarpment and the Atlantic Ocean lies a strip of desert called the Namib that is 80 miles to 100 miles (129 km to 161 km) wide. The Namib runs 1,200 miles (1,931 km) from southern Angola to western South Africa, where it merges with another desert, the Kalahari.

The Kalahari Desert is a vast, sand-covered plateau that sits some 3,000 feet (914 m) above sea level. It is bordered by even higher plateaus. The Kalahari covers much of eastern Namibia and most of Botswana. In some places, long chains of sand dunes rise as much as 200 feet (61 m) high. The sand in some areas is red because of minerals that coat the grains of sand.

South of the Kalahari Desert, much of the rest of Southern Africa is covered by a huge plateau that slopes from about 8,000 feet (2,438 m) in the east to 2,000 feet (610 m) in the west. At the southern tip of this plateau, the Great Escarpment breaks into several small, low mountain ranges. This group of ranges is known as the Cape Ranges. The ranges are separated from each other by dry basins called the Great Karoo and the Little Karoo.

As the Great Escarpment follows South Africa's coastline, it forms the Drakensberg Mountains. This is the most rugged part of the escarpment. Mountain peaks rise to more than 11,000 feet (3,353 m). A narrow coastal plain lies between the mountains and the Indian Ocean.

The Drakensberg is the highest mountain range in all of Southern Africa. The caves and rock shelters contain rock painting made by the San people. The San lived in the area for about 4,000 years.

As rivers spill from one plateau to the next, they create thundering waterfalls, such as the spectacular Victoria Falls.

▶ **CRITICAL THINKING**
Describing Where are the falls located?

The Drakensberg mountains parallel the Indian Ocean coastline for some 700 miles (1,127 km) through Lesotho and Swaziland, two **landlocked** countries in Southern Africa. Near Swaziland, the escarpment pulls back from the coastline to create a broad coastal plain that covers much of Mozambique. Northwestern Mozambique and the neighboring countries of Zimbabwe, Zambia, and Malawi lie at higher elevations west of the escarpment, on the plateau.

Bodies of Water

Three major river systems—the Zambezi, Limpopo, and Orange—drain most of Southern Africa. The Zambezi, which stretches for 2,200 miles (3,541 km), is the region's longest river. On the Zambia-Zimbabwe border, midway through its course, the Zambezi plunges over the spectacular Victoria Falls into a narrow gorge. Roughly a mile (1.6 km) wide and 350 feet (107 m) high, the falls are about twice the width and height of Niagara Falls in North America. Because of the heavy veil of mist that rises from the gorge, the area's indigenous people named the falls *Mosi-oa-Tunya* ("The Smoke That Thunders").

The Orange River is Southern Africa's second-longest river. It begins in the highlands of Lesotho and flows westward to reach the Atlantic Ocean. Its course marks the southern boundary of the Kalahari Desert. The region's third-longest river, the Limpopo, flows eastward in a large arc along South Africa's border with Botswana and Zimbabwe. The river then drops over the Great Escarpment to cross the plains of southern Mozambique to the Indian Ocean.

These three rivers, their tributaries, and Southern Africa's other rivers have carved a **network** of canyons and gorges across the plateaus. Dams have been built in the area to store water. Lake Kariba, Southern Africa's second-largest lake, is really a **reservoir**, or an artificial lake created by a dam.

The region's largest lake—and the third largest in all of Africa—is Lake Malawi (also known as Lake Nyasa), which forms Malawi's border with Mozambique and Tanzania. It is the southernmost lake of the Great Rift Valley which stretches for thousands of miles. Lake Malawi fills a depression, or hollow, that follows one of the rifts, or tears, in Earth's crust. Because of the great depth of the depression, Lake Malawi is one of the deepest lakes in the world.

A number of flat basins, called pans, can be found in Southern Africa. The salt deposits they contain provide nourishment for wild animals. Etosha Pan, in northern Namibia, is an enormous expanse of salt that covers 1,900 square miles (4,921 sq. km). It is the largest pan in Africa, and it is the center of Etosha National Park. The park is home to some of the greatest numbers of lions, elephants, rhinoceroses, and other large animals in the world.

✔ **READING PROGRESS CHECK**

Identifying Which type of landform is common in Southern Africa?

Thinking Like a Geographer

Pans

Pans are believed to be the beds of ancient lakes whose water evaporated over time. They are among the flattest known landforms. Small amounts of rain can flood large areas of their surface. It is this flooding that causes and maintains their flatness. Salt deposits form as rainwater pools slowly evaporate. *Why can a small amount of rain flood a large area of a pan?*

Academic Vocabulary

network a complex, interconnected chain or system of things such as roads, canals, or computers

Zebras are among the many animals that live in the national park that is part of the Etosha Pan.
▶ **CRITICAL THINKING**
Describing What is unique about the Etosha Pan?

Climate

GUIDING QUESTION *What is the climate of Southern Africa?*

Southern Africa has a wide variety of climates, ranging from humid to arid to hot to cool. Nearly all of the region's climates have distinct seasons, with certain seasons receiving most of the rain.

Tropical Zone

The Tropic of Capricorn crosses the middle of Southern Africa. This places the northern half of the region in the Tropics. Northern Angola and northern Mozambique have a tropical wet-dry climate. Each area gets as much as 70 inches (178 cm) of rain per year. Most of it falls in the spring, summer, and fall—from October to May. The high elevation makes temperatures cool.

People struggle to wade through the waters after a heavy rainfall in the coastal city of Maputo in southwestern Mozambique.

▶ CRITICAL THINKING
Describing How does the length of the rainy season in western Mozambique compare with the length in the northern part of the country?

Daily average temperatures range from the upper 60s°F (upper 10s°C) to the upper 70s°F (mid-20s°C). Along the coasts, temperatures are warmer.

Much of northern Mozambique's coastline is watered by rain-bearing winds called monsoons that sweep in from the Indian Ocean during the summer months. More than 70 inches (178 cm) of annual rainfall is common.

Parts of Angola and Mozambique have humid subtropical climates, as do Malawi, Zambia, and northeastern Zimbabwe. The rainy season here is shorter than in the tropical wet/dry zone, and also brings less rainfall. Most places average 24 inches to 40 inches (61cm to 102 cm) per year. Average temperatures are also slightly cooler. Nighttime frosts are not uncommon in July on the high plateaus of Zambia and Malawi. Temperatures on summer days in lowland areas, however, can exceed 100°F (38°C).

Temperate Zones

Much of South Africa, central Namibia, eastern Botswana, and southern Mozambique have temperate, or moderate, climates that are not marked by extremes of temperature. Most of these areas are semiarid. Summer days are warm—from 70°F to 90°F (21°C to 32°C), depending on elevation. Winters are cool, with frosts and sometimes freezing temperatures on the high plateaus.

Annual rainfall varies from 8 inches (20 cm) in some areas to 24 inches (61 cm) in others. Most of the rain falls during the summer, with very little the rest of the year. Droughts are common; in some places, they last for several years.

Lesotho, Swaziland, and eastern South Africa, including the Indian Ocean coastline, are much wetter. Temperatures are like those in the semiarid regions, but ocean currents and moist ocean air bring up to 55 inches (140 cm) of rain annually. Like elsewhere in the region, most of this rain falls in the summer.

Desert Regions

Western South Africa, western Namibia, and much of Botswana are arid. Along the coast, the Namib gets very little rain. In some years, no rain falls. But fog and dew provide small plants with the moisture they need to survive. Temperatures along the coast are mild, however, with daily averages ranging from 48°F to 68°F (9°C to 20°C). The aridity, the fog, and the mild temperatures result from the cold Benguela Current that flows along the coast. This area is sometimes called the "Skeleton Coast" because many ships used to lose their way in the fog and run aground. Once ashore, the sailors rarely survived because of the lack of water in the sandy desert.

In inland areas of the Namib Desert, temperatures are hotter with summer highs from the upper 80s°F to more than 100°F (30°C to 38°C). In winter, freezing temperatures sometimes occur. During wet years, desert grasses and bushes appear. Much of the time, however, the Namib is home to vast areas of barren sand.

The Kalahari's location—farther inland than the Namib—and dry air make its temperatures more extreme than in the Namib. The Kalahari also gets a little more precipitation than the Namib.

☑ **READING PROGRESS CHECK**

Describing Why are temperatures in Southern Africa's tropical countries generally not hot?

Natural Resources

GUIDING QUESTION *What natural resources are found in Southern Africa, and why are they important?*

Southern Africa is the continent's richest region in natural resources. Mineral resources have helped the Republic of South Africa, in particular, to build a strong economy. In other countries, like Angola and Namibia, such resources provide the only source of wealth.

The landscape of the Skeleton Coast is made up of sand dunes, rocky canyons, and mountains. Dense fogs and cool sea breezes are characteristic of the area.

▶ **CRITICAL THINKING**

Describing How did the Skeleton Coast get its name?

More than one-half of the world's diamonds are harvested from mines, such as this one, in Southern Africa. Diamonds were formed deep in Earth thousands of years ago under extreme heat and pressure. Volcanic pressure brings them to Earth's surface.

South Africa's Resources

The Republic of South Africa has some of the largest mineral reserves in the world. It is the world's largest producer of platinum, chromium, and gold, and one of the largest producers of diamonds—both gems and industrial diamonds, or diamonds used to make cutting or grinding tools. These resources, along with important deposits of coal, iron ore, uranium, copper, and other minerals, have created a thriving mining industry. This industry has attracted workers and investments from other countries that have helped South Africa's industries grow.

Energy Resources

The Republic of South Africa, Zimbabwe, Botswana, and Mozambique mine and burn coal from their own deposits to produce most of their electric power. Mozambique has large deposits of natural gas as well, as does Angola. Angola is also one of Africa's leading oil producers. Namibia has oil and natural gas deposits, too, and they are slowly being developed. Oil and gas must be refined, or changed into other products, before they can be used.

The region's rivers are another resource for providing power. Zimbabwe and Zambia get electricity from the huge Kariba Gorge dam on the Zambezi River. Malawi's rivers and falls generate power for that country. Deforestation, however, allows more sediment to enter the rivers, which reduces the water flow and the electricity that the rivers produce. Mozambique, Zimbabwe, and Angola have not made full use of their rivers to provide power. Economic development and the standard of living in those countries have suffered as a result.

Minerals and Other Resources

Namibia is one of Africa's richest countries in mineral resources. It is an important producer of tin, zinc, copper, gold, silver, and uranium. It also ranks with South Africa and Botswana as a leading world supplier of diamonds. In the 1990s, rebels captured Angola's mines and sold the diamonds to continue a 20-year-old civil war against the government. In countries outside Southern Africa, groups have also mined diamonds to pay for rebellions and other violent conflicts. Diamonds used for this purpose are called **blood diamonds**.

Gold is a leading export for Zimbabwe. Mozambique has the world's largest supply of the rare metal tantalite, which is used to make electronic parts and camera lenses. Gold, platinum, and diamonds are mined there too, as are iron ore and copper. Much of Zambia's economy is based on copper and cobalt, although gold, silver, and iron ore are also mined. Zambia has some of the largest emerald deposits in the world. A small amount of rubies, sapphires, and a variety of semiprecious gems are mined in neighboring Malawi.

Malawi's most important natural resource is its fertile soil. The country's economy is based mainly on agriculture. Tobacco is its most important export. Exporting farm products is also a major economic activity in Zimbabwe. Lesotho and Swaziland have few natural resources. Most of their people practice subsistence farming, growing only enough to meet their needs.

Wildlife

Southern Africa is known for its variety of animal life. Wildebeests, lions, zebras, giraffes, and many other animals are found across the region. They live within and outside the many national parks and wildlife reserves that nearly every country has created to protect them. Tourists come from throughout the world to see these animals. **Poaching**, or illegally killing game, is a problem. Poachers shoot elephants for their valuable ivory tusks and rhinoceroses for their horns. Others kill animals to sell their skins and meat and to protect livestock and crops.

Include this lesson's information in your Foldable®.

✓ **READING PROGRESS CHECK**

Describing How does deforestation affect the energy supply in the region?

LESSON 1 REVIEW

Reviewing Vocabulary
1. Why is *poaching* against the law?

Answering the Guiding Questions
2. **Describing** How has damming Southern Africa's rivers benefited the people and countries of the region?
3. **Identifying** What are the rainfall and temperature differences between Southern Africa's tropical, temperate, and arid regions?
4. **Describing** For what resources is Southern Africa known throughout the world?
5. **Narrative Writing** Create a journal entry recording your observations and experiences during one day of a photo safari at Etosha National Park.

networks

There's More Online!

 VIDEO

Reading HELPDESK CCSS

Academic Vocabulary
- exploit
- grant

Content Vocabulary
- apartheid
- civil disobedience
- embargo

TAKING NOTES: *Key Ideas and Details*

Sequence Create a time line like this one. Then list five key events and their dates in the history of the region.

←—+——+——+——+——→

Lesson 2
History of Southern Africa

ESSENTIAL QUESTION • *How do new ideas change the way people live?*

IT MATTERS BECAUSE
Many of Southern Africa's resources have become important parts of the global economy. Political instability and unrest have sometimes disrupted the flow of products to world markets. Much of the instability and unrest is directly or indirectly the result of the region's colonial history.

Rise of Kingdoms

GUIDING QUESTION What major events mark the early history of Southern Africa?

Southern Africa's indigenous people have inhabited the region for thousands of years. Some lived as hunter-gatherers. Others farmed and herded cattle. Trade among the groups flourished. Ivory, gold, copper, and other goods moved from the interior to the east coast. There such goods were exchanged for tools, salt, and luxury items including beads, porcelain, and cloth from China, India, and Persia.

Great Zimbabwe

Around the year A.D. 900, the Shona people built a wealthy and powerful kingdom in what is now Zimbabwe and Mozambique. The capital was a city called Great Zimbabwe. (*Zimbabwe* is a Shona word meaning "stone houses.") As many as 20,000 people lived in the city and the surrounding valley.

Great Zimbabwe was the largest of many similar cities throughout the region. By the 1300s, it had become a great commercial center, collecting gold mined nearby and trading it to Arabs at ports on the Indian Ocean.

Great Zimbabwe was abandoned in the 1400s, possibly because its growing population exhausted its water and food resources. The city's ruins show the Shona's skill as builders. Some structures were more than 30 feet (9 m) high. Their large stones were cut to fit and stay in place without mortar to hold them together.

The Mutapa Empire

In the late 1400s, the Shona conquered the region between the Zambezi and Limpopo rivers from Zimbabwe to the coast of Mozambique. Like Great Zimbabwe, the Mutapa Empire thrived on the gold it mined and traded for goods from China and India.

The Portuguese arrived and took over the coastal trade in the 1500s. They gradually gained control over the empire and forced its people to mine gold for them. In the late 1600s, Mutapa kings allied with the nearby Rozwi kingdom to drive out the Portuguese. Instead, the Rozwi conquered the Mutapa's territory and ruled it until the early 1800s, when it became part of the Zulu Empire.

Other Kingdoms

The Zulu leader Shaka united his people in the early 1800s to form the Zulu Empire in what is now South Africa. He built a powerful army and used it to expand the empire by conquering neighboring people. Shaka was killed in 1828, but his empire survived until the British destroyed it in the Zulu War of 1879.

A series of kingdoms rose and fell on the island of Madagascar from the 1600s to the 1800s. Some of the early kingdoms were influenced by Arab and Muslim culture. In the early 1800s, one king allied with the British on the nearby island of Mauritius to prevent the French from taking control of Madagascar. He eventually conquered most of the island and formed the Kingdom of Madagascar. French troops invaded the kingdom in 1895 and made it a French possession.

✅ **READING PROGRESS CHECK**

Identifying Which outsiders traded with Southern Africans before the Europeans arrived?

Shown are remnants of the walls of the Great Enclosure of the city of Great Zimbabwe. According to historians, houses of the royal family were located within the walls.

Identifying How did Great Zimbabwe become an important center of trade?

European Colonies

GUIDING QUESTION How did Southern Africa come under European control?

Around 1500, Portugal and other European countries began establishing settlements along the African coast. The first settlements were trading posts and supply stations at which ships could stop on their way to and from Asia. As time passed, the Europeans grew interested in **exploiting** Africa's natural resources and, as a source of labor, its people.

Clashes in South Africa

During the 1600s till about the 1800s, Europeans set up trading posts but did not establish colonies, which are large territories with settlers from the home country. One exception was Cape Colony, founded by the Dutch in 1652 at the Cape of Good Hope on the southern tip of what is now South Africa. The Dutch became known as Boers, the Dutch word for farmers. They grew wheat and raised sheep and cattle. Enslaved people from India, Southeast Asia, and other parts of Africa provided much of the labor.

The Africans did not like the Dutch pushing into their land, and soon they started fighting over it. By the late 1700s, the Africans had been defeated. Some fled north into the desert. Others became workers on the colonists' farms.

The Union of South Africa

Wars in Europe gave Britain control of the Cape Colony in the early 1800s. Thousands of British settlers soon arrived. The Boers resented British rule. Many decided to seek new land beyond the reach of British control. Beginning in the 1830s, thousands of Boers left the colony in a migration called the Great Trek and settled north of the Orange River.

In the 1860s, the Boers discovered diamonds in their territory. Then, in 1886, they found the world's largest gold deposits. British efforts to gain these resources led to the Boer War in 1899. The Boers were defeated and again came under British control. In 1910 Britain allowed the Boer colonies to join the Cape Colony in forming an independent country—the Union of South Africa. The small African kingdoms of Lesotho and Swaziland remained under British control.

Three Zulu leaders are shown holding shields and wearing traditional attire. The Zulu built a great empire, but during the 1800s, European settlers took control of their grazing and water resources. The Zulu population is about 9 million today, making them the largest ethnic group in the Republic of South Africa.

Academic Vocabulary

exploit to make use of something, sometimes in an unjust manner for one's own advantage or gain

Colonialism in Other Areas

While the British and the Boers competed for South Africa, other European countries were competing over the rest of Africa. In 1884 representatives of these countries met in Berlin, Germany, to divide the continent among themselves.

In Southern Africa, Britain gained control over what is now Malawi, Zambia, Zimbabwe, and Botswana. The Berlin Conference decided Portugal had rights to Angola and Mozambique. Germany received what is now Namibia, although South Africa seized the colony during World War I. Besides Madagascar, France controlled what is now Comoros. Mauritius and Seychelles were British colonies.

European control in Southern Africa continued for about the next 80 years. Not until the 1960s did the region's colonies begin to gain independence and self-rule.

✓ **READING PROGRESS CHECK**

Analyzing Which European country claimed the most territory in Southern Africa in the 1800s?

Independence and Equal Rights

GUIDING QUESTION *What challenges did Southern Africans face in regaining freedom and self-rule?*

French rule in Madagascar ended in 1960, making it the first Southern African country to gain independence. Britain **granted** independence to Malawi and Zambia in 1964 and to Botswana and Lesotho in 1966. Swaziland and Mauritius gained their freedom in 1968, and Seychelles in 1976. Elsewhere, however, freedom was more difficult to achieve.

Academic Vocabulary

grant to permit as a right, a privilege, or a favor

Boer soldiers fight from trenches at the siege of Mafeking in 1900. The siege, lasting more than 200 days, resulted in an important victory for British forces.

▶ **CRITICAL THINKING**

Describing Who were the Boers? Why were the Boer Wars fought?

Laws in South Africa limited the political rights of black Africans and set up separate parks, beaches, and other public places.

▶ **CRITICAL THINKING**
Describing Who controlled South Africa's government until World War II? Who controlled the government beginning in 1948?

The End of Portuguese Rule

While other European nations gave up their African colonies, Portugal refused to do so. Revolts for independence broke out in Angola in 1961 and in Mozambique in 1964. The thousands of troops Portugal sent to crush these revolts failed to do so.

By 1974, the Portuguese had grown tired of these bloody and expensive wars. Portuguese military leaders overthrew Portugal's government and pulled the troops out of Africa. Angola and Mozambique became independent countries in 1975 as a result. Fighting continued, however, as rebel groups in each country competed for control. Mozambique's long civil war ended when a peace agreement was reached in 1994. Peace was not finally achieved in Angola until 2002.

The Birth of Zimbabwe

After granting Malawi and Zambia independence, Britain prepared to free neighboring Zimbabwe, then called Southern Rhodesia. The colony's white leaders, who controlled the government, instead formed a country they called Rhodesia and continued to rule.

Rhodesia's African population demanded the right to vote. When the government resisted, a guerrilla war began. In 1979 the government finally agreed to hold elections in which all Rhodesians could take part. Rebel leader Robert Mugabe was elected president, and Rhodesia's name was changed to Zimbabwe.

Equal Rights in South Africa

After independence, the growth of South Africa's mining and other industries depended on the labor of black Africans, who greatly outnumbered the country's whites. The white minority government stayed in power by limiting the black population's educational and economic opportunities and political rights.

English South Africans controlled the government until the end of World War II. Then a strike by more than 60,000 black mine workers frightened white voters into electing an Afrikaner government in 1948 that promised to take action. (Afrikaners are the descendants of the Boers. They speak a language called Afrikaans, which gives them their name.)

The new government leaders began enacting laws that created a system called **apartheid**—an Afrikaans word meaning "apartness." Apartheid limited the rights of blacks. For example, laws forced black South Africans to live in separate areas called "homelands." People of non-European background were not even allowed to vote. The African National Congress (ANC), an organization of black South Africans, began a campaign of **civil disobedience**, disobeying certain laws as a means of protest. The government's violent response to peaceful protests caused the ANC to turn to armed conflict. In 1962 ANC leader Nelson Mandela was arrested and sentenced to life in prison.

By the 1970s, apartheid-related events in South Africa had gained world attention. Countries began placing **embargos**, or bans on trade, on South Africa. Meanwhile, the struggle in South Africa grew more intense. In 1989 South Africa's president, P.W. Botha, was forced to resign. In 1990 the government, under Botha's successor, F.W. de Klerk, began repealing the apartheid laws. Mandela was released from prison in 1991. In 1993 a new constitution gave South Africans of all races the right to vote. The ANC easily won elections held in 1994, and Mandela became the country's president.

In 1995 the new government created a truth and reconciliation commission. Its task was to ease racial tensions and heal the country by uncovering the truth about the human rights violations that had occurred under apartheid.

By 1994, South Africa's policy of apartheid was officially over. Nelson Mandela became the first black person to be elected president of South Africa. Mandela is shown voting for the first time in his life on April 27, 1994.

Include this lesson's information in your Foldable®.

✓ **READING PROGRESS CHECK**

Determining Central Ideas Why do you think South Africa's government created the apartheid system?

LESSON 2 REVIEW

Reviewing Vocabulary
1. Why might some people disapprove of *civil disobedience* as a means of protest and of achieving change?

Answering the Guiding Questions
2. *Analyzing* How did some of Southern Africa's early people benefit from the region's natural resources?

3. *Identifying* Name five present-day countries in Southern Africa that were once controlled by Britain.

4. *Determining Central Ideas* Why was gaining independence especially difficult for Angola and Mozambique?

5. *Argument Writing* Write a paragraph explaining whether actions against the governments of Rhodesia and South Africa were justified.

Chapter 13 **393**

networks

There's More Online!

- ☑ **SLIDE SHOW** Cities in Southern Africa
- ☑ **VIDEO**

Reading HELPDESK

Academic Vocabulary
- contact
- trend

Content Vocabulary
- utility
- thatch
- periodic market

TAKING NOTES: *Key Ideas and Details*

Summarize Create a chart like this one. Then list information about Southern Africa on these three topics.

Population	
Culture Groups	
Health Issues	

Lesson 3
Life in Southern Africa

ESSENTIAL QUESTION • *How does geography influence the way people live?*

IT MATTERS BECAUSE
Control over Southern Africa's vast and vital natural resources has been passed on to new leadership. Great economic, political, and social changes and challenges have accompanied this transfer.

The People of the Region

GUIDING QUESTION *Where do people live in Southern Africa?*

The population of Southern Africa is overwhelmingly black African. The largest white minority is in the country of South Africa, where whites represent 10 percent of the population. In almost every other country, whites and Asians make up less than 1 percent of the population. The region's black African population is made up of many different ethnic and culture groups.

Population Patterns

Southern Africa's countries vary widely in population. Fewer than 2 million people live in the small countries of Lesotho and Swaziland. South Africa, which surrounds both of them, has the region's largest population—about 49 million.

Population depends heavily on geography and economics. For example, Botswana and Namibia are much larger than Swaziland and Lesotho, but their populations are only slightly larger. Most Batswana, as the people of Botswana are called, live in the northeast, away from their country's desert areas. Similarly, most Namibians live in the northern part of their country, away from the arid south and west.

South Africa and Angola are about the same size. South Africa, the region's most industrialized nation, has three times as many people. In both countries, most people live in

394

cities. Angola's rural areas are thus much more thinly populated than rural areas in South Africa.

Mozambique, which is slightly smaller than Namibia and much smaller than Angola, has a population greater than those two countries combined. Most of Mozambique's 23 million people are engaged in farming, mainly along the fertile coastal plain.

Zambia is twice as big as Zimbabwe. Zimbabwe, with a population of about 12 million, has only 2 million fewer people. Both countries are largely rural, with only about one-third of their people living in cities. Large parts of Zambia are thinly populated.

Malawi is just one-third the size of Zimbabwe and one-sixth the size of Zambia, yet it exceeds both in population. With some 16 million people living in an area roughly the size of Pennsylvania, it is the region's most densely populated country. On average, every square mile holds more than 250 people.

Surprisingly, Malawi is also Southern Africa's most rural nation. Only 20 percent of its people live in cities. Its small size and large rural population mean that most of its farms are small. Most farm villages are not able to produce much more than what they need. As a result, Malawi is the region's poorest country. The average Malawian earns less than $350 per year.

MAP SKILLS

1. **PLACES AND REGIONS** What do the cities of Johannesburg, Durban, and Cape Town have in common?

2. **THE GEOGRAPHER'S WORLD** In general, which area of Southern Africa is more densely populated: eastern or western?

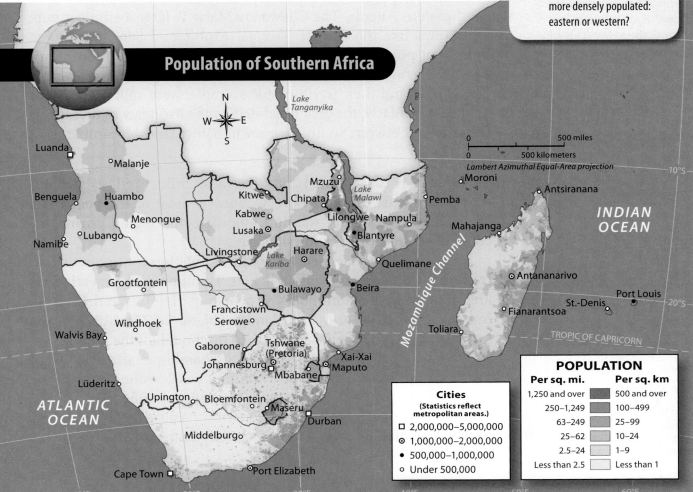

Population of Southern Africa

Ethnic and Culture Groups

Africans are not a single people. Southern Africa is home to many ethnic and cultural groups who speak several different languages. One group, the Shona, makes up more than 80 percent of the population of the country of Zimbabwe. South Africa's 9 million Zulu make up that country's largest ethnic group. More than 7 million Xhosa also live there, as do the Khoekhoe. Some 4.5 million Tsonga people are spread among the countries of South Africa, Zimbabwe, and Mozambique.

About 4 million Tswana form the major population group in Botswana. A similar number of Ovimbundu and 2.5 million Mbundu make up approximately two-thirds of Angola's population. A smaller group, the Ambo, live in Angola and Namibia. About half of Namibia's people belong to this ethnic group. The San, a nomadic people, live mainly in Namibia, Botswana, and southeastern Angola. The Chewa are Malawi's largest ethnic group.

Groups like the Chewa, Tsonga, Ambo, and San illustrate an important point about Southern Africa's history. When Europeans divided the region, they paid little attention to its indigenous people. The Chewa and their territory, for example, were split among four colonies. Similarly, the area inhabited by the Tsonga was divided by the borders between South Africa, Zimbabwe, and Mozambique.

Religion and Languages

Southern Africa's colonial past has also influenced its people's religious beliefs. In almost every country, most of the people are Christians. Christianity was introduced to the region during the colonial era by Christian missionaries.

In Angola, however, nearly half the population continues to hold traditional indigenous religious beliefs. Traditional African religions are followed by large numbers of people in Namibia and Lesotho, too. In Zimbabwe and Swaziland, a blend of Christianity and traditional religious beliefs is followed by about half the population.

Swaziland, Zambia, Malawi, and Mozambique also have large Muslim populations. Most of Mozambique's Muslims live on the coast, where **contact** with Arab traders led long ago to the introduction of Islam. Immigration from Asia explains Zambia's Muslim population, as well as its large Hindu minority.

Members of the Nazareth Baptist Church in South Africa take part in their annual pilgrimage to the mountain of Nhlangakazi. The church is also called the Shembe Church after its founder, Isaiah Shembe.

Think Again

Southern Africa's large island country of Madagascar was settled by African people.

Not true. Most of Madagascar's people speak Malagasy, a language related to those spoken in Indonesia, the Philippines, and islands in the South Pacific. The language of Madagascar indicates that the island's early inhabitants probably came from that part of the world.

Portuguese remains the official language in Angola and Mozambique. English is an official language in most of the former British colonies. Its use, however, is mainly limited to official and business communications; nowhere is it widely spoken by the people. Instead, most speak indigenous languages. South Africa has 10 official languages besides English; Zambia has 7.

Academic Vocabulary
contact communication or interaction with someone

✓ **READING PROGRESS CHECK**

Determining Central Ideas What is the main religion practiced in Southern Africa?

Life in Southern Africa

GUIDING QUESTION *How do the various people of Southern Africa live?*

As in other regions of Africa, life differs from city to countryside. Many rural people continue to follow traditional ways of life. At the same time, urban and economic growth are challenging and changing many of the traditional ways.

Urban Life

Although most people in the region of Southern Africa live in the countryside, migration to cities grows because of job opportunities. Harare, Zimbabwe, has grown to more than 1.5 million, as have Lusaka, Zambia, and Maputo, Mozambique. Luanda, Angola's capital, is even larger: It holds some 4.5 million people. South Africa has four cities—Durban, Ekurhuleni, Cape Town, and Johannesburg—with populations of around 3 million or more.

Shown here is a high-rise building under construction in the city of Luanda in Angola. Luanda is the country's main seaport and government center.

At an outdoor market in Lusaka, vendors come to sell handcrafted items. Food and entertainment are also available.

▶ **CRITICAL THINKING**
Describing What are periodic markets?

Urban Growth and Change

The rapid growth of some cities has strained public **utilities**—services such as trash collection, sewage treatment, and water distribution. Luanda, for example, has had many problems providing enough clean water for its many people. Outbreaks of cholera and other diseases have resulted from drinking polluted water.

The region's cities have a mix of many ethnic groups and cultures. An example is Johannesburg, where the wealth from nearby gold fields helped build one of the most impressive downtowns in all of Africa. Outside the central city are the white neighborhoods where about 20 percent of the city's population live. Some black South Africans have moved into these neighborhoods since the end of apartheid. Most, however, live in "townships" at the city's edge. These areas often have no electricity, clean water, or sewer facilities. Most of the region's large cities have shantytowns.

Johannesburg's role as a mining, manufacturing, and financial center has attracted people from around the world. Every black ethnic group in Southern Africa is present, as well. The white community is mainly English and Afrikaner. Large Portuguese, Greek, Italian, Russian, Polish, and Lebanese populations also live there. Indians, Filipinos, Malays, and Chinese live mainly in the townships. At least 12 languages are heard on city streets.

Family and Traditional Life

People who move to the cities must adjust to new experiences and a different way of life. In the countryside, traditional ways of life remain strong.

Rural villages are often small—consisting of perhaps 20 or 30 houses. Building materials, which vary by ethnic group, include rocks, mud bricks, woven sticks and twigs packed with clay, and **thatch**—straw or other plant material used to cover roofs.

In many cultures, all the people in a village are related by blood or marriage to the village's headman or chief. Men often have more than one wife. They provide a house for each wife and their children. Growing food crops is the main economic activity. Many families raise cattle as well, mainly for milk and as a symbol of wealth.

People in the countryside practice subsistence farming, growing the food they need to survive. Artwork sometimes provides a family with a source of cash. Wood and ivory carving are art forms that are generally practiced by men. Pottery-making is usually a woman's craft. In some cultures, both men and women make baskets. They sell the products in cities or at **periodic markets**—open-air trading markets held regularly at crossroads or in larger towns.

In recent times, more and more men have been leaving their villages to work at jobs in cities or mines. Although the money they send home helps support their families, this **trend** has greatly changed village life. Many villages now consist largely of women, children, and older men. Women have increasingly taken on traditional male roles in herding, family and community leadership, and other activities.

Academic Vocabulary

trend a general tendency or preference

Members of South Africa's Ndebele tribe attend a gathering of traditional leaders from all over the country in November 2009 to honor former President Nelson Mandela.

✓ **READING PROGRESS CHECK**

Citing Text Evidence Where in their countries do most Southern Africans live?

Southern Africa Today

GUIDING QUESTION *What challenges and prospects do the countries of Southern Africa face?*

Southern Africa's wealth of mineral, wildlife, and other resources may be the key to its future. Still, the region faces serious social, economic, and political challenges.

Health Issues

Life expectancy in Southern Africa is low. In the majority of countries, most people do not live beyond age 50 to 55. Lack of good rural health care is one reason, although many countries are trying to build or improve rural clinics.

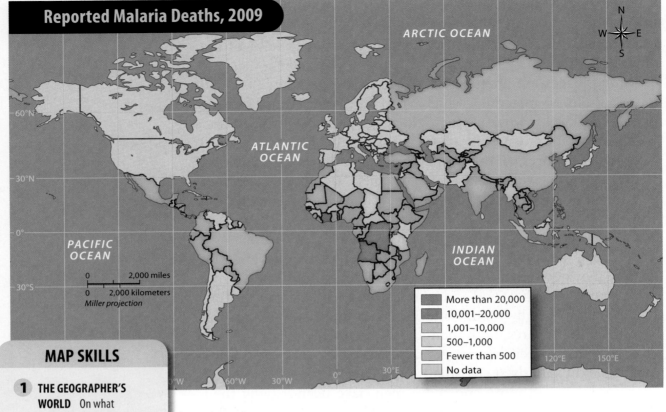

Reported Malaria Deaths, 2009

Legend:
- More than 20,000
- 10,001–20,000
- 1,001–10,000
- 500–1,000
- Fewer than 500
- No data

MAP SKILLS

1. **THE GEOGRAPHER'S WORLD** On what continent has malaria been responsible for the greatest number of deaths?

2. **THE GEOGRAPHER'S WORLD** In what regions of the world has malaria been responsible for the fewest deaths?

Disease

Malaria, a tropical disease carried by mosquitoes, is a problem in several countries. Dysentery and cholera, potentially fatal diseases caused by bacteria in water, are also widespread. So is tuberculosis. Malnutrition is a cause of death for many infants and young children.

Southern Africa has some of the highest rates of infant death in the world. In Angola, Malawi, and Mozambique, about 100 to 120 of every 1,000 children die in infancy. Elsewhere in the region, the figure is 40 to 60 per 1,000. (The infant death rate in the United States is 7 per 1,000.)

A major cause of death in children and adults is HIV/AIDS. Southern Africa has a higher HIV/AIDS rate than any other region in Africa. Swaziland, Botswana, Lesotho, and South Africa have the highest rates in the world. About one of every four adults (25 percent) in these countries is infected with this sexually transmitted disease, which women pass on to their children at birth. In the rest of the region, the adult HIV/AIDS rate averages between 11 and 14 percent. (In the United States, the rate is 0.6 percent.)

The high incidence of HIV/AIDS has disrupted the labor force by depriving countries of needed workers. It has also disrupted families through death, inability to work, or AIDS-related family issues. The disease has created millions of AIDS orphans, children whose mother and father have died from AIDS. The huge number of AIDS orphans is a major social problem.

Progress and Growth

Angola and Mozambique continue to rebuild the cities and towns, industries, railroads, and communications systems that have been damaged or destroyed by years of civil war. Oil exports in Angola and aluminum exports in Mozambique help finance this effort. So does the tourism that peace and stability have brought back to the beautiful beaches and resorts along Mozambique's coast.

Tourism at national parks has grown with the establishment of stable, democratic governments. Zambia and Malawi replaced one-party rule with more democratic forms of government in the 1990s. Botswana and Namibia have been strong democracies, respecting and protecting human rights, since independence. Only Zimbabwe and Swaziland continue to suffer economic decline and political unrest, largely due to repressive leaders.

Help From Other Countries

The United States has used economic aid to strengthen democracy in Southern Africa. Other U.S. programs have provided billions of dollars to pay for medications and care for AIDS sufferers and AIDS orphans.

Other countries and international organizations have also made huge investments in the region. Taiwan's development of a textile industry in Lesotho, for example, is giving some of that poor country's workers an alternative to employment in South Africa's mines.

Foreign investment, workers, and tourists have also returned to South Africa as it continues to recover from the effects of apartheid. South Africa remains the region's most industrial and wealthiest country. It also faces serious economic challenges. Many of its traditional African farming communities struggle in poverty, growing few if any cash crops. Its heavy reliance on the export of mineral and agricultural goods places it at risk if world demand or prices for the goods fall. These problems mirror the challenges that many other countries in Southern Africa also confront.

Include this lesson's information in your Foldable®.

✓ **READING PROGRESS CHECK**

Analyzing Why is life expectancy in Southern Africa so low?

LESSON 3 REVIEW

Reviewing Vocabulary

1. What did rural Southern Africans use clay and *thatch* for?

Answering the Guiding Questions

2. **Determining Central Ideas** How did colonialism and contact with traders influence religious beliefs in Southern Africa?

3. **Describing** What are rural and city life like for Southern Africa's black population?

4. **Analyzing** How and why has Southern Africa benefited from the growth of democracy in the region?

5. **Argument Writing** Write a letter to the editor of a Southern African newspaper explaining whether the region should continue to work for change.

Chapter 13 **401**

Chapter 13 ACTIVITIES CCSS

Directions: Write your answers on a separate piece of paper.

1 Use your FOLDABLES to explore the Essential Question.
INFORMATIVE/EXPLANATORY WRITING Write two paragraphs explaining how Southern Africa's resources place the region in a favorable position to develop trade with other countries.

2 21st Century Skills
DESCRIBING Using information from the text and online, create a brief slide show of Southern Africa's energy resources and how the region uses the resources. Narrate the slide show, identifying the different countries' means of generating power.

3 Thinking Like a Geographer
DETERMINING CENTRAL IDEAS As a geographer, would you favor setting aside more or less land for game preserves in Southern Africa? Use a T-chart to list your pro and con arguments.

4 GEOGRAPHY ACTIVITY

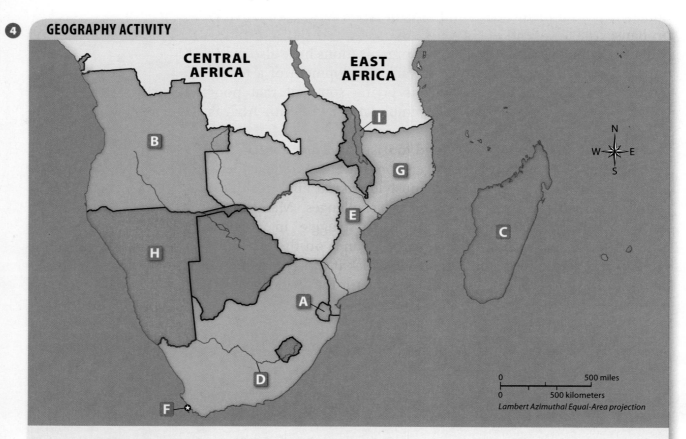

Locating Places
Match the letters on the map with the numbered places listed below.

1. Zambezi River
2. Madagascar
3. Angola
4. Cape Town
5. Orange River
6. Mozambique
7. Namibia
8. Swaziland
9. Lake Malawi (Lake Nyasa)

Chapter 13 Assessment

REVIEW THE GUIDING QUESTIONS

Directions: Choose the best answer for each question.

1. The country of Madagascar is
   A. a large plateau.
   B. Southern Africa's regional capital city.
   C. the world's fourth-largest island.
   D. the world's largest exporter of coconut milk.

2. Which is the longest river in Southern Africa?
   F. Kariba
   G. Congo
   H. Great Karoo
   I. Zambezi

3. Western South Africa, western Namibia, and Botswana have what climate zone in common?
   A. tropical
   B. desert
   C. Mediterranean
   D. steppe

4. The amount of hydroelectric power in this region has been reduced by
   F. deforestation.
   G. droughts.
   H. monsoons.
   I. civil disturbances.

5. South Africa's Afrikaners are descended from which population group?
   A. native Africans
   B. Boers
   C. Portuguese colonists
   D. Zambians

6. Which is the most densely populated country in Southern Africa?
   F. Zambia
   G. the Republic of South Africa
   H. Madagascar
   I. Malawi

Chapter 13 Assessment (continued)

DBQ ANALYZING DOCUMENTS

7 ANALYZING Read the following passage about the area around the Okavango River and the Kalahari Desert.

"*During dry periods [the Okavango Delta] is estimated to cover at least 6,000 square miles, but in wetter years, with a heavy annual flood, the Okavango's waters can spread over 8,500 square miles of the Kalahari's sands. Deep water occurs in only a few channels, while vast areas of reed beds are covered by only a few inches of water.*"

—from Cecil Keen, *Okavango*

As described in the reading, the Okavango is a

A. desert.
B. mountain.
C. river.
D. reed bed.

8 ANALYZING What can you infer about the land of the Kalahari from this passage?

F. It is sandy because it absorbs most of the water fairly quickly.
G. It is fairly flat because more of the water is shallow than deep.
H. It is wet most of the time because it lets the floodwaters stand.
I. It tilts to the west because that is where the deep channels form.

SHORT RESPONSE

"*Discouraged about the lack of results from their nonviolent campaign, Nelson Mandela and others called for an armed uprising . . . that paralleled the nonviolent resistance. That, too, failed to tear down the apartheid system, and in the end a concerted grassroots nonviolent civil resistance movement [together] with international support and sanctions [against the government] forced the white government to negotiate.*"

—from Lester R. Kurtz, "The Anti-Apartheid Struggle in South Africa"

9 DETERMINING CENTRAL IDEAS What were Mandela and others trying to achieve?

10 ANALYZING How did they eventually succeed?

EXTENDED RESPONSE

11 INFORMATIVE/EXPLANATORY WRITING Southern Africa has an abundance of wildlife, including animals, birds, fish, and exotic plant life. Tourists come from all over the world to see the animals, which live on animal preserves and in the wild. Do some research on travel in Southern Africa, then write an essay describing the experience of going on safari. Talk about which areas of the region you visited and what you saw, and what kind of accommodations you had on your safari.

Need Extra Help?

If You've Missed Question	1	2	3	4	5	6	7	8	9	10	11
Review Lesson	1	1	1	1	2	3	1	1	2	2	3

Oceania, Australia, New Zealand, and Antarctica

UNIT **4**

Chapter 14
Australia and New Zealand

Chapter 15
Oceania

Chapter 16
Antarctica

Explore the Continent

This region lies almost entirely in the Southern Hemisphere, reaching from north of the Equator to the South Pole. The countries that make up this region contain an amazing variety of landforms and range in size from tiny islands to large continents.

① NATURAL RESOURCES This region holds abundant natural resources. New Zealand's North Island has good farmland and pasture for grazing sheep and other animals. The location of New Zealand along the Ring of Fire provides geothermal energy. Australia is rich in precious metals, oil and natural gas, and fertile farmland. In Oceania, wind and solar energy are plentiful, as are fish and other seafood.

② ISLANDS AND REEFS Oceania is made up of thousands of islands with different physical features. These islands were formed millions of years ago by underwater volcanoes. Coral islands, called atolls, are made up of reef islands surrounding lagoons. Off Australia's northeastern shore lies the spectacular Great Barrier Reef. By contrast, the water surrounding Antarctica's coasts freezes into thick plains of ice during winter. Other huge ice formations include glaciers and icebergs.

UNIT 4

③ LANDFORMS Physical landscapes vary throughout the region. In New Zealand, the towering mountains of the Southern Alps rise on South Island. Australia is mostly flat. The Central Lowlands, however, contain a massive stone monolith known as Ayers Rock, or *Uluru* to the Aboriginal people of Australia. Antarctica is made up of one large, icy landmass and an archipelago of rocky islands.

Fast FACT

Australia is slightly smaller than the 48 U.S. states.

Unit 4 407

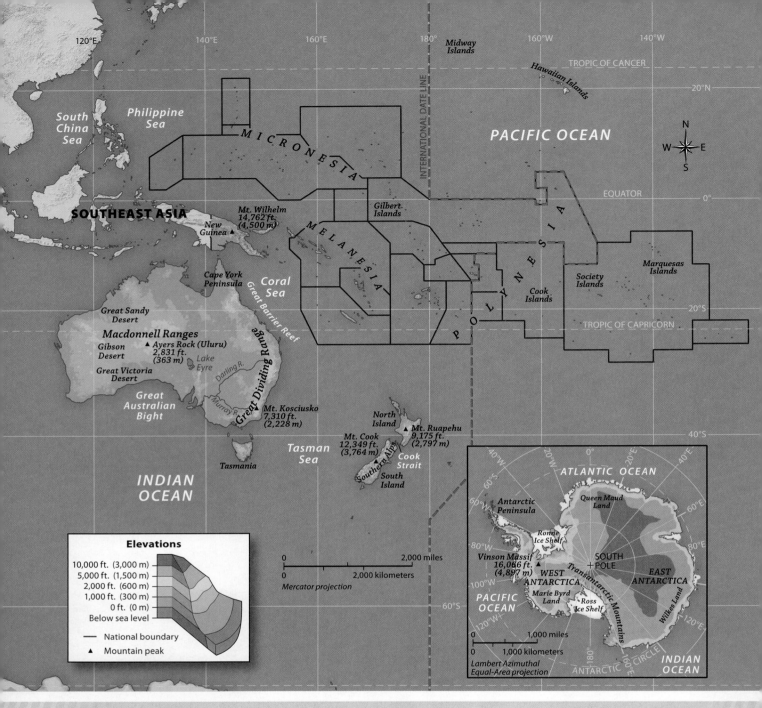

OCEANIA, AUSTRALIA, NEW ZEALAND, AND ANTARCTICA

PHYSICAL

MAP SKILLS

1. **THE GEOGRAPHER'S WORLD** Where are the region's major rivers located?

2. **PLACES AND REGIONS** How is the land elevation in Antarctica different from elevations elsewhere in the region?

3. **PHYSICAL GEOGRAPHY** Which deserts are found in Australia?

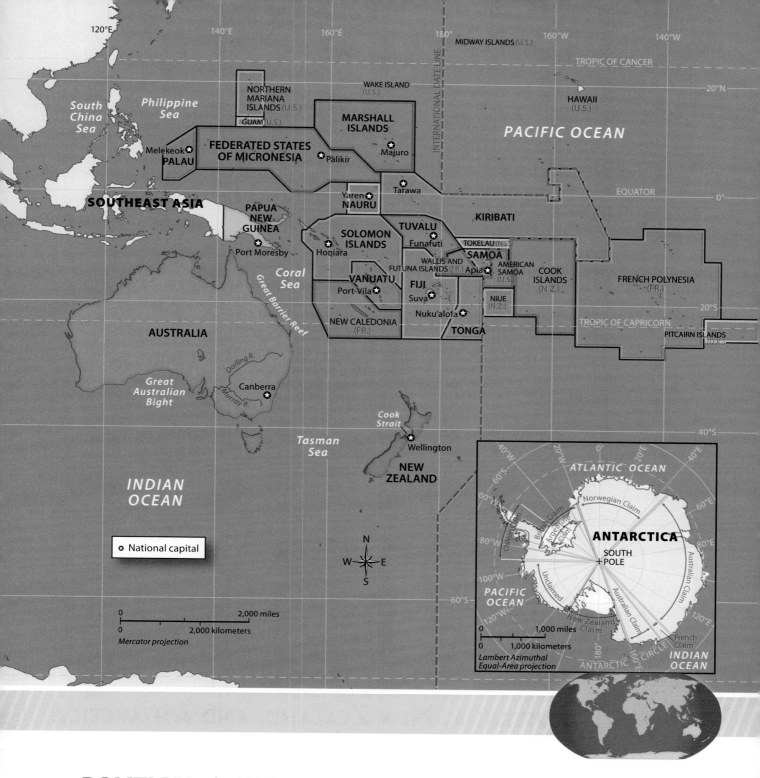

POLITICAL

MAP SKILLS

1. **THE GEOGRAPHER'S WORLD** How far is Wellington, New Zealand, from Canberra, Australia?

2. **PLACES AND REGIONS** Which country controls Guam and the Northern Mariana Islands?

3. **PLACES AND REGIONS** Which country is located on half of a major island?

Unit 4 409

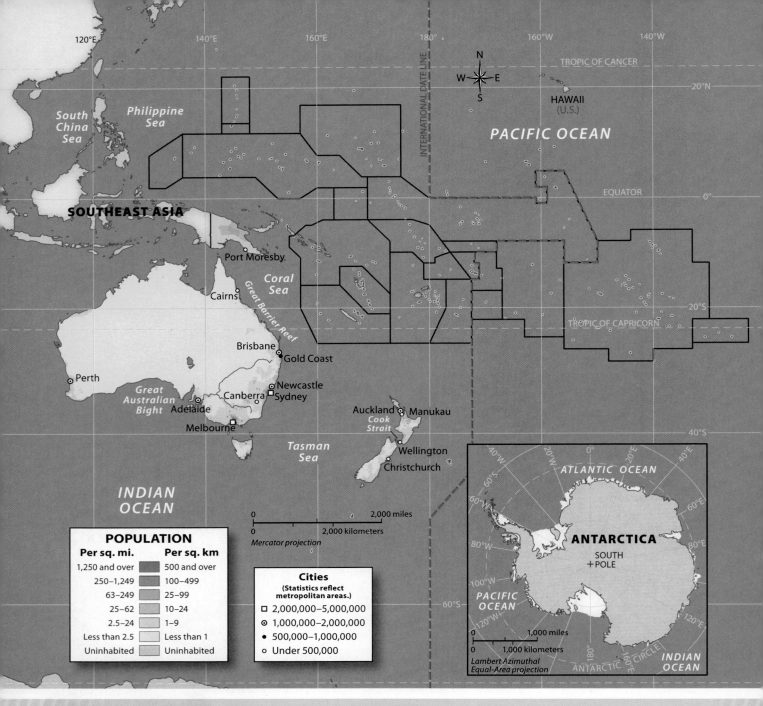

Oceania, Australia, New Zealand, and Antarctica

POPULATION DENSITY

MAP SKILLS

1. **ENVIRONMENT AND SOCIETY** Why do you think most people in Australia live along the country's east coast?

2. **PLACES AND REGIONS** How do population densities compare on New Zealand's South and North Islands?

3. **ENVIRONMENT AND SOCIETY** Why do you think Antarctica has no permanent human residents?

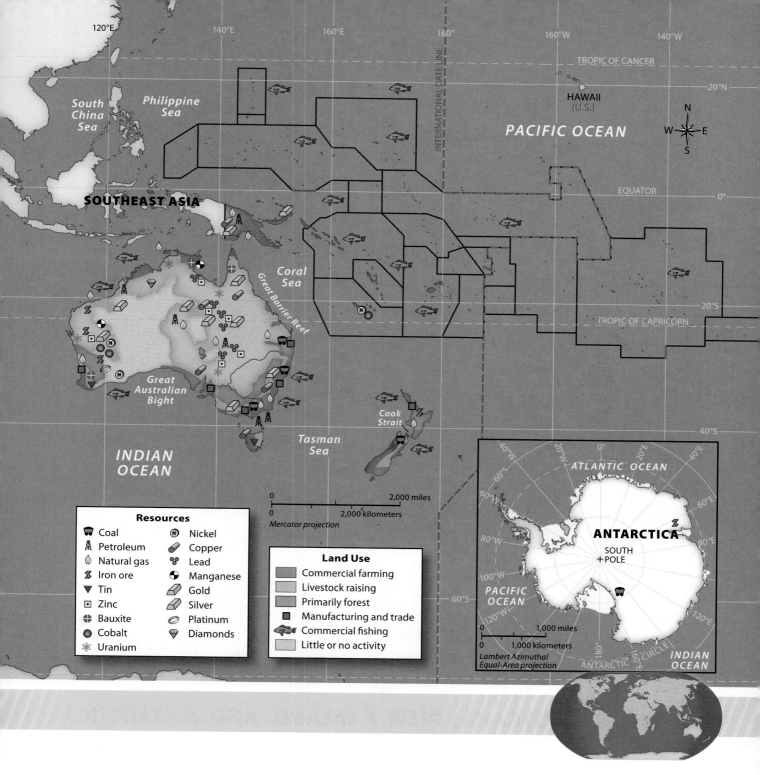

ECONOMIC RESOURCES

MAP SKILLS

1. **ENVIRONMENT AND SOCIETY** How is much of the land in Australia used?

2. **PHYSICAL GEOGRAPHY** What mineral resources are found in New Zealand?

3. **HUMAN GEOGRAPHY** What are the primary economic activities in Oceania north of New Zealand?

Unit 4 411

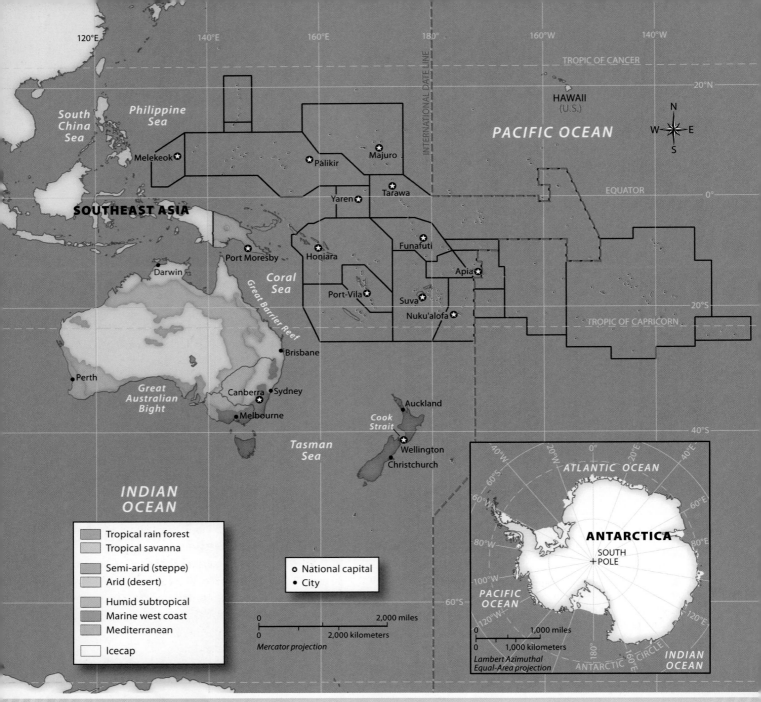

OCEANIA, AUSTRALIA, NEW ZEALAND, AND ANTARCTICA

CLIMATE

MAP SKILLS

1. **PLACES AND REGIONS** What climate zones are found in northern Australia?

2. **PLACES AND REGIONS** What climate zone is found throughout New Zealand?

3. **ENVIRONMENT AND SOCIETY** In what parts of the region could tropical crops be grown?

AUSTRALIA AND NEW ZEALAND

ESSENTIAL QUESTIONS • How does geography influence the way people live? • Why does conflict develop? • What makes a culture unique?

A baby kangaroo nuzzles a young boy. The name "kangaroo" stems from aboriginal language.

networks

There's More Online about Australia and New Zealand.

CHAPTER 14

Lesson 1
Physical Geography

Lesson 2
History of the Region

Lesson 3
Life in Australia and New Zealand

The Story Matters...

Thousands of years ago, Asian and Pacific people began migrating to Australia and New Zealand. People from different European countries later migrated to the region because of its abundant natural resources. These diverse cultures are woven into the fabric of Australia and New Zealand.

Go to the Foldables® library in the back of your book to make a Foldable® that will help you take notes while reading this chapter.

413

Chapter 14
AUSTRALIA AND NEW ZEALAND

As the only place on Earth that is a continent and a country, Australia is unique. Located 1,200 miles (1,931 km) southeast of Australia, two large islands make up most of New Zealand's landmass.

Step Into the Place

MAP FOCUS Use the map to answer the following questions.

1. **PLACES AND REGIONS**
   What city is Australia's national capital?

2. **THE GEOGRAPHER'S WORLD**
   What sea separates Australia and New Zealand?

3. **THE GEOGRAPHER'S WORLD**
   What reef lies off the northeastern coast of Australia?

4. **CRITICAL THINKING** Integrating Visual Information Look at the map. What can you infer from the location of Australia's major cities about where most Australians live?

HARBOR VIEW A view from the harbor's bridge provides a magnificent view of Sydney, Australia. Sydney is the country's largest city in population and a major world port and business center.

MOUNTAIN SPECTACLE A hiker looks toward Mount Cook, New Zealand's highest mountain. Mount Cook lies in the Southern Alps, a mountain range that runs the length of South Island.

Step Into the Time

TIME LINE Choose at least two events from the time line. For each event, write a paragraph describing some of the positive and negative effects that event had on the region.

C. A.D. 800–1300 Maori arrive in New Zealand from Polynesia

B.C. 48,000 Aboriginal people begin migrating to Australia

1642 Explorer Abel Tasman lands on what is now Tasmania

1770 Captain James Cook explores Australian coast

1800

There's More Online!

- ☑ **IMAGE** Australia's Outback
- **SLIDE SHOW** Views of Australia and New Zealand
- **VIDEO**

Reading HELPDESK

Academic Vocabulary
- overall

Content Vocabulary
- Outback
- monolith
- Aboriginal
- coral reef
- hot spring
- geyser
- drought
- marsupial
- eucalyptus

TAKING NOTES: *Key Ideas and Details*

Summarize As you read about the physical geography of Australia and New Zealand, take notes on each section of the lesson using the graphic organizer below.

Heading	Main Idea
The Land	
Climate	
Plants/Animals	

Lesson 1
Physical Geography

ESSENTIAL QUESTION • *How does geography influence the way people live?*

IT MATTERS BECAUSE
Australia and New Zealand have unique landscapes, plants, and wildlife not found in other parts of the world.

The Land of Australia and New Zealand

GUIDING QUESTION *What physical features make Australia and New Zealand unique?*

Australia is nicknamed "The Land Down Under" because it is located south of, or "under," the Equator. On a map, Australia looks like a large island, because it is not attached to any other landmasses. Australia, however, is a continent. It is the world's smallest continent, but it is the world's sixth-largest country. New Zealand is made up of two islands with a variety of landscapes, ecosystems, and climate zones. Australia and New Zealand also have some of the world's most unusual plants and animals.

Australia's Landforms

It might not seem that Australia is a flat continent based on images of its huge rock formations and mountain ranges. **Overall**, however, Australia has low elevation. This means that although the land is high in some places, most of its surface is low compared to the land on other continents. Australia generally has a dry climate. One-third of Australia is covered by deserts. Another one-third is semiarid.

Geographers divide Australia into three main geographic regions: the Western Plateau, the Central Lowlands, and the Eastern Highlands. In general, the Western Plateau is rocky

416

and dry, and the Central Lowlands are flat and rugged. The Eastern Highlands have a variety of high and low areas, forests, and fertile farmlands. In addition, coastal lowlands are found all around the continent, particularly in the north and east.

The Western Plateau makes up the western half of Australia. The plateau is rocky, with few water sources other than small salt lakes. Near the center of the continent are the interior highlands. Here, the Musgrave and Macdonnell mountain ranges rise above a huge expanse of flat plateau.

The Central Lowlands are a rough, dry region. Although a system of rivers runs through the rugged land, much of the region is desert. The most rural and isolated parts of the Central Lowlands are commonly called the **Outback**. Survival is a challenge for humans and animals in the Outback. Strong winds cause harsh dust storms across the desert plains. In many parts of the Outback, water is difficult to find. To get the water they need, ranchers must dig wells deep into the sun-baked earth.

One of Australia's most fascinating landforms is found in the Central Lowlands. A massive, solid stone called a **monolith**, measuring 1,100 feet (335 m) tall and 2.2 miles (3.5 km) long, stands alone on the desolate plain. This amazing monolith has two names. Many Australians know it as Ayers Rock. However, the first humans to live in Australia, the **Aboriginal** people, call it *Uluru*. Uluru is sacred to many Aboriginal people. Small caves along the base of the monolith contain ancient Aboriginal paintings and carvings. Today, Uluru is an official World Heritage site located within a protected national park.

The Eastern Highlands, called the Great Dividing Range, run parallel to Australia's east coast. The mountains and valleys of this range were formed by folding and uplifting movements that occurred in the past within Earth's lithosphere. Today, little movement occurs, and earthquakes are rare.

One of the most spectacular and complex ecosystems on Earth is located in the ocean waters just off Australia's northeastern shore. The Great Barrier Reef is a living **coral reef**, a giant community of marine animals called corals.

Academic Vocabulary

overall as a whole; generally

Ayers Rock, or Uluru, in central Australia, is all that is left of a large mountain range that slowly eroded over millions of years.
▶ **CRITICAL THINKING**
Describing Why is Ayers Rock (Uluru) called a monolith?

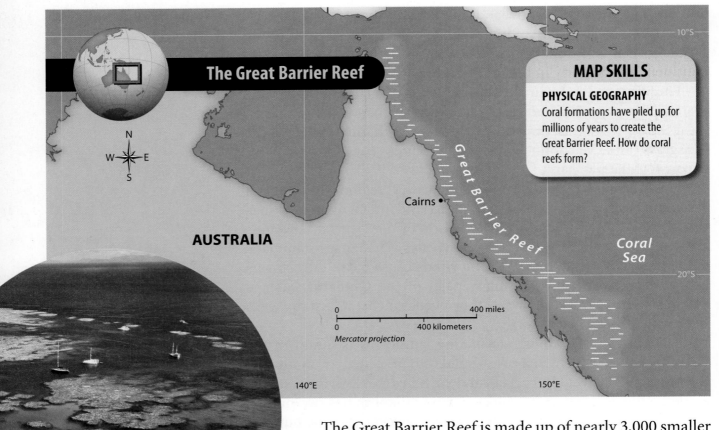

The Great Barrier Reef

MAP SKILLS

PHYSICAL GEOGRAPHY
Coral formations have piled up for millions of years to create the Great Barrier Reef. How do coral reefs form?

Bait Reef in northeastern Australia

The Great Barrier Reef is made up of nearly 3,000 smaller coral reefs extending for more than 1,250 miles (2,012 km). These reefs were formed over millions of years by natural actions of the biosphere. As corals lived, grew, and died, their hard skeletons built layer upon layer, forming large underwater formations called reefs. Coral reefs are held together by algae and tiny bits of plant and animal matter that become stuck in the reefs.

The Great Barrier Reef is teeming with marine life. Brightly colored fish and shellfish search for food and hide among the corals. Sponges, starfish, and anemones cling to the coral reefs, decorating them like living ornaments.

Just south of Australia's southeastern coast is a unique island called Tasmania. Mountains, valleys, and plateaus cover its surface. Some of the world's wildest, unexplored rain forests grow in Tasmania. These are cool temperate rain forests, which differ from the warm tropical rain forests in the northeastern part of Australia. Tasmania's broad Central Plateau is scattered with more than 4,000 shallow lakes.

Waterways of Australia

Most of Australia has a dry climate. Lack of rain means few large rivers flow across the land. Australia's largest rivers flow from the eastern mountains. The important Murray-Darling river system flows through a dry basin in southeastern Australia. The many rivers in the Murray-Darling system bring so much water to the area that the land is green and fertile.

Many of Australia's people, plants, and animals depend on underground aquifers for their water. Nearly one-third of all water used in Australia comes from underground sources. In a dry land such as Australia, water is a precious resource.

New Zealand's Landforms and Waterways

New Zealand is made up of two main islands called North Island and South Island plus many small islands. These islands are located in the southeastern Pacific Ocean, far from any other landmasses. Geographers believe that more than 60 million years ago, movements within the lithosphere pushed the land that would become New Zealand up out of the ocean. Over time, processes within the lithosphere and hydrosphere shaped the land. The movement of ice and liquid water has carved out basins and eroded rocks into unusual shapes. Volcanic eruptions and earthquakes created low hills, fertile valleys and plains, and rows of sharp mountain peaks.

Today, one-third of New Zealand's lands are protected as national parks and nature reserves. The country is known for its beautiful and unusual landscapes. New Zealand's main islands are similar in that both have many forests, mountains, and waterways. Local climates and other types of landforms, however, show the differences between the two islands.

Many of North Island's landforms were created by volcanic activity. A huge volcanic plateau makes up the center of North Island. Lake Taupo, the largest lake in New Zealand, formed millions of years ago in the massive crater left behind by a devastating volcanic eruption. Lake Taupo is surrounded by fertile plains and valleys. This land is enriched by volcanic soil, which makes good farmland and pasture for grazing. North Island's most productive farm and pasture lands are located in a region called Waikato.

Northeast of Lake Taupo is Rotorua, an area famous for its steaming hot springs, bubbling mud pools, and violent geysers. **Hot springs** are pools of hot water that occur naturally. Hot springs form in rocky areas when rainwater seeps into cracks in Earth's surface. Water is exposed to intense heat, and then bubbles back up to gather in surface pools. **Geysers** are hot springs that sometimes shoot hot water out of the ground. The water in hot springs and geysers is warmed by heat energy from deep within Earth.

Pohutu Geyser in New Zealand's North Island erupts about 20 times each day. The water it releases can reach as high as 100 feet (30 m).

▶ CRITICAL THINKING

Describing How has volcanic activity shaped New Zealand's North Island?

The stunning landscape of the Dark Cloud Range stretches across the southern end of New Zealand's South Island.

▶ **CRITICAL THINKING**
Analyzing Why does the southern end of South Island have many deep valleys, fjords, and lakes?

New Zealand's South Island is famous for its spectacular Southern Alps. These towering mountains are higher than the mountains on North Island. The Southern Alps cover hundreds of miles along the western side of South Island.

The southern end of South Island is filled with an incredible variety of landforms and environments. Millions of years ago, the movement of glaciers cut deep valleys into the rocky land. As the glaciers melted, clear, cold lakes formed. Some glacier valleys were so deep they were not completely filled until rising sea levels flooded them with ocean water more than 6,000 years ago. These water-filled landforms are called *fjords*. The fjords of South Island are such fascinating landforms that many are preserved within a huge national park.

The Ring of Fire

New Zealand's location in the southeastern Pacific Ocean puts it within the volcano-studded Ring of Fire. Active volcanoes and the frequent movement of tectonic plates along the Ring of Fire often result in earthquakes. In fact, geographers using special equipment detect more than 15,000 earthquakes in New Zealand every year. Most of these tremors are far too small for humans to feel. At least

100 of these yearly earthquakes, however, are large enough to be noticed by humans. Scientists predict that New Zealand will experience at least one severe earthquake each century.

☑ **READING PROGRESS CHECK**

Citing Text Evidence In what ways are the lands of Australia and New Zealand alike, and in what ways are they different?

Climates of the Region

GUIDING QUESTION *What types of climates and climate zones are found in Australia and New Zealand?*

Australia and New Zealand are located in the Southern Hemisphere, so their seasons are at opposite times of the year from seasons in the United States. For example, June, July, and August are winter months in the region. Summer months are December, January, and February.

Australia's Climates

The climate changes dramatically from one part of Australia to another. The northern third is in a tropical climate zone, and most of the other two-thirds are in a subtropical climate zone. Northern Australia, generally, has a warm, tropical climate. Winter months are dry, and summer months are rainy and hot. Seasonal monsoons can bring damaging winds and heavy rainfall.

Climate in most areas of the western, southern, and eastern regions changes with the seasons. Winters in Queensland, the Northern Territory, and Western Australia are warm and dry. Most rainfall in these regions happens during the long, hot spring and summer seasons. Coastal areas tend to be sunny and dry, with seasonal rains. The climate along the east coast is more humid, and the area receives more rainfall than the rest of the continent.

Much of central and western Australia, such as the Outback, has a desert climate with bands of semiarid steppes to the north, east, and south. These dry areas have extremely hot weather during much of the year. Weather in the Outback can change quickly from one extreme to another. Daytime temperatures in the Outback can reach 122°F (50°C), yet temperatures can fall below freezing at night.

Australia is the driest inhabited continent in the world. Only the icy continent of Antarctica receives less precipitation than Australia. A major problem in the dry areas is that long periods of little or no rain can result in **drought**.

Droughts are common in Australia and threaten the survival of wildlife, livestock, and farm crops. Low water reserves can lead to poor-quality drinking water for humans. Geographers can now predict some droughts by monitoring climate changes caused by El Niño. El Niño occurs every few years when global winds and ocean currents shift, affecting global rainfall patterns.

A cattle driver on a station, or ranch, in Australia's Northern Territory herds Brahman cattle, a breed that is well adapted to the area's climate extremes.

▶ CRITICAL THINKING

Analyzing How does climate in inland Australia compare with climate in coastal areas of the country?

New Zealand's Climates

New Zealand's climate is as varied as its land. Climates range from subantarctic in the south to subtropical in the north. Most areas on North Island and South Island have mild, temperate climates, however, with average daily temperatures between 86°F (30°C) and 50°F (10°C). For most of the year, temperatures on North Island are higher than temperatures on South Island. South Island experiences the hottest summers and the coldest winters in New Zealand. Both islands receive plenty of rain, keeping their forests and grasslands healthy.

Although average temperatures are generally mild and extremes of heat and cold are rare, snow does fall in New Zealand. The central region of South Island experiences the coldest winters of any part of the country. Highland zones, such as the Southern Alps, receive the most snowfall.

☑ **READING PROGRESS CHECK**

Describing What problems are caused by drought in Australia? Why are geographers monitoring climate changes there?

Plant and Animal Life

GUIDING QUESTION *What plants and animals are unique to Australia and New Zealand?*

If you have seen pictures of plants and animals from Australia and New Zealand, you already know that many of them are unusual. Why are plants and animals in this region so different from living things in other parts of the world?

The landmasses that would become Australia and New Zealand separated from other lands millions of years ago. The animals and plants living on these lands developed in isolation, separated from living things on other landmasses. Over millions of years, the plants and animals in Australia and New Zealand adapted to live in their own unique environments.

Plants and Animals of Australia

Bandicoots, kangaroos, and koalas are 3 of the 150 native species of marsupials in Australia. A **marsupial** is a type of mammal that raises its young in a pouch on the mother's body. Marsupials vary in size from tiny kangaroo mice to 6-foot-tall, gray kangaroos.

In the driest regions of Australia, water can be difficult to find. Over time, many native animals learned to get water from plants. The leaves, stems, and roots of desert plants contain water. Koalas adapted to eating only one type of plant—the leaves of the eucalyptus tree. **Eucalyptus** trees are native Australian evergreen trees with stiff, pleasant-smelling leaves.

Plants and Animals of New Zealand

Many of New Zealand's trees, ferns, and flowering plants are unique to New Zealand and cannot be found on other continents. Much of the land is covered in hardwood beech tree forests. Alpine plants such as sundew and edelweiss have adapted to the cold, windy, dry environments of South Island's mountain regions.

Lizards such as geckos are common sights in New Zealand. Other reptiles, such as the chevron skink, have become rare. Long ago, New Zealand was home to two types of flightless birds—the kiwi and the moa. The moa was hunted to extinction, but the kiwi was more fortunate. Today, the kiwi is a common sight on both islands and has become a symbol of New Zealand and its people.

✓ **READING PROGRESS CHECK**

Determining Central Ideas Explain why animals and plants native to Australia and New Zealand are different from living things in other parts of the world.

Road signs in rural Australia often warn motorists about kangaroo crossing areas.
▶ **CRITICAL THINKING**
Citing Text Evidence Why is the kangaroo found in Australia and nowhere else?

Include this lesson's information in your Foldable®.

LESSON 1 REVIEW

Reviewing Vocabulary
1. What kinds of equipment might you need to explore a *coral reef*?

Answering the Guiding Questions
2. *Identifying* List three physical features that make Australia unique and three physical features that make New Zealand unique.

3. *Identifying* Identify one of the two locations in New Zealand that has a cold, snowy climate in winter.

4. *Citing Text Evidence* What native animal has become a well-known symbol of New Zealand and its people?

5. *Informative/Explanatory Writing* Write a paragraph explaining why either Australia or New Zealand would be an interesting place to visit. Use information from the lesson to describe landscapes, ecosystems, climates, and other features of either Australia or New Zealand.

networks

There's More Online!

- ☑ **IMAGES** Aboriginal Rock Paintings
- ☑ **MAP** History of Australia
- ☑ **TIME LINE** Australian Gold Rush
- ☑ **VIDEO**

Reading HELPDESK

Academic Vocabulary
- unify

Content Vocabulary
- boomerang
- dingo
- *tikanga*
- *kapahaka*
- station
- introduced species
- dominion

TAKING NOTES: *Key Ideas and Details*

Determine Cause and Effect As you read about the histories of Australia and New Zealand, use a graphic organizer like this one to take notes about the causes and effects of human settlement in the region.

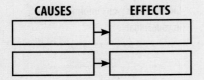

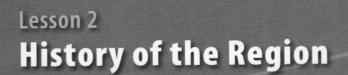

Lesson 2
History of the Region

ESSENTIAL QUESTION • *Why does conflict develop?*

IT MATTERS BECAUSE
Australia and New Zealand have made many advances in democratic government and have close ties to the United States.

First People

GUIDING QUESTION *How and when did the first humans settle in Australia and New Zealand?*

Ancient tools, cave paintings, rock art, and fossilized human remains provide clues that humans traveled to this region from other places on Earth. There is still some debate, however, about where the early people came from and when they first arrived in the region. Most scientists agree that the lands of Australia and New Zealand existed for millions of years before the first human settlers arrived.

Aboriginals of Australia

Fossil evidence shows that humans began migrating to Australia at least 50,000 years ago. At that time, much of Earth's surface water was frozen, and sea levels were hundreds of feet lower than they are today. Areas that are underneath the ocean today were exposed above the surface of the water. This created land bridges and peninsulas over which groups of people left New Guinea and walked across to Australia. Other migrating groups from the Asian mainland may have traveled longer distances by boat. These early people journeyed far and wide across the land, even to the island of Tasmania. In time, humans settled the entire Australian continent.

The first people of Australia are known as Aboriginal people. They generally lived a hunter-gatherer lifestyle. These early people survived by gathering fruit, roots, and other

424

plant parts for food and by hunting animals. They developed a flat, bent, wooden weapon called a **boomerang**. Hunters threw the L-shaped boomerang to stun their prey. If the boomerang missed, it curved and sailed back to the hunter.

Early Settlements

Some hunter-gatherers were nomadic, moving from place to place, following animals or searching for water sources. Nomadic people did not practice farming. The only domesticated animals they owned were dingoes. **Dingoes** are a species of domestic dog first brought to Australia from Asia about 4,000 years ago. Today, wild dingoes roam free in the Outback and other parts of Australia. There is controversy regarding the animal's suitability as a pet.

Other Aboriginal people settled permanently in one location. Some settled in the rain forests of the northeast. Some made their homes in the mountains of the Great Dividing Range. Others traveled as far as the humid lands of the southeast and the island of Tasmania. Some even learned to live in the harsh, dry lands of the Outback. Eventually, Aboriginal settlers began farming the land.

Aboriginal Culture

Traditional Aboriginal culture takes many forms. Aboriginal peoples living in different parts of Australia developed their own languages, religions, traditions, and ways of life. It was difficult for separate tribes of native people to communicate with one another, because about 400 different languages were spoken across Australia. Yet, many of Australia's native people share a number of common beliefs and cultural traditions. Aboriginal culture is closely connected to the natural world. Australia's native cultures have traditional beliefs about the creation of Earth and of plants and animals. They use song, dance, poetry, drama, storytelling, and visual arts to retell the creation story known as "the Dreaming" or "the Dreamtime." The creation story tells how the Spirit Ancestors created the world around them, the universe, and the laws of life, death, and society.

The concept of Dreaming is central to many of Australia's native cultures' social structures and belief systems. Music, dance, storytelling, and other art forms remain an important part of Aboriginal culture.

An Anangu woman prepares wood for carving. In addition to wood carving, Australia's aboriginal artists work in media such as painting on leaves, rock carving, fabric printing, and sandpainting. The Anangu people live in an area extending from Uluru in Northern Territory to the Nullarbor Plain in southwestern Australia.

▶ **CRITICAL THINKING**
Determining Central Ideas What do the Aborigines see as the purpose of their art forms?

The Maori of New Zealand

The first humans to live on the islands of New Zealand were the Maori. The Maori people came to New Zealand much later than the Aboriginal people came to Australia. Historians believe that sometime between A.D. 800 and 1300, humans began traveling in canoes from Polynesian islands such as Tahiti to New Zealand. The Maori built villages and lived in tribal groups led by chiefs. Tribes traded and went to war with other Maori tribes. Nearly all early Maori settlements were located on North Island.

The Maori way of life had minimal impact on the environment. They fished, gathered plants, and hunted wild animals for food. Huge, flightless birds called moa were easy prey and became an important part of the Maori diet. Eventually, the Maori hunted all 10 species of moa to extinction. In time, the Maori introduced food crops such as taro and yams. Boating and diving were important parts of Maori life. Canoes carved from the trunks of massive trees were used for transportation and also as vehicles for warfare.

Maori Culture

The Maori have a spiritual belief system based on the concept that all life in the universe is connected. At the heart of Maori culture is *tikanga*. **Tikanga** are traditional Maori customs and traditions passed down through generations. Maori tradition says that tikanga come from *tika*, the "things that are true," which began with all creation at the dawn of time. According to tikanga, the past is always in front of an individual, there to teach and guide that person. The future is behind the individual, hidden and unknown. Tikanga is part of everyday Maori life, from building homes and preparing food and medicine to social customs and arts, such as *kapahaka*. **Kapahaka** is a traditional Maori art form combining music, dance, singing, and facial expressions.

Maori warriors sail in a war canoe below one of their lookout points on the North Island coast. To get to their battlefields, the Maori built large war canoes called *waka taua*. Each vessel held about 100 people and was up to 130 feet (40 m) in length. Viewed as sacred, the war canoes were elaborately carved with images of deities and ancestors.

▶ **CRITICAL THINKING**
Analyzing Why do you think the early Maori carved sacred images on their war canoes?

One of the most important parts of any culture is language. Before Europeans arrived in New Zealand, the Maori people spoke different dialects of the same language. When New Zealand was colonized by the British, many Maori began speaking English. Over time, fewer and fewer people spoke the traditional Maori language, even in their homes. By the mid-1900s, the language had become so rare it was in danger of dying out. The efforts of Maori leaders and activists brought the Maori language back from the brink of extinction. During the 1970s and 1980s, the Maori language was reborn, as language recovery programs and schools across New Zealand taught younger generations to speak Maori. In 1987 Maori became an official language of New Zealand.

A unique part of Maori culture is the sacred practice of facial tattooing. Maori tattoos are permanent decorations on the body made by cutting designs into the skin, then rubbing black soot into the cuts. Traditionally, Maori boys were tattooed during puberty as a rite of passage into manhood.

Each individual's tattoo was unique, showing his ancestry, status in the tribe, military rank, profession, and family relationships. Maori tattooing is still practiced in New Zealand as an expression of cultural pride and identity. Other Maori art forms such as weaving, painting, and wood carving are also important parts of past and present Maori culture.

A Maori meeting house in New Zealand's Waitangi area displays detailed wooden carvings, all done by hand. Maori meeting houses have long served as centers for important community events and ceremonies.

▶ **CRITICAL THINKING**
Analyzing How do Maori meeting houses reflect Maori beliefs about their society?

☑ **READING PROGRESS CHECK**

Describing How did the early hunter-gatherers live?

Colonial Times

GUIDING QUESTION *What happened when Europeans came to Australia and New Zealand?*

Australia and New Zealand were colonized by the British. However, the British were not the first Europeans to visit the region. Explorers from other European countries had been sailing around Australia and landing on its shores for hundreds of years before the British arrived.

Europeans Come to the Region

During the 1600s, 1700s, and 1800s, Dutch, Spanish, French, and Portuguese explorers visited Oceania. Some of them mapped Australia's coastline. Some even went ashore to explore coastal areas and search for supplies. In 1642 the Dutch East India Company sent Abel Tasman on a mission to sail around the Australian continent. During this voyage of discovery, Tasman circled the island of Tasmania and then sighted the coast of New Zealand. More than a century would pass, however, before Europeans started colonizing the region.

British Australia

Perhaps the most well-known British explorer was the sailor Captain James Cook. He carried out three voyages in the 1760s and 1770s. Following Cook's explorations, the British government prepared to send settlers to the wild, unexplored lands of Australia. Most of the first colonists sent to Australia did not go by choice. In 1788 a group of 11 British ships, known as the First Fleet, landed on Australia's east coast. The crowded ships carried 778 convicted criminals from the British Isles. The First Fleet also included 250 soldiers and government officials. This was the first shipment of about 160,000 convicts sent to Australia during the next 80 years, due to a lack of space in England's prisons. Living conditions were terrible, and punishments were harsh for convicts held in Australia's cruel, filthy prisons. For many years, Australia was known to most of the world as a prison colony.

On April 29, 1770, Captain James Cook made his first landing in Australia at Botany Bay, near present-day Sydney. This imagined view of the landing was painted by E. Phillips Fox, an Australian artist of the early 1900s.

Beginning in the late 1700s, a settlement began to form in Sydney. Colonists were slow to come to the area, but in time, Sydney grew into a busy center of trade and industry. Settling the inland areas of Australia did not happen for many years. Only a few people other than escaping criminals dared to venture into the rough Outback. By the 1880s, however, most of the continent had been explored by Europeans.

In 1851 an English prospector found gold near Bathurst, New South Wales. Soon after, thousands of people from all over Australia were camped in the area, digging for gold. Gold was discovered in other parts of Australia, and word spread across the globe. People from around the world flocked to Australia at a rate of 90,000 per year, all hoping to find a bounty of gold. Prospectors came from as far away as England, Ireland, China, and the United States.

Australia began to grow for other reasons, as well. When resources such as coal, tin, and copper were discovered, workers came for mining jobs. Business owners started and built shops and hotels wherever towns sprang up, and there were people to spend money. Small towns grew into cities. Farmers planted crops in Australia's most fertile areas. Ranchers brought sheep and cattle from overseas, and Australia's ranching industry was born. Vast ranches called **stations** covered millions of acres in the Outback and other areas. Today, millions of sheep and cattle live on ranches all across Australia.

Challenges and Conflict

As humans from other parts of the world moved to Australia, they brought animals with them. Ranchers brought dogs for guarding and herding sheep. Wealthy landowners imported European rabbits to hunt for sport. In an effort to rid sugarcane fields of a destructive beetle, farmers brought in huge, poisonous cane toads. These and other nonnative animal species caused major problems. Rabbits multiplied quickly to a population of 1 billion. They ate so much grass and so many wildflowers that entire areas were left bare. The cane toads also multiplied, crowding out and killing native animal species. Animals that are not native to an area but are brought from other places are called **introduced species**. It is impossible to estimate how much damage has been done to Australia's environment by introduced species. Some introduced species are now under control, but others continue to cause problems for humans and native animals.

British settlers built their homes and farms all over Australia, often forcing native Aboriginal people off their land. Thousands of Aboriginal families and tribes were forced to leave lands where their ancestors had lived for generations.

A gold prospector sits outside a hut he built in the settlement of Gippsland, Australia. During the late 1800s, gold discoveries in southeastern Australia drew many prospectors to the area.

▶ **CRITICAL THINKING**
Describing How did the discovery of gold and other mineral resources contribute to Australia's development?

In both world wars, Australia sent its soldiers to foreign battlefields in support of the British Empire. The heroism and sacrifices of Australian soldiers in these global conflicts attracted the attention of people in other parts of the world.

▶ CRITICAL THINKING

Analyzing How did involvement in both world wars change the way Australians viewed themselves and their country?

European diseases and violence steadily reduced the Aboriginal population. The survivors had no choice but to live on rugged lands that European settlers did not want.

British New Zealand

Captain Cook explored the islands of New Zealand during the early 1770s. Cook reported to the British government that the fertile islands had many valuable natural resources and would be good places to colonize. Soon, British colonists and British, American, and French traders and whalers built settlements on North Island. At first, most relations between the Maori and foreign settlers were peaceful. In 1840 the British government, ruled by Queen Victoria, convinced Maori leaders to sign the Treaty of Waitangi. This treaty gave legal ownership and control of New Zealand to Great Britain, but it guaranteed protection and certain land rights to the Maori.

As Europeans continued to arrive in New Zealand, the Maori saw more and more of their land taken by foreign settlers. Maori society and ways of life weakened when British settlers brought new methods of farming and other features of European culture. Conflict between the Maori and British continued sporadically, until 1872 when many Maori were killed and they lost most of their land to the British.

As was happening in Australia, businesses, industries, farms, and sheep ranches were built across New Zealand. Sheep ranching changed the land, as native scrubland and forests were cleared to make pastures for livestock. Introduced species, such as rabbits, goats, pigs, deer, rodents, and feral cats, began destroying natural habitats. They killed many native animals and threatened the survival of entire species.

Independent Countries

In 1901 the six British colonies set up in Australia took action to **unify** as a federation. This action formed a political alliance between New South Wales, Queensland, Northern Territory, Western Australia, South Australia, and Tasmania. The former colonies set up the Commonwealth of Australia. The new country was a

Academic Vocabulary

unify to unite; to join together; to make into a unit or a whole

dominion, a largely self-governing country within the British Empire. Like Canada, Australia had a form of government that blended a U.S.-style federal system with a British-style parliamentary democracy.

Throughout the 1800s, New Zealand residents pushed independence from Great Britain. The 1852 New Zealand Constitution Act recognized local governments in the six provinces, but New Zealand was still a long way from independence. In 1907 the British government named New Zealand an independent dominion with a British-style parliamentary democracy. Even before independence, New Zealand had made a number of political advances. In 1893 it became the first country in the world to legally recognize women's right to vote.

The Region in Contemporary Times

Australia and New Zealand were pulled into World War I and World War II through their ties to Great Britain. During World War II, Australian, New Zealand, and U.S. soldiers fought together in the Pacific region. This alliance created closer ties among the three countries.

After World War II, Australia became completely independent. Australia loosened ties with Britain and established closer ties to the United States and Asia. In 1951 Australia, New Zealand, and the United States signed a mutual security treaty called the ANZUS Pact. The treaty was meant to guarantee protection and cooperation among the three countries in case of military threats in the Pacific region.

In recent years, a huge increase in Asian immigration has led to more diversity in Australia and New Zealand. Today more people of Asian background live in New Zealand than native Maori. The governments of both countries have also continued to address Aboriginal and Maori rights and social concerns. The native people of Australia and New Zealand and their supporters continue to work for justice and equal rights under the law.

Include this lesson's information in your Foldable®.

☑ **READING PROGRESS CHECK**

Determining Central Ideas How and when did Australia and New Zealand become independent nations?

LESSON 2 REVIEW CCSS

Reviewing Vocabulary
1. Are *dingoes* native to Australia?

Answering the Guiding Questions
2. *Identifying* For whom was the island of Tasmania named?
3. *Identifying* Name one native species and one introduced species in Australia.
4. *Determining Central Ideas* In what ways were the colonization of Australia and the colonization of New Zealand alike?
5. *Narrative Writing* Imagine you are a young Australian Aboriginal or New Zealand Maori living during the time the first Europeans came to your homeland. Write a few paragraphs telling how you feel about the arrival of these foreign settlers. Whenever possible, include details from the lesson in your narrative.

networks

There's More Online!

- ☑ **IMAGE** Wool Processing
- ☑ **SLIDE SHOW** Comparing Football Around the World
- ☑ **VIDEO**

Reading HELPDESK

Academic Vocabulary

- controversy

Content Vocabulary

- bush
- didgeridoo
- action song
- geothermal energy
- kiwifruit
- lawsuit

TAKING NOTES: *Key Ideas and Details*

Summarize As you read the lesson, use a graphic organizer like the one below to write a short summary about each of the topics.

Topic	Australia	New Zealand
Natural Resources		
Economy		

Lesson 3
Life in Australia and New Zealand

ESSENTIAL QUESTION • *What makes a culture unique?*

IT MATTERS BECAUSE
The people of Australia and New Zealand are working to blend diverse populations successfully.

Life in the Region

GUIDING QUESTION *What is it like to live in Australia and New Zealand?*

European culture exercises the most influence in Australia and New Zealand, but indigenous cultures also play an important role. In recent years, Asian influences have increased in the region.

The People of the Region

Australia and New Zealand are multicultural lands. They have diverse human populations where different cultures, languages, and lifestyles are mixed together. New Zealand has had a diverse population for much of its history. Today, New Zealand's population is about 57 percent European, 12 percent Asian and Pacific Islander, and 8 percent Maori. Other groups account for the rest. Australia's population is much less diverse. About 92 percent of Australians are of European descent, 7 percent are Asian, and 1 percent are Aboriginal and other groups.

Religion and Culture

Christianity is the most common religion in the region. Also practiced in the region are Buddhism, Islam, Hinduism, and native religions. In addition, about 30 percent of the people in the region describe themselves as "nonreligious" or did not state a religious affiliation.

English is the official language in Australia. New Zealand has three official languages: English, Maori, and New Zealand sign language. The sign language, the main language used for communicating by members of the deaf community, became an official language in 2006.

The lifestyles in Australia and New Zealand are similar to modern American and British lifestyles. The residents drive cars and use public transportation. In their free time, Australians and New Zealanders shop, go out to eat, watch television, and go to movies. They keep pets such as cats and dogs. Outdoor activities, such as hiking, biking, running, boating, surfing, and swimming, are popular. Watching and playing sports such as football (soccer) and rugby are popular pastimes.

Urban Life

Human populations are unevenly distributed in the region. The vast majority of Australians and New Zealanders live in urban areas. A population map of Australia shows something interesting: The highest populations are concentrated in small land areas, while the smallest populations are scattered throughout the largest land areas. Approximately 89 percent of Australia's people live in cities and suburbs. Most New Zealanders, about 87 percent of the population, live in cities and suburbs.

Australia's largest cities are Sydney, Melbourne, Brisbane, and Perth, all with populations of more than 1 million. New Zealand's largest cities are Auckland, Christchurch, and Wellington. Life in these cities is busy. People who live in cities face everyday challenges—noise, traffic, urbanization and crowding, rising housing prices, pollution, and crime. Cities also offer an endless variety of culture, recreation, shops, restaurants, and entertainment. Urban residents must balance the challenges of city life with its many benefits.

A player from Australia's Queensland Reds charges forward during a rugby match between the Reds and New Zealand's Canterbury Crusaders at Suncorp Stadium in Brisbane, Australia. Sports like football (soccer) and rugby are popular in Australia and New Zealand.

Rural Life

Life in Australia's rural areas moves at a slower pace. Many individuals and families live alone on huge sheep or cattle stations. These people tend to be isolated, far from towns and other people. Farm life can be hard, with work from sunrise to nighttime. The term **bush** means any large, undeveloped area where few people live. The phrase "in the bush" can refer to any location that is wild, unsettled, and rough, such as the Australian Outback.

A student on a remote sheep station in rural Australia takes part in a School of the Air lesson on her home computer. She can see and talk to her teacher by way of a live video cam hookup made possible by a satellite system.

▶ CRITICAL THINKING

Analyzing How has electronic technology improved the education of Australian students living in remote areas?

Rural areas in New Zealand are not as remote as those in Australia, because New Zealand has a much smaller land area. Lush pasture lands can feed herds of sheep and cattle on fewer acres than the dry Australian Outback. As a result, New Zealand farms are located closer together, and farm families have more contact with friends and neighbors. Many people in rural areas live near small towns, where they can shop and interact with others. In recent decades, the populations of New Zealand's small towns have been shrinking as more and more rural people move to cities.

Australian English

Australians are famous for their use of nicknames and slang. Australian English, called *Strine*, has a unique vocabulary made up of Aboriginal words, terms used by early settlers, and slang created by modern Australians. Common slang in Australia uses rhymes and word substitutions. For example, "frog and toad" is slang for "road," and "steak and kidney" is a slang nickname for the city of Sydney. The Australian people are nicknamed "Aussies." New Zealanders also have a common nickname. During World War I, New Zealand soldiers and Australian soldiers served together. The Australian soldiers nicknamed the New Zealanders "Kiwis." The nickname is still used today, even by New Zealanders.

Education

The people of Australia and New Zealand take pride in their educational systems. Both countries are home to well-respected universities that rank as some of the best schools in the world.

Australia is a huge continent with many isolated communities. Many students, especially in the Outback, use modern methods of communication to receive and turn in their lessons. Beginning in the 1950s, classes were conducted via shortwave radio, with students having direct contact with a teacher in town. Previously, students relied on mail service to deliver assignments. Today, the Internet provides quicker and more reliable delivery.

Aboriginal and Maori Culture

Australia and New Zealand have experienced revivals in Aboriginal and Maori cultures. In Australia, some Aboriginal storytellers still use oral tradition to pass down history and myths from one generation to the next. Traditional customs and tools such as the boomerang are still part of daily life. Boomerangs have been used for centuries as tools for hunting, as toys, and as weapons for hand-to-hand combat. Today, boomerangs are used for recreation and contests of skill. Another Aboriginal artifact still used today is a musical instrument called the didgeridoo. The **didgeridoo** is a long wood or bamboo tube that creates an unusual vibrating sound when the player breathes into one end. Aboriginal instruments such as the didgeridoo are part of modern Australian culture, helping to keep traditional music alive.

As part of the modern movement to restore Maori culture in New Zealand, performers created **action songs**. These performances combined body movement with music and singing, often with lyrics that celebrated Maori history and culture.

✓ **READING PROGRESS CHECK**

Identifying Points of View Based on information in the lesson, decide if the nicknames "Aussie" and "Kiwi" are offensive to Australians and New Zealanders, respectively.

Maori young people in the city of Christchurch, New Zealand, perform an action song. Both men and women perform action songs, using tight arm motions with straight, vibrating hands.

▶ **CRITICAL THINKING**
Determining Central Ideas Why do you think the performance of action songs is important to Maori people today?

Natural Resources and Economies

GUIDING QUESTION What resources are important to the economies of Australia and New Zealand?

Australia's Natural Resources

Australia is rich in valuable natural resources. Coal mining is a major industry in eastern and northwestern Australia and also in Tasmania. Iron ore is plentiful in the northwest. Gold discovered in Western Australia in the 1850s spurred a gold rush. Precious metals, including gold and silver, are still mined today. Large offshore oil and natural gas reserves are located in northern Australia and also in the Bass Strait between Australia and Tasmania. Australia exports some of its oil and natural gas.

The fertile farmland in southeastern Australia and other areas is one of the country's valuable natural resources. Wool, food crops, and other agricultural products are raised in many different parts of Australia. Timber and other products come from various species of trees growing across Australia. For example, eucalyptus trees are harvested for their wood, oil, resin, and leaves.

Australia's Economy

Australia's economy relies heavily on exports of its many natural resources. Coal, iron ore, and gold are Australia's three leading exports. Australia's economy also depends on its exports of meat, wool, wheat, and manufactured goods to countries all over the world. Australia's chief trading partners are China, Japan, South Korea, and the United States. Australia's manufacturing sector is not as strong as is the manufacturing sector in several East Asian nations, but Australia has enjoyed robust economic growth in recent years. Unemployment is relatively low. Australia ranks twelfth in world gas reserves and eleventh in world gas exports. Tourism continues to be a vital industry. More than 5 million people visit Australia each year, bringing revenue to local businesses and employing thousands of Australians. During the past few decades, Australia's film industry has grown to international status.

New Zealand's Natural Resources

New Zealand enjoys one important benefit of its location along the Ring of Fire: easy access to geothermal energy. **Geothermal energy** is naturally occurring heat energy produced by extremely hot liquid rock in Earth's upper mantle. As magma rises up through cracks or holes in Earth's crust, it heats the rock and water within the crust. Humans reach this heat and hot water by digging and drilling. The heat energy is used to warm homes, to provide hot water,

An Australian worker empties recently shorn wool into a bin for cleaning. Traditionally, Australian wool sold mostly in Europe and North America. Today, Australian wool suppliers rely increasingly on growing markets in China and other countries of Asia.

Identifying What countries are Australia's major trading partners?

and to generate power. Another benefit of geothermal energy is that it is clean and does not pollute the environment. Hydroelectric power, derived from the energy of moving water, and windmills are other kinds of nonpolluting renewable resources.

Farmland is one of New Zealand's most valuable resources. Almost one-half of all land on the islands is used for farming and livestock grazing. The grass growing on steep hillsides tends to dry out as rainwater drains down to lower pastures. Low-lying lands can become too soaked with rain, which is not good for most crops. The most productive farmland in New Zealand is in places where the soil receives enough rain but is well drained. The dark soil of North Island's volcanic plateau also holds ribbons of mineral deposits. New Zealand's other valuable natural resources include coal, iron ore, natural gas, gold, timber, and limestone.

New Zealand's Economy

During the past two decades, New Zealand's economy has gone through a major transformation. What was once a farm-based economy has become an industrialized, free-market economy. New Zealand's government has plans to continue to increase production of wood and paper products, food products, machinery, and textiles. Agriculture is still a key part of the economy, though. Products such as beef, lamb, fish, wool, wheat, flowering plants, vegetables, and fruits are exported and shipped around the world. One notable food export is the **kiwifruit**, a type of gooseberry fruit that originated in East Asia but has become a symbol of New Zealand.

New Zealand is not as wealthy as Australia, but it has a strong economy: It is forty-eighth in the world in per capita gross domestic product (GDP). New Zealand's chief trading partners are Australia, China, and the United States. Because New Zealand depends on export income, low export demand can badly damage its economy. As in Australia, tourism is a major industry. After the release of the hugely popular *Lord of the Rings* films in the early 2000s, tourists from all over the world flocked to New Zealand. New Zealand's film industry continues to grow, producing top-grossing films that are viewed worldwide.

New Zealand produces specialized food products that are exported around the world. (Top) A fruit grower inspects his kiwifruit grown on supported vines. (Bottom) A factory worker prepares pieces of New Zealand's famous Egmont cheese for wrapping.

✓ **READING PROGRESS CHECK**

Distinguishing Fact From Opinion Is the following statement a fact or an opinion? *Tourism is an important industry in Australia and New Zealand.*

Activists protest the development of a gas project in a coastal area of Western Australia that is rich in petroleum and natural gas. The protestors claim the building of a gas plant and a port will damage offshore reefs and the fossil remains of prehistoric animals.

▶ **CRITICAL THINKING**
Citing Text Evidence What other environmental challenges do Australia and New Zealand face?

Academic Vocabulary

controversy a dispute; a discussion involving opposing views

Current Issues

GUIDING QUESTION *What challenges do the people of Australia and New Zealand face?*

Australia and New Zealand face challenges resulting from their locations, populations, climates, and physical geography. Just as the landforms and wildlife of the region are unique, so are the issues and problems facing Australia and New Zealand.

Indigenous Rights

The concern over Aboriginal people's rights has been an ongoing issue in Australia for well over a century. Aboriginal activist groups in Australia have filed several major lawsuits over land rights and environmental issues. A **lawsuit** is a legal case that is brought before a court of law. Many Aboriginal lawsuits have sparked **controversy** over the rights of the Aboriginal people versus the rights of big businesses to use the land and its resources. Some cases are still being decided. Similar human rights issues continue in New Zealand between New Zealanders of European descent and native Maori.

The Maori own only about 5 percent of the land in New Zealand. They petitioned the government to return their lost land. The government could not return land to the Maori without hurting the people who were living on it. The government agreed to pay the Maori for lost land and lost fishing rights. Payments continue, but they have been slow in coming.

Protecting Natural Resources

Protecting the environment and natural resources is important to many Australians and New Zealanders. Recent environmental issues in Australia are drought and limited water supplies, bushfires, and threats to the survival of the Great Barrier Reef. All of these are affected by global warming. For example, the Great Barrier Reef is affected as water temperatures in the oceans rise, resulting in the death of organisms that need a cooler climate. The corals are also sensitive to climate change and water pollution. Parts of the Great Barrier Reef have already died. Scientists are concerned that if global warming continues, this entire massive reef system will be lost.

In recent decades, many geothermal hot springs and geysers have disappeared in New Zealand, but many of the remaining hot springs and geysers are located in protected areas. Geographers believe human activities such as drilling for hot water and building power stations have destroyed many natural wonders. Such environmental issues are of great concern to Australians and New Zealanders.

Other Issues

Australia and New Zealand have low birthrates. They also have low death rates. Low birthrates result from families having fewer children. Low death rates result from Australia's and New Zealand's high life expectancies. While such rates are generally beneficial for a country, the combination produces a gradually aging population. It also creates a need for more workers to support the older population. People immigrating to Australia and New Zealand are filling some of these positions, which helps meet the need for care providers. Immigration is also changing the region's ethnic makeup.

Australia and New Zealand face many other issues and challenges. Both countries are affected by the overall health of the planet. Their economies depend upon trade with other countries. Their industries could not survive without strong economic relations with one another and without trade with other countries.

Include this lesson's information in your Foldable®.

☑ **READING PROGRESS CHECK**

Describing In your own words, explain why survival of the Great Barrier Reef is threatened.

LESSON 3 REVIEW

Reviewing Vocabulary
1. What is the *bush* like?

Answering the Guiding Questions
2. ***Identifying*** What type of resource is *geothermal energy*?
3. ***Describing*** How is daily life in Australia and New Zealand similar to life in the United States?
4. ***Identifying*** Give three examples of resources found in Australia and three examples of resources found in New Zealand.
5. ***Argument Writing*** Think about the issues facing Australians and New Zealanders today. Write a paragraph in the form of an argument describing which issue you believe is the most important and why.

Chapter 14 **439**

GLOBAL CONNECTIONS

Unfriendly Invaders

Invasive species are animals, plants, and infectious organisms that take over the natural environment of other species. Invasive species often harm the environment and cause the native species to decline in number. They can also harm the health of humans.

Cane Toads Cane toads were brought to Australia in 1935 to control beetles that attacked sugarcane crops. But cane toads contain toxins that poison many native animals that eat them. Cane toads also eat large numbers of honeybees that pollinate plants and crops. The toads carry diseases that can be passed on to other frogs and to fish.

> " Wild rabbits are considered Australia's most widespread and destructive pest. "

A Vast Number Why are cane toads so plentiful? Twice every year, female cane toads produce 8,000 to 35,000 eggs. That's many more eggs than the average frog lays. These eggs quickly hatch and form a school of tiny, black tadpoles. Many of the tadpoles do not survive. Those that survive can live for 10 to 40 years.

Animals That Prey Australia has many wild foxes and feral cats that prey on other animals, and rabbits that devastate vegetation. Their numbers are so great that the government says it cannot eliminate them. The goal is to reduce the damage they cause.

$4 Billion Per Year Invasive alien plants present problems for the economy. The cost of the damage and attempts to control the plants amounts to $4 billion per year.

Did You Know?

The International Union for Conservation of Nature (IUCN) assesses the risk of extinction for species. The four categories are:
- **Extinct**—Species has died out and no longer exists
- **Critically endangered**—Species faces a high risk of becoming extinct
- **Endangered**—Species faces a risk of extinction
- **Vulnerable**—Species is likely to become endangered unless circumstances improve

A worker weighs a cane toad at a collection point in Cairns, Australia. ▶

THERE'S MORE ONLINE

HEAR cane toads • *SEE* invasive plant species • *WATCH* an animation on invasive species

Chapter 14 **441**

GLOBAL CONNECTIONS

These numbers and statistics can help you learn about the invasive species of Australia.

TWENTY-TWO EXTINCT

There are 22 extinct mammals in Australia. *Extinct* means no more are left. Many other animals are in danger of dying out. Australia has more endangered species than any other continent.

The red fox, the feral cat, and the rabbit are probably responsible for the loss of 20 of the 22 extinct marsupials and rodents in Australia. Other animals that have become major agricultural and environmental problems are wild pigs, goats, and deer.

12 pounds

In many parts of Australia, all native mammals weighing up to 12 pounds (5.4 kilograms) are extinct. Nine species of mammals exist only on Australian islands that have no cat or fox population.

1974 Epidemic

Phytophthora root rot is an invasive disease that threatens many important crops and plant species in Australia. For example, a 1974 root rot epidemic destroyed more than one-half of all the avocado trees in eastern Australia.

3,300-MILE FENCE

The dingo looks like a dog and is the largest carnivorous animal in Australia. More than 100 years ago, Australians built a long fence to keep the dingo away from sheep flocks and other animals. At 3,300 miles (5,311 km) long, the dingo fence is the world's longest fence. Rabbit-proof fences were built to protect Western Australian crops and pasture lands.

No. 1

Wild rabbits are considered Australia's most destructive pest. European rabbits were introduced in Australia more than 150 years ago. The rabbit population grew huge because few animals prey on the rabbits. A virus in the 1950s killed many rabbits, but as rabbits built up an immunity to the virus, the rabbit population began to grow again. Today, millions of wild rabbits live in all parts of Australia.

27,000

In the 1800s, a weed called the prickly pear overran large areas, forcing many farmers off their land. Today, more than 27,000 invasive alien plants grow in Australia.

VULNERABLE	ENDANGERED	CRITICALLY ENDANGERED	EXTINCT
Species is likely to become endangered unless circumstances improve	Species faces a risk of extinction in the wild; usually when fewer than 250 mature individuals exist	Species faces a high risk of becoming extinct	Species has died out and no longer exists
African Elephant, Polar Bear, **Great White Shark**	Blue Whale, Giant Panda, **Asian Elephant**	California Condor, Red Wolf, **Mountain Gorilla**	Zanzibar Leopard, Caspian Tiger, **Passenger Pigeon**

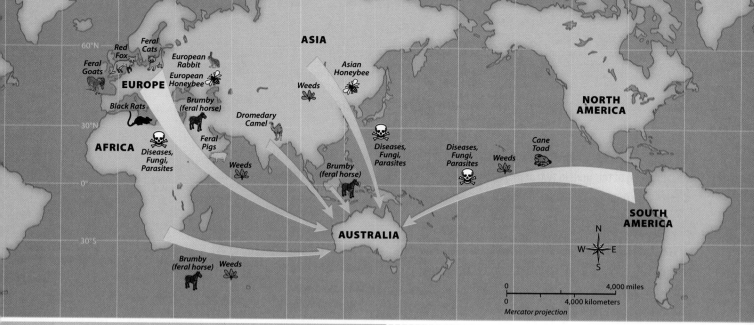

GLOBAL IMPACT

AUSTRALIA'S INVASIVE SPECIES

The map shows the location from where various species were brought to Australia. But why were they brought to Australia? To European settlers, Australia seemed strange. There were no familiar wild animals. In 1859 a rancher brought wild rabbits from England and set them free on his land. As the number of rabbits grew, businesses began to can rabbit meat to sell and used the skins and fur to make clothing and hats. When the rabbit population continued to grow, steps were taken to control them. In the 1950s, a virus was developed that killed most of the rabbits. However, rabbits became resistant to the virus, and the population grew again.

Global Species Extinction

Mass extinctions are time periods in the history of Earth when an extraordinarily large number of species go extinct. Today, many scientists believe the evidence shows a mass extinction is underway.

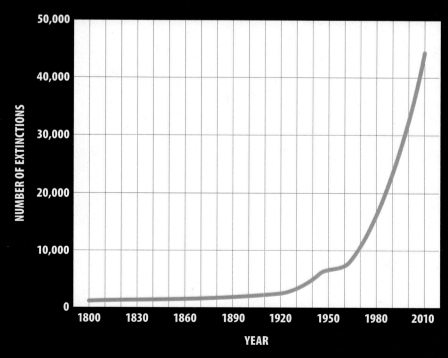

Thinking like a Geographer

1. **Physical Geography** Why was the cane toad introduced into Australia? Why did the cane toad population grow so fast?

2. **The Uses of Geography** Find out what invasive species have been introduced into your area. Prepare a poster to show how to protect native plants and animals against invasive species.

3. **Environment and Society** Research to find information about an endangered species. Write about what is being done to protect these animals.

Chapter 14 ACTIVITIES

Directions: Write your answers on a separate piece of paper.

1 Use your **FOLDABLES** to explore the Essential Question.
INFORMATIVE/EXPLANATORY WRITING Review the physical map and the population map at the beginning of this unit. In two or more paragraphs, explain how Australia's physical geography has affected the country's settlement patterns.

2 21st Century Skills
INTEGRATING VISUAL INFORMATION Work in small groups to research the physical geography, people, and culture of each of Australia's three main geographic regions, New Zealand's North and South Islands, and Tasmania. Present the information as a slide show or a poster.

3 Thinking Like a Geographer
IDENTIFYING Create a two-column chart similar to the one shown here. Label one side Australia and the other side New Zealand. Use the chart to list the natural resources of each country.

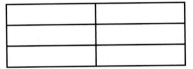

4 GEOGRAPHY ACTIVITY

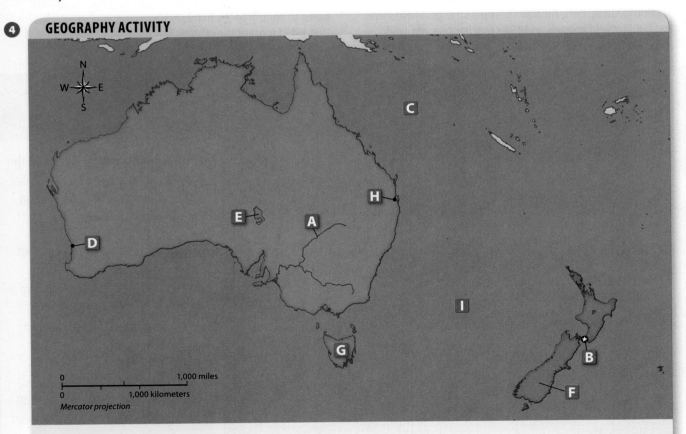

Locating Places
Match the letters on the map with the numbered places listed below.

1. Darling River
2. Wellington
3. Coral Sea
4. South Island
5. Perth
6. Lake Eyre
7. Tasmania
8. Tasman Sea
9. Brisbane

Chapter 14 Assessment

REVIEW THE GUIDING QUESTIONS

Directions: Choose the best answer for each question.

1. New Zealand can best be described as a land of
    A. extremely diverse landscapes, ecosystems, and climate zones.
    B. nomadic people.
    C. harsh, dry deserts.
    D. cold, barren landscapes.

2. How much of Australia is covered by desert?
    F. one-half
    G. one-fourth
    H. two-thirds
    I. one-third

3. The first people to live in Australia are called
    A. Maori.
    B. Indians.
    C. Aboriginal people.
    D. convicts.

4. The first people to inhabit New Zealand came from
    F. Borneo across a land bridge.
    G. Polynesia in canoes.
    H. Australia in sailboats.
    I. Britain in convict ships.

5. What is the basis for New Zealand's economy?
    A. coal
    B. geothermal energy
    C. agriculture
    D. fishing and hunting

6. What is the name of the bamboo instrument used by Aboriginal musicians to make traditional music?
    F. boomerang
    G. kiwi
    H. marsupial
    I. didgeridoo

Chapter 14 ASSESSMENT (continued)

DBQ ANALYZING DOCUMENTS

7. ANALYZING Read the following passage about new construction taking place in Australia:

"Australia is set for an . . . $115 billion infrastructure boom as the nation adds ports and railways to feed China and India's appetite for coal and iron ore. . . . The demand, coupled with economic slowdowns in the U.S. and Europe, has helped make Australia the developed world's fastest-growing construction market."

—from David Fickling, "China Trade Spurs $115 Billion Australia Building Boom: Freight"

What factor spurred this upcoming building boom in Australia?

A. slumping economies in the United States and Europe
B. discovery of new sources of coal and iron in Australia
C. development of new uses for coal and iron ore
D. economic growth in China and India

8. DETERMINING CENTRAL IDEAS What economic trend does the Australian building boom demonstrate?

F. increasing productivity
G. global interdependence
H. growth of high-technology industries
I. growth of service industries

SHORT RESPONSE

"In June 2010, the government signed a new agreement with the Maori over contentious [disputed] foreshore and seabed rights, replacing a 2006 deal that had ended Maori rights to claim customary titles in courts of law. Tribes can now claim customary title to areas proven to have been under continuous indigenous occupation since 1840. Maori tribes that secure a customary title will be granted title deeds, but cannot sell the property or bar public access to the area."

—from "New Zealand," FreedomHouse.org

9. DETERMINING WORD MEANINGS What does the term "continuous indigenous occupation" mean in the agreement?

10. ANALYZING How does the 2010 agreement protect the rights of Maori and non-Maori?

EXTENDED RESPONSE

11. INFORMATIVE/EXPLANATORY WRITING If you had the opportunity to relocate and live for a couple of years in Australia or New Zealand, which country would you choose? Explain your choice in a short essay. Be sure to consider such things as climate, landforms, recreation, cost of living, and employment opportunities in your writing.

Need Extra Help?

If You've Missed Question	1	2	3	4	5	6	7	8	9	10	11
Review Lesson	1	1	2	2	3	3	3	3	2	2	1

OCEANIA

ESSENTIAL QUESTIONS • How does geography influence the way people live?
• What makes a culture unique? • Why do people make economic choices?

networks
There's More Online about Oceania.

CHAPTER 15

Lesson 1
Physical Geography of Oceania

Lesson 2
History and People of Oceania

Lesson 3
Life in Oceania

The Story Matters...

Thousands of islands make up the three sections of Oceania in the Pacific Ocean—Micronesia, Melanesia, and Polynesia. Because of their location, the islands attracted Europeans. Oceania's colonization and occupation during World War II had a tremendous impact on the region. Many countries, including the United States, still have strategic military bases there.

Go to the Foldables® library in the back of your book to make a Foldable® that will help you take notes while reading this chapter.

A man of Papua New Guinea wears an elaborate ceremonial headdress.

447

Chapter 15
OCEANIA

Thousands of islands differing in size and extending across millions of square miles of the Pacific Ocean are located in the region called Oceania.

Step Into the Place

MAP FOCUS Use the map to answer the following questions.

1. **THE GEOGRAPHER'S WORLD** What is Oceania's largest country in land area?

2. **THE GEOGRAPHER'S WORLD** In which direction would you travel to go from the Marshall Islands to the Cook Islands?

3. **PLACES AND REGIONS** What is the capital of Palau?

4. **CRITICAL THINKING** Describing Of the Solomon Islands, Tonga, and Samoa, which are located east of the International Date Line?

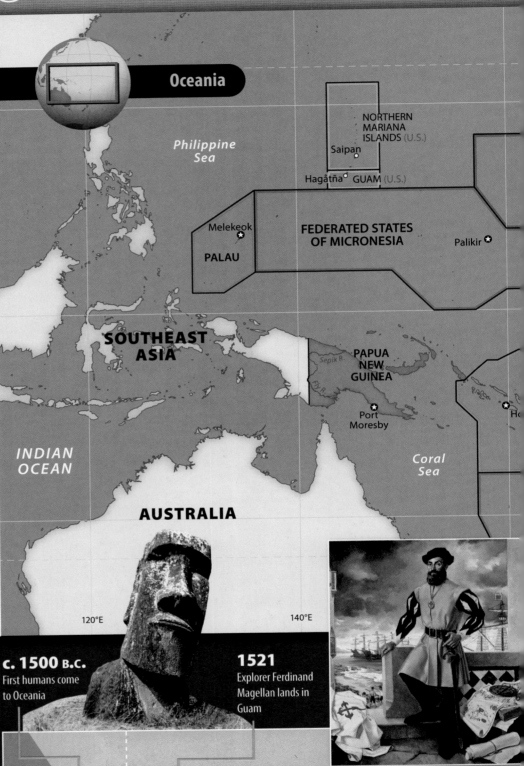

Step Into the Time

TIME LINE Choose two events from the time line to explain why many countries, including the United States, view the strategic location of Oceania's islands as an important resource.

c. 1500 B.C. First humans come to Oceania

1521 Explorer Ferdinand Magellan lands in Guam

1660 Dutch claim possession of New Guinea

1766 French explore Tahiti, Samoa, and the Solomon Islands

1800

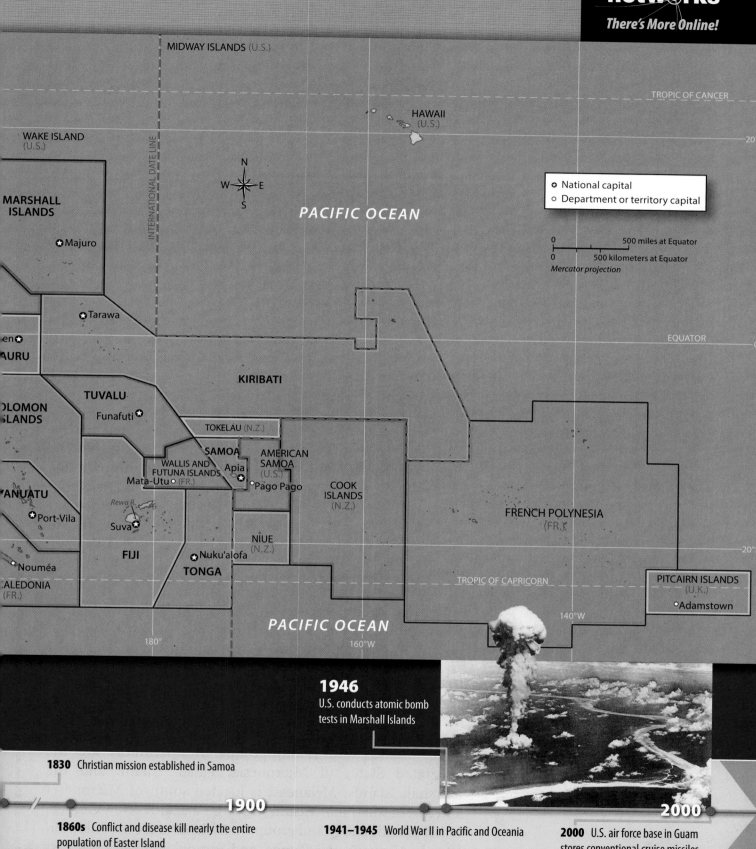

networks

There's More Online!

- ☑ **ANIMATION** How Volcanoes Form Islands
- ☑ **SLIDE SHOW** Views of Oceania
- ☑ **VIDEO**

Reading HELPDESK

Academic Vocabulary
- capable

Content Vocabulary
- continental island
- archipelago
- high island
- low island
- lagoon
- atoll

TAKING NOTES: *Key Ideas and Details*

Organize On a chart like the one below, fill in important facts about the physical geography for each of the areas.

Micronesia	
Melanesia	
Polynesia	

Lesson 1
Physical Geography of Oceania

ESSENTIAL QUESTION • *How does geography influence the way people live?*

IT MATTERS BECAUSE
The United States has many interests in Oceania, including marine resources and important shipping routes across the Pacific.

Landforms of Oceania

GUIDING QUESTION How did thousands of islands appear across the Pacific Ocean?

Oceania covers 3.3 million square miles (8.5 million sq. km) of the Pacific Ocean between Australia, Indonesia, and the Hawaiian Islands. An estimated 10,000 islands make up Oceania. Most of the islands are small, and many are uninhabited. The smallest inhabited island country is Nauru, which measures a mere 8 square miles (21 sq. km) of land area. Some islands in Oceania are located close together in clusters or island chains. Others stand alone, hundreds of miles from their nearest neighbors.

Islands, Nations, and Territories

Geographers divide Oceania into three sections, according to culture and physical location. The sections are called Micronesia, Melanesia, and Polynesia. Micronesia is located in the northwest section of Oceania, east of the Philippines. Major islands and island groups in Micronesia are the Federated States of Micronesia, Palau, Guam, and the Marshall Islands. Melanesia is located south of Micronesia and east of Australia. Melanesia's largest island is New Guinea. Other islands and island groups in Melanesia are the Solomon Islands, Vanuatu, Fiji, Tonga, and Samoa. Polynesia is a vast

(l to r) Images & Stories/Alamy; Friedrich Stark/Alamy; Visual&Written/Newscom

450

section of Oceania located in the central Pacific Ocean. Major islands and island groups in Polynesia include French Polynesia, Kiribati, Niue, the Hawaiian Islands, and the Cook Islands.

The islands of Oceania range in size from New Guinea, which is 303,381 square miles (785,753 sq. km), to tiny rock outcroppings and patches of sand covering less than 1 square mile (2.6 sq. km). Oceania's islands vary in physical characteristics such as landforms and native plants and animals. The islands also differ in their forms of government. Some islands, such as Fiji and Palau, are independent countries. Other islands, such as American Samoa, Guam, and New Caledonia, are overseas territories that are under the jurisdiction of other countries. Australia, England, France, New Zealand, and the United States have island territories in Oceania. One island, New Guinea, is divided politically.

The Divided Island of New Guinea

New Guinea is the largest island in Oceania. New Guinea is a **continental island**, an island that lies on a continental shelf and was once connected to a larger continental landmass. New Guinea lies on the same continental shelf as Australia and was part of that continent during the past when sea levels were lower. New Guinea is now part of the Malay Archipelago. (An **archipelago** is a group of islands clustered together or closely scattered across an area.)

Traditional homes along the Sepik River in Papua New Guinea are thatched houses on stilts. In this swampy and isolated area, the dugout canoe is the only means of travel.

Children play in a highland village in the Owen Stanley Range of Papua New Guinea. The rugged mountain range is the southeastern part of a long mountain chain that stretches across New Guinea, the world's second-largest island after Greenland.

Physically, New Guinea is one island; however, the island is divided into two parts. The western part belongs to Indonesia. The eastern part is an independent country called Papua New Guinea. Both parts of the island have the same types of land features, plants, animals, climate, and resources. Only the eastern part of the island, Papua New Guinea, however, is considered part of Oceania. When describing the physical features of the island, we will use the name of the country, Papua New Guinea.

Papua New Guinea has a great many different landforms. A very large mountain range stretches across the island. This chain of rugged mountain peaks and glaciers dominates the inland areas of the island. Low mountains and fertile river valleys roll across northern Papua New Guinea. Papua New Guinea also has a northern coastal plain. In the south, swampy lowlands lead to the Owen Stanley Range, which is considered the "backbone" of Papua New Guinea.

The Smaller Islands

Oceania includes several physically different types of islands. **High islands** have steep slopes rising from the shore, higher landforms, and diverse plant and animal life. High islands are generally the largest and greenest of Oceania's small islands. The soil that covers high islands is fertile, and the climates are generally humid and rainy. Many have freshwater streams, rivers, and waterfalls. These conditions allow dense rain forests to grow. Tahiti and the Hawaiian islands are examples of high islands.

Low islands are smaller, flatter islands with sandy beaches. Low islands tend to have fewer forests and less diverse plant and animal life. Most of the islands scattered across Micronesia and Polynesia are low islands. The typical "desert island" described in books and seen in films could be classified as a low island. Some desert islands in Oceania are so low they just break the water's surface. The island country of Tuvalu is an example of a low island with an extremely low elevation. Tuvalu is gradually being eroded by ocean waters. With each passing decade, Tuvalu loses more surface area, as its land washes away with the tides. If sea levels continue to rise, Tuvalu and Oceania's other lowest islands will disappear below the water.

The islands of Oceania were formed by processes involving the lithosphere, the hydrosphere, and the biosphere. Most of Oceania's low islands were formed by a gradual process involving volcanic eruptions (lithosphere), erosion by water movement (hydrosphere), and the growth of coral (biosphere). This process began millions of years ago with the eruption of an undersea volcano. As lava from the volcano cooled, it formed a buildup of volcanic rock below the surface of the water. Corals started growing on the volcano, eventually forming large reefs that circled the volcano. Over time, the volcano crumbled and sank, while the coral reefs continued to grow higher and higher. The coral reefs built up layer upon layer until they grew above the water's surface. Waves crashing against the reefs eventually eroded channels, allowing ocean water to flood the center area of the island and form shallow pools called **lagoons**. As the reef aged, crumbled, and died, ocean waves deposited sediment, such as sand and tiny specks of plant and animal life, on the coral remains. The resulting landform is called an **atoll**, a coral island made up of a reef island surrounding a lagoon. If the center of an island does not fill with water, or if the atoll becomes completely covered with sediment over time, a desert island forms.

The majority of the high islands in Oceania were formed by underwater volcanoes. As lava from erupting volcanoes flowed into the ocean waters, it cooled and formed huge mounds of volcanic rock. Eventually the volcanic rock built up above the water's surface, forming high islands.

☑ **READING PROGRESS CHECK**

Identifying What are the names of Oceania's sections?

> **DIAGRAM SKILLS >**

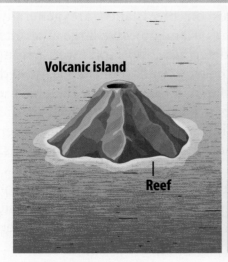

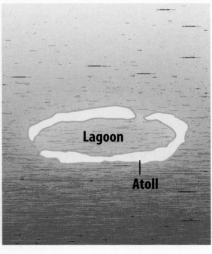

ATOLL FORMATION
South Pacific atolls are known for their beautiful coral reefs and marine life, which draw tourists from around the world. The typical atoll takes about 30 million years to form.

▶ **CRITICAL THINKING**
1. ***Describing*** What two processes are involved in atoll formation?
2. ***Analyzing*** How might tectonic plate movements affect the formation of an atoll?

French Polynesia is known for its black pearl industry. Oysters containing the black pearls are raised in underwater farms. Black pearls make up more than half of French Polynesia's exports.

▶ CRITICAL THINKING

Analyzing Why are other abundant ocean resources, such as fish and seafood, used by Oceania's people but not exported?

Academic Vocabulary

capable having the ability to cause or accomplish an action or an event

Climates of Oceania

GUIDING QUESTION *What factors affect climate in Oceania?*

Nearly all of Oceania's islands are located within the Tropics. Only a few, such as the Midway and Pitcairn Islands, lie north or south of this climate zone. Thus, most of Oceania's islands have warm, humid, tropical climates. The islands located outside the Tropics have mixed tropical and subtropical climates. Some islands experience local climate variations caused by elevation, winds, and ocean currents.

Papua New Guinea Climates

Papua New Guinea has two climate zones: tropical and highland. The lowland areas have warm to hot temperatures. The highland regions are much cooler. For example, the average daily temperature in the lowland coastal plains is 82°F (28°C). The average daily temperature in the highland mountains is 73°F (23°C). Papua New Guinea's climate is wet, with an average of 45 inches (114 cm) of rain falling on the island annually. February is the month with the most rainfall, and July has the least. Monsoon rains are common. Dense rain forests grow well in the warm, wet conditions, and much of the island is covered in trees and other forest plants. Papua New Guinea does, however, experience occasional droughts caused by the El Niño effect.

Climates of the Smaller Islands

The smaller islands scattered throughout Oceania have tropical and subtropical climates. Temperatures on the islands are generally warm throughout the year. Oceania's islands receive large amounts of rain. Seasonal rainfall patterns determine an island's wet season, which is the time of year when the heaviest rains fall. Islands located north of the Equator, such as the Marshall Islands and Palau, have a wet season from May to November. Islands located south of the Equator, such as Samoa and Tonga, have a wet season from December to April. The rainfall patterns occur under normal conditions. Heavy rains can also be caused by storms, such as typhoons. Typhoons cause intense winds and powerful waves **capable** of toppling trees and houses and eroding island shores. Only the island of Yap is affected by monsoon winds, which are brought by weather patterns of the western Pacific and Indian oceans.

☑ READING PROGRESS CHECK

Analyzing The Northern Mariana Islands experience a wet season from May to November. Based on this information, are the Northern Mariana Islands located north or south of the Equator?

Resources of Oceania

GUIDING QUESTION *What natural resources do the islands of Oceania possess?*

Most of Oceania's islands are small. Limited land area limits the amount of natural resources, such as minerals, to be found on land. Still, the islands have some valuable resources. Resources that the islanders trade or sell are essential to the islands' economies.

Papua New Guinea's Resources

The independent nation of Papua New Guinea has natural resources such as gold, copper, timber, fish, petroleum, and natural gas. Compared to the island's size, Papua New Guinea's natural gas reserves, discovered fairly recently, are large. They have the potential to benefit the nation's economy.

Resources of the Smaller Islands

Oceania's smaller islands have few valuable natural resources for trading on the international market. Limited land area is one reason for this; another reason is that low islands, based on coral, do not have rock foundations. It is in deep layers of rock that large deposits of metal ores, such as gold, are found. New Caledonia is also developing wind power by building large stands of wind turbines. This technology is beginning to spread to other islands. People living in rural areas of Kiribati and the Solomon Islands have begun using solar-powered lighting in their homes. Wind and solar energy are renewable, nonpolluting resources. Because of the sunny climates and ocean winds, these resources are plentiful throughout Oceania.

Some of Oceania's high islands have large trees that are used for timber, rubber, and other products. Soil quality varies from island to island. Some islands, such as those with rich volcanic soil, have excellent soil for growing farm crops. Other islands have poor-quality soil, making farming difficult. Fish and other seafood are important resources for the people in Oceania. Most islands use fish only for their own food, not for export.

Include this lesson's information in your Foldable®.

✓ **READING PROGRESS CHECK**

Identifying What are two types of renewable resources in Oceania?

LESSON 1 REVIEW

Reviewing Vocabulary
1. Why is New Guinea called a *continental island*?

Answering the Guiding Questions
2. *Identifying* Which countries have territories in Oceania?
3. *Describing* What is the difference between New Guinea and Papua New Guinea?
4. *Determining Central Ideas* In what ways are low islands and high islands alike and different?
5. *Analyzing* What factors affect the climates of Oceania?
6. *Analyzing* Why would people living in rural Kiribati and the Solomon Islands use solar-powered lights in their homes?
7. *Informative/Explanatory Writing* In your own words, write a paragraph describing how an atoll forms.

networks

There's More Online!

- ☑ **CHART/GRAPH** The People of Oceania
- ☑ **IMAGE** Colonizing Oceania
- ☑ **VIDEO**

Reading HELPDESK

Academic Vocabulary
- distinct

Content Vocabulary
- wayfinding
- trust territory
- possession
- pidgin
- *fale*

TAKING NOTES: *Key Ideas and Details*

Summarize Using a chart like the one shown here, summarize important information from each section of the lesson.

History of Oceania	People of Oceania

Lesson 2
History and People of Oceania

ESSENTIAL QUESTION • *What makes a culture unique?*

IT MATTERS BECAUSE
Oceania has a unique culture but one that faces challenges.

History of Oceania

GUIDING QUESTION *How were the islands of Oceania populated?*

The first humans began settling in Oceania sometime around 1500 B.C. Historians believe the early settlers came from Southeast Asia, using available resources to build large sailing canoes. They filled their sturdy canoes with people, food, plants, and animals. Traveling from west to east and powered only by sails, the settlers crossed the waters of the Pacific Ocean.

The Polynesian Migrations

Their only way of navigating was to use the ancient practice of wayfinding. **Wayfinding** is a method of navigation that relies on careful observation of the natural world. For many thousands of years, humans have charted courses across the open ocean by watching the sun, the stars, and the movement of ocean currents and swells. Wayfinding was practiced long before the invention of navigation instruments, such as compasses and sextants. Even today, navigators practice wayfinding as a way to stay connected to Earth and to keep cultural traditions alive.

Over many centuries, people from areas such as the Philippines and Indonesia sailed from their homelands and settled the islands across Oceania. Many islands were uninhabited until settlers from other islands migrated farther into unexplored areas of Oceania.

The Coming of Europeans

European explorers began sailing through Oceania in the 1500s. They began colonizing the islands in the 1600s. Often, violent conflict broke out between colonists and native people. During the 1800s and 1900s, Christian missionaries came to thousands of islands in the region. Native people did not always welcome the missionaries. Many missionaries succeeded, however, in converting local populations to Christianity. Christian faiths are still widely practiced across Oceania.

Europeans had many reasons for wanting to colonize territories in Oceania. For practical reasons, the locations of many islands made them convenient stops for ships crossing the vast Pacific Ocean. Travelers wanted safe, reliable locations to restock their ships with food, drinking water, and other supplies. European powers were also interested in claiming resources.

Some Europeans mined gold and other precious metals from the islands. Some governments also saw the advantage of building military bases in Oceania. In fact, many islands in the region were occupied by Japanese, German, English, and American forces during World War II. Several battles were fought in the region. Unfortunately, some islands in the Pacific also became testing sites for nuclear weapons.

Contemporary Times

After World War II, many island colonies began to demand independence. Some independence movements involved conflict, but many islands were able to negotiate freedom with their former colonial powers. Some islands negotiated independence by "free association" with foreign powers.

A typical South Pacific canoe lies on a beach in Fiji. Early people, originally from the Asian mainland, sailed and settled in Oceania centuries before the arrival of Europeans in the 1500s.
▶ **CRITICAL THINKING**
Describing How were early people able to sail vast distances to settle Oceania?

The bicycles of worshippers are parked outside a picturesque Catholic church on the Fakarava atoll in the Tuamotu Islands. Built mostly of coral in 1874, the building is the oldest church in Polynesia. Christian faiths are widely practiced in Oceania today.

The island of Palau, for example, is an independent republic, but it has a voluntary free association with the United States. Palau has its own constitution and governs itself. Palau and the United States have an agreement that benefits both: Palau allows the United States to keep military facilities on one of its islands, and in return, the United States provides millions of dollars of aid money to Palau each year.

Other islands in Oceania have agreements of various kinds with foreign governments, including Australia, New Zealand, Great Britain, France, and the United States. The agreements generally involve use of land or other resources in exchange for military protection and economic aid. Relationships between island territories and foreign governments have different levels of political control and responsibility. A **trust territory** is one that has been placed under the governing authority of another country by the Trusteeship Council of the United Nations. The Marshall Islands were a trust territory until they gained independence in 1986. **Possession** is another name for a territory occupied or controlled by a foreign government and its people. French Polynesia can be classified as a possession because it is an overseas territory of France. Trust territories and possessions do not govern themselves but are run by foreign governments.

✓ READING PROGRESS CHECK

Describing Why did the first Europeans come to Oceania?

The People of Oceania

GUIDING QUESTION *What is life like in Oceania?*

Many people imagine the South Sea Islands as tropical paradises. There are many wonderful things about living in this beautiful region of the world. On the other hand, life in Oceania has many challenges.

The People of the Region

Oceania has one of the world's most diverse populations. So many different ethnic groups live on Oceania's thousands of islands that it is impossible to classify them all. Groups have their own **distinct** languages, cultures, and ways of life. Many islands are home to a wide range of ethnic groups. The most amazing diversity is found in Papua New Guinea.

Papua New Guinea has a total human population of more than 6 million. This is by far the largest population of any of Oceania's islands. Papua New Guinea's large population is condensed onto an island about the size of California.

The island country's population is made up of people from many different ethnic and native tribal groups. Natives of Papua New Guinea make up about 84 percent of the total population. The native population includes people from hundreds of tribal groups. The other 16 percent come from various backgrounds, including Polynesian, Chinese, and European.

The different groups have their own lifestyles, cultural traditions, beliefs, and languages. Geographers and language experts believe 860 different languages are spoken in Papua New Guinea. In other words, 10 percent of all languages known to exist are used in Papua New Guinea.

Academic Vocabulary

distinct separate; easily recognized as separate or different

The *USS Bonhomme Richard* pulls into Apra Harbor in Guam. The *Bonhomme Richard* is an amphibious assault ship. These ships are able to land and aid forces on shore during armed conflict.

▶ **CRITICAL THINKING**
Describing What are the terms of agreement between foreign governments and the islands of Oceania?

This village is located on Viti Levu, the largest and most populous of Fiji's more than 300 islands. About 75 percent of Fiji's 600,000 people live on Viti Levu. The island measures about 65 miles (106 km) from north to south and 90 miles (146 km) east to west.

In spite of speaking many different languages, the people of Papua New Guinea are still able to speak to one another. Many of the people speak their own language, as well as a pidgin language. A **pidgin** is a simplified language that is used for communication between people who speak different languages.

Such a wide diversity has both positive and negative effects. For instance, Papua New Guinea's many different ethnic groups create a rich and varied culture on the island. An endless variety of music, foods, clothing, and artwork can be enjoyed. At the same time, serious problems, such as crime, ethnic discrimination, and violent conflicts among tribal groups, are common.

Papua New Guinea's population is denser in some parts of the main island than in others. In general, highland areas have higher population densities than do the lowlands and coastal plains. Only about 12 percent of the population live in urban areas. Isolated towns are hidden in rugged mountain areas. Thousands of distinct tribal groups live in small villages in the islands' many remote locations. A traditional folk saying in Papua New Guinea is, "For every village, a different culture."

The Culture of the Region

Life in the villages of Oceania is based on tradition. Many traditions involve fishing, diving, and celebrating battle victories. Another tradition is an important event in the lives of young people called coming-of-age ceremonies. In Polynesian cultures, ceremonies and celebrations include feasts and dancing. Like their ancestors, Polynesian people practice artistic wood carving and use the carvings to decorate their homes. Many traditional Polynesian and Micronesian cultures practice tatooing. Micronesian people use storytelling to retell history and to keep track of family heritage. The Melanesian people were the only traditional culture of Oceania known to use bows and arrows in hunting.

In the island nation of Samoa, traditional homes called **fales** are common. *Fales* are open structures made of wood poles with thatched roofs. Local trees are used to make the poles, and palm leaves are used for roof thatch. The dwellings have no walls and are used mainly for shade and shelter from frequent rainfall.

Today, Oceania's many island cultures are mixtures of traditional and modern practices, beliefs, and lifestyles. Although they have adopted many Western attitudes, people see the value in continuing some of the traditional ways.

For example, Christianity is widely practiced in island communities, along with elements of traditional religions, such as songs, dances, and ceremonial costumes. Many island people wear Western-style clothing and hairstyles. Cell phones and laptop computers are common. Elements of local traditional cultures, including tribal tattoos, jewelry, and art forms such as wood carving, are common on many islands, as well.

Traditional celebrations are practiced throughout Oceania. Some traditional events, such as the Hawaiian luau, have become more modern in recent decades. Luaus were traditionally ritual ceremonies and feasts celebrating important events, such as victories in battle. Centuries ago, men and women ate in separate areas during luaus, and only chiefs ate certain foods. Some luaus were attended only by men. Today, luaus are banquets of traditional and modern foods eaten on a low table.

Maintaining elements of traditional cultures in their lives is important to the people of Oceania. Respecting and continuing certain traditions keeps cultures alive. Celebrating the traditional culture of their ancestors gives young people a sense of pride and identity. Making the past part of the present keeps people of all ages connected to their cultural heritage.

Include this lesson's information in your Foldable®.

✓ **READING PROGRESS CHECK**

Analyzing Why might using a pidgin language be useful to the population of Papua New Guinea?

LESSON 2 REVIEW

Reviewing Vocabulary
1. For what purpose did the early people of Oceania use *wayfinding*?

Answering the Guiding Questions
2. **Identifying** For approximately how long have humans been living on the islands of Oceania?
3. **Identifying** List three of the many groups of people who came to Oceania from other parts of the world.
4. **Determining Word Meanings** Would a *fale* be an appropriate home for the climate where you live? Why or why not?
5. **Distinguishing Fact From Opinion** Is the following statement about the culture of Oceania a fact or an opinion?

   With more than 860 different spoken languages, Papua New Guinea has one of the most culturally diverse populations in the world.
6. **Narrative Writing** Write a short story from the perspective of a young person living a traditional lifestyle in a small village on one of the islands of Oceania. Include details about your daily life. Describe your home, your family and friends, the foods you eat, the work you do, and what you do for fun.

networks

There's More Online!

- ☑ **IMAGES** Tropical Fish and Sharks in Tahiti
- ☑ **SLIDE SHOW** A Beach in Polynesia
- ☑ **VIDEO**

Reading HELPDESK CCSS

Academic Vocabulary

- collapse

Content Vocabulary

- cash crop
- resort
- remittance
- MIRAB economy

TAKING NOTES: *Key Ideas and Details*

Determine Cause and Effect As you read, use a graphic organizer like the one shown here to identify two important issues and describe the effects of the issue.

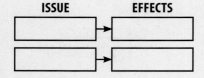

Lesson 3
Life in Oceania

ESSENTIAL QUESTION • *Why do people make economic choices?*

IT MATTERS BECAUSE
The people of Oceania face economic challenges that affect the United States and other countries.

The Economies of Oceania

GUIDING QUESTION *How do the people of Oceania earn their living?*

The islanders of Oceania face difficult economic challenges. With small land areas, few valuable resources, and vast distances between islands, earning a living in Oceania can be difficult. The people of Oceania, however, have found ways to support their families. They also take great care in using the natural resources available to them. Despite challenges, Oceania's island communities have the potential for a bright economic future.

Papua New Guinea's Economy

Papua New Guinea has the most valuable natural resources in Oceania, other than Australia and New Zealand. The challenge is to locate, harvest, and transport the resources through Papua New Guinea's wild and rough terrain.

Gold and copper are Papua New Guinea's most profitable resources. Sales of gold and copper account for about 60 percent of the country's total export income. Other major exports are silver, timber, and agricultural products. **Cash crops** are crops grown or gathered to sell for profit. Papua New Guinea's cash crops include coffee, cacao, coconuts, rubber, and tea.

Mining and farming provide jobs for many people in Papua New Guinea. Industries such as timber processing, palm oil refining, and petroleum refining employ many

462

others. The tourist industry also provides many jobs. The vast majority of Papua New Guinea's people, however, live by subsistence farming. Most families raise their own food crops, such as yams, taro, bananas, and sweet potatoes. Some raise pigs or chickens for meat and eggs. Although the unemployment rate in Papua New Guinea is low, most people earn low incomes. The government of Papua New Guinea plans to increase exports of minerals and petroleum to strengthen the country's economy.

Economies of Small, Independent Countries

Smaller independent island countries throughout Oceania face many obstacles to economic development. With limited land, poor soil quality, and large populations, many islands must import much of their food, fuel, finished goods, and raw materials. Most islands import far more than they export—many import five or six times more goods and materials than they export. On the tiny island country of Tuvalu, for example, import values exceed export values by 200 to 1. People on the smaller islands raise what food they can by subsistence farming.

Some islands in Oceania raise limited cash crops, such as fruits, vegetables, sugar, nuts, coffee, tea, cocoa, and palm and coconut oils. The farms and plantations that produce cash crops employ some island residents. This type of agricultural work can be physically exhausting and usually pays very little.

Fish and other seafood are available across Oceania. Fishing operations on the smaller islands are usually small; fishers catch only enough to feed their families or to sell to local markets. Most island countries do not have the equipment or processing facilities for operating large fishing industries. Some island countries earn revenue by selling fishing rights to other countries. Japan, Taiwan, South Korea, and the United States are some of the foreign lands that pay for access to Oceania's marine resources.

On islands that have minerals and other marketable resources, many people work for industries such as mining, fishing, clothing, and farming. Some people make a living by using local materials, such as shells, wood, and fibers, to create art and to craft tools and artifacts.

Farmers in Oceania often grow taro, a root vegetable, brought from Southeast Asia centuries ago. The leaves and the root of the taro plant are widely used in South Pacific cooking.

▶ **CRITICAL THINKING**
Describing Why is subsistence farming widely practiced in Oceania?

Tourists go snorkeling at a beach resort in French Polynesia. Tourism is now the major industry in Oceania, creating jobs and bringing in money for many people of the region. Most of Oceania's tourists come from Australia, New Zealand, Japan, and other countries of the Pacific area.

Academic Vocabulary

collapse a sudden failure, breakdown, or ruin

Tourism is important to the economies of many small, independent countries in Oceania. Tourists come from all over the world to enjoy the sunshine, warm ocean waters, and panoramic views. **Resorts** provide comfortable lodging, food, recreation, and entertainment. Most resorts are located in beautiful natural areas, such as tropical beaches, mountains, and forests. Tourist businesses employ thousands of people. Without the revenue from tourism, many small island countries would be at risk of economic **collapse**.

Many people on the islands that have less-developed economies depend on income earned by family members living overseas. Thousands of young people have left Oceania in search of jobs in other countries. Most have settled in Australia, New Zealand, and the United States. Young workers support their families in Oceania by sending them the money they earn. Foreign-earned wages, called **remittances**, are vital sources of income for families. Workers employed overseas also send remittances to pay for community projects, such as schools. Workers contribute to their home countries' tourist industries when they return for visits.

Many of the islands of Oceania with less-developed economies are known as **MIRAB economies**. MIRAB is an acronym for *migration, remittances, aid,* and *bureaucracy*. Island countries with MIRAB economies are not able to fully support the needs of their people through their own resources and labor. These countries depend on outside aid and remittances from workers who have migrated to overseas areas. One example of a MIRAB economy is the Polynesian island nation of Tonga. Tonga raises a few export crops, such as squash, vanilla beans, and root vegetables. The tourist industry also brings in some revenue. Still, the most important source of income for most Tongans is remittances. Seventy-five percent of all Tongan families receive money from family members who live and work elsewhere.

Tonga and many other small island countries also depend on aid from foreign governments. The governments of Australia, Great Britain, Japan, New Zealand, and the United States have created international trust funds for many islands in Oceania. Foreign governments also give money directly to various island countries. Governments of MIRAB economies use some of the foreign aid for food, schools, water and sanitation, and other basic needs. Foreign aid also funds economic development by expanding local industries and exports.

Economies of America's Pacific Islands

American Samoa is a territory of the United States. Nearly all of American Samoa's economic activity involves the United States. The island group's chief industry is tuna fishing and processing, which employs 80 percent of its people. American Samoa produces a few cash crops, such as bananas, coconuts, taro, papayas, breadfruit, and yams.

The unemployment rate is high at nearly 30 percent. One cause of unemployment in American Samoa is the lasting impact of a 2009 earthquake and tsunami. The disasters caused terrible damage to the islands and their transportation systems, electrical systems, and businesses. Some industries were completely ruined, causing loss of jobs. Other industries, such as tourism, are starting to regain strength and show promise for future development.

Other U.S. territories in Oceania include Guam and Wake Island. Both islands are home to U.S. military bases. Local people are employed in transportation, housing, maintenance, food service, and other industries that serve the needs of military personnel. Guam also has a well-developed tourist industry.

✓ READING PROGRESS CHECK

Describing In your own words, briefly explain MIRAB economies.

Issues Facing the Region

GUIDING QUESTION *What challenges do the people of Oceania face?*

Despite signs of progress, the future is uncertain for the islands of Oceania. Three major areas of concern in the region are migration, economics, and the environment.

Human Migration

The movement of people from the islands of Oceania to other parts of the world is a major issue. So many young people have left Oceania to seek employment overseas that the populations of some islands have changed dramatically. The island of Tonga is an extreme example: Population estimates show that 50 percent of all Tongan people live abroad in foreign countries.

The major population shift has had negative effects. A large percentage of the people remaining in Tonga are children or older people who are unable to work. Migration has left many young and elderly Tongans without caregivers. The income that migrating workers send home helps.

Most Tongan workers who live abroad maintain strong ties to their homeland. With half of all Tongans now living overseas, however, more and more Tongan children are born in other countries. These children will grow up as natives of their adopted lands, not as native Tongans. Within the next few years, the majority of Tongan people will be born and raised overseas. This situation can weaken cultural ties to the homeland, as more foreign-born Tongans adapt to the cultures where they live.

Economics and Society

Another major issue in Oceania is the need for economic development. Many islands have slow economic development because of lack of resources. New industries cannot be planned and built without money to invest in growth and development. The need is urgent on islands where large numbers of people are migrating from farms and rural villages to cities to try to find work. With fewer people living in rural areas, fewer people are growing their own food. There is a continuing need across Oceania to buy food, energy resources, and raw materials from other countries. Many countries of Oceania import more goods and services than they export. When a less-developed country imports more than it exports, the entire economy is affected. Countries like to export more than they import because this creates jobs and demand for their goods and services.

Trash covers an otherwise-attractive beach on the island nation of Kiribati. Oceania, like other parts of the world, faces serious problems with waste pollution caused by an increasing population, rapid economic development, and the concentration of people in urban centers.

Tensions and Conflict

Although most island nations in Oceania are free of conflict, some unrest occurs. Crime and human rights abuses are problems on some of the islands. In Fiji, tensions between native Fijians and immigrants from India led to conflicts. The Solomon Islands have also seen conflict on issues. Disagreement on the issue of land rights led to conflict between the native people of Guadalcanal and people from the neighboring island of Malaita. Peace was restored with help from the United Nations.

Environmental Issues

The islands of Oceania face serious environmental issues, including climate change, deforestation, pollution, natural disasters, and declining fish populations. Many scientists and island residents view climate change as one of the most urgent and serious issues in the region. With continued global warming, sea levels are already rising. If this continues, many of Oceania's low islands could be completely covered by water. Some of the lowest islands are already experiencing surface flooding from rising oceans and eroding beaches.

Natural disasters such as earthquakes, tsunamis, typhoons, and resulting floods continue to threaten the islands. These events have the potential to destroy homes and claim lives, and they can also damage farmland and natural habitats.

Survival of Marine Animals

Commercial fishing companies have harvested so many fish from some areas that almost no fish remain for local people to catch and eat. When huge numbers of fish are caught at one time, the remaining fish are unable to reproduce quickly enough to restore populations. In time, entire populations of fish will be gone. Ocean pollution also threatens the survival of fish and other marine animals.

Include this lesson's information in your Foldable®.

☑ **READING PROGRESS CHECK**

Citing Text Evidence Why is climate change a serious concern for the islands of Oceania?

LESSON 3 REVIEW

Reviewing Vocabulary
1. What is the major difference between *cash crops* and *subsistence crops*?

Answering the Guiding Questions
2. *Identifying* List three ways the people of Oceania earn a living.
3. *Describing* How does aid from foreign countries benefit islands with MIRAB economies?
4. *Citing Text Evidence* The birthrate in Tonga has increased, but the island's population has decreased. Explain.
5. *Citing Text Evidence* What do you think is the most serious issue or challenge in Oceania today? Use information from the lesson in your explanation.
6. *Argument Writing* Using information from the lesson, write an argument either in favor of or against remittances. Explain why you believe remittances benefit or harm the cultures and economies of Oceania.

Chapter 15 ACTIVITIES CCSS

Directions: Write your answers on a separate piece of paper.

1 Use your **FOLDABLES** to explore the Essential Question.
INFORMATIVE/EXPLANATORY WRITING It is generally believed that inhabitants of Southeast Asia began to populate the islands of Oceania more than 2,500 years ago. Describe their method of navigation known as *wayfinding*.

2 **21st Century Skills**
DETERMINING CENTRAL IDEAS Work in a group to learn more about one of the island countries of Oceania. Put together a travel brochure to appeal to people who are interested in traveling there. Annotate your brochure with information from the text and from online sources.

3 **Thinking Like a Geographer**
ANALYZING How did the South Pacific's physical geography contribute to Oceania's cultural diversity?

4 **GEOGRAPHY ACTIVITY**

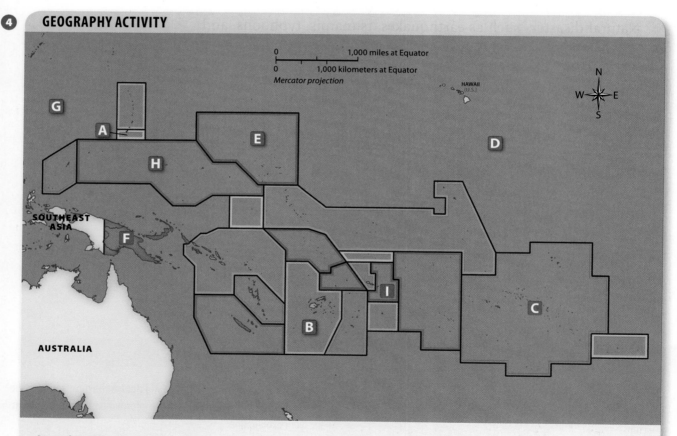

Locating Places
Match the letters on the map with the numbered places listed below.

1. Guam
2. French Polynesia
3. Papua New Guinea
4. Federated States of Micronesia
5. Philippine Sea
6. Marshall Islands
7. Fiji Islands
8. North Pacific Ocean
9. American Samoa

468 *Chapter 15*

Chapter 15 Assessment

REVIEW THE GUIDING QUESTIONS

Directions: Choose the best answer for each question.

1. What is the smallest inhabited island in Oceania?
    A. Micronesia
    B. Nauru
    C. Guam
    D. Papua New Guinea

2. New Guinea is a continental island, which means that
    F. like Australia, it is also a continent.
    G. it is part of an archipelago.
    H. it was at one time connected to a continent.
    I. its people are highly educated.

3. From what region did the original settlers of the islands of Oceania come?
    A. New Zealand
    B. East Africa
    C. Southeast Asia
    D. the Bering Strait

4. Apart from Australia and New Zealand, which island of Oceania has the largest population?
    F. American Samoa
    G. Solomon Islands
    H. Papua New Guinea
    I. Tahiti

5. Why, despite abundant fish and seafood, do the islands of Oceania not export more of these resources?
    A. Oceania does not have enough workers.
    B. The nations do not have enough equipment or processing facilities.
    C. Other countries steal the fish from Oceania's waters.
    D. Mercury poisoning is a major threat.

6. Guam and Wake Island are territories of the United States that are maintained to support
    F. movie filming.
    G. forestry.
    H. military bases.
    I. the MIRAB program.

Chapter 15 Assessment (continued)

DBQ ANALYZING DOCUMENTS

7 DETERMINING CENTRAL IDEAS Read the following passage about Tonga's economy:

> "The remittances of cash and goods from migrants who live and work overseas . . . [keep] the Tongan economy afloat. . . . Remittances . . . accounted in [2002] for about 50 percent of [gross domestic product]. . . . Although individuals and families are the main benefactors, . . . overseas Tongans [also] regularly send back money to their villages and local institutions, . . . funding practical community projects."
>
> —from Cathy A. Small and David L. Dixon, "Tonga: Migration and the Homeland"

What best explains why Tongans leave the island to find work?
- A. Tongans traditionally have loved to travel.
- B. Tonga's climate makes farming difficult.
- C. Tonga's economy does not offer enough jobs.
- D. Wages in Tonga are higher than elsewhere.

8 IDENTIFYING What is an example of a "practical community project"?
- F. equipment for a village health clinic
- G. better housing for a migrant's family
- H. the education of a cousin
- I. wedding presents to a sister

SHORT RESPONSE

> "The isolation created by the mountainous [landscape] is so great that some groups, until recently, were unaware of the existence of neighboring groups only a few kilometers away. The diversity, reflected in a folk saying, 'For each village, a different culture,' is perhaps best shown in the local languages. . . . Over 850 of these languages have been identified; of these, only 350–450 are related."
>
> —from "Papua New Guinea," State Department Background Notes

9 DETERMINING CENTRAL IDEAS Explain the meaning of the folk saying quoted in the passage.

10 ANALYZING What other kinds of environments have a similar effect on groups of people?

EXTENDED RESPONSE

11 INFORMATIVE/EXPLANATORY WRITING The MIRAB system (migration, remittances, aid, bureaucracy) has helped the economies of some of the islands of Oceania. What are the disadvantages of the MIRAB system? In an essay, identify some of its flaws, and predict their future impact on the region.

Need Extra Help?

If You've Missed Question	1	2	3	4	5	6	7	8	9	10	11
Review Lesson	1	1	2	2	3	3	3	3	2	2	3

ANTARCTICA

CHAPTER 16

networks
There's More Online about Antarctica.

ESSENTIAL QUESTIONS • How does physical geography influence the way people live? • How do people adapt to their environment?

Lesson 1
The Physical Geography of Antarctica

Lesson 2
Life in Antarctica

South Korean research team endures icy conditions to gather biological material from sea on southeastern tip of Antarctica.

The Story Matters...

For centuries, explorers from many countries set out on dangerous expeditions, competing to be the first to reach the frozen continent of Antarctica. Antarctica, which means "opposite to the Arctic," is shared today by many nations whose scientists are researching its vital role in maintaining the world's climate balance.

FOLDABLES
Study Organizer

Go to the Foldables® library in the back of your book to make a Foldable® that will help you take notes while reading this chapter.

Chapter 16
ANTARCTICA CCSS

The continent of Antarctica lies at the southern extreme of Earth, with its land lying beneath a massive ice cap. It is larger in size than either Europe or Australia.

Step Into the Place

MAP FOCUS Use the map to answer the following questions.

1. **THE GEOGRAPHER'S WORLD** Which country's claim in Antarctica is larger: the Australian, the Argentine, or the French?

2. **THE GEOGRAPHER'S WORLD** What sea is located between longitudes 100°W and 120°W?

3. **PLACES AND REGIONS** Which country's claim is adjacent to the Ross Sea?

4. **CRITICAL THINKING** *Analyzing* Think about the location of Antarctica in relation to other parts of the world. How might its location affect the development of natural resources?

ICY SUMMIT A climber approaches the summit of Mount Vaughn in the Queen Maud Mountain range. The range runs from Antarctica along the ocean floor and continues into South America as the Andes Mountains.

ANTARCTIC ISLAND Kayakers paddle near the rugged cliffs of Petermann Island off the west side of the Antarctic Peninsula. The island is home to large numbers of penguin colonies.

Step Into the Time

TIME LINE Based on events on the time line, write a paragraph explaining the international importance of Antarctica.

1760s Several nations hunt the seals of the Antarctic seas

1700 — 1800

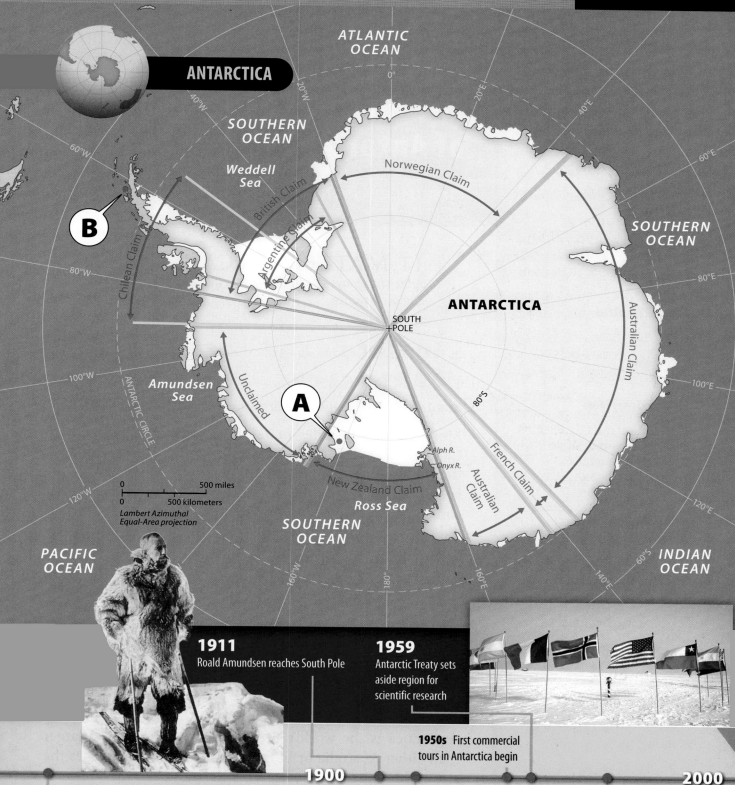

networks

There's More Online!

- **IMAGES** Antarctica
- **ANIMATION** How Icebergs Form
- ☑ **VIDEO**

Reading HELPDESK

Academic Vocabulary
- visible

Content Vocabulary
- ice sheet
- ice shelf
- calving
- iceberg
- katabatic wind
- lichen
- krill
- plankton

TAKING NOTES: Key Ideas and Details

Summarize Using a chart like the one shown here, summarize the information presented in each section of the lesson.

Landforms and Water

Climate and Resources

Lesson 1
The Physical Geography of Antarctica

ESSENTIAL QUESTION • How does physical geography influence the way people live?

IT MATTERS BECAUSE
Antarctica plays an important role in maintaining world climate balance.

Landforms and Waters of Antarctica

GUIDING QUESTION *Is Antarctica the last unknown land region on Earth?*

More than 2,500 miles (4,023 km) south of the Tropic of Capricorn, far from the warm, sunny islands of Oceania, is the coldest, cruelest, darkest land on Earth. Nicknamed "the last continent" and "the bottom of the world," this frozen continent is Antarctica. Throughout most of history, humans did not know that this bitterly cold, windy, barren land even existed. Scholars and scientists had theories and ideas that a large "southern land" might be at the bottom of Earth. There was no proof that Antarctica was real, however, until explorers first sighted its icy shores in 1820. Exploration was, and still is, extremely difficult in this harsh land. It was not until 20 years after the first sighting that geographers confirmed that Antarctica is, in fact, a continent, not a mass of floating ice as some early explorers believed.

Size of Antarctica

Antarctica is the world's fifth-largest continent. It is larger than either Australia or Europe, and it is about 1.5 times the size of the United States. Antarctica is located in the Southern Hemisphere and is surrounded on all sides by the Southern Ocean. Together, Antarctica and the Southern Ocean cover

474

about 7 percent of Earth's surface. However, very few people reside, even for a short time, within these millions of square miles. The South Pole is located near the center of the continent.

Rock, Ice, and Water

Antarctica is called the highest continent because it has the highest overall elevation of any continent. Antarctica's surface is an average of 7,000 feet (2,134 m) above sea level. This figure measures the height of the **visible** surface of the continent, which is mostly solidly packed ice. While some of Antarctica's mountains rise above the ice, most of the land that makes up the foundation of the continent is much lower than the dense ice layer that covers it. In fact, the weight of so much ice has forced some of Antarctica's surface land far below sea level. Entire mountains are completely buried under layers of ice.

It might be difficult to believe, but the ice sheet covering most of Antarctica is 2 miles (3.2 km) thick in some places. An **ice sheet** is a thick layer of ice and compressed snow that forms a solid crust over an area of land. This vast ice sheet covers about 98 percent of Antarctica's surface. Scientists estimate that two-thirds of all freshwater on Earth is frozen in the Antarctic ice sheet. But it is important to remember that land lies beneath this ice sheet. By using high-resolution satellites and other specialized equipment, geographers have learned that Antarctica is one large landmass that also has an archipelago of rocky islands.

Many extremes of high and low elevation can be found across this frozen land. The highest point on the continent is Vinson Massif, which measures an incredible 16,066 feet (4,897 m) high. The Transantarctic Mountains stretch for 2,200 miles (3,541 km) across Antarctica. This enormous mountain range divides the continent into two regions: East Antarctica and West Antarctica.

Academic Vocabulary

visible able to be seen

Tourists bathe in a thermal pool, which is actually an inactive volcanic crater on Deception Island, Antarctica. Volcanic activity creates the thermal pools by heating underground water that comes to the surface.

The lowest point on the continent is deep within the Bentley Subglacial Trench. This low area is also the lowest place on the surface of Earth that is not under seawater.

Ice not only covers the land in ice sheets but also extends out into the ocean in what are called ice shelves. An **ice shelf** is a thick slab of ice that is attached to a coastline but floats on the ocean. It is a seaward extension of the ice sheet that forms a thick, hard plain of ice. During the dark winter months, temperatures fall to unbelievable lows in Antarctica. The air and water get so cold that the seawater next to the ice shelves freezes. This increases the size of the ice shelf. So much coastal water freezes into the ice shelves that Antarctica appears to double in size every winter. As the weather warms during the spring and summer months, parts of the ice shelves farthest away from the land break up into huge, floating islands of ice.

The process of ice breaking free from an ice shelf or glacier is called **calving**. The large chunk of ice that breaks off is called an **iceberg**. Icebergs are made of freshwater, not salt water. These massive bodies of ice slowly drift through frigid ocean waters with only their top portions showing. As much as 90 percent of an iceberg remains underwater. Because most of their mass is hidden below the water's surface, icebergs appear much smaller than they really are. This has led to accidents when ships have come close to what appeared to be small chunks of ice floating on the surface but then crashed into the icebergs' enormous, unseen bulk under the water. Icebergs are classified as chunks of ice larger than 16 feet (5 m) across. While many icebergs are the size of houses and mountains, some are the size of small islands. Satellite images show a huge crack in an Antarctic glacier that could produce an iceberg of 350 square miles (906 sq. km), the size of New York City.

In the middle of summer, a special-purpose ship called an icebreaker moves through an ice pack in the Ross Sea along the coast of Antarctica. The Ross Sea region has permanent year-round icy conditions, and a powerful icebreaker is the best vessel suited for its exploration.

▶ **CRITICAL THINKING**
Describing How do icebergs form?

Researchers battle powerful winds as they carry out their research in the interior of Antarctica.

Antarctica is surrounded by stormy seas. These include the Weddell Sea, the Scotia Sea, the Amundsen Sea, the Ross Sea, and the Davis Sea. These icy waters were named by and for explorers and scientists who have mapped and studied the region. In recent years, hundreds of new species of marine animals have been discovered in the mysterious depths of these cold seas.

☑ **READING PROGRESS CHECK**

Determining Central Ideas Does the continent of Antarctica increase in size during the winter?

Climate and Resources

GUIDING QUESTION *Does any life exist on such a forbidding continent?*

How would you like to live in a place where summer brings a high temperature of only a few degrees above freezing?

A Dry, Frigid Climate

Because of its high latitude, Antarctica never receives direct rays from the sun, even when the Southern Hemisphere is tilted toward the sun. Thus, the sun's energy is spread out over a wider surface area than at lower-latitude places such as the Equator. The result is that very little heat energy reaches the surface of the land. In addition, being south of the Antarctic Circle, most of Antarctica receives no sunlight at all and is completely dark for at least three months of the year.

The high elevation across Antarctica affects its climate, as well. Strong, fast winds blow colder air down from high interior lands toward the coasts. These powerful gusts, called **katabatic winds**, are driven by the force of Earth's gravity.

A gentoo penguin leaps onto an iceberg to join its companion near Antarctica's Gerlache Strait. The gentoo is one of six penguin species that populate the Antarctic Peninsula and the many islands around the frozen continent. An orange beak, white head marking, and longer tail distinguish the gentoo from the other penguin species.

All of these factors combined produce an intensely cold climate. Temperatures are coldest in the interior highlands and warmest near the coasts.

Antarctica also has an extremely dry climate. The continent receives only 2 inches to 4 inches (5 cm to 10 cm) of precipitation each year. The extreme cold makes the climate even drier by limiting the amount of moisture the air can hold. This effect combined with lack of precipitation makes Antarctica the world's driest continent.

Antarctica plays a vital role in maintaining the global climate balance. The millions of square miles of ice covering Antarctica act like a giant reflector. The white, smooth surface of the ice shelf reflects sunlight. An estimated 80 percent of the solar radiation that reaches Antarctica is reflected back into the upper atmosphere. So, this continent-sized reflector reduces the amount of solar energy that is absorbed by Earth. On such a large scale, this effect lowers the overall temperature of the entire planet.

Plants and Wildlife

Antarctica's harsh climate limits the types of plants and animals that can live on its surface. Only 1 percent of Antarctica's land area is suitable for plant life. Plants found in Antarctica include mosses, algae, and **lichens**, which are organisms that usually grow on solid, rocky surfaces and are made up of algae and fungi. Only two species of flowering plants grow in the region: Antarctic hair grass and Antarctic pearlwort. Most plants are found on the Antarctic Peninsula and its surrounding islands. These are the warmest, wettest areas in the region. However, a few plants have been discovered in intensely cold inland areas. Scientists have found hardy species of mosses, lichens, and algae growing in the tiny cracks and pores of rocks in Victoria Land.

Most of Antarctica's land animals are tiny insects and spiderlike mites. Most other land animals are only part-time visitors to the region. Weddell seals are water animals, but they only come to Antarctica's shores during summer months to raise their pups. Many species of seabirds, such as albatross, cormorants, and gulls, visit Antarctica in the summer to breed. Only the emperor penguin stays in the region year-round. These giant penguins are adapted to extremely cold temperatures and cannot survive outside of Antarctica's frozen climate.

Although the land of Antarctica is not home to many living things, the Antarctic seas are filled with life. Many kinds of seals, dolphins, fish, and other marine animals live in waters surrounding Antarctica. Whales come to the Southern Ocean to feed during the summer. Sea mammals and seabirds eat **krill**, tiny crustaceans similar to shrimp. Krill thrive in the waters of the Southern Ocean, feeding on even smaller life-forms called plankton.

Plankton are tiny organisms floating near the water's surface. Some types of plankton are single-celled bacteria; others are plants such as algae. Plankton, krill, fish, sea mammals, seabirds, and land animals are part of Antarctica's ecosystem. These creatures are essential parts of the food chain that supports life on the continent.

Resources of Antarctica

Several factors make it difficult or impossible to explore Antarctica for natural resources. The harsh climate creates dangerous conditions for human workers. The land that could hold mineral resources is buried by the thick ice sheet. Even reaching the interior of the continent with heavy mining equipment could prove impossible. As a result, mineral and energy resources that might exist have not been exploited. Some species of fish are plentiful in the Southern Ocean, and commercial fishing operations harvest fish from Antarctica's waters.

Include this lesson's information in your Foldable®.

✓ **READING PROGRESS CHECK**

Describing Where in Antarctica would you find the most living things? Explain why this is so.

LESSON 1 REVIEW

Reviewing Vocabulary
1. How are *krill* important to the survival of Antarctica's seabirds and sea mammals?

Answering the Guiding Questions
2. **Analyzing** Why do you think Antarctica was nicknamed "The Last Continent"?
3. **Determining Central Ideas** How are ice sheets and ice shelves alike and different?
4. **Identifying** What types of life-forms are found on the Antarctic Peninsula?
5. **Describing** Explain why Antarctica's mineral and energy resources are not commercially mined or harvested.
6. **Informative/Explanatory Writing** In your own words, explain why Antarctica's climate is so cold and dry. Use facts and details from the lesson in your writing.

networks

There's More Online!

- ☑ **IMAGES** Earth's Ozone Layers
- ☑ **MAP** Antarctica Explored
- ☑ **SLIDE SHOW** Life in Antarctica
- ☑ **VIDEO**

Reading HELPDESK

Academic Vocabulary
- acknowledge

Content Vocabulary
- treaty
- research station
- ozone
- remote sensing

TAKING NOTES: *Key Ideas and Details*

Determining Central Ideas Using a chart like this one, take notes on the central idea from each section of the lesson.

Section	Central Idea
Sharing the Land	
The Scientific Continent	

Lesson 2
Life in Antarctica

ESSENTIAL QUESTION • *How do people adapt to their environment?*

IT MATTERS BECAUSE
Antarctica is a region shared by many nations. It is important to scientific research.

Sharing the Land

GUIDING QUESTION *How do people share common resources and protect unique lands?*

In 1911, two explorers, Roald Amundsen and Robert Scott, were in a race to be the first to reach the South Pole. Amundsen was a Norwegian explorer who had experience dealing with cold and snow. Scott was an English explorer who had commanded one previous expedition to Antarctica but had little experience traveling in the cold. Both men were driven to compete by a combination of curiosity and national pride. The two explorers and their teams made incredible journeys across the brutal land, losing men, ponies, and sled dogs to cold and starvation along the way. In the end, it was Amundsen's team that reached the South Pole first. Scott and his team arrived 34 days later.

The remarkable feats of Amundsen and Scott gave the world an insider's view of Antarctica for the first time. Their experiences and discoveries helped the scientists and explorers that came after them. Memorials to both men have been built in Antarctica.

The Antarctic Treaty

Antarctica is the only continent with no native population. Therefore, when the first humans visited the continent, there were no previous claims to the land. Several countries made territorial claims to sections of Antarctica's land, but these

(l to r) ©Deborah Zabarenko/Reuters/Corbis; ©YONHAP/epa/Corbis; Hugh Rose/Danita Delimont/Alamy

claims were set aside by the Antarctic Treaty. A **treaty** is an agreement or a contract between two or more nations, governments, or other political groups. Signed by 12 countries in 1959, the Antarctic Treaty states that the continent of Antarctica should be used only for scientific research and other peaceful purposes. In later years, 33 additional countries signed the Antarctic Treaty. The treaty also sets rules for the management of the land and its resources. The Antarctic Treaty is an effort to protect Antarctica's environment and to prevent weapons testing and other military actions from being carried out on the continent. Other agreements such as the Antarctic Conservation Act relate to protecting the region's wildlife.

Regulating Relations

Nations including Argentina, Australia, Great Britain, New Zealand, and Norway still claim large sections of the continent as territories, but not all governments **acknowledge** these claims. Because Antarctica is not considered a nation or a country and it has no government, issues relating to the continent's use and protection are decided by the Antarctic Treaty System. This is a group of representatives from 48 countries that have interests in Antarctica. Important decisions affecting Antarctica's environment, wildlife, and resident scientists are made by annual meetings of the Antarctic Treaty System.

✓ **READING PROGRESS CHECK**

Citing Text Evidence What is the purpose of the Antarctic Treaty?

Academic Vocabulary

acknowledge to recognize the rights, status, or authority of a person, a thing, or an event

The Amundsen-Scott station is a U.S. scientific research center at the South Pole. The original station was built in 1956 for the International Geophysical Year—a special period of international scientific investigations in Antarctica during 1957 and 1958. The rebuilt station today houses as many as 200 people in summer and about 50 in winter.

Chapter 16 **481**

The Scientific Continent

GUIDING QUESTION *How and why do scientists study Antarctica?*

Because of its harsh climate, Antarctica is the only continent that has no permanent human settlement. Its largely unspoiled environment has made it a favorable place for scientific research.

Living in Antarctica

No one lives in Antarctica full time, but many people stay in this harsh world for months or even years at a time. During the summer, when the sun shines brightly and temperatures are cold but not deadly, about 4,400 people reside in Antarctica. When the temperatures fall and the constant darkness of winter sets in, this temporary population shrinks to about 1,100 people. Most of these people are scientists and support teams living in shelters on land, but some stay aboard ships near the coasts. People from all over the world come to the frozen continent to study climate change, oceanography, geology, and countless other sciences involving Antarctica's landforms, weather, plants, animals, and the ancient rocks that form the continental shelf below the land.

Why do most research scientists stay in Antarctica only during the summer season? Weather conditions during summer months are still cold and windy, but they are not nearly as severe as those in winter. Summer temperatures usually remain below freezing, but the sun shines 24 hours a day. In winter, powerful winds tear across the land from the mountains, creating blizzard-like conditions. Temperatures are so low that without protective clothing, humans can freeze to death in a matter of minutes. The ice shelves that form along the coasts prevent ships from coming anywhere near the land. For these reasons, summer is the best time for scientists to work in Antarctica. Only a small number of brave, patient researchers continue their work through the long winters, staying sheltered indoors as much as possible.

Governments with scientific interests in Antarctica have built research stations in different locations across the continent. A **research station** is a base where scientists live and work. These stations usually consist of a few

A scientist from South Korea collects meteorites on an ice field in the Ross Sea region of Antarctica. Meteorites are rocks from space that have fallen on earth. They provide clues about the history of our solar system. Antarctica is the best place in the world to collect meteorites. There, meteorites fall into ice rather than hard rock and so are damaged less. Also, Antarctica's icy environment makes them more visible and easier to collect.

▶ **CRITICAL THINKING**
Analyzing Why do scientists come to Antarctica despite its climatic conditions?

DIAGRAM SKILLS

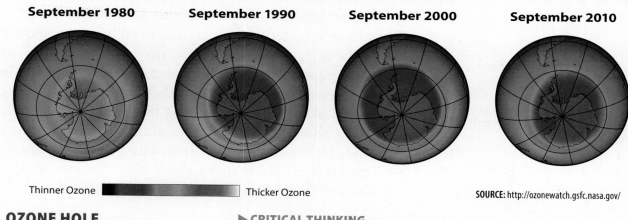

OZONE HOLE
Images show changes in the size of the ozone hole over Antarctica during a 30-year period.

▶ **CRITICAL THINKING**
Analyzing How has the size of the ozone hole changed over time? What do these changes reveal?

SOURCE: http://ozonewatch.gsfc.nasa.gov/

buildings set close together, each built with special insulation against the brutally cold and windy weather. Currently, Antarctica has about 50 research stations. Each station has living quarters for the scientists and other staff, such as medical doctors, cooks, and mechanics. Stations also have laboratories, other types of research facilities, and large storage areas filled with fuel, food, and equipment. Research stations need to have enough food stored away to feed teams of scientists for a year or more. Having extra food is important because violent storms can happen at any time, making it impossible for planes or ships to bring in fresh food and other supplies.

The Ozone Layer

One important area of study focuses on the ozone layer that is high in Earth's atmosphere. **Ozone** is a gas in the atmosphere that absorbs harmful ultraviolet radiation from the sun. Through decades of research and study, scientists discovered that the layer of ozone in our planet's atmosphere had thinned and decreased. This means that less ozone exists to protect Earth and life on Earth from damaging solar radiation.

The two locations on Earth where levels of ozone are the lowest are over the Arctic Circle and over Antarctica and the Southern Ocean. Loss of ozone affects all life on the planet. Scientists in Antarctica and other parts of the world are working to find a way to protect and preserve the ozone layer.

Studying Earth From Above

Have you ever wondered how scientists are able to figure out what lies below the surface of Earth? For example, how do scientists know what landforms lie under the thick Antarctic ice sheet? Scientists are able to study what they cannot see, such as layers of rock under the ice, by using an amazing kind of technology called remote sensing.

Chapter 16

Remote sensing involves using scientific instruments placed onboard weather balloons, airplanes, and satellites to study the planet. Remote-sensing technology is used to collect data, allowing scientists to take images and measurements of objects and places that would be impossible to reach in person. Some remote-sensing equipment emits electromagnetic waves that create detailed images of landforms and bodies of water buried under miles of ice. This technology has allowed scientists to study Antarctica's rock foundation and the floor of the Southern Ocean. Remote sensing has helped geographers learn more about how Antarctica's land, water, and atmosphere interact and affect the global climate system.

Research Yields Clues

The research work done in Antarctica has resulted in many amazing discoveries. Scientists have learned about our planet's birth and evolution over time, and they have found clues about the origins of the universe. By comparing rocks and fossils found in Antarctica to those found on other continents, scientists have determined that the Antarctic continent was once part of a huge landmass called Gondwana. Antarctica, Asia, Africa, and other continents broke away from this supercontinent millions of years ago and drifted across Earth's surface as a result of plate tectonics.

Climate Change

Another major area of research in Antarctica is climate change. Through satellite imagery and local measurements, geographers have learned that the Antarctic ice sheet is shrinking. Many people are concerned that this shrinkage might be caused by global warming. This concern is twofold: First, scientists fear that the impact of climate change will permanently damage Antarctica's environment and ecosystems. Second, Antarctica is seen as a "global barometer," or an indicator of what is happening to the climate of the entire planet. Signs such as melting ice and shrinking icebergs tell geographers and scientists that global temperatures are rising quickly. As massive chunks of ice calve, or break away, from

Tourists enjoy close-up views of a female humpback whale and her calf. Humpback whales receive their name because they raise and bend their backs to begin diving. They gather in groups along the Antarctic coast, where they feed and breed. Because humpback whales are slow swimmers, tourist boats can easily approach them.

▶ **CRITICAL THINKING**
Describing How is climate change affecting the life of Antarctica's sea animals?

Antarctica's glaciers, more icebergs clutter the surrounding waters. This makes traveling by ship in these areas more difficult and dangerous. Even more serious is the possibility that as these icebergs melt, global sea levels will rise. Rising sea levels would make survival impossible in many parts of the world.

Climate change is also affecting the food supplies of Antarctic land and sea animals. Scientists believe that global warming and changing water currents are affecting the amount of plankton and krill in the region. Recent studies have found low populations of these tiny water organisms in the waters around Antarctica. Without enough plankton and krill to eat, fish will die off or leave the area. This leaves seals and penguins without enough fish to eat. This disruption in the local food chain could have terrible consequences for Antarctica's animal life.

Taking a Tour

The bitterly cold, icy shores of Antarctica might not seem like a great place to spend a vacation. Still, about 6,000 tourists visit Antarctica each year. Most visitors are interested in seeing the incredible natural beauty of the area, taking photographs of wildlife, and visiting the research stations. Some are excited by the idea of visiting a harsh, dangerous place that few humans will ever see firsthand. Most tourists choose Antarctica because they want an adventure more than a vacation.

Thanks to the work of the scientists who have studied Antarctica, humans have come a long way in understanding this mysterious continent. However, there is still much to learn. Geographers and other scientists living and working in Antarctica are dedicated to their research. With each passing year, they discover more clues about our planet's past. They also use what they learn to make predictions about our planet's future. Antarctica's secrets are slowly being uncovered, but much remains unknown about this frozen land at the bottom of the world.

Thinking Like a Geographer

Lake Vostok
After spending years drilling through more than two miles (3 km) of solid ice in Antarctica, Russian scientists reached the surface of a gigantic freshwater lake in February 2012. Named Lake Vostok, it is the largest of the hundreds of the continent's subglacial lakes—roughly the size of Lake Ontario in North America. Scientists are taking samples of the water hoping to find living organisms that could provide clues about the unusual environment.

FOLDABLES Study Organizer

Include this lesson's information in your Foldable®.

✓ **READING PROGRESS CHECK**

Determining Central Ideas Why do scientists think Antarctica was part of a larger landmass that included continents such as Africa and Asia?

LESSON 2 REVIEW (CCSS)

Reviewing Vocabulary
1. Why is *ozone* high in the atmosphere important?

Answering the Guiding Questions
2. *Describing* Describe the 1911 race to the South Pole in your own words.

3. *Analyzing* What are some of the advantages of using remote sensing to study Earth?

4. *Identifying* What are the possible causes and effects of Antarctica's shrinking ice sheet?

5. *Narrative Writing* Imagine that you are a young scientist who has just arrived in Antarctica for the summer. Write a fictional journal entry describing your research and daily life at a research station. Whenever possible, use facts, details, and vocabulary terms from the lesson in your narrative.

What Do You Think?

Is Global Warming a Result of Human Activity?

Scientists agree that Earth's climate is changing. They do not agree on what is causing the change. Is it just another natural warming cycle like so many cycles that have occurred in the past? Scientists who support this position cite thousands of years' worth of natural climatic change as evidence. Or is climate change anthropogenic—caused by human activity? Scientists who support this position cite the warming effect of rapidly increasing amounts of greenhouse gases in the atmosphere. Greenhouse gases occur naturally, but they also result from the burning of fossil fuels. Which side's evidence is more convincing?

No!
PRIMARY SOURCE

" The Intergovernmental Panel on Climate Change (IPCC), an agency of the United Nations, claims the warming that has occurred since the mid-twentieth century "is *very likely* due to the observed increase in anthropogenic greenhouse gas concentrations." Many climate scientists disagree with the IPCC on this key issue.

Scientists who study the issue say it is impossible to tell if the recent small warming trend is natural, a continuation of the planet's recovery from the more recent "Little Ice Age," or unnatural, the result of human greenhouse gas emissions. Thousands of peer-reviewed articles point to natural sources of climate variability that could explain some or even all of the warming in the second half of the twentieth century. S. Fred Singer and Dennis Avery documented natural climate cycles of approximately 1,500 years going back hundreds of thousands of years. "

—Joseph Bast and James M. Taylor, "Global Warming: Not a Crisis," The Heartland Institute

A scientist weighs an ice core sample on a glacier in Antarctica. Ice cores provide detailed information about changes in Earth's climate over many centuries.

Afternoon traffic flows along the streets of Beijing, China, on a smog-filled day. Despite efforts to improve air quality, the Chinese capital remains one of the world's most polluted cities.

Yes!

PRIMARY SOURCE

"It is very unlikely that the 20th-century warming can be explained by natural causes.... Palaeoclimatic reconstructions show that the second half of the 20th century was likely the warmest 50-year period in the Northern Hemisphere in the last 1300 years. This rapid warming is consistent with the scientific understanding of how the climate should respond to a rapid increase in greenhouse gases like that which has occurred over the past century, and the warming is inconsistent with the scientific understanding of how the climate should respond to natural external factors such as variability in solar output and volcanic activity. Climate models provide a suitable tool to study the various influences on the Earth's climate. When the effects of increasing levels of greenhouse gases are included in the models, as well as natural external factors, the models produce good simulations of the warming that has occurred over the past century. The models fail to reproduce the observed warming when run using only natural factors."

—Contribution of Working Group I: The Physical Science Basis to the Fourth Assessment Report of the Intergovernmental Panel on Climate Change, 2007

What Do You Think? DBQ

1. **Analyzing** What specific evidence is offered to support the position that warming is caused by human activity?

2. **Identifying Point of View** What specific evidence is offered to support the position that warming results from natural causes?

Critical Thinking

3. **Analyzing** What effect could belief in one viewpoint or the other have on people or governments?

Chapter 16 ACTIVITIES

Directions: Write your answers on a separate piece of paper.

❶ Exploring the Essential Question
INFORMATIVE/EXPLANATORY WRITING In two or more paragraphs, explain why Antarctica can be described as a polar desert.

❷ 21st Century Skills
PEER REVIEW Work with a partner. Each partner will write a paragraph to answer the question: What do you think it's like to live in a cold climate like Antarctica? Check your partner's paper for meaning, grammar, and complete sentences. Discuss the review of your paragraph with your partner. Revise as needed.

❸ Thinking Like a Geographer
LISTING Create a Venn diagram like the one shown here. Label one side West Antarctica and the other side East Antarctica. List some physical geography features that can be found in each. In the center of the diagram, list features found in both.

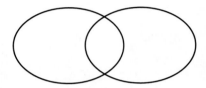

❹ GEOGRAPHY ACTIVITY

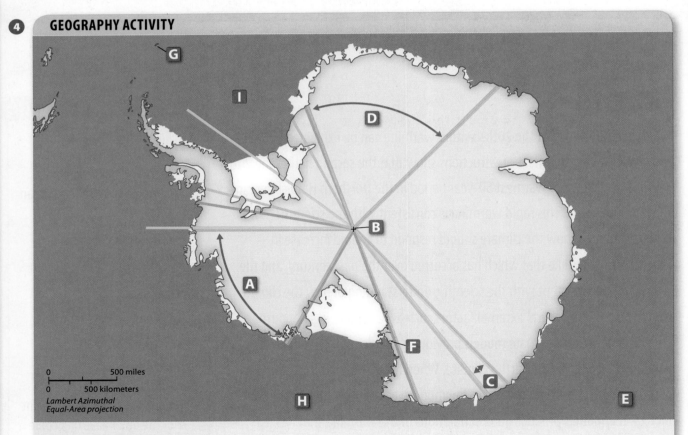

Locating Places
Match the letters on the map with the numbered places or phrases listed below.

1. French Claim
2. Norwegian Claim
3. Southern Ocean
4. Unclaimed
5. Weddell Sea
6. Indian Ocean
7. South Orkney Islands
8. South Pole
9. Onyx River

488 *Chapter 16*

Chapter 16 Assessment

REVIEW THE GUIDING QUESTIONS
Directions: Choose the best answer for each question.

1. How much of Earth's freshwater is frozen in Antarctic ice?
   A. one-half
   B. one-third
   C. one-fourth
   D. two-thirds

2. Who was the first explorer to reach the South Pole?
   F. Ernest Shackleton
   G. Roald Amundsen
   H. Robert Scott
   I. Richard Byrd

3. It gets so cold in Antarctica during the winter that salt water from the surrounding seas freezes into
   A. huge ice sculptures.
   B. ice shelves.
   C. ice cliffs.
   D. icebergs.

4. During the summer, birds, whales, and other sea mammals come to the Southern Ocean around Antarctica to feed on tiny crustaceans called
   F. plankton.
   G. shrimp.
   H. krill.
   I. lichens.

5. How does the Antarctic Treaty protect the continent's environment?
   A. It prohibits drilling for oil in the Southern Ocean.
   B. It restricts the number of people who can be there at any one time.
   C. It prohibits tourism.
   D. It reserves the continent for peaceful scientific research.

6. In which months of the year do researchers in Antarctica have to tolerate months of total darkness, frigid temperatures, and howling winds?
   F. July, August, September
   G. only December
   H. all 12 months
   I. April, May, June

Chapter 16 ASSESSMENT (continued)

DBQ ANALYZING DOCUMENTS

7 DETERMINING CENTRAL IDEAS Read the following passage about sea ice in the Arctic and Antarctic regions:

"Sea ice differs between the Arctic and Antarctic, primarily because of their different geography. The Arctic is . . . almost completely surrounded by land. As a result, the sea ice that forms in the Arctic is not [very] mobile. . . . Antarctica is a land mass surrounded by an ocean. . . . Sea ice is free to float northward into warmer waters."

—from "Arctic vs. Antarctic," National Snow & Ice Data Center

What causes the differences between the two kinds of sea ice?
A. the different geographies of the two regions
B. the colder temperatures of Antarctica
C. different impacts of global warming
D. coldness of Arctic waters

8 ANALYZING Which of these events is more likely to occur to Antarctic sea ice than Arctic sea ice?
F. growing larger over time
G. piling up in high ice jams
H. slowly melting over time
I. being packed tightly with other ice

SHORT RESPONSE

"Lars-Eric Lindblad led the first traveler's expedition to Antarctica in 1966. Lindblad once said, "You can't protect what you don't know." He believed that by providing a first-hand experience to tourists you would . . . promote a greater understanding of the earth's resources and the important role of Antarctica in the global environment."

—from "Tourism Overview," International Association of Antarctic Tour Operators

9 CITING TEXT EVIDENCE According to Lindblad, how is experience connected to environmental preservation?

10 IDENTIFYING POINT OF VIEW Do you agree with Lindblad? Why or why not?

EXTENDED RESPONSE

11 INFORMATIVE/EXPLANATORY WRITING Conduct research and then write a paper on what it is like to live and work on a research station in Antarctica. To find out what it was like to build these research stations under such harsh conditions and how the stations are equipped to handle the weather and protect their inhabitants from the coldest weather on Earth.

Need Extra Help?

If You've Missed Question	1	2	3	4	5	6	7	8	9	10	11
Review Lesson	1	2	1	1	2	2	1	1	2	2	2

Using **FOLDABLES** is a great way to organize notes, remember information, and prepare for tests. Follow these easy directions to create a Foldable® for the chapter you are studying.

CHAPTER 1: STUDYING EARTH'S LAND, PEOPLE, AND ENVIRONMENTS

Describing Make this Foldable and label the top *Geographer's View* and the bottom *Geographer's Tools*. Under the top fold, describe three ways you experience geography every day. Under the bottom fold, list and describe the tools of geography and explain how a map is a tool. In your mind, form an image of a map of the world. Sketch and label what you visualize on the back of your shutter fold.

Step 1
Bend a sheet of paper in half to find the midpoint.

Step 2
Fold the outer edges of the paper to meet at the midpoint.

CHAPTER 2: THE PHYSICAL WORLD

Identifying Make this Foldable and label the four tabs *Processes*, *Forces*, *Land*, and *Water*. Under *Processes*, identify and describe the processes that operate above and below Earth's surface. Include specific examples. Under *Forces*, give examples of how forces are changing Earth's surface where you live. Finally, under *Land* and *Water*, identify land and water features within 100 miles (161 km) of your community and explain how they influence your life.

Step 1
Fold the outer edges of the paper to meet at the midpoint. Crease well.

Step 2
Fold the paper in half from side to side.

Step 3
Open and cut along the inside fold lines to form four tabs.

Step 4
Label the tabs as shown.

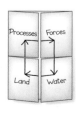

Foldables® Library **491**

CHAPTER 3: THE HUMAN WORLD

Analyzing Create this Foldable, and then label the tabs *Adaptations*, *Cultural Views*, and *Basic Needs*. Under *Adaptations*, describe how humans have adapted to life in two different geographic regions and describe population trends in each. Under *Cultural Views*, analyze what makes two different cultures unique. Finally, under *Basic Needs*, describe how your basic needs might be met in two different economic systems.

Step 1
Fold a sheet of paper in half, leaving a ½-inch tab along one edge.

Step 2
Then fold the paper into three equal sections.

Step 3
Cut along the folds on the top sheet of paper to create three tabs.

Step 4
Label your Foldable as shown.

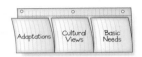

CHAPTER 4: EAST ASIA

Analyzing Create the Foldable below. Write the chapter title on the front and label the tabs *Mainland and Islands*, *Cultural Influences*, and *Economic Growth*. Under *Mainland and Islands*, describe the physical environment of the mainland and the islands. Under *Cultural Influences*, draw a three-circle Venn diagram and label the circles *China*, *Japan*, and *Korea*. Use the diagram to analyze similarities and differences in the history of these countries. Under *Economic Growth*, summarize the present economy of the region.

Step 1
Stack two sheets of paper so that the back sheet is 1 inch higher than the front sheet.

Step 2
Fold the paper to form four equal tabs.

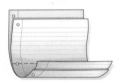

Step 3
When all tabs are an equal distance apart, fold the papers and crease well.

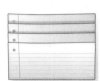

Step 4
Open the papers, and then glue or staple them along the fold.

FOLDABLES

CHAPTER 5: SOUTHEAST ASIA

Identifying Sketch a map of Southeast Asia on the back of the Foldable and label the 11 countries discussed in the text. Label the top of each of the three columns: *Geography*, *History*, and *Cultural Diversity*. Use the sections to identify three geographic features that make Southeast Asia unique, make a time line of events that occurred during the history of the spice trade, and explain why Southeast Asia is such a culturally diverse region.

Step 1
Fold a sheet of paper into thirds to form three equal columns.

Step 2
Label your Foldable as shown.

CHAPTER 6: SOUTH ASIA

Organizing Make the Foldable below. Cut notebook paper into eighths to make small note cards that fit in the pockets. Label the pockets *Geography*, *History*, and *Economy*. Use small note cards to record information on major geographic features in this region, to record important historical events, and to document economic and political events that have affected the region over the last century.

Step 1
Fold the bottom edge of a piece of paper up 2 inches to create a flap.

Step 2
Fold the paper into thirds.

Step 3
Glue the flap on both edges and at both fold lines to form pockets. Label as shown.

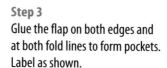

Foldables® Library **493**

Foldables

CHAPTER 7: Central Asia, the Caucasus, and Siberian Russia

Identifying Sketch an outline of Central Asia, the Caucasus, and Siberian Russia on the back of the Foldable and label the physical features. Label the top two tabs *Waterways—Landforms,* and list examples of each that form borders between countries within this region. Label the two middle tabs *Conquest—Independence,* and list examples of empires from the region and countries that are currently independent. Label the bottom two tabs *Rural—Urban*. Then describe and differentiate between rural and urban life in the region.

Step 1
Fold the outer edges of the paper to meet at the midpoint. Crease well.

Step 2
Open and cut three equal tabs from the outer edge to the crease on each side.

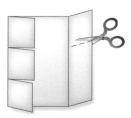

Step 3
Label the tabs as shown.

CHAPTER 8: Southwest Asia

Describing On your Foldable, label the three tabs *Water, Civilization and Religion,* and *Oil and Water*. Under *Water,* describe the role of freshwater and salt water in the development of Southwest Asia. Under *Civilization and Religion,* explain why this region in called the "cradle of civilization." Under *Oil and Water,* describe and compare the importance of oil and water to the economy of the region.

Step 1
Fold a sheet of paper in half, leaving a ½-inch tab along one edge.

Step 2
Then fold the paper into three equal sections.

Step 3
Cut along the folds on the top sheet of paper to create three tabs.

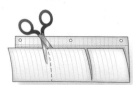

Step 4
Label your Foldable as shown.

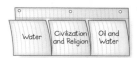

CHAPTER 9: NORTH AFRICA

Organizing Label the rows *Geography*, *History*, and *Economy*. Label the columns *Know* and *Learned*. Use the table to record what you know and what you learn about the geography, history, and economy of North Africa.

Step 1
Fold the paper into three equal columns. Crease well.

Step 2
Open the paper and then fold it into four equal rows. Crease well. Unfold and label as shown.

CHAPTER 10: EAST AFRICA

Analyzing Write the chapter title on the cover and label the tabs *"Great" Things; Past Affects Present;* and *Daily Life, Literacy, and Health*. Under the first tab, describe the Great Rift Valley and the Great Migration and their impact on the economy of the region. Under the second tab, give examples of historical events that affect the region's current economy and politics. Under the third tab, compare two countries in East Africa using literacy rates and life expectancy.

Step 1
Stack two sheets of paper so that the back sheet is 1 inch higher than the front sheet.

Step 2
Fold the paper to form four equal tabs.

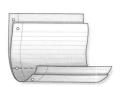

Step 3
When all tabs are an equal distance apart, fold the papers and crease well.

Step 4
Open the papers and then glue them along the fold.

Foldables

CHAPTER 11: CENTRAL AFRICA

Describing Label the four tabs *Rain Forests*, *Savannas*, *Triangular Trade*, and *Rural vs. Urban*. Under the *Rain Forests* and *Savannas* tabs, differentiate between the climate and vegetation of rain forests and savannas. Under *Triangular Trade*, describe the three stages of the triangular trade route and explain why each was profitable for merchants. Under *Rural vs. Urban,* explain why you think the capital cities of many of the countries within this region are becoming huge metropolises.

Step 1
Fold the outer edges of the paper to meet at the midpoint. Crease well.

Step 2
Fold the paper in half from side to side.

Step 3
Open and cut along the inside fold lines to form four tabs.

Step 4
Label the tabs as shown.

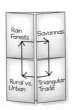

CHAPTER 12: WEST AFRICA

Analyzing Label the first tab *Natural Resources,* and identify which resources you think should be protected and which should be developed. Label the second tab *Trade*. Summarize the importance of trade to the region and list three valuable trade goods. Finally, label the third tab *Population Growth*. Describe how and why rapid population growth is negatively affecting the economy of the region.

Step 1
Fold a sheet of paper in half, leaving a ½-inch tab along one edge.

Step 2
Then fold the paper into three equal sections.

Step 3
Cut along the folds on the top sheet of paper to create three tabs.

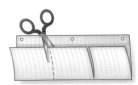

Step 4
Label your Foldable as shown.

CHAPTER 13: SOUTHERN AFRICA

Organizing Cut notebook paper into eighths to make small note cards that fit in the pockets. Sketch an outline of Southern Africa on the back of the Foldable and label geographic features of the region. On the front, label the pockets *Geography*, *History*, and *Economy*. On the note cards, record information about major geographic features, important historical events, and economic and political events that occurred in the region.

Step 1
Fold the bottom edge of a piece of paper up 2 inches to create a flap.

Step 2
Fold the paper into thirds.

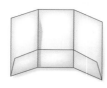

Step 3
Glue the flap on both edges and at both fold lines to form pockets. Label as shown.

CHAPTER 14: AUSTRALIA AND NEW ZEALAND

Describing Label the top section *Australia* and the bottom section *New Zealand*. Under the top-left tab, describe three geographic features of Australia. Under the bottom-left tab, describe the geography of New Zealand. Under the middle tabs, describe the ethnic groups in the regions. Under the top- and bottom-right tabs, describe issues facing Australia and New Zealand.

Step 1
Fold the outer edges of the paper to meet at the midpoint. Crease well.

Step 2
Label the tabs as shown.

Step 3
Open and cut three equal tabs from the outer edge to the crease on each side.

Foldables

CHAPTER 15: OCEANIA

Identifying Write the chapter title on the cover tab, and label the three small tabs *Resources; Past Affects Present;* and *Daily Life, Literacy, and Health*. Under *Resources*, explain the impact of limited resources on Oceania's economy. Under *Past Affects Present*, explain how past colonization affects the economy and government of two countries in Oceania. Under *Daily Life, Literacy, and Health*, explain how the migration of young adults from Oceania to other countries is affecting the region.

Step 1
Stack two sheets of paper so that the back sheet is 1 inch higher than the front sheet.

Step 2
Fold the paper to form four equal tabs.

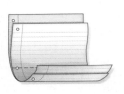

Step 3
When all tabs are an equal distance apart, fold the papers and crease well.

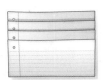

Step 4
Open the papers, and then glue or staple them along the fold.

CHAPTER 16: ANTARCTICA

Describing Label the top three columns of your Foldable *Extremes, The Antarctic Treaty,* and *Global Importance*. Under *Extremes*, describe the geographic and climate extremes that make this continent unique. Under *The Antarctic Treaty* and *Global Importance*, summarize the history of Antarctica and the Antarctic Treaty, and explain the global importance of this protected continent.

Step 1
Fold a sheet of paper into thirds to form three equal columns.

Step 2
Label your Foldable as shown.

GAZETTEER

Gazetteer

A gazetteer (ga•zuh•TIHR) is a geographic index or dictionary. It shows latitude and longitude for cities and certain other places. Latitude and longitude are shown in this way: 48°N 2°E, or 48 degrees north latitude and two degrees east longitude. This Gazetteer lists many important geographic features and most of the world's largest independent countries and their capitals. The page numbers tell where each entry can be found on a map in this book. As an aid to pronunciation, most entries are spelled phonetically.

A

Abidjan [AH•BEE•JAHN] Capital of Côte d'Ivoire. 5°N 4°W (p. RA22)
Abu Dhabi [AH•BOO DAH•bee] Capital of the United Arab Emirates. 24°N 54°E (p. RA24)
Abuja [ah•BOO•jah] Capital of Nigeria. 8°N 9°E (p. RA22)
Accra [ah•KRUH] Capital of Ghana. 6°N 0° longitude (p. RA22)
Addis Ababa [AHD•dihs AH•bah•BAH] Capital of Ethiopia. 9°N 39°E (p. RA22)
Adriatic [AY•dree•A•tihk] **Sea** Arm of the Mediterranean Sea between the Balkan Peninsula and Italy. (p. RA20)
Afghanistan [af•GA•nuh•STAN] Central Asian country west of Pakistan. (p. RA25)
Albania [al•BAY•nee•uh] Country on the Adriatic Sea, south of Serbia. (p. RA18)
Algeria [al•JIHR•ee•uh] North African country east of Morocco. (p. RA22)
Algiers [al•JIHRZ] Capital of Algeria. 37°N 3°E (p. RA22)
Alps [ALPS] Mountain ranges extending through central Europe. (p. RA20)
Amazon [A•muh•ZAHN] **River** Largest river in the world by volume and second-largest in length. (p. RA17)
Amman [a•MAHN] Capital of Jordan. 32°N 36°E (p. RA24)
Amsterdam [AHM•stuhr•DAHM] Capital of the Netherlands. 52°N 5°E (p. RA18)
Andes [AN•DEEZ] Mountain system extending north and south along the western side of South America. (p. RA17)
Andorra [an•DAWR•uh] Small country in southern Europe between France and Spain. 43°N 2°E (p. RA18)
Angola [ang•GOH•luh] Southern African country north of Namibia. (p. RA22)
Ankara [AHNG•kuh•ruh] Capital of Turkey. 40°N 33°E (p. RA24)
Antananarivo [AHN•tah•NAH•nah•REE•voh] Capital of Madagascar. 19°S 48°E (p. RA22)
Arabian [uh•RAY•bee•uhn] **Peninsula** Large peninsula extending into the Arabian Sea. (p. RA25)
Argentina [AHR•juhn•TEE•nuh] South American country east of Chile. (p. RA16)
Armenia [ahr•MEE•nee•uh] European-Asian country between the Black and Caspian Seas. 40°N 45°E (p. RA26)
Ashkhabad [AHSH•gah•BAHD] Capital of Turkmenistan. 38°N 58°E (p. RA25)
Asmara [az•MAHR•uh] Capital of Eritrea. 16°N 39°E (p. RA22)
Astana Capital of Kazakhstan. 51°N 72°E (p. RA26)
Asunción [ah•SOON•see•OHN] Capital of Paraguay. 25°S 58°W (p. RA16)
Athens Capital of Greece. 38°N 24°E (p. RA19)
Atlas [AT•luhs] **Mountains** Mountain range on the northern edge of the Sahara. (p. RA23)
Australia [aw•STRAYL•yuh] Country and continent in Southern Hemisphere. (p. RA30)
Austria [AWS•tree•uh] Western European country east of Switzerland and south of Germany and the Czech Republic. (p. RA18)
Azerbaijan [A•zuhr•BY•JAHN] European-Asian country on the Caspian Sea. (p. RA25)

B

Baghdad Capital of Iraq. 33°N 44°E (p. RA25)
Bahamas [buh•HAH•muhz] Country made up of many islands between Cuba and the United States. (p. RA15)
Bahrain [bah•RAYN] Country located on the Persian Gulf. 26°N 51°E (p. RA25)
Baku [bah•KOO] Capital of Azerbaijan. 40°N 50°E (p. RA25)
Balkan [BAWL•kuhn] **Peninsula** Peninsula in southeastern Europe. (p. RA21)
Baltic [BAWL•tihk] **Sea** Sea in northern Europe that is connected to the North Sea. (p. RA20)
Bamako [BAH•mah•KOH] Capital of Mali. 13°N 8°W (p. RA22)
Bangkok [BANG•KAHK] Capital of Thailand. 14°N 100°E (p. RA27)
Bangladesh [BAHNG•gluh•DEHSH] South Asian country bordered by India and Myanmar. (p. RA27)
Bangui [BAHNG•GEE] Capital of the Central African Republic. 4°N 19°E (p. RA22)
Banjul [BAHN•JOOL] Capital of Gambia. 13°N 17°W (p. RA22)
Barbados [bahr•BAY•duhs] Island country between the Atlantic Ocean and the Caribbean Sea. 14°N 59°W (p. RA15)
Beijing [BAY•JIHNG] Capital of China. 40°N 116°E (p. RA27)
Beirut [bay•ROOT] Capital of Lebanon. 34°N 36°E (p. RA24)

GAZETTEER

Belarus [BEE·luh·ROOS] Eastern European country west of Russia. 54°N 28°E (p. RA19)

Belgium [BEHL·juhm] Western European country south of the Netherlands. (p. RA18)

Belgrade [BEHL·GRAYD] Capital of Serbia. 45°N 21°E (p. RA19)

Belize [buh·LEEZ] Central American country east of Guatemala. (p. RA14)

Belmopan [BEHL·moh·PAHN] Capital of Belize. 17°N 89°W (p. RA14)

Benin [buh·NEEN] West African country west of Nigeria. (p. RA22)

Berlin [behr·LEEN] Capital of Germany. 53°N 13°E (p. RA18)

Bern Capital of Switzerland. 47°N 7°E (p. RA18)

Bhutan [boo·TAHN] South Asian country northeast of India. (p. RA27)

Bishkek [bihsh·KEHK] Capital of Kyrgyzstan. 43°N 75°E (p. RA26)

Bissau [bihs·SOW] Capital of Guinea-Bissau. 12°N 16°W (p. RA22)

Black Sea Large sea between Europe and Asia. (p. RA21)

Bloemfontein [BLOOM·FAHN·TAYN] Judicial capital of South Africa. 26°E 29°S (p. RA22)

Bogotá [BOH·GOH·TAH] Capital of Colombia. 5°N 74°W (p. RA16)

Bolivia [buh·LIHV·ee·uh] Country in the central part of South America, north of Argentina. (p. RA16)

Bosnia and Herzegovina [BAHZ·nee·uh HEHRT·seh·GAW·vee·nuh] Southeastern European country bordered by Croatia, Serbia, and Montenegro. (p. RA18)

Botswana [bawt·SWAH·nah] Southern African country north of the Republic of South Africa. (p. RA22)

Brasília [brah·ZEEL·yuh] Capital of Brazil. 16°S 48°W (p. RA16)

Bratislava [BRAH·tih·SLAH·vuh] Capital of Slovakia. 48°N 17°E (p. RA18)

Brazil [bruh·ZIHL] Largest country in South America. (p. RA16)

Brazzaville [BRAH·zuh·VEEL] Capital of Congo. 4°S 15°E (p. RA22)

Brunei [bru·NY] Southeast Asian country on northern coast of the island of Borneo. (p. RA27)

Brussels [BRUH·suhlz] Capital of Belgium. 51°N 4°E (p. RA18)

Bucharest [BOO·kuh·REHST] Capital of Romania. 44°N 26°E (p. RA19)

Budapest [BOO·duh·PEHST] Capital of Hungary. 48°N 19°E (p. RA18)

Buenos Aires [BWAY·nuhs AR·eez] Capital of Argentina. 34°S 58°W (p. RA16)

Bujumbura [BOO·juhm·BUR·uh] Capital of Burundi. 3°S 29°E (p. RA22)

Bulgaria [BUHL·GAR·ee·uh] Southeastern European country south of Romania. (p. RA19)

Burkina Faso [bur·KEE·nuh FAH·soh] West African country south of Mali. (p. RA22)

Burundi [bu·ROON·dee] East African country at the northern end of Lake Tanganyika. 3°S 30°E (p. RA22)

C

Cairo [KY·roh] Capital of Egypt. 31°N 32°E (p. RA24)

Cambodia [kam·BOH·dee·uh] Southeast Asian country south of Thailand and Laos. (p. RA27)

Cameroon [KA·muh·ROON] Central African country on the northeast shore of the Gulf of Guinea. (p. RA22)

Canada [KA·nuh·duh] Northernmost country in North America. (p. RA6)

Canberra [KAN·BEHR·uh] Capital of Australia. 35°S 149°E (p. RA30)

Cape Town Legislative capital of the Republic of South Africa. 34°S 18°E (p. RA22)

Cape Verde [VUHRD] Island country off the coast of western Africa in the Atlantic Ocean. 15°N 24°W (p. RA22)

Caracas [kah·RAH·kahs] Capital of Venezuela. 11°N 67°W (p. RA16)

Caribbean [KAR·uh·BEE·uhn] **Islands** Islands in the Caribbean Sea between North America and South America, also known as West Indies. (p. RA15)

Caribbean Sea Part of the Atlantic Ocean bordered by the West Indies, South America, and Central America. (p. RA15)

Caspian [KAS·pee·uhn] **Sea** Salt lake between Europe and Asia that is the world's largest inland body of water. (p. RA21)

Caucasus [KAW·kuh·suhs] **Mountains** Mountain range between the Black and Caspian Seas. (p. RA21)

Central African Republic Central African country south of Chad. (p. RA22)

Chad [CHAD] Country west of Sudan in the African Sahel. (p. RA22)

Chang Jiang [CHAHNG jee·AHNG] Principal river of China that begins in Tibet and flows into the East China Sea near Shanghai; also known as the Yangtze River. (p. RA29)

Chile [CHEE·lay] South American country west of Argentina. (p. RA16)

China [CHY·nuh] Country in eastern and central Asia, known officially as the People's Republic of China. (p. RA27)

Chişinău [KEE·shee·NOW] Capital of Moldova. 47°N 29°E (p. RA19)

Colombia [kuh·LUHM·bee·uh] South American country west of Venezuela. (p. RA16)

Colombo [kuh·LUHM·boh] Capital of Sri Lanka. 7°N 80°E (p. RA26)

Comoros [KAH·muh·ROHZ] Small island country in Indian Ocean between the island of Madagascar and the southeast African mainland. 13°S 43°E (p. RA22)

Conakry [KAH·nuh·kree] Capital of Guinea. 10°N 14°W (p. RA22)

Congo [KAHNG•goh] Central African country east of the Democratic Republic of the Congo. 3°S 14°E (p. RA22)

Congo, Democratic Republic of the Central African country north of Zambia and Angola. 1°S 22°E (p. RA22)

Copenhagen [KOH•puhn•HAY•guhn] Capital of Denmark. 56°N 12°E (p. RA18)

Costa Rica [KAWS•tah REE•kah] Central American country south of Nicaragua. (p. RA15)

Côte d'Ivoire [KOHT dee•VWAHR] West African country south of Mali. (p. RA22)

Croatia [kroh•AY•shuh] Southeastern European country on the Adriatic Sea. (p. RA18)

Cuba [KYOO•buh] Island country in the Caribbean Sea. (p. RA15)

Cyprus [SY•pruhs] Island country in the eastern Mediterranean Sea, south of Turkey. (p. RA19)

Czech [CHEHK] **Republic** Eastern European country north of Austria. (p. RA18)

D

Dakar [dah•KAHR] Capital of Senegal. 15°N 17°W (p. RA22)

Damascus [duh•MAS•kuhs] Capital of Syria. 34°N 36°E (p. RA24)

Dar es Salaam [DAHR EHS sah•LAHM] Commercial capital of Tanzania. 7°S 39°E (p. RA22)

Denmark Northern European country between the Baltic and North Seas. (p. RA18)

Dhaka [DA•kuh] Capital of Bangladesh. 24°N 90°E (p. RA27)

Djibouti [jih•BOO•tee] East African country on the Gulf of Aden. 12°N 43°E (p. RA22)

Dodoma [doh•DOH•mah] Political capital of Tanzania. 6°S 36°E (p. RA22)

Doha [DOH•huh] Capital of Qatar. 25°N 51°E (p. RA25)

Dominican [duh•MIH•nih•kuhn] **Republic** Country in the Caribbean Sea on the eastern part of the island of Hispaniola. (p. RA15)

Dublin [DUH•blihn] Capital of Ireland. 53°N 6°W (p. RA18)

Dushanbe [doo•SHAM•buh] Capital of Tajikistan. 39°N 69°E (p. RA25)

E

East Timor [TEE•MOHR] Previous province of Indonesia, now under UN administration. 10°S 127°E (p. RA27)

Ecuador [EH•kwuh•dawr] South American country southwest of Colombia. (p. RA16)

Egypt [EE•jihpt] North African country on the Mediterranean Sea. (p. RA24)

El Salvador [ehl SAL•vuh•dawr] Central American country southwest of Honduras. (p. RA14)

Equatorial Guinea [EE•kwuh•TOHR•ee•uhl GIH•nee] Central African country south of Cameroon. (p. RA22)

Eritrea [EHR•uh•TREE•uh] East African country north of Ethiopia. (p. RA22)

Estonia [eh•STOH•nee•uh] Eastern European country on the Baltic Sea. (p. RA19)

Ethiopia [EE•thee•OH•pee•uh] East African country north of Somalia and Kenya. (p. RA22)

Euphrates [yu•FRAY•teez] **River** River in southwestern Asia that flows through Syria and Iraq and joins the Tigris River. (p. RA25)

F

Fiji [FEE•jee] **Islands** Country comprised of an island group in the southwest Pacific Ocean. 19°S 175°E (p. RA30)

Finland [FIHN•luhnd] Northern European country east of Sweden. (p. RA19)

France [FRANS] Western European country south of the United Kingdom. (p. RA18)

Freetown Capital of Sierra Leone. (p. RA22)

French Guiana [gee•A•nuh] French-owned territory in northern South America. (p. RA16)

G

Gabon [ga•BOHN] Central African country on the Atlantic Ocean. (p. RA22)

Gaborone [GAH•boh•ROH•nay] Capital of Botswana. (p. RA22)

Gambia [GAM•bee•uh] West African country along the Gambia River. (p. RA22)

Georgetown [JAWRJ•town] Capital of Guyana. 8°N 58°W (p. RA16)

Georgia [JAWR•juh] European-Asian country bordering the Black Sea south of Russia. (p. RA26)

Germany [JUHR•muh•nee] Western European country south of Denmark, officially called the Federal Republic of Germany. (p. RA18)

Ghana [GAH•nuh] West African country on the Gulf of Guinea. (p. RA22)

Great Plains The continental slope extending through the United States and Canada. (p. RA7)

Greece [GREES] Southern European country on the Balkan Peninsula. (p. RA19)

Greenland [GREEN•luhnd] Island in northwestern Atlantic Ocean and the largest island in the world. (p. RA6)

Guatemala [GWAH•tay•MAH•lah] Central American country south of Mexico. (p. RA14)

Guatemala Capital of Guatemala. 15°N 91°W (p. RA14)

Guinea [GIH•nee] West African country on the Atlantic coast. (p. RA22)

Guinea-Bissau [GIH•nee bih•SOW] West African country on the Atlantic coast. (p. RA22)

Gulf of Mexico Gulf on part of the southern coast of North America. (p. RA7)

Gazetteer **501**

GAZETTEER

Guyana [gy·AH·nuh] South American country between Venezuela and Suriname. (p. RA16)

H

Haiti [HAY·tee] Country in the Caribbean Sea on the western part of the island of Hispaniola. (p. RA15)
Hanoi [ha·NOY] Capital of Vietnam. 21°N 106°E (p. RA27)
Harare [hah·RAH·RAY] Capital of Zimbabwe. 18°S 31°E (p. RA22)
Havana [huh·VA·nuh] Capital of Cuba. 23°N 82°W (p. RA15)
Helsinki [HEHL·SIHNG·kee] Capital of Finland. 60°N 24°E (p. RA19)
Himalaya [HI·muh·LAY·uh] Mountain ranges in southern Asia, bordering the Indian subcontinent on the north. (p. RA28)
Honduras [hahn·DUR·uhs] Central American country on the Caribbean Sea. (p. RA14)
Hong Kong Port and industrial center in southern China. 22°N 115°E (p. RA27)
Huang He [HWAHNG HUH] River in northern and eastern China, also known as the Yellow River. (p. RA29)
Hungary [HUHNG·guh·ree] Eastern European country south of Slovakia. (p. RA18)

I

Iberian [eye·BIHR·ee·uhn] **Peninsula** Peninsula in southwest Europe, occupied by Spain and Portugal. (p. RA20)
Iceland Island country between the North Atlantic and Arctic Oceans. (p. RA18)
India [IHN·dee·uh] South Asian country south of China and Nepal. (p. RA26)
Indonesia [IHN·duh·NEE·zhuh] Southeast Asian island country known as the Republic of Indonesia. (p. RA27)
Indus [IHN·duhs] **River** River in Asia that begins in Tibet and flows through Pakistan to the Arabian Sea. (p. RA28)
Iran [ih·RAN] Southwest Asian country that was formerly named Persia. (p. RA25)
Iraq [ih·RAHK] Southwest Asian country west of Iran. (p. RA25)
Ireland [EYER·luhnd] Island west of Great Britain occupied by the Republic of Ireland and Northern Ireland. (p. RA18)
Islamabad [ihs·LAH·muh·BAHD] Capital of Pakistan. 34°N 73°E (p. RA26)
Israel [IHZ·ree·uhl] Southwest Asian country south of Lebanon. (p. RA24)
Italy [IHT·uhl·ee] Southern European country south of Switzerland and east of France. (p. RA18)

J

Jakarta [juh·KAHR·tuh] Capital of Indonesia. 6°S 107°E (p. RA27)
Jamaica [juh·MAY·kuh] Island country in the Caribbean Sea. (p. RA15)
Japan [juh·PAN] East Asian country consisting of the four large islands of Hokkaido, Honshu, Shikoku, and Kyushu, plus thousands of small islands. (p. RA27)
Jerusalem [juh·ROO·suh·luhm] Capital of Israel and a holy city for Christians, Jews, and Muslims. 32°N 35°E (p. RA24)
Jordan [JAWRD·uhn] Southwest Asian country south of Syria. (p. RA24)
Juba [JU·buh] Capital of South Sudan. 5°N 31°E (p. RA22)

K

Kabul [KAH·buhl] Capital of Afghanistan. 35°N 69°E (p. RA25)
Kampala [kahm·PAH·lah] Capital of Uganda. 0° latitude 32°E (p. RA22)
Kathmandu [KAT·MAN·DOO] Capital of Nepal. 28°N 85°E (p. RA26)
Kazakhstan [kuh·ZAHK·STAHN] Large Asian country south of Russia and bordering the Caspian Sea. (p. RA26)
Kenya [KEHN·yuh] East African country south of Ethiopia. (p. RA22)
Khartoum [kahr·TOOM] Capital of Sudan. 16°N 33°E (p. RA22)
Kigali [kee·GAH·lee] Capital of Rwanda. 2°S 30°E (p. RA22)
Kingston [KIHNG·stuhn] Capital of Jamaica. 18°N 77°W (p. RA15)
Kinshasa [kihn·SHAH·suh] Capital of the Democratic Republic of the Congo. 4°S 15°E (p. RA22)
Kuala Lumpur [KWAH·luh LUM·PUR] Capital of Malaysia. 3°N 102°E (p. RA27)
Kuwait [ku·WAYT] Country on the Persian Gulf between Saudi Arabia and Iraq. (p. RA25)
Kyiv (Kiev) [KEE·ihf] Capital of Ukraine. 50°N 31°E (p. RA19)
Kyrgyzstan [S·gih·STAN] Central Asian country on China's western border. (p. RA26)

L

Laos [LOWS] Southeast Asian country south of China and west of Vietnam. (p. RA27)
La Paz [lah PAHS] Administrative capital of Bolivia, and the highest capital in the world. 17°S 68°W (p. RA16)
Latvia [LAT·vee·uh] Eastern European country west of Russia on the Baltic Sea. (p. RA19)
Lebanon [LEH·buh·nuhn] Country south of Syria on the Mediterranean Sea. (p. RA24)

Lesotho [luh•SOH•TOH] Southern African country within the borders of the Republic of South Africa. (p. RA22)

Liberia [ly•BIHR•ee•uh] West African country south of Guinea. (p. RA22)

Libreville [LEE•bruh•VIHL] Capital of Gabon. 1°N 9°E (p. RA22)

Libya [LIH•bee•uh] North African country west of Egypt on the Mediterranean Sea. (p. RA22)

Liechtenstein [LIHKT•uhn•SHTYN] Small country in central Europe between Switzerland and Austria. 47°N 10°E (p. RA18)

Lilongwe [lih•LAWNG•GWAY] Capital of Malawi. 14°S 34°E (p. RA22)

Lima [LEE•mah] Capital of Peru. 12°S 77°W (p. RA16)

Lisbon [LIHZ•buhn] Capital of Portugal. 39°N 9°W (p. RA18)

Lithuania [LIH•thuh•WAY•nee•uh] Eastern European country northwest of Belarus on the Baltic Sea. (p. RA21)

Ljubljana [lee•oo•blee•AH•nuh] Capital of Slovenia. 46°N 14°E (p. RA18)

Lomé [loh•MAY] Capital of Togo. 6°N 1°E (p. RA22)

London Capital of the United Kingdom, on the Thames River. 52°N 0° longitude (p. RA18)

Luanda [lu•AHN•duh] Capital of Angola. 9°S 13°E (p. RA22)

Lusaka [loo•SAH•kah] Capital of Zambia. 15°S 28°E (p. RA22)

Luxembourg [LUHK•suhm•BUHRG] Small European country bordered by France, Belgium, and Germany. 50°N 7°E (p. RA18)

M

Macao [muh•KOW] Port in southern China. 22°N 113°E (p. RA27)

Macedonia [ma•suh•DOH•nee•uh] Southeastern European country north of Greece. (p. RA19). Macedonia also refers to a geographic region covering northern Greece, the country Macedonia, and part of Bulgaria.

Madagascar [MA•duh•GAS•kuhr] Island in the Indian Ocean off the southeastern coast of Africa. (p. RA22)

Madrid Capital of Spain. 41°N 4°W (p. RA18)

Malabo [mah•LAH•boh] Capital of Equatorial Guinea. 4°N 9°E (p. RA22)

Malawi [mah•LAH•wee] Southern African country south of Tanzania and east of Zambia. (p. RA22)

Malaysia [muh•LAY•zhuh] Southeast Asian country with land on the Malay Peninsula and on the island of Borneo. (p. RA27)

Maldives [MAWL•DEEVZ] Island country southwest of India in the Indian Ocean. (p. RA26)

Mali [MAH•lee] West African country east of Mauritania. (p. RA22)

Managua [mah•NAH•gwah] Capital of Nicaragua. (p. RA15)

Manila [muh•NIH•luh] Capital of the Philippines. 15°N 121°E (p. RA27)

Maputo [mah•POO•toh] Capital of Mozambique. 26°S 33°E (p. RA22)

Maseru [MA•zuh•ROO] Capital of Lesotho. 29°S 27°E (p. RA22)

Masqat [MUHS•KAHT] Capital of Oman. 23°N 59°E (p. RA25)

Mauritania [MAWR•uh•TAY•nee•uh] West African country north of Senegal. (p. RA22)

Mauritius [maw•RIH•shuhs] Island country in the Indian Ocean east of Madagascar. 21°S 58°E (p. RA3)

Mbabane [uhm•bah•BAH•nay] Capital of Swaziland. 26°S 31°E (p. RA22)

Mediterranean [MEH•duh•tuh•RAY•nee•uhn] **Sea** Large inland sea surrounded by Europe, Asia, and Africa. (p. RA20)

Mekong [MAY•KAWNG] **River** River in southeastern Asia that begins in Tibet and empties into the South China Sea. (p. RA29)

Mexico [MEHK•sih•KOH] North American country south of the United States. (p. RA14)

Mexico City Capital of Mexico. 19°N 99°W (p. RA14)

Minsk [MIHNSK] Capital of Belarus. 54°N 28°E (p. RA19)

Mississippi [MIH•suh•SIH•pee] **River** Large river system in the central United States that flows southward into the Gulf of Mexico. (p. RA11)

Mogadishu [MOH•guh•DEE•shoo] Capital of Somalia. 2°N 45°E (p. RA22)

Moldova [mawl•DAW•vuh] Small European country between Ukraine and Romania. (p. RA19)

Monaco [MAH•nuh•KOH] Small country in southern Europe on the French Mediterranean coast. 44°N 8°E (p. RA18)

Mongolia [mahn•GOHL•yuh] Country in Asia between Russia and China. (p. RA23)

Monrovia [muhn•ROH•vee•uh] Capital of Liberia. 6°N 11°W (p. RA22)

Montenegro [MAHN•tuh•NEE•groh] Eastern European country. (p. RA18)

Montevideo [MAHN•tuh•vuh•DAY•oh] Capital of Uruguay. 35°S 56°W (p. RA16)

Morocco [muh•RAH•KOH] North African country on the Mediterranean Sea and the Atlantic Ocean. (p. RA22)

Moscow [MAHS•KOW] Capital of Russia. 56°N 38°E (p. RA19)

Mount Everest [EHV•ruhst] Highest mountain in the world, in the Himalaya between Nepal and Tibet. (p. RA28)

Mozambique [MOH•zahm•BEEK] Southern African country south of Tanzania. (p. RA22)

Myanmar [MYAHN•MAHR] Southeast Asian country south of China and India, formerly called Burma. (p. RA27)

GAZETTEER

N

Nairobi [ny•ROH•bee] Capital of Kenya. 1°S 37°E (p. RA22)
Namibia [nuh•MIH•bee•uh] Southern African country south of Angola on the Atlantic Ocean. 20°S 16°E (p. RA22)
Nassau [NA•SAW] Capital of the Bahamas. 25°N 77°W (p. RA15)
N'Djamena [uhn•jah•MAY•nah] Capital of Chad. 12°N 15°E (p. RA22)
Nepal [NAY•PAHL] Mountain country between India and China. (p. RA26)
Netherlands [NEH•thuhr•lundz] Western European country north of Belgium. (p. RA18)
New Delhi [NOO DEH•lee] Capital of India. 29°N 77°E (p. RA26)
New Zealand [NOO ZEE•luhnd] Major island country southeast of Australia in the South Pacific. (p. RA30)
Niamey [nee•AHM•ay] Capital of Niger. 14°N 2°E (p. RA22)
Nicaragua [NIH•kuh•RAH•gwuh] Central American country south of Honduras. (p. RA15)
Nicosia [NIH•kuh•SEE•uh] Capital of Cyprus. 35°N 33°E (p. RA19)
Niger [NY•juhr] West African country north of Nigeria. (p. RA22)
Nigeria [ny•JIHR•ee•uh] West African country along the Gulf of Guinea. (p. RA22)
Nile [NYL] **River** Longest river in the world, flowing north through eastern Africa. (p. RA23)
North Korea [kuh•REE•uh] East Asian country in the northernmost part of the Korean Peninsula. (p. RA27)
Norway [NAWR•way] Northern European country on the Scandinavian peninsula. (p. RA18)
Nouakchott [nu•AHK•SHAHT] Capital of Mauritania. 18°N 16°W (p. RA22)

O

Oman [oh•MAHN] Country on the Arabian Sea and the Gulf of Oman. (p. RA25)
Oslo [AHZ•loh] Capital of Norway. 60°N 11°E (p. RA18)
Ottawa [AH•tuh•wuh] Capital of Canada. 45°N 76°W (p. RA13)
Ouagadougou [WAH•gah•DOO•goo] Capital of Burkina Faso. 12°N 2°W (p. RA22)

P

Pakistan [PA•kih•STAN] South Asian country northwest of India on the Arabian Sea. (p. RA26)
Palau [puh•LOW) Island country in the Pacific Ocean. 7°N 135°E (p. RA30)
Panama [PA•nuh•MAH] Central American country on the Isthmus of Panama. (p. RA15)
Panama Capital of Panama. 9°N 79°W (p. RA15)
Papua New Guinea [PA•pyu•wuh NOO GIH•nee] Island country in the Pacific Ocean north of Australia. 7°S 142°E (p. RA30)
Paraguay [PAR•uh•GWY] South American country northeast of Argentina. (p. RA16)
Paramaribo [PAH•rah•MAH•ree•boh] Capital of Suriname. 6°N 55°W (p. RA16)
Paris Capital of France. 49°N 2°E (p. RA18)
Persian [PUHR•zhuhn] **Gulf** Arm of the Arabian Sea between Iran and Saudi Arabia. (p. RA25)
Peru [puh•ROO] South American country south of Ecuador and Colombia. (p. RA16)
Philippines [FIH•luh•PEENZ] Island country in the Pacific Ocean southeast of China. (p. RA27)
Phnom Penh [puh•NAWM PEHN] Capital of Cambodia. 12°N 106°E (p. RA27)
Poland [POH•luhnd] Eastern European country on the Baltic Sea. (p. RA18)
Port-au-Prince [POHRT•oh•PRIHNS] Capital of Haiti. 19°N 72°W (p. RA15)
Port Moresby [MOHRZ•bee] Capital of Papua New Guinea. 10°S 147°E (p. RA30)
Port-of-Spain [SPAYN] Capital of Trinidad and Tobago. 11°N 62°W (p. RA15)
Porto-Novo [POHR•toh•NOH•voh] Capital of Benin. 7°N 3°E (p. RA22)
Portugal [POHR•chih•guhl] Country west of Spain on the Iberian Peninsula. (p. RA18)
Prague [PRAHG] Capital of the Czech Republic. 51°N 15°E (p. RA18)
Puerto Rico [PWEHR•toh REE•koh] Island in the Caribbean Sea; U.S. Commonwealth. (p. RA15)
P'yŏngyang [pee•AWNG•YAHNG] Capital of North Korea. 39°N 126°E (p. RA27)

Q

Qatar [KAH•tuhr] Country on the southwestern shore of the Persian Gulf. (p. RA25)
Quito [KEE•toh] Capital of Ecuador. 0° latitude 79°W (p. RA16)

R

Rabat [ruh•BAHT] Capital of Morocco. 34°N 7°W (p. RA22)
Reykjavík [RAY•kyah•VEEK] Capital of Iceland. 64°N 22°W (p. RA18)
Rhine [RYN] **River** River in western Europe that flows into the North Sea. (p. RA20)
Riga [REE•guh] Capital of Latvia. 57°N 24°E (p. RA19)
Rio Grande [REE•oh GRAND] River that forms part of the boundary between the United States and Mexico. (p. RA10)
Riyadh [ree•YAHD] Capital of Saudi Arabia. 25°N 47°E (p. RA25)

Rocky Mountains Mountain system in western North America. (p. RA7)

Romania [ru•MAY•nee•uh] Eastern European country east of Hungary. (p. RA19)

Rome Capital of Italy. 42°N 13°E (p. RA18)

Russia [RUH•shuh] Largest country in the world, covering parts of Europe and Asia. (pp. RA19, RA27)

Rwanda [ruh•WAHN•duh] East African country south of Uganda. 2°S 30°E (p. RA22)

S

Sahara [suh•HAR•uh] Desert region in northern Africa that is the largest hot desert in the world. (p. RA23)

Saint Lawrence [LAWR•uhns] River River that flows from Lake Ontario to the Atlantic Ocean and forms part of the boundary between the United States and Canada. (p. RA13)

Sanaa [sahn•AH] Capital of Yemen. 15°N 44°E (p. RA25)

San José [SAN hoh•ZAY] Capital of Costa Rica. 10°N 84°W (p. RA15)

San Marino [SAN muh•REE•noh] Small European country located on the Italian Peninsula. 44°N 13°E (p. RA18)

San Salvador [SAN SAL•vuh•DAWR] Capital of El Salvador. 14°N 89°W (p. RA14)

Santiago [SAN•tee•AH•goh] Capital of Chile. 33°S 71°W (p. RA16)

Santo Domingo [SAN•toh duh•MIHNG•goh] Capital of the Dominican Republic. 19°N 70°W (p. RA15)

São Tomé and Príncipe [sow too•MAY PREEN•see•pee] Small island country in the Gulf of Guinea off the coast of central Africa. 1°N 7°E (p. RA22)

Sarajevo [SAR•uh•YAY•voh] Capital of Bosnia and Herzegovina. 43°N 18°E (p. RA18)

Saudi Arabia [SOW•dee uh•RAY•bee•uh] Country on the Arabian Peninsula. (p. RA25)

Senegal [SEH•nih•GAWL] West African country on the Atlantic coast. (p. RA22)

Seoul [SOHL] Capital of South Korea. 38°N 127°E (p. RA27)

Serbia [SUHR•bee•uh] Eastern European country south of Hungary. (p. RA18)

Seychelles [say•SHEHL] Small island country in the Indian Ocean off eastern Africa. 6°S 56°E (p. RA22)

Sierra Leone [see•EHR•uh lee•OHN] West African country south of Guinea. (p. RA22)

Singapore [SIHNG•uh•POHR] Southeast Asian island country near tip of the Malay Peninsula. (p. RA27)

Skopje [SKAW•PYAY] Capital of the country of Macedonia. 42°N 21°E (p. RA19)

Slovakia [sloh•VAH•kee•uh] Eastern European country south of Poland. (p. RA18)

Slovenia [sloh•VEE•nee•uh] Southeastern European country south of Austria on the Adriatic Sea. (p. RA18)

Sofia [SOH•fee•uh] Capital of Bulgaria. 43°N 23°E (p. RA19)

Solomon [SAH•luh•muhn] Islands Island country in the Pacific Ocean northeast of Australia. (p. RA30)

Somalia [soh•MAH•lee•uh] East African country on the Gulf of Aden and the Indian Ocean. (p. RA22)

South Africa [A•frih•kuh] Country at the southern tip of Africa, officially the Republic of South Africa. (p. RA22)

South Korea [kuh•REE•uh] East Asian country on the Korean Peninsula between the Yellow Sea and the Sea of Japan. (p. RA27)

South Sudan [soo•DAN] East African country south of Sudan. (p. RA22)

Spain [SPAYN] Southern European country on the Iberian Peninsula. (p. RA18)

Sri Lanka [SREE LAHNG•kuh] Country in the Indian Ocean south of India, formerly called Ceylon. (p. RA26)

Stockholm [STAHK•HOHLM] Capital of Sweden. 59°N 18°E (p. RA18)

Sucre [SOO•kray] Constitutional capital of Bolivia. 19°S 65°W (p. RA16)

Sudan [soo•DAN] East African country south of Egypt. (p. RA22)

Suriname [SUR•uh•NAH•muh] South American country between Guyana and French Guiana. (p. RA16)

Suva [SOO•vah] Capital of the Fiji Islands. 18°S 177°E (p. RA30)

Swaziland [SWAH•zee•land] Southern African country west of Mozambique, almost entirely within the Republic of South Africa. (p. RA22)

Sweden Northern European country on the eastern side of the Scandinavian Peninsula. (p. RA18)

Switzerland [SWIHT•suhr•luhnd] European country in the Alps south of Germany. (p. RA18)

Syria [SIHR•ee•uh] Southwest Asian country on the east side of the Mediterranean Sea. (p. RA24)

T

Taipei [TY•PAY] Capital of Taiwan. 25°N 122°E (p. RA27)

Taiwan [TY•WAHN] Island country off the southeast coast of China; the seat of the Chinese Nationalist government. (p. RA27)

Tajikistan [tah•JIH•kih•STAN] Central Asian country east of Turkmenistan. (p. RA26)

Tallinn [TA•luhn] Capital of Estonia. 59°N 25°E (p. RA19)

Tanzania [TAN•zuh•NEE•uh] East African country south of Kenya. (p. RA22)

Tashkent [tash•KEHNT] Capital of Uzbekistan. 41°N 69°E (p. RA26)

Tbilisi [tuh•bih•LEE•see] Capital of the Republic of Georgia. 42°N 45°E (p. RA26)

Tegucigalpa [tay•GOO•see•GAHL•pah] Capital of Honduras. 14°N 87°W (p. RA14)

Tehran [TAY•uh•RAN] Capital of Iran. 36°N 52°E (p. RA25)

GAZETTEER

Thailand [TY•LAND] Southeast Asian country east of Myanmar. 17°N 101°E (p. RA27)

Thimphu [thihm•POO] Capital of Bhutan. 28°N 90°E (p. RA27)

Tigris [TY•gruhs] **River** River in southeastern Turkey and Iraq that merges with the Euphrates River. (p. RA25)

Tiranë [tih•RAH•nuh] Capital of Albania. 42°N 20°E (p. RA18)

Togo [TOH•goh] West African country between Benin and Ghana on the Gulf of Guinea. (p. RA22)

Tokyo [TOH•kee•OH] Capital of Japan. 36°N 140°E (p. RA27)

Trinidad and Tobago [TRIH•nuh•DAD tuh•BAY•goh] Island country near Venezuela between the Atlantic Ocean and the Caribbean Sea. (p. RA15)

Tripoli [TRIH•puh•lee] Capital of Libya. 33°N 13°E (p. RA22)

Tshwane [ch•WAH•nay] Executive capital of South Africa. 26°S 28°E (p. RA22)

Tunis [TOO•nuhs] Capital of Tunisia. 37°N 10°E (p. RA22)

Tunisia [too•NEE•zhuh] North African country on the Mediterranean Sea between Libya and Algeria. (p. RA22)

Turkey [TUHR•kee] Country in southeastern Europe and western Asia. (p. RA24)

Turkmenistan [tuhrk•MEH•nuh•STAN] Central Asian country on the Caspian Sea. (p. RA25)

U

Uganda [yoo•GAHN•dah] East African country south of Sudan. (p. RA22)

Ukraine [yoo•KRAYN] Eastern European country west of Russia on the Black Sea. (p. RA25)

Ulaanbaatar [oo•LAHN•BAH•TAWR] Capital of Mongolia. 48°N 107°E (p. RA27)

United Arab Emirates [EH•muh•ruhts] Country made up of seven states on the eastern side of the Arabian Peninsula. (p. RA25)

United Kingdom Western European island country made up of England, Scotland, Wales, and Northern Ireland. (p. RA18)

United States of America Country in North America made up of 50 states, mostly between Canada and Mexico. (p. RA8)

Uruguay [YUR•uh•GWAY] South American country south of Brazil on the Atlantic Ocean. (p. RA16)

Uzbekistan [uz•BEH•kih•STAN] Central Asian country south of Kazakhstan. (p. RA25)

V

Vanuatu [VAN•WAH•TOO] Country made up of islands in the Pacific Ocean east of Australia. (p. RA30)

Vatican [VA•tih•kuhn] **City** Headquarters of the Roman Catholic Church, located in the city of Rome in Italy. 42°N 13°E (p. RA18)

Venezuela [VEH•nuh•ZWAY•luh] South American country on the Caribbean Sea between Colombia and Guyana. (p. RA16)

Vienna [vee•EH•nuh] Capital of Austria. 48°N 16°E (p. RA18)

Vientiane [vyehn•TYAHN] Capital of Laos. 18°N 103°E (p. RA27)

Vietnam [vee•EHT•NAHM] Southeast Asian country east of Laos and Cambodia. (p. RA27)

Vilnius [VIL•nee•uhs] Capital of Lithuania. 55°N 25°E (p. RA19)

W

Warsaw Capital of Poland. 52°N 21°E (p. RA19)

Washington, D.C. Capital of the United States, in the District of Columbia. 39°N 77°W (p. RA8)

Wellington [WEH•lihng•tuhn] Capital of New Zealand. 41°S 175°E (p. RA30)

West Indies Caribbean islands between North America and South America. (p. RA15)

Windhoek [VIHNT•HUK] Capital of Namibia. 22°S 17°E (p. RA22)

Y

Yamoussoukro [YAH•MOO•SOO•kroh] Second capital of Côte d'Ivoire. 7°N 6°W (p. RA22)

Yangon [YAHNG•GOHN] City in Myanmar; formerly called Rangoon. 17°N 96°E (p. RA27)

Yaoundé [yown•DAY] Capital of Cameroon. 4°N 12°E (p. RA22)

Yemen [YEH•muhn] Country south of Saudi Arabia on the Arabian Peninsula. (p. RA25)

Yerevan [YEHR•uh•VAHN] Capital of Armenia. 40°N 44°E (p. RA25)

Z

Zagreb [ZAH•GREHB] Capital of Croatia. 46°N 16°E (p. RA18)

Zambia [ZAM•bee•uh] Southern African country north of Zimbabwe. (p. RA22)

Zimbabwe [zihm•BAH•bway] Southern African country northeast of Botswana. (p. RA22)

GLOSSARY/GLOSARIO

- Content vocabulary words are words that relate to geography content.
- Words that have an asterisk (*) are academic vocabulary. They help you understand your school subjects.
- All vocabulary words are **boldfaced** or **highlighted in yellow** in your textbook.

Aboriginal • axis

ENGLISH — A — ESPAÑOL

Aboriginal the first people to live in Australia (p. 417)

Aborigen el primer pueblo que habitó en Australia (pág. 417)

absolute location the exact location of something (p. 21)

localización absoluta ubicación exacta de algo (pág. 21)

absolute monarchy a system of government in which the ruler has complete control (p. 161)

monarquía absoluta sistema de gobierno en el cual el gobernante detenta el control absoluto (pág. 161)

*****accurate** without mistakes or errors (p. 44)

*****exacto** sin faltas o errores (pág. 44)

acid rain rain that contains harmful amounts of poisons due to pollution (p. 65)

lluvia ácida lluvia que contiene cantidades nocivas de venenos debido a la polución (pág. 65)

*****acknowledge** to recognize the rights, status, or authority of a person, thing, or event (p. 481)

*****admitir** reconocer los derechos, el estatus o la autoridad de una persona, cosa o suceso (pág. 481)

action song a song that combines singing and dancing to celebrate Maori history and culture in order to instill pride among Maori people (p. 435)

canción de acción canción que combina el canto y la danza para honrar la historia y cultura maoríes e inculcar orgullo entre el pueblo maorí (pág. 435)

alluvial plain an area built up by rich fertile soil left by river floods (pp. 177; 228)

llanura aluvial área formada por los sedimentos fértiles que dejan las inundaciones fluviales (págs. 177; 228)

animist a person who believes in spirits that can exist apart from bodies (p. 370)

animista persona que cree en espíritus que viven por fuera del cuerpo (pág. 370)

*****annual** yearly or each year (p. 178)

*****annual** cada año (pág. 178)

apartheid the system of laws in South Africa aimed at separating the races (p. 393)

apartheid sistema jurídico de Sudáfrica que establecía la segregación racial (pág. 393)

aquifer an underground layer of rock through which water flows (p. 271)

acuífero estrato rocoso subterráneo por donde corre el agua (pág. 271)

archipelago a group of islands (pp. 118; 451)

archipiélago grupo de islas (págs. 118; 451)

atmosphere the layer of gases surrounding Earth (p. 44)

atmósfera capa de gases que rodea la Tierra (pág. 44)

atoll a circular-shaped island made of coral (pp. 177; 453)

atolón isla coralina de forma anular (págs. 177; 453)

axis an imaginary line that runs through Earth's center from the North Pole to the South Pole (p. 42)

eje línea imaginaria que atraviesa el centro de la Tierra desde el Polo Norte hasta el Polo Sur (pág. 42)

507

B

basin an area of land that is drained by a river and its tributaries (p. 355)

cuenca area de terreno drenada por un río y sus afluentes (pág. 355)

*behalf in the interest of (p. 87)

*a favor de en beneficio de (pág. 87)

biodiversity the wide variety of life on Earth (p. 332)

biodiversidad la amplia variedad de vida terrestre (pág. 332)

birthrate the number of babies born compared to the total number of people in a population at a given time (p. 72)

tasa de natalidad número de nacimientos comparado con el número total de habitantes de una población en un tiempo determinado (pág. 72)

blood diamonds diamonds that are sold on the black market, with the proceeds going to provide guns and ammunition for violent conflicts (p. 387)

diamantes sangrientos diamantes que se venden en el mercado negro y cuyas ganancias se utilizan para adquirir armas y municiones en conflictos violentos (pág. 387)

boomerang the flat, bent wooden tool of the Australian Aborigines that is thrown to stun prey when it strikes them and that sails back to the hunter if it misses its target (p.425)

búmeran utensilio de madera curvo y plano de los aborígenes australianos, que se lanza para aturdir a las presas cuando las golpea y regresa al cazador en caso de fallar el blanco (pág. 425)

boycott to refuse to buy items from a particular country or company (p. 186)

boicotear rehusarse a comprar los artículos de un país o compañía en particular (pág. 186)

bush a rural area in Australia (p. 433)

brezal área rural de Australia (pág. 433)

C

caliph the successor to Muhammad (p. 276)

califa sucesor de Mahoma (pág. 276)

calving the process in which a section of ice breaks off the edge of a glacier (p. 476)

ablación proceso en el que un bloque de hielo se desprende del borde de un glaciar (pág. 476)

*capable having the ability to cause or accomplish an action or an event (p. 454)

*capaz que tiene habilidad para provocar o llevar a cabo una acción o un suceso (pág. 454)

cash crops a farm product grown for sale (p. 462)

cultivo comercial producto agrícola que se cultiva para la venta (pág. 462)

cassava a tuberous plant that has edible roots (p. 336)

yuca planta tuberosa de raíces comestibles (pág. 336)

caste the social class a person is born into and cannot change (p. 183)

casta clase social en la que nace una persona y no puede cambiar (pág. 183)

*channel a course for a river to flow through (p. 266)

*canal curso artificial por donde fluye un río (pág. 266)

*characteristic a quality or an aspect (p. 343)

*característica cualidad o aspecto (pág. 343)

civil disobedience the use of nonviolent protests to challenge a government or its laws (pp. 187; 393)

desobediencia civil rebatir un gobierno o sus leyes mediante protestas no violentas (págs. 187; 393)

civil war a fight between opposing groups for control of a country's government (p. 279)

guerra civil lucha entre grupos opositores por el control del gobierno de un país (pág. 279)

clan a large group of people who have a common ancestor in the far past (p. 311)

climate the average weather in an area over a long period of time (pp. 23; 48)

*****collapse** a sudden failure, breakdown, or ruin (pp. 235; 464)

collective a farm that is owned by the government but run by a group of farmers who work together (p. 208)

colonialism a policy based on control of one country by another (p. 337)

command economy an economy in which the means of production are publicly owned (p. 96)

*****commodity** a material, resource, or product that is bought and sold (p. 152)

communism a system of government in which the government controls the ways of producing goods (p. 129)

compass rose the feature on a map that shows direction (p. 28)

*****complex** highly developed (p. 210)

*****component** a part of something (p. 23)

condensation the result of water vapor changing to a liquid or a solid state (p. 64)

*****consist** to be made up of (p. 294–95)

constitution a document setting forth the structure and powers of a government and the rights of people in a country (p. 286)

constitutional monarchy a form of government in which a monarch is the head of state but elected officials run the government (p. 163)

*****contact** communication or interaction with someone (p. 396–97)

continent a large, unbroken mass of land (p. 52)

continental island an island formed centuries ago by the rising and folding of the ocean floor due to tectonic activity (p. 451)

continental shelf the part of a continent that extends into the ocean in a plateau, then drops sharply to the ocean floor (p. 60)

*****controversy** a dispute; a discussion involving opposing views (p. 438)

*****convert** to change from one thing to another (pp. 27; 276)

clan agrupación extensa de personas que tienen un ancestro común en el pasado remoto (pág. 311)

clima tiempo atmosférico promedio en una zona durante un periodo largo (págs. 23; 48)

*****colapso** bancarrota, caída o ruina súbita (págs. 235; 464)

colectiva granja de propiedad del gobierno que administra una cooperativa de granjeros (pág. 208)

colonialism política que se basa en el control o dominio de un país sobre otro (pág. 337)

economía planificada sistema económico en el que los medios de producción son de propiedad pública (pág. 96)

*****mercancía** materia prima, bien o producto que se compra y se vende (pág. 152)

comunismo forma de gobierno en la que el gobierno controla los modos de producción de los bienes (pág. 129)

rosa de los vientos convención de un mapa que señala la dirección (pág. 28)

*****complejo** muy desarrollado (pág. 210)

*****componente** parte de algo (pág. 23)

condensación cambio del vapor de agua a un estado líquido o sólido (pág. 64)

*****consistir** estar hecho de (pág. 294–95)

constitución documento que establece la estructura y los poderes de un gobierno así como los derechos de las personas en un país (pág. 286)

monarquía constitucional sistema de gobierno en el que un monarca ostenta la jefatura del Estado pero el gobierno lo administran funcionarios elegidos (pág. 163)

*****contacto** comunicación o interacción (pág. 396–97)

continente extensión de tierra grande e ininterrumpida (pág. 52)

isla continental isla formada siglos atrás por el levantamiento y plegamiento del fondo oceánico resultantes de la actividad tectónica (pág. 451)

plataforma continental parte de un continente que se adentra en el océano en forma de meseta y luego desciende abruptamente hasta el fondo oceánico (pág. 60)

*****controversia** disputa; discusión que involucra puntos de vista opuestos (pág. 438)

*****convertir** cambiar de una cosa a otra (págs. 27; 276)

coral reef · **desalinization**

coral reef a long, undersea structure formed by the tiny skeletons of coral, a kind of sea life (p. 417)

arrecife coralino extensa estructura submarina formada por los diminutos esqueletos de los corales, una especie de vida marina (pág. 417)

cottage industry a home- or village-based industry in which people make simple goods using their own equipment (p. 192)

industria artesanal industria doméstica o aldeana en la cual las personas elaboran bienes sencillos utilizando sus propios equipos (pág. 192)

coup an action in which a group of individuals seize control of a government (p. 339)

golpe (de Estado) acción mediante la cual un grupo de individuos se apodera del control de un gobierno (pág. 339)

couscous a small, round grain used in North African and Southwest Asian cooking (p. 282)

cuscús cereal pequeño y redondo que se utiliza en la cocina de África del Norte y el Sudeste Asiático (pág. 282)

Creole two or more languages that blend and become the language of the region (p. 369)

criollo dos o más lenguas que se mezclan y convierten en la lengua de una región (pág. 369)

cultural region a geographic area in which people have certain traits in common (p. 86)

región cultural área geográfica donde las personas tienen ciertos rasgos comunes (pág. 86)

culture the set of beliefs, behaviors, and traits shared by a group of people (p. 82)

cultura conjunto de creencias, comportamientos y rasgos compartidos por un grupo de personas (pág. 82)

***currency** the paper money and coins in circulation (p. 101)

***moneda** dinero en billetes y monedas en circulación (pág. 101)

cyclone a storm with high winds and heavy rains (p. 178)

ciclón tormenta con vientos huracanados y lluvias torrenciales (pág. 178)

D

dalit the lowest caste of Indian society; also called the "untouchables" (p. 193)

paria casta más baja de la sociedad india; también se le denomina "los intocables" (pág. 193)

de facto actually; in reality (p. 116)

de facto de hecho; en la realidad (pág. 116)

death rate the number of deaths compared to the total number of people in a population at a given time (p. 72)

tasa de mortalidad número de defunciones comparado con el número total de habitantes de una población en un tiempo determinado (pág. 72)

deciduous describing trees that shed their leaves in the autumn (p. 205)

caducifolios árboles que pierden sus hojas en el otoño (pág. 205)

***define** to describe the nature or extent of something (p. 201)

***definir** describir la naturaleza o el alcance de algo (pág. 201)

delta an area where sand, silt, clay, or gravel is dropped at the mouth of a river (pp. 62; 177; 266)

delta área donde se deposita arena, sedimento, lodo o gravilla en la desembocadura de un río (págs. 62; 177; 266)

democracy a type of government run by the people (p. 86)

democracia tipo de gobierno dirigido por el pueblo (pág. 86)

***demonstrate** to show (p. 274)

***demostrar** probar (pág. 274)

***depict** to describe or show (p. 344)

***representar** describer o mostrar (pág. 344)

desalinization a process that makes salt water safe to drink (p. 61)

desalinización proceso que elimina la sal del agua para hacerla potable (pág. 61)

desertification the process by which an area turns into a desert (p. 299)

dialect a regional variety of a language with unique features, such as vocabulary, grammar, or pronunciation (p. 83)

dictatorship a form of government in which one person has absolute power to rule and control the government, the people, and the economy (p. 87)

didgeridoo a large, bamboo musical instrument of the Australian aboriginal people (p. 435)

dingoes wild dogs of Australia (p. 425)

*****displace** to take over a place or position of others (p. 361)

*****distinct** separate; easily recognized as separate or different (p. 459)

*****distort** to change something so it is no longer accurate (p. 27)

*****diverse** composed of many distinct and different parts (pp. 311; 368–69)

diversified increased variety to achieve a balance (p. 284)

*****dominate** to have the greatest importance (p. 117)

dominion a largely self-governing country within the British Empire (p. 431)

doubling time the number of years it takes a population to double in size based on its current growth rate (p. 73)

drought long period of time without rainfall (p. 421)

*****dynamic** always changing (p. 20)

dynasty a line of rulers from a single family that holds power for a long time (p. 124)

desertización proceso por el cual un área se transforma en un desierto (pág. 299)

dialecto variedad regional de una lengua con características únicas, como vocabulario, gramática o pronunciación (pág. 83)

dictadura forma de gobierno en la que una persona detenta el poder absoluto para mandar y controlar al gobierno, el pueblo y la economía (pág. 87)

diyiridú instrumento musical de bambú, de gran tamaño, de los aborígenes australianos (pág. 435)

dingos perros salvajes de Australia (pág. 425)

*****desplazar** tomar el lugar o la posición de otros (pág. 361)

*****distinto** separado; fácilmente reconocible como separado o diferente (pág. 459)

*****distorsionar** cambiar algo de modo que ya no es correcto (pág. 27)

*****diverso** compuesto de muchas partes distintivas y diferentes (págs. 311; 368–69)

diversificado variedad incrementada para lograr un equilibrio (pág. 284)

*****dominar** tener la mayor importancia (pág. 117)

dominio país autónomo dentro del Imperio británico (pág. 431)

tiempo de duplicación número de años que le toma a una población doblar su tamaño con base en la tasa de crecimiento actual (pág. 73)

sequía periodo largo sin lluvias (pág. 421)

*****dinámico** en permanente cambio (pág. 20)

dinastía serie de gobernantes de una sola familia que detentan el poder por mucho tiempo (pág. 124)

E

earthquake an event in which the ground shakes or trembles, brought about by the collision of tectonic plates (p. 54)

economic system how a society decides on the ownership and distribution of its economic resources (p. 96)

ecotourism a type of tourism in which people visit a country to enjoy its natural wonders (p. 168)

*****element** an important part or characteristic (p. 361)

terremoto suceso en el cual el suelo se agita o tiembla como consecuencia de la colisión de placas tectónicas (pág. 54)

sistema económico la forma en que una sociedad decide la propiedad y distribución de sus recursos económicos (pág. 96)

ecoturismo tipo de turismo en el cual las personas visitan un país para disfrutar de sus maravillas naturales (pág. 168)

*****elemento** parte o característica importante (pág. 361)

English	Spanish
elevation the measurement of how much above or below sea level a place is (p. 29)	**elevación** medida de cuánto más alto o más bajo está un lugar respecto del nivel del mar (pág. 29)
embargo a ban on trade with a particular country (p. 393)	**embargo** prohibición de comerciar con un país específico (pág. 393)
emigrate to leave one's home to live in another place (p. 78)	**emigrar** abandonar el hogar propio para vivir en otro lugar (pág. 78)
*****emphasis** an expression that shows the importance of something (p. 283)	*****énfasis** expresión que muestra la importancia de algo (pág. 283)
endemic specific to a particular place or people (p. 155)	**endémico** específico de un lugar o una persona en particular (pág. 155)
environment the natural surroundings of a place (p. 23)	**medioambiente** entorno natural de un lugar (pág. 23)
Equator a line of latitude that runs around the middle of Earth (p. 21)	**ecuador** línea de latitud que atraviesa la mitad de la Tierra (pág. 21)
equinox one of two days each year when the sun is directly overhead at the Equator (p. 46)	**equinoccio** uno de dos días al año cuando el sol se halla situado directamente sobre el ecuador (pág. 46)
erg a large area of sand (p. 268)	**erg** zona extensa de arena (pág. 268)
erosion the process by which weathered bits of rock are moved elsewhere by water, wind, or ice (p. 55)	**erosión** proceso por el cual fragmentos desgastados de rocas son llevados a otra parte por acción del agua, el viento o el hielo (pág. 55)
escarpment a steep cliff at the edge of a plateau with a lowland area below (p. 381)	**escarpado** acantilado pendiente, al borde de una meseta, que tiene debajo un área de tierras bajas (pág. 381)
*****establish** to start (p. 190)	*****establecer** comenzar (pág. 190)
estuary an area where river currents and the ocean tide meet (p. 329)	**estuario** área donde convergen corrientes fluviales y la marea oceánica (pág. 329)
ethnic group a group of people with a common racial, national, tribal, religious, or cultural background (p. 83)	**grupo étnico** grupo de personas con un antecedente racial, nacional, tribal, religioso o cultural común (pág. 83)
eucalyptus a tree found only in Australia and nearby islands that is well suited to dry conditions with leathery leaves, deep roots, and the ability to survive when rivers flood (p. 423)	**eucalipto** árbol de hojas carnosas y raíces profundas que solo se encuentra en Australia e islas adyacentes. Se adapta bien a las condiciones de sequía y puede sobrevivir a las inundaciones fluviales (pág. 423)
evaporation the change of liquid water to water vapor (p. 63)	**evaporación** cambio del agua en estado líquido a vapor (pág. 63)
*****expand** to spread out; to grow larger (p. 234)	*****expandir** extender; agrandar (pág. 234)
*****exploit** to use a person, resource, or situation unfairly and selfishly (pp. 169; 390)	*****explotar** utilizar a una persona, un recurso o una situación de manera injusta y egoísta (págs. 169; 390)
export to send a product produced in one country to another country (p. 99)	**exportar** enviar un bien producido en un país a otro país (pág. 99)
extended family a unit of related people made up of several generations, including grandparents, parents, and children (p. 370)	**familia extensa** unidad de personas emparentadas conformada por varias generaciones, incluidos abuelos, padres e hijos (pág. 370)

F

***factor** a cause (p. 285)

fale a traditional Samoan home that has no walls, leaving the inside open to cooling ocean breezes (p. 461)

fault a place where two tectonic plates grind against each other (p. 54)

fauna the animal life in a particular environment (p. 155)

fellaheen the peasant farmers of Egypt who rent small plots of land (p. 281)

flora the plant life in a particular environment (p. 155)

fossil water water that fell as rain thousands of years ago and is now trapped deep below ground (p. 247)

foundation the basis of something (p. 335)

free trade arrangement whereby a group of countries decides to set little or no tariffs on quotas (p. 100)

fundamentalist a person who believes in the strict interpretation of religious laws (p. 279)

*factor causa (pág. 285)

fale casa tradicional de Samoa que carece de paredes, de manera que el interior queda abierto a las frescas brisas del océano (pág. 461)

falla lugar donde dos placas tectónicas chocan entre sí (pág. 54)

fauna vida animal en un medioambiente específico (pág. 155)

fellaheen campesinos de Egipto que arriendan pequeñas parcelas (pág. 281)

flora vida vegetal en un medioambiente específico (pág. 155)

agua fósil agua que cayó en forma de lluvia hace miles de años y ahora se encuentra atrapada en las profundidades del subsuelo (pág. 247)

fundamento la base de algo (pág. 335)

libre comercio acuerdo por el cual un grupo de países decide imponer aranceles bajos a las cuotas o no fija ningún arancel (pág. 100)

fundamentalista persona que cree en la interpretación estricta de las leyes religiosas (pág. 279)

G

genocide the mass murder of people from a particular ethnic group (p. 308)

geography the study of Earth and its peoples, places, and environments (p. 18)

geothermal energy the electricity produced by natural, underground sources of steam (pp. 301; 436)

geyser a spring of water heated by molten rock inside Earth that, from time to time, shoots hot water into the air (p. 419)

glacier a large body of ice that moves slowly across land (p. 56)

globalization the process by which nations, cultures, and economies become mixed (p. 89)

*grant to allow as a right, privilege, or favor (p. 391)

genocidio asesinato masivo de personas de un grupo étnico específico (pág. 308)

geografía estudio de la Tierra y de sus gentes, lugares y entornos (pág. 18)

energía geotérmica electricidad producida por fuentes naturales de vapor subterráneas (págs. 301; 436)

géiser fuente de agua calentada por rocas fundidas en el interior de la Tierra que, de vez en cuando, expulsa agua caliente al aire (pág. 419)

glaciar masa de hielo enorme que se mueve lentamente sobre la tierra (pág. 56)

globalización proceso mediante el cual naciones, culturas y economías se integran (pág. 89)

*conceder permitir como un derecho, privilegio o favor (pág. 391)

green revolution the effort to use modern techniques and science to increase food production in poorer countries (p. 192)

gross domestic product (GDP) the total dollar value of all final goods and services produced in a country during a single year (p. 98)

groundwater the water contained inside Earth's crust (p. 61)

revolución verde esfuerzo por utilizar técnicas modernas y la ciencia para aumentar la producción de alimentos en los países más pobres (pág. 192)

producto interno bruto (PIB) valor total en dólares de todos los bienes y servicios finales producidos en un país durante un año (pág. 98)

agua subterránea agua contenida en el interior de la corteza terrestre (pág. 61)

H

harmattan a wind off the Atlantic coast of Africa that blows from the northeast to the south, carrying large amounts of dust (p. 357)

hemisphere each half of Earth (p. 26)

hieroglyphics the system of writing that uses small pictures to represent sounds or words (p. 274)

high island a type of island in the Pacific Ocean formed many centuries ago by volcanoes and still having mountainous areas (p. 452)

homogeneous made up of many things that are the same (p. 216)

hot springs places where naturally heated water rises out of the ground (p. 419)

human rights the rights belonging to all individuals (p. 87)

hydroelectric power the electricity that is created by flowing water (p. 300)

hydropolitics the politics surrounding water access and usage rights (p. 247)

harmattan viento de la costa atlántica de África que sopla del nordeste al sur, arrastrando consigo grandes cantidades de polvo (pág. 357)

hemisferio cada mitad de la Tierra (pág. 26)

jeroglífico sistema de escritura que representa sonidos o palabras con dibujos pequeños (pág. 274)

isla alta tipo de isla del océano Pacífico formada hace muchos siglos por volcanes y que aún tiene zonas montañosas (pág. 452)

homogéneo compuesto por muchas cosas iguales (pág. 216)

fuentes termales lugares donde agua calentada por medios naturales brota del suelo (pág. 419)

derechos humanos los derechos que tienen todos los individuos (pág. 87)

energía hidroeléctrica electricidad producida por agua en movimiento (pág. 300)

hidropolítica política relativa al acceso al agua y los derechos de su uso (pág. 247)

I

ice sheet a large, thick area of ice that covers a region (p. 475)

ice shelf a thick layer of ice that extends above the water (p. 476)

iceberg a huge piece of floating ice that broke off from an ice shelf or glacier and fell into the sea (p. 476)

immigrate to enter and live in a new country (p. 78)

manto de hielo área extensa y gruesa de hielo que cubre una región (pág. 475)

plataforma de hielo capa gruesa de hielo que se extiende sobre la superficie del agua (pág. 476)

iceberg témpano gigante de hielo flotante que se desprendió de una plataforma de hielo o glaciar y cayó al mar (pág. 476)

inmigrar entrar a un nuevo país y vivir allí (pág. 78)

impact an effect or influence (p. 305)

imperialism a policy by which a country increases its power by gaining control over other areas of the world (pp. 305; 365)

import when a country brings in a product from another country (p. 99)

infrastructure a system of roads and railroads that allow the transport of materials (p. 372)

***inhibit** to limit (p. 207)

insular separate from other countries (p. 150)

***intense** strong (p. 54)

***intertwine** to become closely connected or involved (p. 127)

introduced species a nonnative species that is brought to a new environment (p. 429)

irrigate to supply land with water through ditches or pipes (p. 209)

isthmus a narrow strip of land that connects two larger land areas (p. 59)

impacto efecto o influencia (pág. 305)

imperialismo política mediante la cual un país aumenta su poder ejerciendo control sobre otras áreas del mundo (págs. 305; 365)

importación cuando un país ingresa un producto de otro país (pág. 99)

infraestructura sistema de carreteras y ferrocarriles que permite el transporte de materiales (pág. 372)

***inhibir** limitar (pág. 207)

insular separado de otros países (pág. 150)

***intenso** poderoso (pág. 54)

***entretejer** unirse o envolverse estrechamente (pág. 127)

especie introducida especie foránea que se lleva a un medioambiente nuevo (pág. 429)

irrigar suministrar agua a un terreno por medio de zanjas o ductos (pág. 209)

istmo franja estrecha de tierra que conecta dos áreas de tierra más grandes (pág. 59)

K

kapahaka a traditional art form of the Maori people that combines music, dance, singing, and facial expressions (p. 426)

katabatic winds the strong, fast, cold winds that blow down from the interior of Antarctica (p. 477)

kente the colorful, handwoven cloth produced in Ghana (p. 371)

key the feature on a map that explains the symbols, colors, and lines used on the map (p. 28)

kiwifruit a small, fuzzy, brownish-colored fruit with bright green flesh (p. 437)

krill the tiny, shrimplike sea creatures that are eaten by whales and many other sea creatures (p. 479)

kapahaka manifestación artística tradicional del pueblo maorí que combina música, danza, canto y expresiones faciales (pág. 426)

vientos catabáticos vientos fuertes, rápidos y fríos que soplan desde el interior de la Antártida (pág. 477)

kente tela tejida de vivos colores que se fabrica en Ghana (pág. 371)

clave elemento de un mapa que explica los símbolos, colores y líneas usados en este (pág. 28)

kiwi fruta pequeña y vellosa de tono marrón cuya carne es verde brillante (pág. 437)

krill crustáceo diminuto, similar al camarón, que comen las ballenas y otras criaturas marinas (pág. 479)

L

lagoon a shallow pond near a larger body of water (p. 453)

laguna pozo poco profundo cercano a una masa de agua mayor (pág. 453)

landform a natural feature found on land (p. 23)

accidente geográfico formación natural que se encuentra sobre la tierra (pág. 23)

landlocked having no border with an ocean or a sea (p. 382)

sin salida al mar que no limita con un océano o un mar (pág. 382)

landscape the portions of Earth's surface that can be viewed at one time from a location (p. 19)

paisaje partes de la superficie terrestre que se pueden observar a un mismo tiempo desde una ubicación (pág. 19)

latitude the lines on a map that run east to west (p. 21)

latitud líneas sobre un mapa que van de este a oeste (pág. 21)

lawsuit a legal action in which people ask for relief from some damage done to them by someone else (p. 438)

demanda acción legal en la que las personas exigen una indemnización por algún daño que les causó un tercero (pág. 438)

lichen tiny, sturdy plants that grow in rocky areas (p. 478)

líquenes plantas pequeñas y resistentes que crecen en las zonas rocosas (pág. 478)

loess a fine-grained, fertile soil deposited by the wind (p. 119)

loes suelo fértil y de granos finos depositado por el viento (pág. 119)

longitude the lines on a map that run north to south (p. 21)

longitud líneas sobre un mapa que van de norte a sur (pág. 21)

low island a type of island in the Pacific Ocean formed by the buildup of coral (p. 452)

isla baja tipo de isla del océano Pacífico formada por acumulaciones de coral (pág. 452)

M

map projection one of several systems used to represent the round Earth on a flat map (p. 28)

proyección cartográfica uno de los varios sistemas que se usan para representar la esfera terrestre en un mapa plano (pág. 28)

*margin an edge (p. 265)

*margen borde (pág. 265)

market economy an economy in which most of the means of production are privately owned (p. 96)

economía de mercado economía en la cual la mayoría de los medios de producción son de propiedad privada (pág. 96)

marsupial a type of mammal that carries its young in a pouch (p. 423)

marsupial tipo de mamífero que carga a su cría en una bolsa (pág. 423)

*mature fully grown and developed as an adult; also refers to older adults (p. 73)

*maduro adulto plenamente crecido y desarrollado; también se refiere a los adultos mayores (pág. 73)

megalopolis a huge city or cluster of cities with an extremely large population (pp. 80; 133)

megalópolis ciudad enorme o cúmulo de ciudades que tienen una población extremadamente grande (págs. 80; 133)

millennium a period of a thousand years (p. 233)

milenio period de mil años (pág. 233)

millet a grass that produces edible seeds (p. 335)

mijo especie de pasto que produce semillas comestibles (pág. 335)

minority a group of people that is different from most of the population (p. 166)

minoría grupo de personas diferente a la mayoría de la población (pág. 166)

MIRAB economy a lesser-developed economy that depends on aid from foreign countries and remittances from former residents working elsewhere (p. 464)

economía MIRAB economía menos desarrollada que depende de la ayuda de países extranjeros y de las remesas de antiguos residentes que trabajan en otro lugar (pág. 464)

missionary someone who tries to convert others to a certain religion (p. 338)

mixed economy an economy in which parts of the economy are privately owned and parts are owned by the government (p. 96)

monarchy the system of government in which a country is ruled by a king or queen (p. 87)

monolith a single standing stone (p. 417)

monotheism the belief in one god (pp. 233; 276)

monsoon a seasonal wind that blows steadily from the same direction for several months at a time but changes directions at other times of the year (p. 178)

myrrh a sweet perfume used as medicine in ancient times (p. 273)

misionario persona que trata de convertir a otras a una religión específica (pág. 338)

economía mixta economía en la cual unos sectores son de propiedad privada y otros son de propiedad del gobierno (pág. 96)

monarquía sistema de gobierno en el que un rey o una reina gobiernan un país (pág. 87)

monolito piedra erguida de una sola pieza (pág. 417)

monoteísmo creencia en un solo dios (págs. 233; 276)

monzón viento estacional que sopla regularmente desde la misma dirección durante varios meses pero cambia de dirección en otras épocas del año (pág. 178)

mirra perfume dulce que antiguamente se empleaba como medicamento (pág. 273)

N

***network** a complex, interconnected chain or system of things such as roads, canals, or computers (p. 383)

nomad a person who lives by moving from place to place to follow and hunt herds of migrating animals or to lead herds of grazing animals to fresh pasture (p. 269)

nonrenewable resources the resources that cannot be totally replaced (p. 95)

nuclear family the family group that includes only parents and their children (p. 371)

nuclear proliferation the spread of control of nuclear power, particularly the knowledge of how to construct nuclear weapons (p. 187)

***red** cadena o sistema complejo e interconectado de carreteras, canales o computadoras, entre otros (pág. 383)

nómada persona que vive trasladándose de un lugar a otro para seguir y cazar manadas de animales migratorios o para conducir rebaños de animales de pastoreo hacia pastos frescos (pág. 269)

recursos no renovables recursos que no se pueden reponer por completo (pág. 95)

familia nuclear grupo familiar que solo incluye a padres e hijos (pág. 371)

proliferación nuclear expansión del dominio de la energía nuclear, en particular el conocimiento para construir armas nucleares (pág. 187)

O

oasis a fertile area that rises in a desert wherever water is regularly available (p. 213)

oral tradition the process of passing stories by word of mouth from generation to generation (p. 314)

orbit to circle around something (p. 42)

Outback the inland areas of Australia west of the Great Dividing Range (p. 417)

oasis área fértil que se desarrolla en un desierto cuando hay una fuente regular de agua (pág. 213)

tradición oral forma de transmitir historias de generación en generación, mediante la palabra hablada (pág. 314)

orbitar moverse en círculo alrededor de algo (pág. 42)

Outback zonas del interior de Australia ubicadas al oeste de la Gran Cordillera Divisoria (pág. 417)

outsourcing hiring workers in other countries to do a set of jobs (p. 192)

*overall as a whole; generally (p. 416–17)

ozone the certain kind of oxygen that forms a layer around Earth in the atmosphere; it blocks out many of the most harmful rays from the sun (p. 483)

subcontratar contratar trabajadores en otros países para que hagan una serie de trabajos (pág. 192)

*global como un todo; generalizado (pág. 416–17)

ozono tipo de oxígeno que forma una capa alrededor de la Tierra en la atmósfera; bloquea el paso de los rayos más dañinos del sol (pág. 483)

P

Pacific Rim the countries bordering the Pacific Ocean, particularly Asian countries (p. 167)

palm oil an oil that is used in cooking (p. 335)

pastoral describing a society based on herding animals (p. 206)

periodic market an open-air trading market that spring ups at a crossroads or in larger towns (p. 399)

permafrost the permanently frozen, lower layers of soil found in the tundra and subarctic climate zones (p. 202)

pharaoh the name for a powerful ruler in ancient Egypt (p. 272)

phosphate a chemical salt used to make fertilizer (p. 271)

pidgin a language formed by combining parts of several different languages (pp. 369, 460)

plain a large expanse of land that can be flat or have a gentle roll (p. 58)

plankton plants or animals that ride along with water currents (p. 479)

plantation a large farm (p. 161)

plateau a flat area that rises above the surrounding land (p. 58)

poaching illegal fishing or hunting (pp. 317, 387)

*policy a plan or course of action (p. 187)

polytheism the belief in more than one god (p. 233)

population density the average number of people living within a square mile or a square kilometer (pp. 76; 310)

population distribution the geographic pattern of where people live (p. 76)

Cuenca del Pacífico países que bordean el océano Pacífico, en particular los países asiáticos (pág. 167)

aceite de palma un aceite que se usa para cocinar (pág. 335)

pastoril sociedad que vive del pastoreo de animales (pág. 206)

mercado ambulante mercado al aire libre que se instala en una aldea o en pueblos más grandes (pág. 399)

permacongelamiento capas bajas del suelo, permanentemente congeladas, que se encuentran en la tundra y las zonas de clima subártico (pág. 202)

faraón nombre dado a un poderoso gobernante en el Antiguo Egipto (pág. 272)

fosfato sal química utilizada para producir fertilizantes (pág. 271)

pidgin lengua formada por la combinación de partes de varias lenguas distintas (págs. 369, 460)

llanura gran extensión de tierra plana o con ligeras ondulaciones (pág. 58)

plancton plantas o animales que se desplazan con las corrientes de agua (pág. 479)

plantación granja grande (pág. 161)

meseta área plana que se eleva por encima del terreno circundante (pág. 58)

caza furtiva pesca o caza ilegal (págs. 317, 387)

*política plan o curso de acción (pág. 187)

politeísmo creencia en más de un dios (pág. 233)

densidad de población número promedio de personas que habitan en una milla cuadrada o un kilómetro cuadrado (págs. 76; 310)

distribución de la población patrón geográfico que muestra dónde habita la gente (pág. 76)

possession an area or a region that is controlled by another country (p. 458)

***potential** the possibility (p. 333)

precipitation the water that falls on the ground as rain, snow, sleet, hail, or mist (p. 48)

primate city a country's main city that is so large and influential that it dominates the rest of the country (p. 165)

Prime Meridian the starting point for measuring longitude (p. 21)

productivity the measurement of what is produced and what is required to produce it (p. 98)

***project** a planned activity (pp. 272–73)

posesión zona o región controlada por otro país (pág. 458)

***potencial** posibilidad (pág. 333)

precipitación agua que cae al suelo en forma de lluvia, nieve, aguanieve, granizo o rocío (pág. 48)

ciudad principal la ciudad más importante de un país, tan grande e influyente que domina el resto del país (pág. 165)

primer meridiano punto de partida para medir la longitud (pág. 21)

productividad medición de lo que se produce y lo que se requiere para producirlo (pág. 98)

***proyecto** actividad planificada (págs. 272–73)

R

rain shadow an area that receives reduced rainfall because it is on the side of a mountain facing away from the ocean (p. 49)

Raj the period of time in which Great Britain controlled India as a part of the British Empire (p. 186)

refugee a person who flees a country because of violence, war, persecution, or disaster (pp. 78; 309; 341)

regime a government (p. 279)

region a group of places that are close to one another and that share some characteristics (p. 22)

reincarnation the belief in Hinduism that after a person dies, his or her soul is reborn into another body (p. 184)

relative location the location of one place compared to another place (p. 20)

relief the difference between the elevation of one feature and the elevation of another feature near it (p. 29)

remittance the money sent back to the homeland by people who have gone somewhere else to work (p. 464)

remote sensing the method of getting information from far away, such as deep below the ground (pp. 32, 484)

renewable resources a resource that can be totally replaced or is always available naturally (p. 95)

sombra pluviométrica zona que recibe pocas precipitaciones porque se halla en la ladera de una montaña que está en el lado contrario al océano (pág. 49)

Raj periodo durante el cual Gran Bretaña controló India como parte del Imperio británico (pág. 186)

refugiado persona que huye de un país por la violencia, una guerra, una persecución o un desastre (págs. 78; 309; 341)

régimen gobierno (pág. 279)

región agrupación de lugares cercanos que comparten algunas características (pág. 22)

reencarnación creencia del hinduismo según la cual el espíritu de una persona muerta renace en otro cuerpo (pág. 184)

localización relativa la ubicación de un lugar comparada con la de otro (pág. 20)

relieve diferencia entre la elevación de una formación y la de otra formación cercana (pág. 29)

remesa dinero enviado al país de origen por personas que se han ido a trabajar a otro lugar (pág. 464)

detección remota método para obtener información muy lejana, como de las profundidades del subsuelo (págs. 32; 484)

recursos renovables recursos que pueden reponerse totalmente o siempre se encuentran disponibles en la naturaleza (pág. 95)

representative democracy a form of democracy in which citizens elect government leaders to represent the people (p. 87)

democracia representativa forma de democracia en la que los ciudadanos eligen líderes de gobierno para que representen al pueblo (pág. 87)

research station a base for scientific research and observation, often in a remote location (p. 482)

estación de investigación base para la investigación y observación científicas, por lo general ubicada en un sitio remoto (pág. 482)

reservoir an artificial lake created by a dam (p. 283)

embalse lago artificial creado por una presa (pág. 283)

resort a vacation place where people go to relax (p. 464)

centro vacacional lugar de vacaciones donde las personas van a descansar (pág. 464)

resource a material that can be used to produce crops or other products (p. 23)

recurso materia prima que se puede utilizar para obtener cultivos u otros productos (pág. 23)

*****revenue** the income generated by a business (p. 364)

*****renta** ingresos generados por un negocio (pág. 364)

revolution a complete trip of Earth around the sun (p. 42)

revolución recorrido completo de la Tierra alrededor del Sol (pág. 42)

rift to separate two pieces from one another (p. 294)

escindir separar dos partes entre sí (pág. 294)

Ring of Fire a long, narrow band of volcanoes surrounding the Pacific Ocean (p. 54)

Cinturón de Fuego banda larga y estrecha de volcanes que rodean el océano Pacífico (pág. 54)

rural describes an area that is lightly populated (p. 77)

rural área poco poblada (pág. 77)

S

samurai a powerful, land-owning warrior in Japan (p. 128)

samurái poderoso guerrero japonés dueño de tierras (pág. 128)

scale the relationship between distances on the map and on Earth (p. 29)

escala relación entre distancias en un mapa y en la Tierra (pág. 29)

scale bar the feature on a map that tells how a measured space on the map relates to the actual distance on Earth (p. 28)

escala numérica elemento cartográfico que muestra la relación entre un espacio medido sobre el mapa y la distancia real sobre la Tierra (pág. 28)

secede to withdraw from a group or a country (p. 367)

separarse retirarse de un grupo o un país (pág. 367)

semiarid having lower temperatures and cooler nights than hot, dry deserts (p. 230)

semiárido que tiene temperaturas más bajas y noches más frías que los desiertos cálidos y secos (pág. 230)

shogun a military leader who ruled Japan in early times (p. 128)

sogún líder militar que gobernaba Japón antiguamente (pág. 128)

*****significant** important (p. 214)

*****significativo** importante (pág. 214)

silt small particles of rich soil (p. 266)

limo pequeñas partículas de suelo fértil (pág. 266)

sitar a long-necked instrument with 7 strings on the outside and 10 inside the neck that provides Indian music with a distinctive sound (p. 191)

cítara instrumento de cuello largo con 7 cuerdas en la parte exterior y 10 dentro del cuello, que da a la música india un sonido distintivo (pág. 191)

slash-and-burn agriculture a method of farming that involves cutting down trees and underbrush and burning the area to create a field for crops (p. 331)

agricultura de tala y quema método agrícola que consiste en talar árboles y rastrojos y quemar el área despejada para crear un campo de cultivo (pág. 331)

solstice one of two days of the year when the sun reaches its northernmost or southernmost point (p. 45)

solsticio uno de dos días al año cuando el sol alcanza su máxima declinación norte o sur (pág. 45)

souk a large, open-air market in North African and Southwest Asian countries (p. 281)

zoco mercado grande al aire libre propio de África del Norte y los países del Sudoeste Asiático (pág. 281)

spatial Earth's features in terms of their places, shapes, and relationships to one another (p. 18)

espaciales características de la Tierra en cuanto a sus lugares, formas y relaciones entre sí (pág. 18)

***sphere** a round shape like a ball (p. 26–27)

***esfera** figura redonda como una pelota (pág. 26–27)

sphere of influence an area of a country where a single foreign power has been granted exclusive trading rights (p. 128)

esfera de influencia área de un país donde se le ha concedido a una sola potencia extranjera derechos comerciales exclusivos (pág. 128)

standard of living the level at which a person, group, or nation lives as measured by the extent to which it meets its needs (p. 98)

estándar de vida nivel en que vive una persona, grupo o nación, medido según la capacidad de satisfacer sus necesidades (pág. 98)

station a cattle or sheep ranch in rural Australia (p. 429)

estación rancho de ganado vacuno o lanar del área rural de Australia (pág.429)

steppe a partly dry grassland often found on the edge of a desert (p. 202)

estepa pradera parcialmente seca que se encuentra con frecuencia al borde de un desierto (pág. 202)

***structure** an arrangement of parts (p. 135)

***estructura** organización de las partes (pág. 135)

subcontinent a large landmass that is part of a continent (p. 176)

subcontinente gran masa de tierra que forma parte de un continente (pág. 176)

subsistence farming a type of farming in which the farmer produces only enough to feed his or her family (pp. 168; 313)

agricultura de subsistencia tipo de agricultura en el que los granjeros producen apenas lo suficiente para alimentar a su familia (págs. 168; 313)

sultan the ruler of a Muslim country (p. 159)

sultán gobernante de un país musulmán (pág. 159)

sustainability the economic principle by which a country works to create conditions where all the natural resources for meeting the needs of society are available (p. 101)

sostenibilidad principio económico según el cual un país crea condiciones para que estén disponibles todos los recursos naturales que satisfacen las necesidades de la sociedad (pág. 101)

T

taiga a large coniferous forest (p. 202)

taiga gran bosque de coníferas (pág. 202)

technology any way that scientific discoveries are applied to practical use (p. 30)

tecnología cualquier forma en que los descubrimientos científicos se aplican para un uso práctico (pág. 30)

tectonic plate one of the 16 pieces of Earth's crust (p. 53)

placa tectónica uno de las 16 partes de la corteza terrestre (pág. 53)

thatch a bundle of twigs, grass, and bark (p. 399)

thematic map a map that shows specialized information (p. 30)

tikanga Maori customs and traditions passed down through generations (p. 426)

trade deficit a situation that occurs when the value of a country's imports is higher than the value of its exports (p. 138)

trade language a common language that emerges when countries trade with each other (p. 342)

trade surplus a situation that occurs when the value of a country's exports is higher than the value of its imports (p. 138)

traditional economy an economy where resources are distributed mainly through families (p. 96)

*****transform** to change something completely (p. 64)

treaty an official agreement, negotiated and signed by each party (p. 481)

trench a long, narrow, steep-sided cut on the ocean floor (p. 60)

*****trend** a general tendency or preference (p. 399)

tribute money paid by one country to another in surrender or for protection (p. 305)

trust territory an area temporarily placed under control of another country (p. 458)

tsunami a giant ocean wave caused by volcanic eruptions or movement of the earth under the ocean floor (pp. 54; 118)

tundra a flat, treeless plain with permanently frozen ground (p. 202)

techo de paja armazón de ramas, pasto y corteza (pág. 399)

mapa temático mapa que muestra información especializada (pág. 30)

tikanga costumbres y tradiciones maoríes transmitidas de generación en generación (pág. 426)

déficit comercial situación que ocurre cuando el valor de las importaciones de un país es superior al valor de sus exportaciones (pág. 138)

lenguaje comercial lenguaje común que surge cuando los países comercian entre sí (pág. 342)

superávit comercial situación que ocurre cuando el valor de las exportaciones de un país es superior al valor de sus importaciones (pág. 138)

economía tradicional economía en la que los recursos se distribuyen principalmente entre las familias (pág. 96)

*****transformar** cambiar algo por completo (pág. 64)

tratado acuerdo oficial, negociado y firmado por las partes (pág. 481)

fosa depresión larga, estrecha y profunda del fondo oceánico (pág. 60)

*****tendencia** inclinación o preferencia general (pág. 399)

tributo dinero que un país paga a otro por sometimiento o para obtener protección (pág. 305)

territorio en fideicomiso área puesta transitoriamente bajo el control de otro país (pág. 458)

tsunami gigantesca ola oceánica provocada por erupciones volcánicas o movimientos de la tierra bajo el lecho oceánico (págs. 54; 118)

tundra llanura plana y sin vegetación cuyo suelo permanece helado (pág. 202)

U

*****ultimate** most extreme or greatest (p. 161)

*****unify** to unite; to join together; to make into a unit or a whole (p. 430)

urban describes an area that is densely populated (p. 77)

*****supremo** extreme o mayor (pág. 161)

*****unificar** unir; juntar; integrar en una unidad o totalidad (pág. 430)

urbana área densamente poblada (pág. 77)

urbanization when a city grows larger and spreads into nearby areas (pp. 80; 133)

utilities the infrastructure provided by companies or governments such as electricity, water, and trash removal (p. 398)

urbanización cuando una ciudad crece y se expande hacia las áreas adyacentes (págs. 80; 133)

servicios públicos infraestructura que proveen algunas compañías o gobiernos, como electricidad, agua y recolección de basuras (pág. 398)

V

*****vary** to show differences between things (p. 230)

*****visible** able to be seen (p. 475)

*****volume** an amount (p. 355)

*****variar** mostrar diferencias entre cosas (pág. 230)

*****visible** que se puede ver (pág. 475)

*****volumen** antidad (pág. 355)

W

wadi a dry riverbed that fills with water when rare rains fall in a desert (pp. 230; 268)

water cycle the process in which water is used and reused on Earth, including precipitation, collection, evaporation, and condensation (p. 63)

watershed land drained by a river and its tributaries (p. 328)

wayfinding the system of navigating a foreign place through observation of natural phenomena, such as the movement of the sun and stars (p. 456)

weathering the process by which Earth's surface is worn away by natural forces (p. 55)

*****widespread** commonly occurring (p. 242)

vado lecho seco de un río que se llena de agua cuando ocasionalmente llueve en un desierto (págs. 230; 268)

ciclo del agua proceso en el cual el agua se usa y reutiliza en la Tierra; incluye la precipitación, recolección, evaporación y condensación (pág. 63)

cuenca territorio cuyas aguas afluyen a un mismo río y sus afluentes (pág. 328)

orientación sistema que consiste en navegar por un lugar desconocido con la ayuda de la observación de fenómenos naturales, como el movimiento del sol y las estrellas (pág. 456)

meteorización proceso mediante el cual la superficie terrestre se deteriora por la acción de fuerzas naturales (pág. 55)

*****extendido** que ocurre con frecuencia (pág. 242)

Y

yurt a large, circular structure made of animal skins that can be packed up and moved from place to place (p. 216)

yurta estructura grande y circular, hecha de pieles de animales, que se puede empacar y trasladar de un lugar a otro (pág. 216)

INDEX

The following abbreviations are used in the index: *m*=map, *c*=chart, *p*=photograph or picture, *g*=graph, *crt*=cartoon, *ptg*=painting, *q*=quote, *i*=infographic, *d*=diagram

Abkhaz Republic, 216
Aboriginal: in Australia, 424–25, 432; British settlers vs., 429; culture of, 85; defined, 417; rights of, 438. *See also* Native Americans
absolute location, 21
absolute monarchy, 161
Abu Dhabi, 241
Abuja, Nigeria, 370
Achebe, Chinua, 372
acid rain, 65
action songs, 435
active layer, *d*203
Addis Ababa, Ethiopia, 310
Adebajo, Adekeye, *q*346
Adulis, 303
Aegean coast, climate of, 230
Afghani people, 243
Afghanistan, *m*225; civil wars in, 238; climate of, 230; Hindu Kush mountain range, 226, 230; Khyber Pass and, 176; Kunduz, outdoor market in, *p*97; languages of, 243; as poor country, 245; Taliban and, 239; where people live in, 241. *See also* Southwest Asia
Africa, *m*30; as continent, 52; emigration from, 78; forests in, 51; Gondwana land mass, 484; Islam religion and, 234; population growth in, 73, 75; slaves from, 390; urbanization in, 81. *See also* Central Africa; East Africa; North Africa; Southern Africa; West Africa.
African National Congress (ANC), 393
Africans, The (Lamb), *q*350
African enslaved people. *See slave trade*
Afrikaners, in South Africa, 393, 398
Afrikaans language, 393
Age of Discovery, Southeast Asia and, 159
Agra, India, *p*185
agricultural revolution, 334
agriculture: development of, in Central Africa, 332, 334–35; farming methods, 233; India, 189; industry, 97; Mesopotamia, 232–33; prehistoric societies, 156–57; slash-and-burn, *p*74, 331–32; Tanzania, 316. *See also* farming
Ahmad, Muhammad, 306
AIDS-related deaths: in Southern Africa, 400–1; in West Africa, 373. *See also* HIV/AIDS
Aimak people, 243
Air Massif, 354

Akbar the Great, 185
Akosombo Dam, 356
Aksum, 302–3
Al-Azhar University, 285
Alexander the Great, 208
Alexandria, Egypt, 269, 303
Algeria, *m*262; aquifer in, 271; cooking in, 282; economic issues in, 283–84; film industry in, 282–83; independence of, 278; location of, 264; mountains in, 265; Muslim fundamentalists and, 286; natural gas in, 270; North Africa and, 277; rock formations in, *p*55. *See also* North Africa
Algiers, Algeria, 281
Allah, 234
alluvial plain, 177, 228
Alma-Ata, 213
al-Mawahib, Abu, 304
Almoravids, 362
al-Qaddafi, Muammar, *p*263, 278–79, *p*278, 283
al-Qaeda, 239
Amazon rain forest. *See* **rain forest; tropical rain forest**
Ambo peoples, 396
American Pacific islands, economies of, 465
American Somoa, 451, 465
Amin, Idi, 308
Amman, Jordan, 81
Amnok River, 120
Amu Dar'ya River, 204, 209, 213, 219
Amundsen, Roald, 480
Amundsen Sea, 477
Amur River, 204
Anabon, 329
Anatolian Peninsula, 235, 236
Anatolian Plateau, 227
ANC. *See* **African National Congress (ANC)**
ancient Egypt, 272–75, *m*273; expansion of Egypt, 273; influence of, 274–75; religion and culture in, 273–74; rise of, 272. *See also* Egypt
Andaman Sea, 153
Angkor Thom, 158
Angkor Wat, 158; Hindu temple at, *p*157
Angola, *m*379; colonialism in, 391; escarpment in, 381; ethnic and culture groups, 396; hard times in, 346; independence for, 392; infant death in, 400; as oil producer, 386, 401; population of, 394–95; Portuguese language in, 397; progress and growth in, 401; in Southern Africa, 380. *See also* Southern Africa
animals. *See* **wildlife**

anime characters, *p*136
animism, 166, 370
Annamese Cordillera, 151
Antarctica, 471–90; climate of, 477–78; as continent, 52; freshwater in ice caps of, 61; in Gondwana land mass, 484; ice sheets covering, 56; living in, 482–83; ozone levels over, 484; physical geography of, 474–79; plants and wildlife, 478–79; researcher in, *p*471; resources of, 479; results of research, 484–85; rock, ice, and water on, 475–77; as scientific continent, 482–85; size of, 474–75; surrounded by water, 59; timeline, *c*472–73; tourism to, 485
Antarctic Circle, 477
Antarctic Conservation Act, 481
Antarctic Treaty, 480–81
ANZUS Pact, 431
apartheid, 393, 401. *See also* human rights violation
aquifer, 271, 419
Arabian Desert, 228, 229
Arabian Peninsula, 227, 294; Arabian Desert and, 228; as arid region, 229; birth of Islam and, 276; religion of, 234, 243
Arabian Sea, 177, 180, 227, 294
Arabic language, 242, 283, 303, 312
Arab-Israeli conflict, 237–38, 246
Arabs, 243, 280; language of, 242; settlement of East African coast of Indian Ocean, 303
Arab Spring, 246, 279, 284, 286
Arab traders, 396
Aral Sea, 204; shrinking, 218–19, *p*218
Ararat Plain, Armenia, 215
Ararat volcano, Turkey, 227
archipelago, 118, 451–52
Arctic Circle, 202; ozone levels over, 484
Arctic climate, 203. *See also* polar climate
Arctic Ocean: freshwater in ice caps of, 61; *Rossiya* breaks ice in, *p*204; as smallest ocean, 62
Argentina: Antarctica and, 481
arid climate: in Central Asia, 203; in North Africa, 268; in Southwest Asia, 229–31; of West Africa, 356–57
Armenia, *p*215; in the Caucasus region, 201; climate of, 203; ethnic unrest in, 216; as independent state, 210, 211; language and religion of, 215; people of, 214. *See also* Caucasus region
Armenian Apostolic Church, *p*215
Armstrong, Neil, *c*17
arts/culture, 85; Central Africa, 344; East Africa, 314–15; East Asia, 134–37; Maori people,

427, 435, *p435*; North Africa, 282–283; South Asia, 191; Southeast Asia, 166; Southwest Asia, 243–44; West Africa, 371–72. *See also* culture

Aryan civilization, 183, 184

Ascension, 354

ASEAN. *See* **Association of Southeast Asian Nations (ASEAN)**

Ashoka, 185

Asia: as continent, 52; emigration from, 78; forests in, 51; immigration from, 396; as part of Gondwana land mass, 484; population growth in, 73, 75. *See also* Central Asia; East Asia; South Asia; South East Asia; Southwest Asia

Association of Southeast Asian Nations (ASEAN), 101, 168

astronomy, 274

Aswam High Dam, 267

Atatürk ("Father of the Turks"), 236

Atlantic Ocean: Central African coast on, 329; relative size of, 62; Southern African coast on, 380

Atlas Mountains, 265, 268, 270, *p265*

atmosphere, 44; pollution of Earth's, 74

atoll, 177, 453, *p178*; formation of, *d453*

Auckland, New Zealand, 433

Aussies, 434, 435. *See also* Australia

Australia, 413–46, *m415*; Aboriginals of, 424–25; agreements with Oceania, 458; Antarctica and, 481; ANZUS Pact, 431; climates of, 421; as continent, 52; culture in, 435; economy of, 436; humans migrating to, 424; international trust funds for islands in Oceania, 465; landforms, 416–18; life in, 432–39; migration from Oceania to, 464; natural resources, 436, 438; people of, 432; plants and animals of, 423; rural areas of, 433; surrounded by water, 59; territories in Oceania, 451; timeline, *c414*–15; trading partner with New Zealand, 437; waterways of, 418–19

Australian Outback, 417, 421, 433

axis, 42

Ayers Rock, 417, *p417*

Azerbaijan, *m198*; in the Caucasus region, 201; climate of, 203; ethnic unrest in, 216; as independent state, 211; as oil and gas producer, 205; people of, 214; religion in, 215; Soviet legacy in, 218; territorial issues in, 219

Azeris people, 214

Babylonians, 232–33

Baghdad, Iraq, 241; woman after voting, *p223*

Bahrain: Arab Spring and, 246; border of Persian Gulf, 228; democracy movements in, 239

Baloch people, 243

Bamako, Mali, 370

Bambuti people, 341

Bamileke people, 342

bananas, 361

Banda Aceh tsunami, *p153*

Bangalore, India, 189

Bangladesh, 190, *m175*; British rule of, 186; population, 188. *See also* South Asia

Bangla language, 190

Bangui, Central African Republic, 344

Bani Wadi Khalid riverbed, *p229*

Bantu language, 303, 341, 342, 361

barracoons, 336

basin, 355

Bass Strait, 436

Bast, Joseph, *q486*

Bathurst, New South Wales, 428

Battle of Adwa, 306

Battle of Omduman (Sudan), *p305*

Battuta, Ibn, 304

bay, 62

Bay of Bengal, 177, 178

Bedouin peoples, 229, 241

Beijing, China, 121

Belgium: as colonizing country, 338, 339; in West Africa, 365

Bengali language, 190

Bengaluru, India, 189

***benga* music,** 314

Benghazi, Libya, young people protest Qaddafi in, *p286*

Benguela ("Skeleton Coast"), 385

Benin, 359, 365, *m353*. *See also* West Africa

Benue River, 355

Berber people, 276, 282, 361; language of, 283

Berlin Conference of 1884–1885, 365, 391

Bhutan, *m175*. *See also* South Asia

Biafra, 367

biodiversity, 332

Bioko, 329

biome, 50

biosphere, 44, 74; Australia, 439; New Zealand, 439; Oceania island formation, 453

Black Sea: in Caucasus region, 201, 203; connected to Atlantic Ocean, 228; Turkey on, 227

Black Volta River, 356

blood diamonds, 387

Blue Nile, 266, 296, *p296*

Boers (Dutch) peoples, 390

Boer War (1899), 390, *p391*

Bokassa, Jean-Bédel, 339

"Bollywood," 191

Bombay, 191

boomerang, 435

Borneo, 151, 156

Botswana, *m379*; British control of, 391; climate in, 384; democratic governments in, 401; energy resources in, 386; ethnic and culture groups, 396; HIV/AIDS, 400; infant death in, 400; Kalahari desert covering, 381; population of, 394; in Southern Africa, 380. *See also* Southern Africa

boycott, 186

Brahman, 183

Brahmaputra River, 176–78

Brazzaville, Republic of the Congo, 341

Brisbane, Australia, 433

Britain. *See* **Great Britain; United Kingdom**

British rule: in Australia, 427–30; in New Zealand, 430; in Southern Africa, 390

British territory: in modern South Asia, 186–87, *m186*. *See also* colonial history

Brunei, 151, *m149*. *See also* Southeast Asia

Bryce Canyon National Park, 55

Buddha, 184, *p184*

Buddhism, *c84*, 184–185; in Australia, 432; in China, 126, 135; in Korea, 127; in New Zealand, 432; in South Asia, 191; in Southeast Asia, 166; spread of, 157

Buddhist monastery: Thimphu, Bhutan, *p174*

Buddhist monk: in Cambodia, *p161*

Burkina Faso, *m353*; in Empire of Mali, 363; as French colony, 366; rivers in, 356; in West Africa, 354; young girl from, *p69*. *See also* West Africa

Burma. *See* **Myanmar**

Burundi, 307, 308, 347

bush, 433

cacao tree, 366

CAFTA-DR. *See* **Dominican Republic-Central America Free Trade Agreement**

Cairo, Egypt, 276, 281; craftsperson in, *p283*; as primate city, 81; spices sold at souk in, *p281*

Calcutta, India, 189, 191

caliph, 234, 276

calligraphy, 135

call to prayer, 280

calving, 476

Cambodia, 150, *m149*; Buddhist monk in, *p161*; economy of, 163; independence of, 162; literacy rate in, 167; Mekong River through, 153; as poor country, 168; shadow-puppet theater in, 166; Tonle Sap in, 158; Vietnamese forces invade, 163. *See also* Southeast Asia

Cameroon: ethnic groups in, 341; German rule of, 338; as independent nation, 339, *m338*; Lake Chad in, 356; mineral resources in, 333; urban population in, 342

Canaan, 233

Canada: farmland in rural, *p77*; population doubling time in, 73

canals, 57

Cantonese cuisine, 136

Cantonese language, 134

canyons, 383

Cape Colony, 390

Cape of Good Hope, 380, 390

Cape Ranges, 381

Cape Town, South Africa, 23, 397, *p378*

Cape Verde, 354

capital: as factor of production, 97

carbon dioxide, 181
Carthage, 275
cartographer, 29
cartography, p30
Casablanca, Morocco, 281
cash crops, 462–63
Caspian Sea: in Caucasus, 201; desert land near, 209; Kazakhstan along, 202; as landlocked, 228; oil fields and, 205; territorial issues, 219
cassava, 336
Catholic Church. *See* **Christianity; Roman Catholic Church**
Caucasus Mountains, 201, 203
Caucasus region, 197–222, m198; climates in, 203; countries of, 201; early history of, 210; electricity production in, 205; energy and mineral resources, 205; ethnic unrest in, 216; forests in, 205; history of, 210–11; Islam and, 210; language and religion of, 215; musicians performing in, p201; natural resources, 204–5; people and places of, 214–15; population of, m214; Russian and Soviet rule, 210–11; Soviet legacy in, 217–18; Soviet Union annex of, 211; waterways in, 204
Central Africa, 325–50, m327; agriculture, development of, 334–35; city and country, 341–42; climate zones, 330; colonialism, 337–39; crops, adoption of new, 336–37; culture and arts, 344; daily life, 343–44; early settlement, 334–35; European contact and afterward, 335–38; growth and environment, 345; history, 334–39; how people live, 343–44; independence of, m338; independent nations, 339; landforms, 328–29; language, 342–43; life in, 340–45; mineral resources, 335; people of, 340–43; population, 340–41; rain forest, 330–31; religion, 343; resources, 332–33; savannas, 331–32; slave trade, 335–36; timeline, c326–27; waterways, 329–30; woman carrying produce on head, p335
Central African Republic: city-dwellers in, 342; French rule over, 338; as independent nation, 339; as landlocked, 329, 333; life expectancy in, 340; modernization and, 344
Central Asia, 197–222, m198–99; climates in, 203; countries of, 200–1; energy and mineral resources, 205; ethnic unrest, 216; forests in, 205; history of, 208–10; Islam religion and, 208–9, 234; landforms in, 201–2; mountains areas of, 203; natural resources in, 204–5; oil and gas resources in, 205; people and places, 212–14; plains and deserts in, 202; population, 213–14, m214; public health in, 218; regions of, 200–1; Russian rule and after, 209–10; Soviet legacy in, 217–18; territorial issues in, 219; timeline, c198–99; waterways in, 204. *See also* entries for individual countries
central Australia, 422

central highlands, South Asia, 177
Central Lowlands, 416–17
Central Rift, 311
Central Siberian Plateau, 205
Chad, 354–58, m353. *See also* West Africa
Chaggas peoples, 314
Chandragupta, 185
Chang Jiang River, 119, 123
Chao Phraya River, 153
Chennai, India, 189
cherry trees, Mount Fuji and, p118
Chewa ethnic group, 396
Chiang Kai-shek, 128
China, 200–1, 204, m115; as Australia trading partner, 436; climate of southeastern, 121; coal deposits in, 123; conquest of Red River delta, 157; Cultural Revolution, 130; dynasties of, 125, c126; early, 124–25; on East Asia mainland, 117; human rights in, 139; modern, 129–30; as New Zealand trading partner, 437; "one-child" policy in, 132; population concentration along, 76; population of, 188; Qinzang rail system in, p117; religions in, 135; Republic of (see Taiwan); rivers in, 119; Siberia and, 208; Sichuan Province, 136; as Southern Africa trading partner, 388; Three Gorges Dam in, p56; "two Chinas," 129; U.S. trade deficit with, 138, g139. *See also* East Asia
Chinese New Year, 79
Chinese peoples, 206, 208
cholera, 400
Choson dynasty, 127
Christchurch, New Zealand, 433
Christian Armenians, 210–11
Christian Bible, 244
Christianity, c84; Age of Discovery and, 159; Aksum kings adoption of, 303; in Armenia, 210, p215; in Australia and New Zealand, 432; birth of, 233; in East Africa, 313; in Georgia, 210; in Kazakhstan, 215; in Kenya, 313; in Lebanon, 243; in North Africa, 276; in Oceania, 457, 461; palm procession in Jerusalem, p234; in Siberia, 215; in Siberian Russia, Central Asia, and Caucasus, 215; in South Asia, 191; in Southern Africa, 396; in Syria, 243; in Tanzania, 313; in West Africa, 367, 369, 370. *See also* Christian missionaries
Christian missionaries: in Africa, 338; in Japan, 129; in Oceania, 457. *See also* missionaries
Church of St. George: Lalibela, Ethiopia, p312
CIA World Factbook: on Libya, q290
civil disobedience, 187, 393
civil war: defined, 279; Somalia, 308–9; in Southwest Asia, 238–39; in West Africa, 367
clan, 311
climate: Australia, 416, 421; Central Africa, 330; Central Asia, Siberia, and Caucasus, 203; changes to, 51; climate zones, 49–51, p49; defined, 23, 48; East Africa, 298–99; East Asia, 120–121; elements affecting, 46–49; elevation and, 47; landforms and, 48–49; Mediterranean, 230; New Zealand, 422; North Africa, 267–70; Oceania, 454; South Asia, 178–79; Southern Africa, 383–85; Southwest Asia, 229–30; weather vs., 48; wind and ocean currents and, 47–48. *See also* entries for specific countries
climate change: Antarctica and, 484–85; worldwide, 51. *See also* global warming
Clinton, Hillary, 284
coal mining: in Australia, 429; in Russian Siberia, 207; in Southwest Asia, 231
coastal area, 62
coastal plain, 56
cocoa, 366
Cold Temperate climate, 49–50
collectives, 208
colonial history, 191
colonialism: in Central Africa, 337–39; in Southeast Asia, 159–61; in Southern Africa, 390–91
colonization. *See* **European colonization**
command economy, 96
commercial fishing. *See* **fish and fishing**
communication system, 192
communism, 129
Communist government: in Myanmar, 163; in North Vietnam, 162; People's Republic of China, 129
Comoros, 380, 391
compass rose, 28
component, 23
condensation, 64
Confucianism, 126, 127, 135
Confucius, 125
Congo. *See* **Democratic Republic of the Congo; Republic of the Congo**
Congo Free State, 337, 338
Congo River, 328, 329, 341
coniferous tree, 205
constitution, 286
constitutional monarchy, 163
continent, 52
continental island, 451
continental shelf, 60
continents, 59, m16–17
converted, 276
Cook, James, 428, p428, 430
Cook Islands, 451
Coptic Christian church, 280, 287
coral, growth of, 453
coral reef, 418
core, Earth's, 43, d43
Côte d'Ivoire, m353, 365, 366. *See also* West Africa

cottage industry, 192
cotton cloth trade, m365
countries, boundaries of, 29
coup, 339
couscous, 282
Couscous and Other Good Food from Morocco **(Wolfert),** 290
craftsperson: in Cairo Egypt, p283
creole language, 369
Crusades, 235
crust, Earth's, 43, d43
Cuba: command economy in, 96
cultural blending, 79, 89
cultural region, 86
Cultural Revolution, in China, 130
culture, 82–86; Central Africa, 344; cultural changes, 88–89; cultural regions, 86; customs and, 84–85; defined, 82; East Africa, 314–15; East Asia, 134–37; economy and, 86; elements of, 82–86; ethnic groups and, 83; global, 89; government and, 85–89; history and, 85; language and, 83; Oceania, 460–61; population growth and, 75. *See also* arts/culture
culture groups: Aboriginal, 435; Australia, 435; Maori, 435; New Zealand, 435; North Africa, 280–83; Oceania, 460–61; Southern Africa, 396. *See also* ethnic groups *and entries for individual groups*
cuneiform, 233
currency, 101; Indian, p190
customs, cultural, 84–85
cyclone, 155, 178
daily life: Central Africa, 343–44; East Africa, 313–14; East Asia, 136–37; Kazakhstan, 216; North Africa, 281–82; Southeast Asia, 166–67; Southwest Asia, 244–45; Tajikistan, 217; Uzbekistan, 216–17; West Africa, 371–72
dalits, 193
dam: Akosombo Dam, 356; Aswām High, 267; in Central Africa, 330; Kariba Gorge, 386; Sayano-Shushenskaya, p209; in South Asia, 180
Damascus, 241
Danakil Plain, 295
dance: in Maori, 427, 435, p435; in South Asia, 191; in West Africa, 372. *See also* arts/culture
Daoism, 125, 126, 135
Dar es Salaam, Tanzania, 310
Darfur, 318, 320
Davis Sea, 477
Dead Sea, 61, 228, 295, p224
death rate, 72–73
de Blij, H.J., q38
debt relief, 373
Deccan Plateau, 177, 178, 179
deciduous trees, 205
de facto state, 116
deforestation, 193, 333, 386
Dehejia, Vidya, q196

Delhi, India, 80
delta, 62, 177, 178
democracy, 86–87; Arab Spring and, 246; in Australia, 430–31; in Ghana, 366; in North Africa, 285; in Southern Africa, 401; in Southwest Asia, 239
Democratic Republic of the Congo (DRC), 296, 332, 333; population of, 340; religion in, 343; Savoia glacier on border of, p298; subsistence farming in, 341
Denakil plain, 295
Denver, Colorado: as "mile High City", 22
desalinization, 61, 247
desert: as biome, 50; in Central Asia, 202–3; formation of, 43; in North Africa, 268; in Siberia, 202–3; in Southern Africa, 385; of Southwest Asia, 228–31. *See also* dry zone *and entries for individual desert regions*
desert climate, 49–50
desertification, 299, 360, 376
"desert island," 452
developed countries, 98
dialect, 83
diamonds: industrial, from DRC, 332; in South Africa, 364, 386, 387
dictatorship, 87
didgeridoo, 435
Dien, Albert, q222
dingoes, 425
disease: population growth and, 74–75; in Southern Africa, 400, p400
Disney World, p22
diverse, 311–12, 368. *See also* ethnic groups
diversified economy, 284
Dixon, David L, q470
Djibouti, 294, 296; Arabic and French languages spoken in, 312; drought in, 299; *futa* garments, 314; geothermal energy and, 301; mineral resources of, 300; population, 310; temperature of, 298
Dominican Republic-Central America Free Trade Agreement (CAFTA-DR), 101
dominion, 430–31
Dong Son, 157
Douala, Cameroon, 342
doubling time, 73
Drakensberg Mountains, 381
DRC. *See* **Democratic Republic of the Congo**
"Dreaming, the," 425
drought: Australia, 421; Ethiopia, 308–9; health issues and, 317; land abuse during, 376; northwestern India, p180; Papua New Guinea, 454; Somalia, 299; South Asia, 183; Southern Africa, 384
drugs: pharmaceutical, making new, i95
drumming, 344

dry zone: South Asia, 178; West Africa, 357–58
Dubai, 241
Durban, South Africa, 397
Dutch East India Company, 428
Dutch settlement, in South Africa, 364
dynamic, 20
dynasty, 124. *See also* China, dynasties of
dysentery, 400
Earth, 42–51, p44; atmosphere of, 51; climate on, 46–51 (*See also* climate); crust of, 53; deepest location on, 60; effects of human actions on, 56; erosion and, 55; forces of change, 52–57; inside, 43; landforms on, 56, 58–60; layers of, 43–44, d43; orbit of, 42, 45–46; physical systems of, 44; plate movements on, 53–54; population growth on, 72–75; saltwater vs. freshwater on, i60; seasons and tilt of, d45; seasons on, 45–46; sun and, 42–43; surface of, 52; tilt of, 45, d45; water on, 60–65; weathering of, 55
earthquake: American Samoa, 465; Caucasus region, 201; defined, 54; Japan, 118, 142; New Zealand, 419, 421; Oceania, 467; Philippines, p54; plate movement and, 54; South Asia, 183; Turkey (1999), p227
East Africa, 291–324, m36, m293; Aksum, 303; Ancient Nubia, 302; anthropology and ecology, 315; Bantu people in, 361; bodies of water, 296–97; climates of, 298–99; colonization of, 304–7; daily life, 313–14; economic development in, 315–16; environmental issues, 316–17; health issues, 317; history of, 302–9; independence, 307–9; land and wildlife, 301; landforms, 294–96; life in, 310–17; national park system, 315; people of, 310–13; physical geography of, 294–301; regions, 294 (*See also entries for individual regions*); religion in, 312–13; resources of, 300–1; Sahel in, 357; timeline, c292–93; trade cities, 303–4
East African Rift, Kenya, p295
East African Rift Valley, 294–95
East Antarctica, 475
East Asia, 113–46, m115; art forms in, 135; bodies of water, 119–20; challenges facing, 138–39; in change, 128–29; climate of, 120–21; culture in, 134–37; daily life in, 136–37; economies and environments, 137–38; education in, 136; energy resources, 122–23; forests in, 123; history of, 124–31; landforms of, 116–19; mainland of, 117; minerals in, 122; modern, 129–31; natural resources of, 121–23; physical geography of, 116–23; population of, 132–33; regional overview, 116–17; Silk Road, m127; timeline, c114–15; trade with, 138; urbanization in, 133
East China Sea, 119
Eastern Ghats, 177, 179, 181
Eastern Hemisphere, 26, c27
Eastern Highlands, 417

East Jerusalem, 238
East Sea, 118
East Timor: independence of, 162; literacy rate in, 167; religion in, 166
economic systems, 96
economy: American Pacific islands, 465; Australia, 436; basic economic question, 94–96; culture and, 86; East Asia, 137–38; economic activities, 97; economic organizations, 100–1; economic performance, 98; economic systems, 96–98; factors of production, 97, *d98*; global, 99–101; Japanese, 131; Oceania, 462–65; Papua New Guinea, 462–63; Southeast Asia, 168–69; three economic questions, *i95*; types of national economies, 98–99; West Africa, 372–73
ecotourism, 168, 192, 317
Edo peoples, 366
Egypt, *m263*; Ancient (see Ancient Egypt); break with Muslim nations, 287; farms in, 281; independence of, 278; location of, 264; natural gas in, 271; new constitution in, 286; Nile River in, 264, 266, 271, 296; population concentration in, 76; population growth in, 284; rainfall in, 269; Rome's defeat of, 275; water in, 271. *See also* North Africa
***Eid al-Fitr* (Festival of Breaking Fast),** 245
Ekurhuleni, South Africa, 397
electricity: in Caucasus, 205; in Southern Africa, 386
elephant, 383
elevation, 58; climate and, 47; on maps, 29
El Niño, 421, 454
embargo, 393
emigrate, 78
Emi Koussi, 355
Empire of Mali, *m362*, 363
endemic, 155
Energy Information Administration: Saudi Arabia, *q252*
energy resources, 74; South Asia, 180–81; Southern Africa, 386; West Africa, 359
engineering, 274
England. *See* **Great Britain; United Kingdom**
English language: in Australia, 433; in East Africa, 312; by Maori peoples, 427; in New Zealand, 433; in North Africa, 283; in South Asia, 190; in West Africa, 369
English South Africans, 393, 398
Ennedi, 355
environment, the: Australia and New Zealand, 438–39; Central Africa, 345; climate zone and, 50; defined, 23; East Africa, 316–17; East Asia, 137–38; Oceania, 467; population growth's effects on, 74; society and, *c24*; in South Africa, 23; Southeast Asia, 169

environmental disaster, Aral Sea, 218–19
environmental hazards, 23
Environmental Protection Agency, "Smart Growth and Schools," *q104*
Epic of Gilgamesh, 233
Equator, 21, *c21*, 26; Central Africa on, 330; climate around, 46; warm air masses near, 47; water currents and, 48
Equatorial Guinea, 329, 333, 338–40, 344
equinoxes, 46
ergs, 268, 269
Eritrea, 294, 306
erosion, 55, 452; in West Africa, 354
escarpment, 381. *See also* Great Escarpment
estuary, 329
Ethiopia, 294; Blue Nile in, 266; Danakil Plain in, 295; drought/famine (in 1980s), *p41*; drought in, 299; earliest known humans from, 315; ethnic groups in, 311; HIV/AIDS, 317; independent, 306–7, 309; manufacturing in, 316; Menelik II, *p306*; mineral resources of, 300; population of, 310–11; religion in, 312; tsunami in, 140. *See also* East Africa
Ethiopian Orthodox Church, 312
ethnic groups: Caucasus, 216; Central Africa, 340–41; defined, 83; East Africa, 310–12; East Asia, 134; Oceania, 459–60; South Asia, 190; Southeast Asia, 166; Southern Africa, 396; Southwest Asia, 242–43; West Africa, 368–69
Etosha National Park, 383
Etosha Pan, 383
EU. *See* **European Union**
eucalyptus tree, 423, 436
Euphrates River, 228, 232, 247
Eurasia, 235
Eurasian Plate, 152
Europe, *m30*; as continent, 52; emigration from, 78; urbanization in, 81
European colonization: in Africa, 305–6; in Australia and New Zealand, 427–28; in East Africa, 304–6; in Southern Africa, 390–91. *See also* colonialism
European Union (EU), 101
evaporation, 63
experience, perspective of, 20
exports, 99
extended families, 370
factors of production, 97, *d98*
fales, 461
family: in Southern Africa, 399. *See also* daily life
famine, 317
Fang people, 341
farming, *p246*; Australia, 429, 434; Egypt, 281; Libya, 282; Morocco, 282; New Zealand, 434, 437; Nile River and, *p266*; Papua New Guinea, 463; Southeast Asia, 168; Tanzania, 314. *See also* agriculture

Farsi language, 242
Fatamids, 276
fault, 54
fauna, 155
Federated States of Micronesia, 450
felafel, 282
fellaheen, 281
Festival of Breaking Fast (*Eid al-Fitr*), 245
feudalism: in Japan, 128
Fickling, David, *q446*
Fiji, 451, *p457*, 467
film industry: in Algeria, 282–83; in New Zealand, 437
First Fleet, 428
fish and fishing: American Samoa, 465; Antarctica, 479; Central Africa, 333; Lake Victoria, *p297*; Oceania, 463; overfishing, 467; Papua New Guinea, 455; South Asia, 192; Southeast Asia, 168
fishing grounds, 271
fjords, 420
floods and flooding, 178; of Niger River (West Africa), 355; Oceania, 467
flora, 155
food: East Asia, 136; in North Africa, 282; in West Africa, 361
foreign rule, of North Africa, 277–78
forests: cutting and burning, *p74*, 331–32; in Siberia, 204–5; in South Asia, 181
fossil fuels: effects on water supply, 65
fossil water, 247
Four Noble Truths, 184
Fouta Djallon, 355, 356
France: agreements with Oceania, 458; colonies in Southern Africa, 391; as colonizing country, 338, 339, 364–65; Morocco and, 278; North Africa and, 277, 278; possession of Madagascar, 389; territories in Oceania, 451; trading in West Africa and, 364; West Africa and, 365–66
Freetown, 364
free trade, 100
French language, 283, 369; in Djibouti, 312
French Polynesia, 451, 458
French Sudan, 366
freshwater, 61; in Antarctic ice sheet, 475; salt water vs., *i60*
Fukushima Daiichi nuclear power plant, 142
Fulani people, 342, 344
fundamentalist, 279
futa, 314
Gabon: city-dwellers in, 342; independence of, *m338*; as independent nation, 339; minerals in, 332; slave trade in, 336
Gambia, 365, 366
Gandhi, Indira, 193, *p175*

Gandhi, Mohandas, *p175*, 186, *p187*; on Indian currency, *p190*
Gandhi, Rajiv, 193
Ganges Delta, 180
Ganges River, 176–78, *p177*
gas. *See* **natural gas**
Gautama, Siddhārtha, 184
Gaza Strip, 238, 246
GDP. *See* **gross domestic product (GDP)**
Gebel Katherina, 265
gems. *See* **mineral resources**
genocide, 308, 346
geographer's tools, 26–33; geospatial technologies, 30–33; globes, 26; maps, 27–30
geographic information system (GIS), 31–32
geography: defined, 18; five themes of, 20–24; human, 70–104; physical, 39–68; six essential elements of, 24, *c24*; skill building, 25; uses of, *c24*
Geography (de Blij and Muller), *q38*
Geography Activity: Antarctica, *m488*; Australia and New Zealand, *m444*; Central Africa, *m348*; Central Asia and Siberian Russia, *m220*; continents, *m66*; East Africa, 322; East Asia, *m144*; India, *m102*; Kenya, *m36*; North Africa, *m288*; Oceania, *m468*; South Africa, *m402*; South Asia, *m194*; Southeast Asia, *m170*; Southwest Asia, *m250*; West Africa, *m374*
Georgia (Caucasus region), 201; climate of, 203; ethnic unrest in, 216; independence of, 210–11, *p210*; population of, 214; religion in, 215; Soviet legacy in, 218
geospatial technologies, 30–33; geographic information system (GIS), 31–32; Global Positioning System (GPS), 31; limits of, 33; satellites and sensors, 32–33
geothermal energy, 301, 436–37, 439
Germany, 211; Cameroon and, 338, 339; invasion of Caucasus, 211; invasion of Russia, World War II, 208; occupation of islands in Oceania during World War II, 457; West Africa and, 365
geysers, 419, 439
Ghana, *m353*; economic challenges for, 373; energy resources in, 358–59; people of, 376; rivers in, 356. *See also* West Africa
Ghana Empire, 362–63, *m362*
Gio people, 367
GIS. *See* **geographic information system (GIS)**
glacier, 56, 61, 420
global economy, 99–101; trade, 99–100. *See also* globalization
globalization: defined, 89; and indigenous cultures, 34–35. *See also* global economy

Global Positioning System (GPS), 31; satellites, *p32*
global warming, 439; debate over, 486–87; defined, 51; Oceania and, 467; ozone, 483, *d483*, 484
globes, 26, 27
Gobi Desert, 121
Golan Heights, 238
gold and gold trade: Afghanistan, 231; Ancient Egypt, 273; Australia, 428–29, 436; Kazakhstan, 205; New Caledonia, 455; Papua New Guinea, 455, 462; Russia, 205; salt and, 361, *p361*, 363; Southern Africa, 387, 388, 390; Uzbekistan, 205; West Africa, 359, 361, 366, 372
Gold Coast, 359, 366
Gold-for-salt trade, 361, *p361*, 363
Gondwana, 484
gorges, 382–83
government: Caucasus and Central Asia, 217–18; China, 135; forms of, 86–87; Myanmar, 169; Southern Africa, 401; Southwest Asia, 246; West Africa, 372–73
GPS. *See* **Global Positioning System (GPS)**
Grand Canyon, 55
Grand Mosque, *p234*
graphite, 180
graph skills, GDP comparison, *g99*
grassland, 50
Great Barrier Reef, 418, 439, *p418*
Great Britain: agreements with Oceania, 458; Antarctica and, 481; Australia and, 430–31; control of New Zealand, 430; control of Sudan, 306; international trust funds for islands in Oceania, 465; North Africa and, 278; occupation of islands in Oceania during World War II, 457; territories in Oceania, 451; trade in West Africa, 364; trading in West Africa, 366. *See also* United Kingdom
Great Dividing Range, 417, 425
Great Escarpment, 381, 382
Great Karoo, 381
Great Leap Forward, 130
Great Migration, 301
Great Rift System, 328
Great Rift Valley, 294–95, 301, 328, 383
Great Salt Lake, 61
Great Trek, 390
Great Wall of China, 125, *p125*
Great Zimbabwe, 388–89. *See also* Zimbabwe
Greek Empire, 210
Greenland, 52; ice sheets covering, 56
green revolution, 192
Greenwich, England, 21, 213, *m213*
griot, 371
gross domestic product (GDP), 98; comparing, *g99*

groundwater, 61, 419
Guam, 450, 451, *p459*, 465
Guangdong Province, 136
Guinea Highlands, 355
Guinea: bodies of water in, 355–56; Empire of Mali in, *m362*, 363; as French possession, 366; tropical rain forest in, 358
Guinea-Bissau, *m352*
Gujarati language, 190
gulf, 62
Gulf of Mexico: surrounded by landmasses, 62
Guptas, 185
haiku, 135
Hainan, China, rubbish-covered beach in, *p138*
hamadas, 269
Han, People of, 126
Han Chinese, 83
Han dynasty, 125, 126, *c126*, 157
Han ethnic group, 134
Hanoi, Vietnam, tourism in, *p168*
Han River, 120
Harappa, 182
Harare, Zimbabwe, 397
harmattan winds, 357
Hausa people, 367, 368
Hawaiian Islands, 451, 452; luau, 461
Hazari people, 243
health issues: in East Africa, 317; in Southern Africa, 399–401
heat energy, 436–37
Hebrew Bible, 244, *p234*
Hebrew language, 242
hemisphere, 26
hieroglyphics, 274
high islands, of Oceania, 452
highland countries, 308
High Mountain climate, 49–50
Hillary, Edmund, *p17*
hills, as landforms, 23
Himalaya Mountains: high altitudes in, *p181*; Kashmir region in, 187; Plateau of Tibet and, 117, 121; temperate zone and, 179
Hinduism, *c84*; in Australia, 432; birthplace of, 184; Ganges River and, 180, *p177*; growth of, 184; in India, 185; Islam and, 187; man in traditional clothes, *p173*; in New Zealand, 432; Sanskrit as language of, 183; in South Asia, 191; in Southeast Asia, 166; spread of, 157
Hindu Kush mountain range, 176, 226, 230
Hindu temple, *p184*; at Angkor Wat, *p157*
history: Australia and New Zealand, 424–31; Central Asia and Siberian Russia, 206–11; culture and, 85; East Africa, 302–9; East Asia, 124–31; New Zealand, 424–31; North Africa, 272–79; Oceania, 456–61; South Asia, 182–87; Southeast Asia, 156–63; Southern Africa,

388–93; Southwest Asia, 232–39; West Africa, 272–79

HIV/AIDS, 317; in Southern Africa, 400, 401; in West Africa, 373

holidays, in East Asia, 137

Holland, spice trade and, 160

Holocaust, 237

homelands, 393

homogenous, 216

Hong Kong: anime characters on billboards in, *p136*; China, Temple Street Night Market in, *p114*; high-rise buildings in, *p133*; languages in, 134; population of, 133

Honshū Island, Japan, 117, *p118*

Horn of Africa, 308–9

hot springs, 419, *p52*

Hot Springs, in Mount Hakone, Japan, *p53*

Huang He River, 119, 120, *p120*. *See also* Yellow River

human actions: damage to water supply, 65; effects on Earth's surface, 57

human-environment interaction, 23

human geography, 70–104; culture, 82–89; population growth, 72–75; population movement, 77–81; population patterns, 76–77; timeline, *c70*–71

human migration, from Oceania, 464, 465–66

human rights, 87

human rights violation, 393; Holocaust, 237; Liberia, 367; in New Zealand, 438; Oceania, 467; Rwanda, 308; Sudan, 318–21; Uganda, 308; United Nations and, 346–347

human systems, *c24*

Humid Temperate climate, 49–50

Huns, 206, 208

hunter-gatherers, 156; Aboriginals as, 424–25; in Central Africa, 334; in Siberia, 206

hurricanes: defined, 155

Hussein, Saddam, 239

Hyderabad, India, 189

hydroelectric power: Central Africa, 330, 333; East Africa, 300–1; Mekong River, 169; Togo and Nigeria, 359; West Africa, 356

hydropolitics, 247

hydrosphere, 44, 419, 453

Ice Age, 56

iceberg, 476

ice caps, 56, 61

ice sheet, 56, 61, 475

ice shelf, 476

Igbo peoples, 367

IMF. *See* **International Monetary Fund (IMF)**

immigrate, 78

immigration: to Australia and New Zealand, 439; to Zambia, 397

imperialism, 305, 365

import, 99

inactive volcano, *p292*

India, 80, *m175*; British rule of, 186, 187; climates in, 178, *m179*; conflicts with Pakistan, 193; drought in, *p180*; influence on early Southeast Asia, 157; man in traditional clothes for Hindu festival, *p173*; population of, *m102*, 188; slaves from, in South Africa, 390; trading with Southern Africa, 388; tsunami in, 140; water in streets after rain in, *p46*. *See also* South Asia

Indian currency note, *p190*

Indian National Congress, 186–87

Indian Ocean: Arabian Sea as part of, 227; East African cities on, 310; East Asia and, 119; moisture from, 121, 178; relative size of, 62; Southeast Asia, 153; Southern Africa and, 380–82; tsunami in, 140, 142

Indian Plate, 152

indigenous peoples, *q34*–*q35*. *See also* entries for individual indigenous groups

indigenous rights, in New Zealand and Australia, 438

Indo-Australian Plate, 152

Indonesia, *m149*; Banda Aceh tsunami, *p153*; climate in, 23, 155; fishing in, 168; independence of, 162; invasion of Timor-Leste, 162; languages in, 166; latex plantation, *p154*; manufacturing in, 163; Mount Lokon volcano, *p53*; natural resources in, 152; plantations in, 168; shadow-puppet theater in, 166; Sunda Isles, 153; tin mining in, 169; tsunamis in, 140, 142, 152; volcanoes in, 152. *See also* Southeast Asia

Indus River, 176–77, 182, 193

industrialized countries, 98

Industrial Revolution, 81

industry: cottage, 192; defined, 97. *See also* entries for individual industries

Indus Valley, 182–83

Indus Valley civilization, 182–83

infant death, in Southern Africa, 400–1

infrastructure, 372

inland delta, 355

insular, 150

Intergovernmental Panel on Climate Change (IPCC), 486, *q487*

International Association of Antarctic Tour Operators, "Tourism Overview," *q490*

International Date Line, 26, 213, *m213*

International Forum on Globalization (IFG), *q35*

International Monetary Fund (IMF), 101, 373

international trade, *p88*, 99–101, *p100*

introduced species, 429, 440–43

Inya Lake, Myanmar, floating market in, *p148*

IPCC. *See* **Intergovernmental Panel on Climate Change (IPCC)**

Iran, *m225*; border of Persian Gulf, 228; carpets from, 244; civil wars in, 238–39; coal deposits in, 231; command economy in, 96; ethnic groups, 242; oil in, 231; plateau and mountain ranges in, 227; population of, 240–41; Shias of, 243; temperatures in mountainous areas of, 230; territorial issues in, 219. *See also* Southwest Asia

Iranian peoples, 206

Iraq, *m225*; civil wars, 238–39; ethnic groups, 242; oil in, 231; oil well fires in Kuwait, *p245*; Persian Gulf, 228; population of, 241; Tigris-Euphrates River system in, 247; tribal identity in, 242–43. *See also* Southwest Asia

Ireland, 429

iron artifact, *p326*

iron ore, 333, 350; in Australia, 436

iron-smelting technology, 361

iron tools, 335

Irrawaddy delta, 158

Irrawaddy River, 153

irrigation, 209; methods, 232–33; in Saudi Arabia, 247

Irtysh River, 204

Islam, *c84*; Africa, 86; Australia, 432; Central Asia, 208; daily life in, 244; modern, 287; monotheistic religion, 276; Mozambique, 396–97; in New Zealand, 432; North Africa, 369; prophet Muhammad, 233; Quran, 244; rise of, 276; South Asia, 191; Southeast Asia, 158–59, 166; Southwest Asia, 86; Soviet leaders and, 209; spread of, 234–35, *m235*, 243; Sunnis and Shia, 243; Tanzania, 313; Uzbekistan mosque, *p208*; West Africa, 363, 370. *See also* Islamic expansion; Muslims; North Africa, culture of

Islamic expansion, 234–36, *m235*

Islamic fundamentalism, 286

Islamic republic, 239

Island: formation of, 53; as landmass, 59; in Southeast Asia, 150–51

Israel, *m237*; area, 233; coast of, 227; Dead Sea and, 228; ethnic groups, 242; Muslim peace treaty with, 287; population of, 240–41. *See also* Arab-Israeli conflict

Israelites, 233

Istanbul, Turkey, 241

isthmus, 59

Italy: Carthage rule in, 275; Eritrea colony, 306; Libya and, 277

ivory, 387; trade, 388

Jaffa, Israel, *p224*

Jainism, 184, 185, 191

Jakarta, Indonesia, 165; climate in, 23; skyscrapers, *p148*; urbanization, 169

Japan, *m115*; access to Oceania's marine resources, 463; birthrate in, 133; climate of, 121; cultured pearls harvested offshore, 122; economic growth of, 131; feudal system in, 128; forests in, 123; history of, 127–28; Honshu Island, Mt. Fuji on, *p118*; international trust funds for islands in Oceania, 465; Kuril Islands dispute with Russia, 139, 219; mineral resources in, 122; modern, 131; Mount Hakone in, *p53*; occupation of islands in Oceania during World War II, 457; religion in, 135; rise of, 129; rivers in, 120; Sea of Japan and, 119; trading partner with Australia, 436; tsunami in (2011), 142; World War II and, 131. *See also* East Asia

Japanese folding screen, *p128*
Java, 151, 152, 156
Javanese people, 166
Jerusalem: as holy city, 233; three religions in, *p234*
Jesus of Nazareth, 233
Jews, 242. *See also* Judaism
Jingxi, 135
Johannesburg, South Africa, 23, 398
Johor, Malaysia, *p88*
Jonglei Canal, 297
Jordan, *m225*; Arab Spring and, 246; coast of, 227; Dead Sea and, 228; ethnic groups, 242; Great Rift in, 295. *See also* Southwest Asia
Jordan River, 295
Jos Plateau, 355
Judaism, 233, 243, *c84*; Hebrew Bible scroll, *p234*
Kaaba, *p234*
Kabul, 227
Kalahari, 381, 385, 404
Kamba people, 311
Kamchatka, 219
Kannada language, 190
Kanuri people, 342
kapahaka, 426
Kaplan, Gordon, *q248*
Karakoram, 176, 187
Kara-Kum, 202
Kara-Kum Canal, 209
Kariba Gorge dam, 386
karma, 184
Karnow, Stanley, *q172*
Kashmir, 193
katabatic winds, 477
Kazakhstan, *m198*; in Central Asia, 200; coal in, 205; daily life in, 216; language and religion of, 215; mineral resources in, 205; mountain range in, *p202*; oil and gas resources in, 205; plains and deserts in, 202; population of, 213; Russian rule of, 209; territorial issues in, 219; USAID for, 222. *See also* Central Asia
Keen, Cecil, *q404*
Kemal, Mustapha (General), *p236*
kente, 371
Kenya, East Africa, 294; benga music, 314; climate of, 298; drought in, 299; earliest known humans in, 315; East African Rift in, 295, *p295*; energy resources in, 300–1; ethnic groups in, 311; geothermal energy and, 301; HIV/AIDS, 317; independence of, 307; as independent trading state, 303; Lake Victoria and, 297; languages spoken in, 312; life expectancy, 317; map of, *m36*, *m293*; Masai Mara National Reserve in, 301, 315; Masai nomadic lifestyle, 314; mineral resources of, 300; Mombasa, *p313*; Nairobi (capital), 310, *p311*, 313; oral tradition in, 314; rainfall in, 298; refugees from Somalia in, 309; religion in, 313; Samburu National Reserve, 315; temperatures in, 298; wildlife reserves in, 301

Kenya National Assembly, 312
Kenya National Museum, *p315*
Kenyatta, Jomo, 307, *p307*
key, map, 28
Khalkha Mongolian language, 134
Khan, Azmat, *q196*
Khan, Genghis, 235
Khan el-Khalili, spices sold at, *p281*
Khartoum, Sudan, *p296*, 302, 306, 309
Khmer Empire, 158
Khmer Rouge, 163
Khyber Pass, 176, 227, *p174*
Kikuyu people, 307, 311
Kilamanjaro, farming in plains around, 314
Kilwa trade city, 304
Kimbanguist Church, 343
Kingdom of Madagascar, 389. *See also* Madagascar
Kinshasa, Democratic Republic of the Congo, 341, 344
Kirghiz people, 216
Kiribati Island, 451, 455
kiwifruit, 437
Kiwis, 434, 435. *See also* New Zealand
knowledge, advances in, 274–75
Kolkata, India, 189, 191
Kongo people, 344
Köppen, Wladimir, 49
Korea, *m130*; divided, 130–31; history of, 126–27; Japanese control of, 129; religion in, 135; rivers in, 120. *See also* Korean Peninsula; North Korea; South Korea
Korean language, 134
Korean Peninsula, 118; forests in, 123; rivers on, 120; Yellow Sea and, 119
Korean War, 131
Koryo dynasty, 127
Kpong, 359
K-pop, 135
kraal, 314
Krahn people, 367
Krakatoa volcano, 152

krill, 479, 485
Kunduz, outdoor market in, *p97*
Kunlun Shan mountain range, 117
Kurdish families, traveling by cart, *p238*
Kurdish language, 242
Kurds, 238, 242, 243
Kuril Islands, 139, 219
Kurtz, Lester, *q404*
Kush, 302; pyramids of ancient, *p303*
Kushites, 302
Kuwait, *m225*; border of Persian Gulf, 228; civil wars in, 239; Iraq's invasion of, 245, *p245*; oil in, 231; oil wells on fire in, *p245*; population of, 240–41; as wealthy country, 245
Kyrgyzstan, *m198*; climate in, 203; in Central Asia, 200, 201; mineral resources in, 205; people of, 213, *m214*. *See also* Central Asia
Kyzyl Kum, 202
labor, as factor of production, 97
laborers and peasants, in South Asia, 183
lagoon, 453
Lagos, Nigeria, 370
Lake Albert, 295, 296
Lake Assal, 300; workers collect salt at, *p300*
Lake Baikal, 204
Lake Chad, 356, *p356*
Lake Malawi, 295, 297, 329, 383
Lake Nyasa, 329
lakes, 61; freshwater, *i60*, 61, 62; pollution of, *p65*
Lake Tanganyika, 295, 297, 329
Lake Taupo, 411
Lake Victoria: fishers in, *p297*; as largest lake in Africa, 297; Luhya people and, 311; Nile River begins at, 266, 296; Sukuma people and, 314
Lake Volta, 356
Lake Vostok, 485
Lalibela, Ethiopia: Church of St. George, *p312*
Lamb, David, *q350*
Lamu, 303
land: on Earth's surface, 44; as factor of production, 97, *d98*; surface features on, 58–59
landforms: bodies of water and, 59–60; Central Africa, 328–29; changing, 52–54; classification of, 58–59; climate and, 48–51; creation of, 53–57; defined, 23; East Africa, 294–96; ocean floor, 59–60; Southern Africa, 381–82; types of, 44; weathering of, 55; West Africa, 354–55
landlocked, 382
landmasses, shifting, 52–54
landscapes, 19; formation of, 43
languages and language groups: Arabic, 276; Australia, 433; Bantu, 361; Berber, 283; Caucasus, 215; Central Africa, 342–43; Central Asia, 215; culture and, 83; East Africa, 312; East Asia, 134; English, 283; French, 283; Maori, 427; New Zealand, 433; North Africa, 283; Oceania, 459–60; Papua New Guinea,

Index **531**

Laos, 460; Siberia, 215; slang, in Australia, 434; South Asia, 190; Southeast Asia, 166; Southern Africa, 396–97; Southwest Asia, 242–43; trade, 342; West Africa, 369. *See also entries for individual languages*

Laos, m149; collapse of government in, 163; independence of, 162; literacy rate in, 167; Mekong River through, 153; outdoor market in, p78; as poor country, 168; on Southeast Asia peninsula, 150. *See also* Southeast Asia

Laozi, 125

La Perouse, Jean-Francois, 457

Latex plantation, Indonesia, p154

Latin America. *See* Mexico; South America

latitude, 21, c21, c27

lawsuit, 438

leap year, 43

Lebanon, m225; in Canaan, 233; Christians in, 243; civil wars in, 238; coast of, 227. *See also* Southwest Asia

Lena River, 204, 205, 212

Leopold II (King of Belgium), 337, 338, p337

Leptis Magna, p275

Lesotho, m379; British control of, 390; climate of, 384; HIV/AIDS, 400; independence of, 391; Orange River in, 382; population in, 394; religion in, 396; subsistence farming in, 387; Taiwan's textile industry in, 401. *See also* Southern Africa

Liberia, 358, 364, 367

libraries, earliest, 275

Libya, m263; aquifers and, 271; Arab Spring and, 279; economic issues in, 283; farms in, 282; Italy seizure of, 277; Muammar al-Qaddafi in, 278, p278; in North Africa, 264; oil in, 270; population growth in, 284; poverty in, 283; rainfall in, 269; self-rule, 286; urbanization, 280. *See also* North Africa

Libyan Desert, 302, p268

Libyan refugees, p79

lichens, 202, 478, p202

life expectancy: in East Africa, 317; in West Africa, 373

Limpopo River, 382, 389

lions, 383

literacy: among North African women, 284–85; in East Africa, 316; in Libya, 284–85

lithosphere, 44, 54, 419, 453

Little Karoo, 381

Little League teams, in East Asia, 137, p137

Lituya Bay, Alaska, tsunami in, 140

Livingstone, David, 304–5

location, types of, 20–21

loess, 119

London, England: port security worker in, p248–49

longitude, 21, c21, c27

Lord of the Rings **(Tolkien),** 437

low islands, of Oceania, 452

Lowland, of North Africa, 265

Luanda, Angola, 397, 398

luau, 461

Luhya people, 311

Luo people, 311; benga music and, 314

Lusaka, Zambia, 397

Luzon, 151

Macau, China, religious shrines in, p134

MacDonnell mountain range, 417

Madagascar, m379; early kingdoms of, 389; French-controlled, 391; land area, 380; language of, 396; tsunami in, 140. *See also* Southern Africa

Madras, India, 189

Magellan, Ferdinand, 160

magma, 43, 53

Mahabharata, 191

Mahatma, 186. *See also* Gandhi, Mohandas

Mahfouz, Naguib, 283

maize (corn), 336–37

Makkah (Mecca), Saudi Arabia, 233, 234, p234, 363

Malabo, Equatorial Guinea, 344

Malacca, Malaysia, 159; mosque on stilts near, p158; Portuguese conquest of, 160

Malagasy language, 396

malaria, 400

Malawi, m379; democratic governments in, 401; independence for, 391–92; infant death in, 400; Lake Malawi in, 297, 383; Malawian income, 395; Muslim population in, 396; population of, 395. *See also* Southern Africa

Malay Archipelago, 151, 158, 451

Malay Peninsula, 151, 153; climate in, 154–55; Malacca on, 159; religion in, 166

Malaysia, 150, 151, m149; fishing in, 168; independence of, 162; international trade, p88; manufacturing in, 163; natural resources in, 152; plantations in, 168; shadow-puppet theater in, 166; tin mining in, 152, 168, 169. *See also* Southeast Asia

Maldives, 176, 177. *See also* South Asia

Mali, m353; climate of, 356–57; Empire of, 363; as French colony, 366; mineral resources in, 359; Niger River in, 355–56; North African culture and, 371. *See also* Empire of Mali; West Africa

Malindi, 303

malnutrition, 400; in East Africa, 317

Manchuria, China, 206

Mandarin Chinese, 134

Mandela, Nelson, 393, p393, 404

Manila, Philippines: slum areas near, p165; urbanization in, 165, 169

Mano people, 367

Mansa Musa, 363, p363

mantle, Earth's, 43, d43

manufacturing: in Ethiopia, 316; in South Africa, 398. *See also* industry

Maori language, 427

Maori tattoo, 427

Maori tribe, 83, 426–27; population of, 432; rights of, 438, 446

Mao Zedong, 129, 130

map projections, 28–29

maps, 27–30; cartography as science of making, p30; distortion of, 27; information on, 27; large- vs. small-scale, p28, 29; map projections, 28–29; map scale, 29; parts of, 28; physical, 29–30; political, 29; relief on, 29, 30; types of, 29–30

Maputo, Mozambique, 397

Marathi language, 190

Margherita Peak, 328

Mariana Trench, 60

market economy, 96

Marshall Islands, 450, 454, 458

Masai Mara National Reserve, 301, 315

Masai people, 311; mother and son with mud home, p313; as nomads, 314

Massif, 354

Matadi, 329

mathematics, 274

Matmata, Tunisia: family meal in, p282

mature, 73

Mauritania: climate of, 356–57; Empire of Mali in, m362, 363; as French colony, 366; independence of, 366; North African culture and, 371; Sahara desert in, 354

Mauritius, 380, 389, 391

Mauryas, 185

Maya people: cassava cultivated by, 336; organized team sports and, 85

Mbundu peoples, 396

Mbuti people, 341

McDougall, Leanne, q222

Mecca. *See* Makkah (Mecca), Saudi Arabia

Mediterranean Sea: description of "sea" fits, 228; as link to North Africa, 266; Nile River empties to, 296; in Southwest Asia, 227; Suez Canal and, 267, 364

megalopolis, 80, 133, 165

Mekong Delta, Vietnam War and, p162

Mekong River, 153; hydroelectric power from, 169

Mekong River Commission, 169

Melanesia, 447, 450–51, m448–49. *See also* Oceania

Melbourne, Australia, 433

Menelik II (Ethiopian King), 306–7, p306

Mercado Camon del Sur, 101

merchant class, in South Asia, 183

MERCOSUR, 101
meridians, 21
Meru people, 311
Mesopotamia: early history in, 232–33; as "land between the rivers", 228; location of, 232; population density in, 241; social advances in, 232–33; Ziggurat of Ur in, p233
Mexico, population concentration along, 76
Mexico City, as megalopolis, 81
Micronesia, 447, m448–49; culture of, 460–61; location of, 450; low islands of, 452. See also Oceania
Mid-Atlantic Ridge, 60
Middle Ages, 275–76
Middle East, 239. See also Southwest Asia
migration: causes of, 78–79; to cities, 80; effects of, 79; human, from Oceania, 465–66
millet, 335
Mindanao, 151
mineral resources: Central Africa, 332–33, 335; Central Asia and Siberian Russia, 205; East Africa, 300; East Asia, 121–22; New Zealand, 437; North Africa, p284; Oceania, 455; South Asia, 180–81; Southeast Asia, 152; Southern Africa, 386, 387, 390; Southwest Asia, 231; West Africa, 359
Ming dynasty, 125, C126
mining, Papua New Guinea, 462; South Africa, 398. See also mineral resources
Min language, 134
minorities, 166
Minott, Geoff, q222
MIRAB economies (migration, remittances, aid, and bureaucracy), 464–65
missionaries, 338. See also Christian missionaries
mixed economy, 96
Mogadishu, Somalia, 303, 310; war-torn buildings in, p292
Mohenjo-Daro, 182, p183. See also Pakistan
Mombasa, 303, 310
monarchy, 87
Mongolia, m115, 202; on East Asia mainland, 117; population density in, 133. See also East Asia
Mongols: Central Asia conquest, 208, 212; invasion of China, 127; Persian conquest, 235; as Siberian invaders, 206
monolith, 417
monotheism, 191, 233, 276
monsoon, 178, 384, 454
Mopti, 355
Moroccans, 363
Morocco, m262; economic issues in, 284; farms in, 282; fishing grounds of, 271; food from, 290; freedom from France, 278; French influence of, 280; location of, 264; mountains in, 265, p265; Muslim fundamentalists and, 286; rainfall in, 269
Moscow, Russia, 200
Mosi-oa-Tunya **("The Smoke That Thunders"),** 382
mosque, 281; in Samarkand, Uzbekistan, p208
mountains: Central Asia, 203; elevation of, 58–59; formation of, 43, 53; as landforms, 23; South Asia, 176–77, 180; tunnels through, 57; underwater, 60. See also entries for specific mountains
mountain spring, p229
Mount Everest, 31, 117, 176
Mount Fuji, p118
Mount Hakone, p53
Mount Kenya, 295, 329
Mount Kilimanjaro, p292, 295, 298, 329
Mount Lokon, p53
Mount Mayon, p152
Mount Toubkal, 265
movies, 191, 282, 436, 437
Mozambique, m379; aluminum in, 401; climate, 383, 384; escarpment, 382; ethnic groups in, 396; Great Rift Valley, 295; infant death in, 400; mineral resources in, 387; Muslim population in, 396; population of, 395; Portuguese language in, 397; Portuguese rule, 391, 392; size of, 380; tourism in, 401. See also Southern Africa
Mugabe, Robert, 392
Mughals and Mughal Empire, 185, 186
Muhammad, 233, 276
Muller, Peter O, q38
Mumbai, India, p93, 181, 189, 191
Murray-Darling River system, 418
Musgrave mountain range, 417
Music: Caucasus region, musicians in, p201; Central Africa, p326, 344, p344; East African, 314; Maori, 426, 435; North Africa, 281, 372; South Asia, 191; West African, 371–72. See also arts/culture
Muslim Brotherhood, 286
Muslims, 86; Central Africa, 342; daily life of, 244–45; East Africa, 312–13; Georgia, 215; Grand Mosque in Makkah, Saudi Arabia, p234; Hindus and, 187; Kashmir territory, 193; Kazakhstan, 215; Mughal rules as, 185; North Africa, 280, 367; Pakistan, 190; rulers of North Africa, 276; Siberia, 215; South Asia, 191; Sudan, 309; Sunni *vs.* Shia, 243; West Africa, 363, 369. See also Islam
Mutapa Empire, 389
Myanmar, 151, 158, m149; communist government in, 163; independence of, 162; Inya Lake in, p148; Mekong River through, 153; political challenges in, 169; as poor country, 168. See also Southeast Asia
NAFTA. See **North American Free Trade Agreement (NAFTA)**

Nagorno-Karabakh, ethnic unrest in, 216
Nagoya, Japan, 133
Nairobi, Kenya, p311; Kenya National Museum, p315; population of, 313
Nejd, Saudi Arabia, p246
Namib Desert, 381, 385
Namibia, m379; climate of, 384; democratic governments in, 401; desert region in, 385; escarpment in, 381; ethnic and, 396; natural resources in, 386–87; population of, 394, 395; relative land area of, 380; religion in, 396. See also Southern Africa
Narmada River project, 180
national identity, 243
Nationalists, Taiwanese, 129
National Map-Hazards and Disasters, The (U.S. Geological Service), q38
National Snow & Ice Data Center, "Arctic vs. Antarctic," q490
Native Americans: as ethnic group, 83. See also entries for individual groups
natural disaster: emigration after, 78; in Indus Valley, 183; in South Asia, 183. See also entries for individual natural disasters
natural gas: Australia, 436; North Africa, 270–71; Oceania, 455; Papua New Guinea, 455; South Asia, 181; Southern Africa, 386; Southwest Asia, 231
natural resources: mineral resources: Australia, 436, 438, 539; Caucasus, 204–5; Central Asia, 204–5; New Zealand, 436–39; Oceania, 455; Papua New Guinea, 462–63; Siberia, 204–5; South Asia, 179–81; Southwest Asia, 231. See also entries for specific resources, countries, or areas
Nauru, 450
navigable, 329
Nazi Germany, 237
Nehru, Jawaharlal, 187, p187, 193
nenets peoples, p202
Nepal, 31, m175, 180, p181, 193. See also South Asia
Netherlands, the. See Holland
New Caledonia, 451, 455
New Delhi, India, 81, 189
New Guinea, 151, 451–52
newly industrialized countries (NICs), 98–99
New Orleans, Louisiana: as "the Crescent City", 22
New South Wales, 430
New York City: Spanish Harlem in, 86
New Zealand, 413–46, m415; agreements with Oceania, 458; Antarctica and, 481; ANZUS Pact, 431; British in, 430; climates of, 422; culture in, 432–34, 435; economy of, 436–37; international trust funds for islands in Oceania,

465; landforms and waterways, 419–21; life in, 432–39; Maori of, 83, 426–27, 435, 438; migration from Oceania to, 464; natural resources in, 436–39; people of, 432–33; plants and animals of, 423; rural areas of, 434; territories in Oceania, 450–51; timeline, c414–15; trading partners, 437

"New Zealand" (FreedomHouse.org), q446

New Zealand Constitution Act (1852), 431

Nguema, Francisco Macías, 339

NICs. *See* **newly industrialized countries**

Niger, m353; Air Massif in, 354; delta region, 359; in Empire of Mali, m362, 363; as French colony, 366; independence of, 366; North African influence in, 371. *See also* West Africa

Nigeria, m353; climate of, 357; debt relief for, 373; eastern region secedes from, 367; ethnic groups in, 368–69; Jos Plateau in, 355; kingdom of Benin in, 364; as new country, 366–67; Niger River in, 355; Songhai Empire in, m362, 363. *See also* West Africa

Niger River, 355

Nile Basin, 296

Nile River: Ancient Nubia and, 302; deep gorge formed by, 265; delta at mouth of, 266; in East Africa, 296; farmland along, p266; importance to Egypt, 264, 266, 271; population concentration along, 76; rise of Egypt and, 272

Nile River valley, 261, 302

Niue Island, 451

Nkrumah, Kwame, 366, p366

nomadic peoples, p202, 269; Australia, 425; Bedouins, 229, 241; Central Asia, 208; desertification and, 376; East Africa, 314; Siberia, 206

nonrenewable resources, 95

Norgay, Tenzing, p17

Norlisk, Russia, 23

North Africa, 261–90, m262–63; art, 282–83; climate in, 267–70; coastal plains and mountains, 265, p265, p269; countries of, 264–65; culture of, 280–83; daily life, 281–82; early trading kingdom, 362; economic issues in, 283–84; food, 282; future of, 285–87; history of, 272–79; independence of, m277; languages and literature, 283; life in, 280–87; lowland of, 265; in Middle East *vs.* Southwest Asia, 239; natural resources, 270–71; people of, 280; physical geography of, 264–71; religion in, 276, 280, 369; Rome's control of, 275–76; size of, 265; society in, 280–87; timeline, c262–63; water pump in, p270; waterways, 266–67

North America, as continent, 52. *See also* Canada; Mexico; United States

North American Free Trade Agreement (NAFTA), 101

North China Plain, 119, 125

Northern Angola, 383

Northern Hemisphere, 26, c27; seasons in, 45, d45

Northern Territory (Australia), 421, p422, 430

North Island, 419, 420, 423, 430

North Korea, m115; Korean Peninsula; command economy in, 96; as communist nation, 130–31; on Korean Peninsula, 118, 120. *See also* East Asia

North Pole, 21; climate in, 46; Earth's axis and, 42; region around, 52; seasons and, 45

North Vietnam, 162

Northwestern Mozambique, 382

Norway, Antarctica and, 481

Nubia, 302

nuclear family, 371

nuclear proliferation, 187

nuclear weapon, 187

Nyika people, 311

oasis, 213, 229, 271

Obama, Barack (U.S. President), q376

Ob River, 204, 205

ocean currents: climate and, 47–48; effect of, 56

ocean floor: landforms on, 59–60; mining of, 63; plate movements and, 53

Oceania, 447–70, m448–49; climates of, 454; conflicts with colonists, 457; countries paying for marine resources of, 463; culture of, 460–61; economics and society, 466; economies of, 462–65; European explorers in, 457–58; first humans settling in, 456; formation of island, 453, d453; history of, 456–58; human migration from, 464, 465–66; landforms of, 450–53; life in, 462–67; location of, 450; MIRAB economies, 464–65; modern times, 457–58; natural resources of, 455; people of, 459–61; physical geography of, 450–55; Polynesian migrations, 456; population growth, 465–66; resources of, 455; smaller islands of, 452–53; timeline, c448–49

oceans, 61–63

oil, 180–81; Angola, 401; Australia, 436; dependency on, Southwest Asia and, 245, 246; North Africa, 270–1; Oceania, 455; Southern Africa, 386; Southwest Asia, 231; West Africa, 358

oil consumption, i230

oil tanker, in Strait of Hormuz, p228

Okavango Delta, 404

Olduvai Gorge, 315

Oman, m225; in Arabian Peninsula, 227; Arab Spring and, 246; border of Persian Gulf, 228. *See also* Southwest Asia

Omotic language, 312

oral tradition, 314

Orange River, 382, 390

orbit, 42

Organization for the Development of the Senegal River, 359

Oriya language, 190

Orlando, Florida, Disney World in, p22

Osaka, Japan, 133

Ottoman Empire: decline of, 236; North Africa in, 277; Southwest Asia in, 235; World War I and, 211

Ottoman Turks, 210

Outback, 417, 421, 433, 435

outer space, 44

outsourcing, 192

overfishing, 467

Ovimbundu peoples, 396

Owens Stanley Range, 452

oxygen, 181

ozone, 484; hole in, 483, d483

Pacific Islander, 432

Pacific Islands, Aboriginal peoples of, 85

Pacific Ocean: East Asia, 119; Magellan exploration crosses, 160; Mariana Trench, 60; New Zealand, 419; Oceania, 450; relative size of, 62; Ring of Fire in (see Ring of Fire); Siberian Russia, 200; Southeast Asia, 153; typhoons over, 155

Pacific Plate, 152

Pacific Rim, 167

Padaung woman, in Myanmar, p166

Pagan kingdom, 158

Pakistan, 201, m175, p183, m198; British rule of, 186, 187; conflicts with India, 193; Hindu Kush mountain range, 226; Khyber Pass in, 176, 227; language in, 190; population, 188; religion in, 190; Thar Desert in, 178. *See also* South Asia

Pakistanis, 190

Palau Island, 450, 451, 454, 458

Palestine, 237, m237. *See also* Arab-Israeli conflict; Palestinian territory

Palestinian territory, 233

palm oil, 335

palm oil trade, 364, 463

Pamirs, 201

Panama: as isthmus, 59

Panama Canal, 59

Pana people, 343

pans, 383

Papua New Guinea: drought in, 454; economy of, 462–63; in Oceania, 450–52; population of, 459–60; resources of, 455

Paris, France, as primate city, 81

Paschal, G. Zachary, q324

Pashto people, 243

Pashtun peoples, 243

pastoral, 206
Paul, Ron, q249
PBC (Peacebuilding Commission), 347
Peking opera, 135
peninsula(s), 59, 118, p119; in Southeast Asia, 150–51
People's Republic of China, 129
periodic market, 399
permafrost, 202, d203
Perry, Matthew C, 129, p129
Persia: trade with Southern Africa, 388. *See also* Iran
Persian Empire, 210
Persian Gulf: oil spill on fire in, p245; petroleum in, 231; Strait of Hormuz and, 228
Persian Gulf War, 239, p245
Persian people, 208, 243
Perth, Australia, 433
Peshawar, Pakistan, 227
Peter the Great (Russian Czar), 209
petroleum, 245; in Papua New Guinea, 455, 463; in Southwest Asia, 231, 239; in West Africa, 358, 372. *See also* oil
pharaohs, 272, 273, p273
Philippines, 151, m149; bridge collapse after earthquake in, p54; climate in, 155; fishing in, 168; independence of, 161–62; manufacturing in, 163; Mount Mayon volcano in, p152; religion in, 166. *See also* Southeast Asia
Philippine Sea, 153
phosphate mine, in Tunisia, p284
physical geography, 39–68; changing Earth, 52–57; land, 58–60; planet Earth, 42–51; water, 60–65. *See also* entries for individual regions
physical map, 29
physical systems, c24; Earth's, 44
pidgin language, 369, 460
Pillars of Islam, 234
pine tree, 205
pita bread, 282
place, c24; as geography theme, 22; perspective of, 19–20
plains, 58; Central Asia, 202; creation of, 56; Siberia, 202–3
plankton, 479, 485
plantation agriculture, 161
plant life, in salt water, 61
plateau: in Arabian Desert, 228; Deccan, 177; defined, 58; in Southern Africa, 381–83; in Southwest Asia, 226–27
Plateau of Tibet, 117, 119, 121, 153
plate movements, 53–54
poaching, 317, 387; burning elephant tusks in Kenya, p316
polar climate, 49–50
polar ice caps, 61

policy, 187
political map, 29
pollution: effects on Earth, 57; of lakes and rivers, p65; population growth and, 74–75
Polynesia, 447, 450, m448–49; culture of, 460–61; islands of, 451; low islands of, 452. *See also* Oceania
Polynesian migrations, 456
polytheism, 233, 271
population: Aboriginals in Australia, 429–30; aging, in Australia and New Zealand, 439; America, 188; Australia, 432; Central Africa, 340–41; East Africa, 310–12; Maoris in New Zealand, 432; Nairobi, Kenya, 313; New Zealand, 432–33; North Africa, 280; Oceania, 459–60, 465–67; Papua New Guinea, 460; South Africa's four cities, 397; South Asia, 188, 189, 193; Southern Africa, 394–95; Southwest Asia, 240; West Africa, 370, 373. *See also entries for specific regions*
population density, 76–77; defined, 190, 310; Southern Africa, m395
population distribution, 76
population growth, 72–75; causes of, 72–73; challenges of, 74; as East Asian challenge, 138; effects on environment, 74; rates of, 73, 75
population movement, 77–81
population patterns, 76–77
population pyramid, g73
Portugal: colonization of East Africa, 305; in Ghana, 359; Islam religion, 234; North Africa and, 277; rule of Timor-Leste, 162; slave trade, 335; Southern Africa and, 390, 392; trading in West Africa, 366; West African kingdoms and, 365
Portuguese explorers, in Southeast Asia, 160
Portuguese language, 369
Portuguese peoples, 389
possession, 458
poverty: population growth and, 74–75
precipitation: defined, 48; evaporation and, 64, d64
primate city, 81, 165
Prime Meridian, 21, c21, 26, c27, 213, m213
Principe, 329, 338–40
prison colony (Australia), 428
productivity, 98
Ptolemy, p16
Puncak Jaya, 154
Punjabi language, 190
Push-pull factors for migration, 78–79
pyramids: ancient Egypt, 274; ancient Kush (Sudan), p303; making of, i274
Qaddafi. *See* al-Qaddafi, Muammar
Qatar: Persian Gulf border, 228; population of, 240; as wealthy country, 245

Qattara Depression, 265
Qing dynasty, 125, c126
Qinzang rail system, p117
Queen Elizabeth National Park, 315
Queensland, 421, 430
Queen Victoria, 430
Quran, 244, p283
railroad: Qinzang rail system in China, p117; Trans-Siberian, 207, m207, 208, 212
rainfall: Australia, 421; Central Africa, 330; East Africa, 298–99; Georgia, 203; North Africa, 264, 269–70; Oceania, 452, 454; Sierra Leone and Liberia, 358; South Asia, m179; Southern Africa, 384; Southwest Asia desert, 230; West Africa, 358. *See also* monsoon
rain forest, Amazon, 33, 90; as biome, 50; in Central Africa, 330–31; climate and, 51; GIS and, 33; loss of, 57; resources and, 90–93. *See also* tropical rain forest
rain forest research, p93
rain shadow, 49, g48
rainwater, 64
Raj, 186
Ramadan, 234, p244, 245
Ramayana, 191
Rangoon. *See* Yangon
recycling, of water supply, 63–64
Red River, 153
Red River delta, 157, 158
Red Sea: Adulis on, 303; Ancient Egyptian control along, 273; Ancient Nubia and, 302; Eritrea, 306; Great Rift Valley and, 295; mountains on shores of (Egypt), 265; in Southwest Asia, 227–28; Suez Canal and, 267
refugees, 318–21, g321; in Central Africa, 341; defined, 78; Somolian, 309; Sudanese, 318, p319, 320, 321
regime, 279
region, 22, c24
reincarnation, 184
relative location, 20–21
relief, on physical maps, 29
religion: ancient Egypt, 273–74; Australia, 432; birthplace of, 233; Central Africa, 343; culture and, 84; East Asia, 135; major world, c84; New Zealand, 432; population growth and, 75; rivalry in West Africa, 367; Siberia, Central Asian, and Caucasus, 215; South Asia, 182–85, 191; Southeast Asia, 166; Southern Africa, 396–97; Southwest Asia, 243–44. *See also entries for individual religions*
remittance, 464, 470
remote sensing, 32, 483–84
renewable resources, 95
representative democracy, 87
Republic of China, 129
Republic of Ghana, 366

Republic of Mali, 366
Republic of South Africa, 386. *See also* Southern Africa
Republic of the Congo, 338; as independent nation, 339; State Department Background Notes about, *q350*
research station, 482–83
reservoir, 383
resort, 464
resources: defined, 23; nonrenewable, 95; renewable, 95; wants and, 94–95. *See also* energy resources; mineral resources; natural resources
revolution: of Earth around the sun, 42–43
Rig Veda, 184
Rhodesia, 392. *See also* Zimbabwe
rice: agricultural societies in Southeast Asia and, 158; Asian cultures and, 120; crop in Philippines, *p152*; in prehistoric Southeast Asia, 156; in Southeast Asia, 168
Richter scale, 140
rifted, 294–95
Ring of Fire: defined, 54; Japanese islands in, 118; New Zealand in, 436; Southeast Asian islands in, 151; tsunamis and, 140
rivers: freshwater, 62; liquid water in, 61; mouth of, 62; in South Asia, 176–77. *See also* entries for individual rivers
Riyadh, 241
Roman bath, *p275*
Roman Catholic Church, in Southeast Asia, 166
Roman Empire, 210, 275–76
Rossiya, p204
Ross Sea, 477
Rotorua, 421
Roughing It **(Twain),** *q68*
Rowan, Chris *q146*
Royal Thai guard, *p147*
Rozwi kingdom, 389
Rub' al-Khali (Empty Quarter), 229
rubber, *p154*, 168, 337, 455
rural areas, 77
Russia: Amur River and, 204; borders with Siberia and Mongolia, 202; climate in, 23; Kuril Islands dispute with Japan, 139; territorial issues in, 219. *See also* Siberian Russia
Russian Empire, 207, 211; German invasion of, 208
Russian Republic, 207
Russian Siberia, 206–7. *See also* Siberia
Ruwenzori Mountains, 296, 298, 329
Rwanda: ethnic tensions in, 308; genocide in, 346; independence of, 307; as landlocked country, 296; life expectancy, 317; mineral resources of, 300; population density of, 310; Volcanoes National Park, 315

Sahara desert, *p355*, harmattan winds in, 356–57; landscapes in, *p268*; the Sahel and, 299; size of, 268; temperatures in, 269; trade across, 361; in West Africa, 354–55, 360
Sahel, 299, 357
salt and salt trade, *p361*; in Central Africa, 335; gold-for-salt (Mali), 363; in Lake Assal, 300; in Southern Africa, 383
salt water, 61; freshwater *vs.*, *i60*
salt water sea, 228
Salween River, 153
Samarkand, Uzbekistan, 208, *p208*
Samburu National Reserve, 315
Samburu people, 311
Samoa, 450
samurai, 128
sand dunes, 268, *p268*. *See also* desert
San people, 396
Sanskrit language, 183, 190
Sao people, 342
Sao Tomé: independence of, 339; life expectancy in, 340; location of, 329; Portuguese colony on, 335, 338
Satellites: remote sensing, 32
Saudi Arabia, *m225*; in Arabian Peninsula, 227; border of Persian Gulf, 228; command economy in, 96; ethnic groups, 242; irrigation in Nejd region, *p246*; Makkah (Mecca), 233; oil in, 231; population of, 240–41. *See also* Southwest Asia
savanna, 354–57; in Central Africa, 331–32
Sayano-Shushenskaya Dam, *p209*
scale, map, 29
scale bar, on map, 28
Schlein, Lisa, *q324*
Scotia Sea, 477
Scott, Robert, 480
Sea of Japan, 118, 119, 212
seas, characteristics of, 228
seasons: Earth's orbit and, 45–46; tilt of Earth and, *d45*
seceded, 367
Second Continental Congress, *p86*
Second Persian Gulf War, 239
semiarid, 121, 230, 357
Seminomadic peoples, 311, 360
Senegal, 357, 366
Senegal River, 356, 359
Senghor, Léopold, 372
Seoul, South Korea, 120; population density in, *p76*
sepaktakraw, 167
Serengeti National Park, 301, 315
Serengeti Plain, 301
service industries, 97
settlement patterns, in West Africa, 370
Seychelles, 380, 391
Shang dynasty, 125, *c126*

Shanghai, China, 119, 133
Shankar, Ravi, 191
Shia Muslims, 243
Shinto, 128, 135
shogun, 128
Shona peoples, 389
Siam, absolute monarchy in, 161
Siberia: below ground in, *d203*; climates in, 203; energy and mineral resources, 205; history of, 206–8; language and religion of, 215; nomadic and nenets people of, *p202*; plains and deserts in, 202; revolution and development, 207–8; Russian, 206–7; settlement, invasion and conquest, 206; waterways, 204; woman of northwestern, *p197*. *See also* Siberian Russia
Siberian Russia, 197–222, *m198*–99; climates in, 203; landforms in, 201–2; people and places of, 212; population of, 212; timeline, *c198*–99; waterways in, 204. *See also* Siberia
Sichuan Province, 136
Sidney, Australia, 433, 434
Sierra Leone: British colony in, 364, 365; civil war in, 367; Peacebuilding Commission (PBC) and, 347; tropical rain forests in, 358
Sikhism, *c84*; in South Asia, 191
Silk Road, 126, *m127*, 208, 222
Silla, the, 126
silt, 266
silver: in Australia, 436; in Namibia, 387
Sinai Peninsula, 238, 264, 265
Singapore: cargo containers in port of, *p249*; Chinatown district in, *p165*; economy of, 163; independence of, 162; off Malay Peninsula, 151
Sinhalese, 190
sitar, 191
Skaka, 389
"Skeleton Coast" (Benguela), 385
skill building, interpreting visuals, 25
slash-and-burn agriculture, *p74*, 331–32
slavery, in Southern Africa, 390
slave trade, 364, *m365*; in Central Africa, 335–36
slums, 75
Small, Cathy A, *q470*
soccer, *p85*
Sogdians, 222
solar panels, *p75*
solar-powered lighting, 455
Solomon Islands, 450, 455, 457, 467
solstice, 45
Somalia, 82; civil war, 308–9; clan, attachment to, 312; drought in, 299; in East Africa, 294; independence of, 307; as independent trading state, 303; Somali language spoken in, 82, 312; temperature of, 298; United Nations and, 346, *p346*; where people live, 310
Somali language, 82, 312
Song dynasty, *c126*

Songhai, m362, 363
souk, 281
South Africa, m379. See also Southern Africa
South America: birthrates in, 75; as continent, 52
South Asia, 173–96; British in, 186; climates, 178–79; cultures, 191; daily life in, 191; history of, 182–87; homes in, 188–89; Independence, achieving, 186–87; Indian empires, 185; Islam religion and, 234–35; islands of, 177; issues in, 187, 192–93; modern, 185–87; natural resources, 179–81; northern mountains and plains, 176–77; people, 190; physical geography of, 176–81; population, 188, 189, 190; rainfall, m179; religion and arts, 191; timeline, c174–75
South Australia, 430
South China Sea, 119, 123, 153; rubbish-covered beach along, p138
Southeast Asia, 147–72, m149; arts in, 166; Association of Southeast Asian Nations (ASEAN), 168; bodies of water in, 152–53; climate in, 154–55; colonies/independence, m160, 161–63; daily life in, 166–67; early societies in, 156–59; earning a living in, 168; economic and environmental challenges for, 169; ethnic and language groups, 166; European traders in, 159–60; four plates meet in, 152; green islands in, p151; history of, 156–63; independent nations in, 161–63; Islam and, 158–59, 234; landforms in, 150–52; modern, 163; mountains and volcanoes in, 151–52; natural resources of, 152; peninsulas and islands in, 150–51; people and cultures, 165–67; plants and animals, 155; population in, 164–65; prehistoric cultures in, 156–57; religion in, 166; slaves from, in South Africa, 390; spices in, 159, p159; sports in, 167; timeline, c148–49; in transition, 167; Western colonization of, 159–61
Southern Africa, 377–404, m379; Bantu people in, 361; bodies of water, 382–83; clashes in, 390; climate, 383–85; desert regions, 384–85; Dutch settling in, 364; energy resources in, 386; environmental characteristics in, 23; equal rights in, 392–93; ethnic and, 396; European colonies, 390–91; family and traditional life in, 399; health issues today, 399–401; history of, 388–93; independence, 391–92; landforms, 381–82; languages of, 396–97; life in, 394–401; physical geography of, 380–87; population density, m395; population patterns, 394–95; progress and growth, 401; region of, 380; religion and languages, 396–97; resources, 385–87; rise of kingdoms, 388–89; timeline, c378–79; urban life in, 397–98; wildlife, 387. See also Union of South Africa
Southern Alps, 420, 422

Southern Hemisphere, 26, c27; seasons in, 45, d45; tropics in, 154–55
Southern Ocean, 62, 483, 484
South Island, 419, 420, 422
South Korea, m115; access to Oceania's marine resources, 463; climate of, 121; K-pop in, 135; as peninsula, 118; urbanization of, 133; U.S. support of, 130–31. See also East Asia; Korean Peninsula
South Ossetia, 216
South Pole, 21, 480; climate in, 46; Earth's axis and, 42
South Sudan, m293, m321; in East Africa, 294; energy resources in, 300; independence of, 309; 318; villagers in, p308
Southwest Asia, 223–52, m225; bodies of water in, 227–28; civil wars in, 238–39; climates, 229–30; as cradle of religions, 233; daily life, 244–45; early, 232–35; ethnic and language groups, 242–43; Islamic expansion, 234–35; issues in, 245–47; life in, 240–47; Mesopotamia, 232–33; modern, 236–39; natural resources, 231; oil reserves and consumption, i230; physical features, 226–29; physical geography of, 226–29; population profile, 240; religion and art, 243–44; timeline, c224–25; water scarcity in, 247; where people live, 241
Soviet legacy: in Caucasus and Central Asia, 217–18
Soviet Union, m198; Caucasus and, 210–11; collapse of, 210. See also Union of Soviet Socialist Republics (USSR)
Sovoia glacier, p298
space, perspective of, 18–20
***Space, Time, Infinity* (Trefil),** q68
Spain: Islam religion and, 234; North Africa and, 277; trade with West African kingdoms, 364
Spanish-American War, 161
Spanish explorers, in Oceania, 457
Spanish Harlem, 86
spatial, 18
sphere of influence, 128
spheres, 26, 27
spices: at souk in Cario, Egypt, p281; from Southeast Asia, 159, p159
sports: culture and, 85; first organized team, 85
Sri Lanka: civil war in, 193; climate of, 178; language in, 190; location of, 177; mineral resources in, 180; in South Asia, 176; timber resources in, 181. See also South Asia
Srivijaya empire, 157–58
St. Helena, 354, 365
Stalin, Joseph, 211
standard of living, 98
Stanley, Henry Morton, 304–5
stations, 429

steppe, 202
Strait of Hormuz, 228, p228
Strait of Malacca, 153, 157, 159, 160
streams, freshwater, 62
subcontinent, 176
subsistence agriculture, 168, 313, 341, 343
Sudan, m293, m321; Ancient Nubia in, 302; Battle of Omduman, p305; as British colony, 306; in East Africa, 294; independence, 309; landforms in, 295–96; life expectancy, 317; president's palace, p296; refugees and displacement, 318–21; statistics about people of, 320; temperature of, 298
Sudd, 296–97
Suez Canal: building of, 277, 365; completion of, 236; container ship in, p267; importance of, 267, 278; in Southwest Asia, 227
Sukuma farm, 314
Sulawesi, 151
sultans, 159
Sumatra: earthquake in, 142, 152; Islamic kingdoms in, 158; prehistoric, 156; in Southeast Asia, 151; Srivijaya kingdom, 157–58
Sumerians, 232–33
sun, Earth and, 42–43
Sunda Isles, 153
Sunni Muslims, 243
sustainability, 101
Swahili language, 303, 312
Swaziland, m379; British control of, 390; climate of, 384; HIV/AIDS in, 400; independence for, 391; as landlocked, 382; Muslim population in, 396; political unrest in, 401; population of, 394; resources of, 387. See also Southern Africa
Syr Dar'ya River, 204, 219
Syria, m225; alluvial plain, 228; Arab Spring and, 246; Christians in, 243; coast of, 227; democracy movements in, 239; ethnic groups, 242; on Mediterranean Sea, 227; religion in, 243; Tigris-Euphrates River system through, 247. See also Southwest Asia
Szechuan Province, 136
Tagalog people, 166
Tahiti, 452
taiga, 202, 205
Taiwan, m115, 118, 119; access to Oceania's marine resources, 463; China's view of, 139; climate of, 121; forests in, 123; Japanese control of, 129; language in, 134; limited mineral reserves in, 122; population of, 132–33; textile industry in Lesotho, South Africa, 401. See also East Asia
Tajikistan, m198, 200, 201; civil war in, 216; coal in, 205; culture and daily life, 217; people of, 214; Soviet legacy in, 218. See also Central Asia
Tajik people, 243
Taj Mahal, p19, p185
Taklimakan Desert, 121, 123

Taliban, 239
Tamil language, 190
Tamil Nadu, 190
Tanganyika, 308
Tang dynasty, 126, *p126*
Tanzania, East Africa, *m293*, 294; agriculture in, 316; Denakil Plain in, 295; energy resources in, 300; ethnic groups in, 311; farming in, 313; HIV/AIDS, 317; independence of, 307–8; as independent trading state, 303, 304; Lake Victoria in, 297; language of, 312; Masai nomadic lifestyle and, 314; mineral resources of, 300; rainfall in, 298–99; religion in, 313; Serengeti National Park in, 301, 315; *tarab* music, 314; wildlife reserves in, 301. *See also* East Africa
tarab **orchestra,** *p314*
taro, 361
Tartars, 206
Tashkent, Uzbekistan, 213
Tasman, Abel, 428
Tasmania, *m415*, 418; early people in, 425; European explorers in, 428; as independent nation, 430–31; natural resources in, 436
Taylor, Charles, 367
technology, 30–31; globalization and, 89; widespread use of, *p89*
tectonic plates, 53; boundaries, *m40–41*
Tehran, 241
Telugu language, 190
textiles, 314
Thailand, *m149*, 151; absolute monarchy in, 161; as constitutional monarchy, 163; fishing in, 168; manufacturing in, 163; Mekong River through, 153; natural resources in, 152; plantations in, 168; rice exports, 168; tin mining in, 152, 168, 169
Thar Desert, 178. *See also* Southeast Asia
thatch, 399, 461
thematic maps, 30
thermal image, *p51*
Thimphu, Bhutan: Buddhist monastery in, *p174*
Things Fall Apart **(Achebe),** 372
thinking spatially, 18–19
Thiong'o, Ngugi wa, 314
Thousand and One Nights, The, 244
Three Gorges Dam, *p56*, 123
Through the Dark Continent **(Stanley),** 304–5
Tian Shan, 201
Tibesti Mountains, 354–55
Tibet: China's treatment of, 139; nomadic shepherd in, *p114*
Tigris River, 228, 232, 247
Tikanga, 426
Tikar people, 342

Timbuktu, 355, 363
timelines: Antarctica, *c472–73*; Australia, *c414–15*; Caucasus, *c198–99*; Central Africa, *c326–27*; Central Asia, *c198–99*; Earth's land, people, environment, *c16–17*; East Africa, *c292–93*; East Asia, *c114–15*; human geography, *c70–71*; New Zealand, *c414–15*; North Africa, *c262–63*; Oceania, *c448–49*; physical world, *c40–41*; Siberian Russia, *c198–99*; South Asia, *c174–75*; Southeast Asia, *c148–49*; Southern Africa, *c378–79*; Southwest Asia, *c224–25*; West Africa, *c352–53*
time zones, 213, *m213*
Timor, 151
Timor-Leste, 162
Timur, 208
Togo, *m353*, 359, 365. *See also* West Africa
Tokyo, Japan, 133
tombs, in ancient Egypt, 274, *i274*
Tonga: aid from foreign governments, 465; European explorers in, 457; migration from, 465–66; MIRAB economy in, 465; in Oceania, 450; remittances to, 464, 470
Tonle Sap, 158
tourism: Antarctica, 485; Australia, 436; Oceania, 464; Vietnam, *p168*
townships, in South Africa, 398
trade deficit, 138
trade languages, 342
trade restrictions, 248–49
trade surplus, 138
trading market, *p398*, 399
traditional economy, 96
traditional music: of Central Africa, *p326*; in East Africa, 314, *p314*
Transantarctic Mountains, 475
Trans-Siberian Railroad, 207, *m207*, 208, 212
treaty, 481
Treaty of Waitangi, 430
Trefil, James S, *q68*
trench, 60
"triangular trade," 336
tribal identity, 243
tribute, 305
Tripoli, Libya, *p278*
Tristan da Cunha, 354
tropical climate, 49–50; South Asia, 178; Southeast Asia as, 154–55; Southern Africa, 383–84
tropical rain forest: in Central Africa, 330–31. *See also* rain forest
Tropic of Cancer, 45, 46
Tropic of Capricorn, 45, 46, 383, 474
Trusteeship Council of the United Nations, 458
trust territory, 458

Tsaatan people, *p121*
Tsonga peoples, 396
tsunami, 54, 140–43; America Samoa, 465; Banda Aceh, *p153*; Krakatoa eruption triggered, 152; statistics about 2011, 142
Tswana population, 396
Tuareg people, 354; with livestock, *p355*
tuberculosis, 400
tundra: in biome, 50; defined, 202; Siberian, 203, 205
Tunis, Tunisia, 281
Tunisia, 239, *m263*; aquifer in, 271; Arab Spring and, 279; cooking in, 282; freedom from France, 277, 278; French influence of, 280; iron ore and phosphates in, 271, *p284*; location of, 264; mountains in, 265; Muslim Brotherhood and, 286; women's rights and, 287. *See also* Carthage
Turkana people, 311
Turkey, *m225*; Ararat volcano, 227; birth of, 236; civil wars in, 238; coal deposits in, 231; coasts of, 227; earthquake in (1999), *p227*; ethnic groups, 242; ethnic unrest in, 216; Islam religion spreading in, 234; Kurdish families in, 238, *p238*; Mediterranean climate, 230; population of, 240–41; reforms in, 239; temperatures in mountainous areas of, 230; Tigris-Euphrates River system and, 228, 247. *See also* Southwest Asia
Turkic people, 206, 212
Turkish language, 242
Turkish people, 210
Turkish republic, creation of, *p236*
Turkmenistan, *m198*; in Central Asia, 200; Kara-Kum desert in, 202; oil and gas resources in, 205; population of, 213; relative size of, 201; Soviet legacy in, 218; territorial issues in, 219. *See also* Central Asia
Turkmen people, 243
Turkomans, ethnic, 213
Turks, 243
Tutankhamen, *p273*
Tuvalu, 452, 463
Twain, Mark, *q68*
typhoon, 155, 454, 467
UAE. *See* **United Arab Emirates**
Uganda: in East Africa, 294; energy resources in, 300; independence for, 308; Lake Victoria in, 266; mineral resources of, 300; Queen Elizabeth National Park in, 315; Ruwenzori Mountains in, 296; Savoia glacier on border of, *p298*; temperatures in, 298
Uluru, 417
UN. *See* **United Nations**
underwater volcano, 453
unfrozen soil, *d203*
Union of South Africa, 390

Union of Soviet Socialist Republics (USSR): formation of, 207. *See also* Soviet Union

United Arab Emirates (UAE), m225; border of Persian Gulf, 228; oil in, 231. *See also* Southwest Asia

United Kingdom: monarchy in, 87, p87; Second Persian Gulf War and, 239

United Nations (UN): Arab-Israeli conflict and, 237; Darfur estimates, 320; Department of Economic and Social Affairs, on population growth, 73; effectiveness regarding Africa, 346–47; Korean War and, 131; land rights in Solomon Islands and, 467; Security Council, 346, 347, q347; World Refugee Day, 318

United Nations Development Programme (UNDP), q34

United States: access to Oceania's marine resourcs, 463; agreements with Oceania, 458; al-Qaddafi, Muammar and, 279; al-Qaeda and, 239; Declaration of Independence approval, p86; early immigrants to, 78; economic activity of American Samoa and, 465; gold prospectors to Australia from, 429; international trust funds for islands in Oceania, 465; life expectancy, 317; migration from Oceania to, 464; Muslim fundamentalists and, 287; Myanmar and, 169; occupation of islands in Oceania during World War II, 458; Persian Gulf Wars, 239, p245; representative democracy, 87; territories in Oceania, 451; trade deficit with China, 138, g139; trade with Australia, 436; trade with New Zealand, 437; voluntary free association with Palau Island, 458. *See also* entries for individual American territories

United States Geological Survey: "Desertification," q376

"untouchables," 193

Ural Mountains: as Siberian Russia border, 200, 202

uranium, 181, 387

urban areas, 77

urbanization, 80–81, 133, 165, 169

Urdu language, 190

U.S. Geological Service, *The National Map—Hazards and Disasters,* q38

U.S. State Department: Background Notes: "Israel," q252; "Singapore: History," q172

USSR. *See* **Russian Empire; Union of Soviet Socialist Republics (USSR)**

utilities, 398

Uzbekistan, m198, 200–2, p208, 209; coal in, 205; daily life in, 216–17; mineral resources in, 205; oil and gas resources in, 205; population of, 213–14; village women weaving in, p216. *See also* Central Asia

Uzbek people, 243; in Tajikistan, 214

valleys, 56, 58–59

Vanautu, 450
Van Sant, Shannon, q146
Varanasi, India, p177
varna, 183
Vedas, 183, 184
vegetation, 178
Victoria Land, 478
Vietnam, m149, 151; Communist forces in, 162; Dong Son culture in, 157; independence of, 162; literacy rate in, 167; manufacturing in, 163; Mekong River through, 153; as poor country, 168; rice exports, 168; tourist, 168; Vietnam War, 162–63, 162. *See also* Southeast Asia
***Vietnam: A History* (Karnow),** q172
Vietnam War, 162–63; Mekong Delta during, p162
Vinson Massif, 475
visuals, interpreting, 25
Vladivostok, 212
volcanoes, p178; Ararat, in Turkey, 227; Caucasus region, 201; Emi Koussi (West Africa), 355; eruption of, 43, 453; inactive, p292; lava from, p295; Mount Kilamanjaro, p292; New Zealand, 419; Oceania, 453; plate movement and, 54; Southeast Asia, 152; underwater, 60, 453
Volcanoes National Park, 315
wadis, 230, 268
Waikato, 419
Wake Island, 465
wants, resources and, 94–95
warriors, in South Asia, 183
Washington, D.C: absolute location of, 21
water: earning a living on, p62; Earth's surface, 44; freshwater *vs.* salt, i60, 61; locating, 61–63; North Africa, 271; recreational activities on, p62; resources for, in North Africa, 271; three states of matter, 61; use to humans, p62, 63. *See also* rainfall *and entries for specific bodies of water*
water, scarcity of: Australia, 423; North Africa, 270, 271; population growth and, 271; Southwest Asia, 231
water cycle, 63–65, d64
watershed, 328
waterways: Australia, 418–19; Central Africa, 329–30; Central Asia, Caucasus, and Siberia, 204; New Zealand, 419–21; North Africa, 266–67; Southwest Asia, 227–28; Suez Canal, 236, 365
wattle-and-daub houses, 343
waves, ocean, 56
wayfinding, 456
weather, 23; climate *vs.*, 48
weathering, 55
Weddell Sea, 477
***Weep Not, Child* (Ngugi),** 314
Wellington, New Zealand, 433
West Africa, 351–76, m352–53; ancient herders, 360; arts, 371–72; bodies of water, 355–56;

challenges, 372–73; civil war, 367; climate of, 356–58; coastal kingdoms, 363; daily life in, 371; energy in, 358–59, i358; ethnic groups, 368–69; first trading kingdom, 362; health and education, 373; history of, 360–67; kingdoms of, 363; landforms, 354–55; languages, 369; life in, 368–73; new countries, 365–67; people, movement of, 361; physical geography of, 354–59; religions in, 369–70; resources in, 358–59, 372; settlement patterns, 370; timeline, c352–53; wet zones, 357

West Antarctica, 476
West Bank, 238, 246
Western Australia, 422, 430
Western Ghats, 177–79
Western Hemisphere, 26, c27
Western Plateau, 416–17
Western Sahara, m262, 264
Western Wall, in Jerusalem, p234
west Siberian plain, 203
wetland ecosystems, 56
wet zones, in West Africa, 357–58
White Nile, 296, p296, 297
White Volta River, 356
wildlife, p315; East Africa, 301, 317; reserve, 301; South Asia, 50, 181; Southern Africa, 383, 387
wind: climate and, 47–48. *See also* typhoon
wind patterns, m47
wind power, 455
Wolfert, Paula, q290
women: in male roles, in Southern Africa, 399; rights of, in North Africa, 285, p285, 287
woodblock printing, 126
world: changing, 20; interconnected, 24; in spatial terms, c24
World Bank, 100–1, 373
World Heritage site, 417
World Refugee Day, 318
World Trade Organization, 100
World War I: Ottoman Empire and, 236
World War II: Australia post-, 431; Holocaust and, 237; Japan controls Southeast Asia during, 161–62; Japan's defeat in, 131
written record, in early South Asia, 182
Xhosa people, 396
Yaka people, 344
Yalu River, 120
yams, 361
Yangon, Padaung woman in car near, p166
Yaoundé, Cameroon, 342
Yap Island, 454
Yellow River, 119, 120, p120. *See also* Huang He River
Yellow Sea, 118, 119
Yemen, m225; in Arabian Peninsula, 227; Arab Spring and, 246; civil wars in, 238; democracy movements in, 239; Eritrea military conflict, 309; population of, 240, 241. *See also* Southwest Asia

Yenisey River, 202, 204, 205, 212
Yerevan, Armenia, 215
Yokohama, Japan, 133; Matthew Perry in, 129, *p129*
Yoritomo, Minamoto, 128
Yoruba, 367, 369
Yuan dynasty, 127
Yucatán Peninsula, 336
yurt, 216, *p216*
Zambezi River, 382, 386, 389

Zambia, 380, 382, *m379*; British control of, 391; democratic governments in, 401; electricity source for, 386; independence for, 392; languages of, 397; mineral resources in, 387; Muslim population in, 396; population of, 395. *See also* Southern Africa
Zanzibar: Tanganyika merged with, 308; tarab orchestra in, *p314*; trade city of Kilwa near, 304
Zheng He, 125
Zhou dynasty, 125
Ziggurat of Ur, *p233*

Zimbabwe, *m379*; birth of, 392; coastal plain of, 382; colonialism in, 391; energy resources in, 386; ethnic and, 396; infant death in, 400; mineral resources in, 387; political unrest in, 401
zinc, in Southern Africa, 387. *See also* Great Zimbabwe; Southern Africa
zones, climate, 49–51, *p49*
Zoroastrianism, 222
Zulu Empire, 389
Zulu peoples, 396
Zulu War of 1879, 389

McGraw-Hill Networks™ meets you anywhere—takes you everywhere. Go online at connected.mcgraw-hill.com.

Circle the globe, travel across time. How do you access networks?

1. Log on to the internet and go to connected.mcgraw-hill.com
2. Get your User Name and Password from your teacher and enter them.
3. Click on your networks book.
4. Select your chapter and lesson. Start networking.